Physics with Excel and Python

Dieter Mergel

Physics with Excel and Python

Using the Same Data Structure
Volume I: Basics, Exercises and Tasks

Springer

Dieter Mergel
Fakultät für Physik
Universität Duisburg-Essen
Duisburg, Nordrhein-Westfalen, Germany

ISBN 978-3-030-82324-5 ISBN 978-3-030-82325-2 (eBook)
https://doi.org/10.1007/978-3-030-82325-2

Based on the German language edition: *Physik mit Excel und Visual Basic* by Dieter Mergel, © 2017 2017. Published by Springer-Spectrum. All Rights Reserved, and extended with Python solutions.

This Springer imprint is published by the registered company Springer Nature Switzerland AG
The registered company address is: Gewerbestrasse 11, 6330 Cham, Switzerland

Preface

Ψ EXCEL*—as powerful as necessary,* `Python`*—as simple as possible.*

This book treats a series of physics exercises that were developed for courses at the University of Duisburg-Essen for students training to become physics/mathematics teachers or physics engineers. The exercises were intended to introduce computational physics based on spreadsheet calculation combined with simple VBA macros and also to broaden the beginner's knowledge of physics. This approach garnered positive reactions from practitioners and resulted in a textbook in German.[1] Furthermore, the methods developed turned out to be powerful enough to treat a broad range of topics in undergraduate physics, resulting in a second volume with exercises on particles, waves, fields, and random processes.[2]

Referees found the exercises interesting and ambitious enough to serve for undergraduate education. However, concerns arose that spreadsheet techniques, although useful in the business world, might be a dead end for students who would be required to use scientific computing in their future research work. Therefore, the concept has been changed for the present English version. Programming in `Python` is now included from the beginning, while the same topics are addressed as in the German textbook.

The key to all of the exercises is data structures, developed in introductory sections that explain the physical problems. They serve as an interface to both EXCEL and `Python`, and potentially also to other applications for scientific computation. To enable this approach, the EXCEL solutions in this edition use vectorized code and matrix formulas to mimic broadcasting, an essential `Python` technique for creating new arrays.

We feel that this approach is suitable as a low-threshold introduction to scientific computing as early as high school all the way up to undergraduate physics

[1] Dieter Mergel, Physik mit Excel und Visual Basic Grundlagen, Beispiele und Aufgaben, Springer Spektrum (2017), https://doi.org/10.1007/978-3-642-37857-7.

[2] Dieter Mergel, Physik lernen mit Excel und Visual Basic, Anwendungen auf Teilchen, Wellen, Felder und Zufallsprozesse, Springer Spektrum (2018), https://doi.org/10.1007/978-3-662-57513-0.

classes at the university and may also be a good start for students who later choose to specialize in computational physics.

Our approach is intended to make the student fit for a computer-oriented world, be it for spreadsheet calculations in business, scientific computing in research, or mathematics and physics teaching in high school. We take into account that not all students have the same attitude towards programming; some have to be encouraged to venture into a new world, whereas others have to be cautioned not to rush into blind programming.

Duisburg, Germany Dieter Mergel

Contents

About the Author

Dieter Mergel studied physics in Göttingen, obtained his doctorate at the Technical University of Clausthal in the field of solid-state physics, and worked 11 years in the Philips Research Laboratories Hamburg/Aachen on automatic speech recognition and optical data storage. Since 1993, he is Professor of Technical Physics at the University of Duisburg-Essen. His professional activities include research in the field of solid-state layers and lectures for students in teaching and medical professions.

Introduction

1

Possible errata and corrections in the internet at: go.sn.pub/Ob4vCR.

For every chapter, solutions for each two exercises in Excel and Python can be found at internet adresses.

1.1 A Two-Track Didactical Approach

History

The exercises in this book arise from a German textbook that emerged from courses for prospective teachers and students of Technical Physics at the University of Duisburg-Essen with the intention to prepare the students for a computerized world. The participants in the courses had already been studying physics for at least one year. However, the explanations of the exercises are so explicit that they should also be suitable for beginners.

Said courses are based on EXCEL and Visual Basic (VBA). The current English version includes `Python` from the very beginning so as to make it more generally useful for students who later choose to dive deeper into Scientific Computation.

Exercises

The subject matter is presented in nine chapters as a series of exercises. Every exercise consists of three steps:

1. The physical concept is introduced with mathematical equations and diagrams.
2. An adequate data structure is set up independent of the implementation in a particular programming platform, but taking care that the same nomenclature can be used in both mathematical equations and programming. This serves as an interface to any programming application.
3. Solutions in EXCEL and `Python` are designed so that a solution in one application can directly be translated into the other one.

D. Mergel, *Physics with Excel and Python*,
https://doi.org/10.1007/978-3-030-82325-2_1

To enable this approach, training in EXCEL emphasizes vectorized code, matrix formulas, and constructs that allow for broadcasting in the same way as `Python`. Furthermore, programming VBA macros interacting with spreadsheets introduces looping, logical queries, and functions.

Didactical advantages of the two-track system
We strive to combine the didactical advantages of both programming applications:

- Spreadsheets are interactive; charts react immediately to changes, and data structures (but not formulas) are immediately visible in spreadsheets.
- In `Python`, formulas (but not data structures) are immediately visible.
- The VBA interpreter allows us to run the program step by step and to watch intermediate results in a spreadsheet, or read the content of variables by mouse-over in the code.

Our guideline: spreadsheet calculations as powerful as necessary, `Python` programming as simple as possible.

Emphasis is not on mere computational techniques, but on exercises that may be regarded as small projects so that project-related difficulties manifest and can be addressed, e.g., by checking the consistency of equations and numerical solutions with limiting cases that can be solved analytically.

We maintain that this approach is not only suitable as a low-threshold introduction to scientific computing as early as High School and up to undergraduate Physics classes at University. It should also be a good start for students who later choose to specialize in Computational Physics or, more generally, for professional use.

1.2 What Can You Expect?

What can you, dear reader, expect from this book?

You can expect to be introduced to the world of `Python` and EXCEL by:

- training to work with `numpy` arrays, list slicing and broadcasting in `Python`,
- working with similar constructs, vector structures, and matrix operations also in EXCEL,
- learning how to write programs with looping, logical queries, and functions in `Python` and VBA for EXCEL,
- training how to lay out spreadsheets clearly so that they are apt for simple scientific computing,
- developing VBA macros that exchange data with spreadsheets,
- applying standard mathematical methods numerically.

and, with that,

- getting a better understanding of certain mathematical and physical concepts.

After having successfully completed the exercises, you should have gained so much self-confidence that you can answer the question "Programming practice?" with an enthusiastic "Yes!".

1.3 What Do You Need?

You will need a Physics textbook (anyone will do, e.g., the one you have at hand during your studies anyway) and two more books on programming as indicated below.

EXCEL
To work with EXCEL, you only need a computer in which EXCEL has been implemented (any version; the exercises in this book have been checked in EXCEL 2010 and EXCEL 2019) and an introduction to EXCEL (do not buy one before having done the basic exercise in Sect. 2.3). In particular, you do not need a special development environment for VISUAL BASIC, because it is included in all versions of EXCEL.

Python
You will need to install Anaconda, a free and open-source distribution of the `Python` and *R* programming languages that also comprises the `Jupyter` Notebook by default. The examples in this book were obtained with `Python 3.7` in `Jupyter`. You are advised to use both a book and internet courses to broaden your training systematically. Make your choice after having gone through Exercises 2.4 and 2.5.

1.4 Tim, Alac, and Mag

You will soon meet two types of students and a tutor who will accompany us throughout this book. The character named *Tim* (which stands for "timidus" or "timida", meaning shy)) represents those students who are somewhat hesitant, fearing that they may fall short of the requirements, although they study hard. The character named *Alac* (which stands for "alacer", meaning alacritous, high-flying) is typical of those vehemently self-confident students (men are generally over-represented) who believe that they already have a superior overview and do not have to deal with what they consider mere bits and pieces. *Mag* (for Magister/Magistra, i.e., the tutor who runs the course) tries to engage with both characters, encouraging Tim and cautioning Alac, and clarifying that both approaches are valuable and that every Physics student should venture into the Computer world.

Tim: Computers are not my thing

▶ **Tim** I see how well some fellow students are juggling programming tools, but I'd rather stand back. I prefer to learn the stuff from textbooks.

▶ **Mag** This course is not intended to turn you into a computer nerd. You will not learn any cool tricks. We restrict ourselves to some basic techniques that are practiced again and again. The computational techniques do not stand by themselves, but are always taught in connection with physical problems.

▶ **Tim** But I have often heard that programming is a black art for which you have to be specially talented.

▶ **Mag** Here, you will learn the most basic computer techniques that every scientist, engineer, and science teacher must master to succeed in their profession.

▶ **Alac** Why EXCEL and `Python`?

▶ **Mag** All algorithm-oriented computer languages have a similar structure. Knowledge of specific commands is not the most important thing. You have to learn how to translate physical and technical problems into a computational structure. Furthermore, the mistakes that beginners make are always the same in all computer languages. The most important thing is to track, correct, and, finally, avoid them. Anyhow, as we shall have spreadsheet solutions and `Python` programs in parallel, you will be sensitized more towards common structures than the peculiarities of specific software.

Alac: How do I become a master programmer?

▶ **Mag** A master can be recognized by how he/she deals with errors. Any unnoticed error in spreadsheet formulas and programs can lead to disaster. It is essential that you gain experience with data structures and programming constructs.

▶ **Alac** And that is what this course will accomplish?

▶ **Mag** Yes! By using data structures in spreadsheets and Python programs and setting up graphical representations that are comprehensible, even when you look at them after some time. And by developing simple procedures that control the program flow.

▶ **Tim** Data structures, procedures, controlling; that sounds pretty challenging. How can I learn all of that?

▶ **Mag** Let's compare this course with learning a foreign language. How do you learn foreign languages?

▶ **Alac** Learning? For foreign languages, academic learning is useless in the long run. You simply have to go abroad; the rest follows by itself.

▶ **Tim** Oh, I couldn't learn like that. I couldn't form a proper sentence in a foreign language without profound foreknowledge. I would have to learn the correct grammar and vocabulary first before I would dare to speak.

A good balance

▶ **Mag** We are trying to find a good balance. You will learn the most straightforward "sentence" structures, but will also be "sent abroad" right off, and you will have to make your way there. If you pass this test, you can be confident of being able to learn the more complex "grammar" if necessary.

▶ **Tim** Is that thorough enough?

▶ **Alac** Will I learn the more tricky constructions too late?

▶ **Mag** Don't worry! Working through this book will make you fit for a computer-oriented world, be it for spreadsheet calculations in business or scientific computing in research. This can be tedious, but it will be worthwhile, whether it be as early as learning at school or working for a Bachelor's or Master's, or even as late as working on a Ph.D. thesis.

▶ **Tim** Can I manage this in addition to my studies in Physics?

▶ **Mag** I think so. Anyway, this course is about physics and will help you to pass your exams.

1.5 Didactic Concept

Workshop atmosphere
Having cleared up the doubts harbored by Tim and Alac, we now explain the didactic concept of this book.

In the courses at the University of Duisburg-Essen on *Physics with Excel and visual basic*, learning was mostly done in a workshop, such as in physics labs for beginners. The students dealt with the tasks alone or in pairs while in a computer lab, ideally also helping each other out across groups and consulting the supervisor when needed. Students could continue to work on their tasks outside of attendance time so that everyone could work according to their learning progress.

Experience shows that the students enjoy the tasks, and the learning progress is fastest when all three aspects—programming, physics, and mathematics—are combined. The systematic practice of various isolated spreadsheet and programming techniques is often perceived as too dull. The combination of calculations and graphs, realized in nearly all exercises in this book, proved to be particularly instructive.

Courses with 30 attendance hours

At the University of Duisburg-Essen, two EXCEL-based courses were offered, each with 30 attendance hours:

- a basic course for beginners, in which two tasks from each of the six Chaps. 2, 3, 4, 8, 9 and 10 were worked on and had to be presented to the supervisor;
- an advanced course with two tasks each from Chaps. 5, 6 and 7, and one task that had not yet been worked on from the chapters of the beginner's course. Sometimes, two short exercises were combined into one task.

Subjects from a one-year physics course

The exercises rely on the subject matter from the first year of undergraduate physics. We do not intend to repeat physics that can be found in standard textbooks. Therefore, the introductions to the tasks are concise, but the solutions are presented in great detail. Experience shows that this creates the risk that the students might work through the exercises mechanically without caring about the physics context. To counteract this tendency, simple questions regarding physics and programming are asked in the middle of the text and answered in footnotes.

In Chaps. 8, 9 and 10, statistical concepts are illustrated in greater detail through simulations, because many students lack basic knowledge in this area. Although no theorems are logically derived, their structure should become clear, because the simulations follow the mathematical ideas.

Simple solutions preferred

The material is presented in nine chapters, each featuring about five detailed exercises. The aim is to pursue clear and simple solutions in which the physical justification for each step is traceable. To achieve this, suboptimal solutions, suboptimal with respect to computational efficiency and numerical precision, are often presented instead of solutions that are perfect from the outset. It has proven to be didactically more efficient to point out the shortcomings of this first approach and give the reader tools for improvements.

Broom rules

To many beginners, spreadsheet calculations and, especially, computer programming seem like witchcraft. We like to address this idea by setting up "broom rules" that the students hopefully will not forget so easily. Some examples: "Ψ *Half, half, full; the halves count twice*" (Runge–Kutta of the 4th order) or "Ψ *Mostly, not always.* ("fundamental rule" of statistical reasoning, no statement is 100% sure).

In addition, *Mag* puts stumbling blocks along the learning path, in talks with the two student characters, *Tim,* who learns the material from the beginner's course diligently, and *Alac,* who does not hesitate to implement premature ad-hoc solutions. It is important to emphasize that both attitudes have their advantages and shortcomings, and neither student should feel denigrated. It is just that some students have to be encouraged to venture into the programming world, whereas others have to be cautioned against rushing too quickly into coding.

Exam questions
At the end of every chapter, a collection of rehearsals and tasks is presented, typically requested in written and oral examinations.

1.6 Subject Matter

Block A, Fundamentals (Chaps. 2, 3 and 4)
Ψ The dollar makes it absolute.

The student will learn how to organize spreadsheets, design charts reasonably, and implement simple programming procedures. The necessary computational techniques are embedded in Physics tasks and trained with clearly arranged formula calculations, presentations, interpretations of curves, and simple mathematics. The reader should consult, in parallel, systematic introductions to `Excel` and `Python` for help.

Block B, Physics and Mathematics
Ψ Half, half, whole, the halves count twice.

In Chap. 5, the reader will find exercises for analysis and vector and matrix calculations in the form of a spreadsheet-specific introduction to mathematics with parallel `Python` programs.

In Chap. 6, the knowledge gained in Chaps. 2 to 5 is applied to the kinematic superposition of movements, including animated charts.

In Chap. 7, we deal with various methods for solving Newton's equation of motion and apply them to one-dimensional motions, such as a jump from the stratosphere, Exercise 7.4, or a bungee jump, Exercise 7.6.

Block C, Simulation and analysis of experiments
Ψ If in doubt, count!

Evaluation of measurements is regarded as a critical skill to be exercised at the beginning of studies in Physics. Therefore, this block is particularly detailed and illustrated with simulations based on chance, because, as experience shows, many students' most significant knowledge gaps are in the field of probability and statistics. Furthermore, statistics is the branch of mathematics that is most important outside of technical professions.

We intend to develop a good understanding of concepts through statistical experiments with random numbers without going deep into formal mathematics. Statistical rules are intended to be illustrated and checked through multiple repetitions of simulations designed to test the hit rate ("Does the error range capture the true value?"). For this purpose, random number generators are introduced in Chap. 8, e.g., for normal distributions.

The student will learn how to analyze and graphically represent measurements (Chap. 9), emphasizing the precise meaning of error ranges ("C-spec errors" related to confidence levels). Before this can be done, the measurement process must be simulated realistically to obtain data that can be evaluated. Our tools for simulation are random numbers generated according to the desired distribution.

With linear regression, mathematical functions are fitted to sets of measured values to get trend lines through data points (Chap. 10). Furthermore, an introduction to the important technique of non-linear regression with solver functions will be given, again, in both EXCEL and `Python`.

▶ **Follow-up book** A follow-up book, "Physics with Excel and Python, Using the Same Data Structure. Applications", is being prepared, dealing, in the same style, with advanced topics, structured according to physical and mathematical aspects, such as:

- properties of oscillations,
- motions in the plane,
- the steady-state Schrödinger equation,
- partial differential equations,
- Monte Carlo methods,
- wave optics,
- statistical physics, and
- variational calculus.

1.7 Getting Started with Excel

1.7.1 Start Menu

In Fig. 1.1, you see the start menu of EXCEL 2019, where the main tab FORMULAS has been activated, and the cursor has been positioned over the group

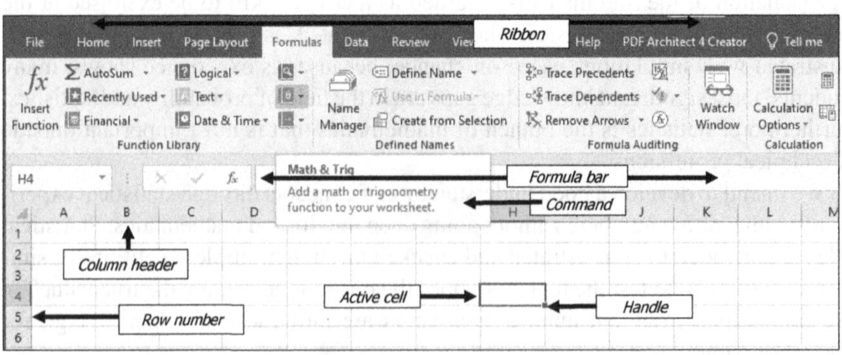

Fig. 1.1 The start menu of EXCEL, with the main tab FORMULAS activated

FUNCTION LIBRARY to show the command MATH&TRIG. Arrows indicate the different elements of the start menu, namely ribbon, tab, group, formula bar, and command, as well as elements of the working area, column header, row number, active cell, and handle of a cell. To indicate a "click path" in the text, we write a sequence TAB/GROUP/COMMAND/FUNCTION, e.g., FORMULAS/FUNCTION LIBRARY/MATH&TRIG/COS to call the cosine function.

Throughout this book, we will take screenshots from EXCEL 2019. Experience has shown that students can work with these instructions in every version of EXCEL without major difficulties.

1.7.2 Spreadsheet Presentation

Γ structure of a spreadsheet

In Fig. 1.2 (S), a spreadsheet organization in the Γ structure, often employed in our exercises, is shown. With Γ ("gamma"), we refer to the straight lines above C14:G14 and to the left of C14: C174.

Above Γ:

- the parameters of the task are defined in the range C2:C6,
- these cells get the names in B2:B6, with which they can be called in formulas,
- the most informative parameters of the exercise are integrated into a text, here, in cell E4 (with the formula in E5) that can be taken as a legend in a figure.

Left of Γ:

- the values for the independent variable *t* are in B14: B174.

	A	B	C	D	E	F	G	H
1	**Prespecifications**							
2	Amplitude of pendulum	**Ap**	1.50					
3	Period of pendulum	**Tp**	1.20					
4	Period of rotation	**Tr**	9.00		**Tp=1.2; Tr=9**			
5	Time interval	**dt**	0.0173		="Tp="&Tp&"; Tr="&Tr			
6	Suspension point vs. rot. axis	**xSh**	0.00					
7	**Calculated therefrom**							
8	Angular frequency pendulum	**wP**	5.24	=2*PI()/Tp				
9	Angular frequency rotating disc	**wR**	-0.70	=-2*PI()/Td				
10								
11			Pendulum	Trace pend. on rot. disc		Trace stylo on rot. disc		
12		=B14+dt	=Ap*COS(wP*t)+xSh	=xP*COS(wR*t)	=xP*SIN(wR*t)	=Ap*COS(wR*t)	=Ap*SIN(wR*t)	
13		**t**	**xP**	**xT**	**yT**	**xSt**	**ySt**	
14		0.000	1.50	1.50	0.00	1.50	0.00	
15		**0.017**	1.49	1.49	-0.02	1.50	-0.02	
174		2.768	-0.52	0.18	0.49	-0.53	-1.40	

Fig. 1.2 (S) Typical Γ structure of a spreadsheet, here, for the calculation of the trace of a Foucault pendulum; rows 16–173 are hidden

```
Sub Protoc()
r2 = 13
Cells(r2, 10) = "Tp"
Cells(r2, 11) = "xT"
Cells(r2, 12) = "yT"
r2 = r2 + 1
For Tp = 1 To 9
    Range("C3") = Tp
    Cells(r2, 10) = Tp
    Cells(r2, 11) = Range("D174")
    Cells(r2, 12) = Range("E174")
    r2 = r2 + 1
Next Tp
End Sub
```

Fig. 1.3 (P) Log procedure, changes the period of the rotational motion in Fig. 1.2 (S) and logs the values of x_T and y_T at the last instant of the calculation period

Below Γ:

- values are calculated from the parameters and independent variables,
- the range C14:G174 contains five columnar vectors of length 171 with the names in row 13,
- row 12 includes, in oblique orientation and in italics, the text of the formulas in the bold-printed cells of the column below. If no cell is printed in bold, the formula applies to the entire column.

Nomenclature

`Python`-typical terms are printed in the `Courier` font.

When EXCEL-typical terms are referred to in the text, e.g., function names, they are set in SMALL CAPS; examples: IF(CONDITION; THEN; ELSE) . Spreadsheet formulas are given in the form B15 = [=B14 + d*t*]. The expression in rectangular brackets corresponds exactly to the entry in the cell, including the equal sign. The equal sign specifies that it is a formula that is in the cell.

Three types of figure are distinguished, two of which are denoted by suffixes, (S) for spreadsheets, e.g., Fig. 1.2 (S), and (P) for the code of Visual Basic programs, e.g., Fig. 1.3 (P). Figures without a suffix are line drawings or screenshots, e.g., Fig. 1.1.

Names given by the programmer are printed in the text in italics, e.g., *f, d*. Sometimes in EXCEL, names are used that contain a dot, e.g., "T.1" or "x.2". This is because T1 and X2 are reserved for cell addresses. The associated variables are referenced in the text without a dot, but with subscripts, i.e., as T_1 and x_2.

Physical units

Sometimes, no physical units are specified in the axis labels of the figures. They can then be deduced from the physical units of the parameters.

1.8 Getting Started with Python

You first have to install `Anaconda` with `Python`. There are many instructions on the Internet as to how to achieve that, e.g., https://docs.anaconda.com/anaconda/install/windows/ or https://www.jcchouinard.com/install-python-with-anaconda-on-windows/ (2020-09-02).

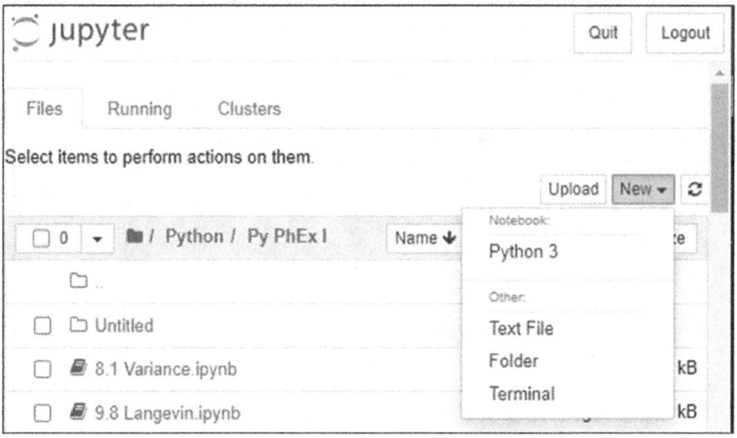

Fig. 1.4 Window opened to create a new program file

When the `Jupyter` Notebook is opened, an overview of the filers and single files on the localhost is shown. The programs used for this book are in a sub-filer "Py PhExI" of the main filer "Python". When we click *Python/Py PhEx I*, the window in Fig. 1.4 pops up. To edit an already existing file, we have to click on that file.

To create a new file, we open the list "New" and click on "Python 3". A new window pops up, opening a new file "`Untitled22`" with an empty program cell "`In [ ]`". The version in Fig. 1.5 is displayed after a small program has been written into that cell. "`In [5]`" indicates that the 5th version of the code is shown. This program is executed by clicking the button "`Run`". The result of the instruction `print[x]` is displayed in the output cell created automatically below the program cell.

1.9 Skills to Be Trained

The different programming techniques are distributed over various exercises. For the purpose of learning about them and how to revise them, the following lists of keywords and broom rules have been compiled. They are meant to assist the readers with the revision of subjects and, of course, their preparation for examinations.

Spreadsheet operations (Exercise 2.3)

- Using cell addressing, absolute, relative, indirect
- Ψ *The dollar makes it absolute.*
- Naming cell ranges and using the names in formulas

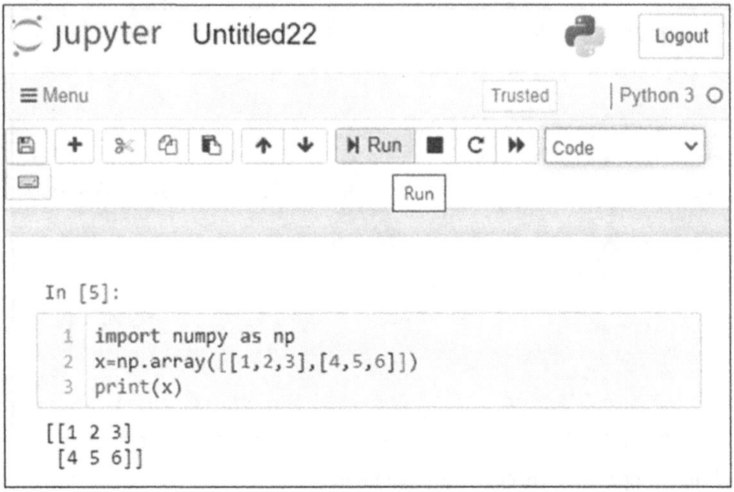

Fig. 1.5 A program creating an output just below the program cell

- Using sliders to change cell contents
- Scaling and formatting *XY* scatter diagrams
- Creating smart legends by linking text and variables, Ψ *"Text" & Variables.*
- Gamma structure of tables (Sect. 1.7.2)
- Ψ *Empty lines separate curves.*
- Ψ *Ctrl + Shift + Enter*. Magic "chord" to complete the entry of matrix functions in EXCEL (Exercise 2.6).

Python constructs (explained mainly in Exercises 2.4 and 2.5)

- Use of `numpy` arrays
- Ab-initio constructors `np.arange`, `np.linspace`
- Creating one and two-dimensional arrays with `np.array` (row vectors, column vectors, matrices)
- Broadcasting row vectors, column vectors, and matrices together in algebraic operations
- Slicing of lists
- List comprehension
- Creating smart legends by linking text and variables
- Applying a standard function to create scatter diagrams
- Creating animated figures (Exercise 6.2).

VBA-macros and Python instructions (Chap. 4)
The terms 'routines', 'programs', and 'procedures' are all used synonymously here. The term 'macro' is for VBA only.

- *For, if, Sub/def*, basic structures of programming: loops, logical queries, subroutines (Exercises 4.4 and 4.5)
- *Loop2i*, loops with a loop index and a running index incremented within the loop (Exercises 4.2 and 4.8)
- Systematically modifying parameters and recording the results of the spreadsheet calculations with *rep-log procedures*
- Processing and decoding texts for evaluation of the protocols of measuring instruments (Exercises 4.8 and 4.9)
- Writing formulas into cells with procedures (Exercises 4.1 and 4.2)
- Creating user-defined functions and using them in spreadsheet calculations (Exercise 4.9)
- Linking macros to control elements (command buttons, sliders) in spreadsheets (Sect. 4.3.3).

Mathematical techniques

Ψ *Imaging equation for lenses with plus and minus!* (Exercise 3.2).

- Using the line equation constructively (Exercise 3.2)
- Calculating with vectors in the plane and displaying them in diagrams (Exercises 5.5, 5.6, 5.7)
- Calculating with matrices (Exercise 5.9)
- Converting polar coordinates and Cartesian coordinates into one another (Chap. 6)
- Ψ *Doppler effect with plus and minus* (Exercise 3.3)
- Differentiating (Exercise 5.3) and integrating (Exercise 5.4) numerically
- Weighted sum (Exercises 5.8 and 6.5) and weighted mean (Exercise 6.5).

Functions

- Properties of the exponential function (Exercise 3.4)
- Ψ *First, the tangent at* $x = 0$*!* (Exercise 3.4)
- Ψ *Plus 1 becomes times e.* (Exercise 3.4)
- Use of the logarithm function for different computing tasks (Exercise 9.3)
- Addition of sines and cosines: overtones and beats (Exercise 2.7)
- Ψ *Cos plus Cos equals mean value times half the difference.* (Sum rule of cosines, Exercise 2.7).

Integration of Newton's equation of motion (Chap. 7)

- Ψ *Approximated average value instead of exact integral*
- Four numerical methods (Sect. 7.1.2, Exercise 7.2):
- Euler
- Progress with look-ahead (our standard procedure in a spreadsheet calculation)
- Half-step
- Runge-Kutta of fourth order, Ψ *half, half, whole, the halves count twice.* (our standard procedure as a `Python` function).

Statistics (mainly Chap. 8)

- Ψ *Mostly, not always*. Fundamental rule of statistical reasoning
- Ψ *If in doubt, count!*
- Ψ *Come to a decision! You may be wrong.* (Exercise 9.6)
- Generating random numbers with specified distribution (Exercises 8.5, 8.6, 8.7)
- Ψ *Chance is blind and checkered.* (Exercise 8.3)
- Empirical frequency distribution (Exercise 8.2), in EXCEL: Ψ *Always one more! Yes, but of what and than what?* In `Python`: Ψ *Always one less!*
- Chi^2 test for comparing theoretical and empirical frequency distributions (Exercises 8.2, 8.8)
- Multiple repetitions of random experiments to test for uniform distribution of the results of Chi^2 tests (Exercise 8.2).

Evaluation of measurements (mainly Chaps. 9, 10)

- Ψ *We know everything and play stupid.*
- Simulating measuring processes and evaluating the generated data sets statistically (Exercises 9.2, 9.9).
- Mean, standard deviation, Ψ *Two within, one out of (the standard error range)* (Exercises 9.4 to 9.8)
- Specifying measurement uncertainty (Exercises 9.2, 9.7 and 9.8)
- Multiple repetition of random experiments to test for error rate (Exercises 9.4, 9.5 and 10.3)
- For only a few repetitions of measurements, taking the t-value into account (Exercise 9.5).
- Ψ *Twice as good with four times the effort* (Exercise 9.4).
- Error propagation (Exercises 9.7, 9.8), Ψ *Calculate with variances, report the C-spec error!*
- Ψ *From variance to confidence with Student's t value* (Exercise 9.8).
- Reducing measurement uncertainty by combining measurement series (Exercise 9.6).
- Ψ *Worse makes good even better.*
- Linear regression, trend lines, coefficients with uncertainty (Sect. 10.2, 10.3 and 10.4).
- Applying non-linear regression using SOLVER in EXCEL and `curve_fit` in `Python` (Exercises 10.5, 10.6 and 10.7).
- Decoding textual logs of measuring instruments and writing the relevant data into tables using VBA macros or `Python` programs (Exercise 4.8).

Data Structures, Excel and Python Basics

2

This chapter aims to develop computational solutions for physics problems, parallel in EXCEL and Python based on the same data structure and the same type of list processing. In this way, EXCEL may serve as a low-threshold entry into scientific computation with a smooth transition to professional platforms. We proceed in three steps: (1) Basic spreadsheet techniques are introduced: absolute and relative addressing and naming of cells, creating charts, and using sliders (scroll bars), with the didactic goal of addressing variables in functions by their names, just as in mathematical formulas. (2) Python basics are explained, with emphasis on the manipulation of Numpy arrays essential for scientific computation. (3) Matrix operations are introduced in EXCEL, equivalent to those in Python. Finally, in one exercise, a set of four parabolas and, in another exercise, a group of four cosines (to simulate acoustic phenomena) are calculated and displayed in parallel in both applications.

2.1 Introduction: Named Ranges in Excel, Arrays in Numpy

Solutions of Exercises 2.3 (Excel), 2.4 (Python), 2.5 (Excel), and 2.6 (Python) can be found at the internet address: go.sn.pub/9Rtzxi.

Spreadsheet technology
This chapter is about how cells are addressed, figures are created and formatted, and sliders are used to change cell contents. This will be easy for you if you are already familiar with EXCEL and know how to write formulas into cells. If you are less experienced, you will first have to go through the basic exercise step by step. If necessary, consult the EXCEL help guide, and, finally, after having gone through the

D. Mergel, *Physics with Excel and Python*,
https://doi.org/10.1007/978-3-030-82325-2_2

basic exercise, find a textbook about EXCEL techniques that is best suited for your learning style.

Required and practiced EXCEL techniques are:

- relative and absolute cell addressing,
- direct and indirect cell addressing,
- naming of cells and cell ranges,
- creation of diagrams,
- application of sliders.

We first exercise different types of cell addressing. Our goal, however, is to write formulas as mathematically as possible, i.e., with letters representing variables. Then, they will be identical to formulas in `Python`. For this purpose, individual cells, ranges of cells within rows (row vectors) or columns (column vectors), and two-dimensional cell ranges (matrices) are to be designated by names. All of these techniques will be introduced step by step in the individual exercises and summarized again in Sect. 2.3.5. The systematic use of vectors and matrices is the reason why spreadsheet calculations can be translated nearly literally into `Python`.

Figures representing spreadsheets are characterized by the supplement (S), e.g., Fig. 2.2 (S).

Formulas in spreadsheets

Formulas in spreadsheets are reported in italic and often in oblique orientation, valid for a cell in the neighborhood in bold font. They will be written in the text in brackets; e.g., for the content of cell A11, we write A11 = [=A10 + dx]. We have to distinguish whether or not an equality sign is written in the cell. For the expressions A9 = [x] and A10 = [3], no equal sign is written in the cell; [x] is thus interpreted as text and [3] as a number.

Python constructs

We will learn `Python` programming by working with the program cell structure of the Jupyter notebook, first dealing with list processing and then focusing on operations on arrays in `Numpy`. The explanations are less technically detailed than for EXCEL, because list processing is the core business of `Python` and has been well described in numerous textbooks and online courses. However, our examples are designed so that the essence of the definitions and procedures should become obvious.

The `Python` constructs for list generation (`np.linspace` and `np.arange` in the `numpy` library) will be used to define vectors that are later transformed with standard functions into other vectors using list comprehension and broadcasting. To mimic the column vectors in spreadsheets, a two-dimensional list with only one row has to be introduced and transposed.

Matrix operations
Having provided EXCEL with the necessary matrix formulas, we can finally demonstrate broadcasting for algebraic operations and some operations of linear algebra parallel in EXCEL and `Python` (Exercise 2.5).

Applications
We practice our newly acquired knowledge in an exercise on four parabolas and their upper envelope. The chapter concludes with a physically meaningful exercise treating the sum of four cosines so as to demonstrate overtones, beats, and the addition theorem of cosines.

2.2 Characteristics of a Parabola

Starting from its vertex form, we set up a data structure to tabulate and plot a parabola, together with its characteristic features *focus* and *directrix*. The data structure set up here is to be used in Exercise 2.3 for a single parabola and in Exercise 2.6 for a set of 4 parabolas.

2.2.1 Different Definitions of a Parabola

Parabola from vertex
A parabola is to be presented in a diagram. Its standard form is defined as

$$y = a + b \cdot x + c \cdot x^2 \tag{2.1}$$

It is, however, more intuitive to start from its vertex form

$$y = y_V + c \cdot (x - x_V)^2 \tag{2.2}$$

because its shape is immediately clear: its vertex is at (x_V, y_V), and its curvature is proportional to c (positive or negative). Transforming into the standard form yields

$$a = y_V + c \cdot x_V^2 \text{ and } b = -2 \cdot c * x_V \tag{2.3}$$

In Fig. 2.1a, a parabola is shown with its vertex marked with a diamond.

Question

What are the coordinates (x_V, y_V) of the vertex and the value of c in Fig. 2.1a?[1]

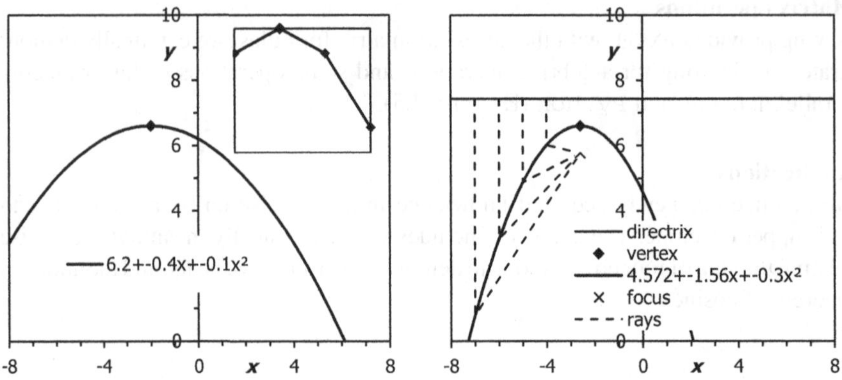

Fig. 2.1 **a** (left) A parabola with its maximum at (−2.6, 6.6). **b** (right) Connecting the parabola with its focus and its directrix; compare Fig. 2.10

Focus and directrix

A parabola may also be defined as the locus with equal distance to the focus (x_F, y_F) and to the directrix ($y = y_D$), both of which are depicted in Fig. 2.1b. The coordinates of the focus and the directrix are:

$$x_F = x_V \quad y_F = y_V + \frac{c}{4} \quad y_D = y_V - \frac{c}{4} \tag{2.4}$$

Ignoring reflection, all rays incident parallel to the symmetry axis of the parabola have the same length up to the directrix, e.g., when calculated from the line $y = 0$. By definition, they all have the same length up to the focus when reflected at the parabola. This is why light incident parallel to the optical axis is focused in the focus of a parabolic mirror.

Calculation points of a curve

Equations 2.1 and 2.2 represent continuous functions. In numerical calculations, however, the function values y are calculated on only a finite number of discrete x-values, x_i (see the inset picture in Fig. 2.1a). The points (x_i, y_i) are called the calculation points of the function. They are the vertices of the polyline representing the graph. In most of our exercises, the adjacent calculation points are equidistant on the x axis , i.e., they all have the same distance to their respective neighbors, usually specified in a parameter dx.

[1] $x_V \approx -2$, $y_V \approx 6.5$ from visual inspection in the coordinate system; $c = -0.1$ from the legend in the figure.

2.2.2 Data Structure and Nomenclature

(x_V, y_V)	coordinates of the vertex of the parabola
a, b, c	coefficients of the parabola in standard form
y_D	ordinate value of the directrix of the parabola
(x_F, y_F)	focus of the parabola
$\mathrm{d}x$	distance between adjacent x values of calculation points
x	sequence of x values
y_P	values of the polynomial for x
y_A, y_B, y_C, y_D	values of the same polynomial but calculated in Excel with column names
y_{Max}	upper envelope of y_A, y_B, y_C, y_D.

2.3 Basic Exercise in Spreadsheet Calculation

With the example of tabulating a parabola and displaying it in a diagram, we train the basic spreadsheet techniques: absolute and relative cell addressing, providing cells with names, connecting text and numeric variables to get informative legends for diagrams, applying scroll bars to change cell contents without typing numbers. After performing these exercises, the reader should be able to select a more detailed textbook on EXCEL techniques that is best suited to her/his taste and needs. At the end of the exercise, an analogous `Python` program with the same data structure is presented.

2.3.1 Cell Addressing

Spreadsheet layout

A spreadsheet layout for generating a parabola $y_P = f(x)$ is displayed in Fig. 2.2 (S). We are going to review three regions successively: A1:B6 (to explain cell addressing), D1:H3 (to name cells and apply sliders), and A8:E169 (to name cell ranges). All relevant cells and the ranges x, y_P, and y_A get names from the beginning. The formulas in cells are printed in neighboring cells in italic to keep track of the calculations.

Relative and absolute cell addressing

The coefficients of the vertex form of the parabola, x_V, y_V, and c, are defined in B1:B3 and recalculated in B4:B5 into b and a of the standard form, with the formulas in C4:C5. The value in B6 specifies the horizontal distance $\mathrm{d}x$ between the calculation points.

	A	B	C	D	E	F	G	H
1	xV	-2	vertex	=(E1-40)/5	30			
2	yV	6.6	vertex	=E2/10	66			
3	c_	-0.1		=(E3-50)/10	49			
4	b	-0.4	=-2*B3*B1	=-2*c_*xV				
5	a	6.2	=B2+B3*B1^2	=yV+c_*xV^2				
6	dx	0.1		6.2+-0.4x+-0.1x^2		=a&"+"&b&"x+"&c_&"x^2"		
7	=A9+$B$6	=$B$5+$B$4*A10+$B$3*A10^2			=a+b*x+c_*x^2			
8	x	yP			yA			
9	-8.0	3.0			3.0			
10	**-7.9**	**3.1**			3.1			
169	8.0	-3.4			-3.4			

Fig. 2.2 (S) Tabulation of a parabola with the parameters *c*, *b*, *a* in B3:B5; F1:H3, sliders to determine the parameters x_V, y_V, *c*; E8:E169, alternative tabulation of the parabola using the named cells *a*, *b*, *c_* and the column range *x*

The independent variables *x* (left of Γ) are in A9:A169. They are obtained by entering the initial value, here −8.0, into the first cell A9 of the range and the formula A10 = A9 + $B$6 or A10 = [=A9 + dx] into the next cell and copying this formula into the whole range. Copying is done by seizing the handle of A9 (see the small black square at the lower right edge in the subpicture in column C) with the pointer and dragging it down to A169.

▶ **Task** Change the contents of cell B6, named "d*x*"! All *x* values in cells A10:A169 should adjust themselves immediately.

The values for y_P are obtained by entering the formula reported in B7 into B10 and dragging it up to B9 and down to B169. The formula is most conveniently obtained by first entering [=], and then by clicking on the corresponding cells and continuing with the operators * for multiplication and ^ for potentiation, resulting in B10 = [=B5 + B4*A10 + B3*A10^2]. In the last term, the variable taken from A10 must be squared. This is done with the power operator ^. You have to press the button with the ^-sign and then the desired power, "2" in our case. Only after the second step does the operator [^2] appear in the cell.

When this formula is copied into another cell, the cell addresses change accordingly. Copying into C11 would yield C11 = [=C6 + C5*B11-C4*B11^2), realizing *relative* addressing but not giving the desired result, because we would like to keep the cells with the coefficients constant. This is achieved by making the references to B3:B5 *absolute*, with dollars as prefixes: $B$5, $B$4, $B$3, either by introducing the $ sign explicitly before the column letter or the row number or by pressing the function key F4, resulting in the formula reported in B7.

Having now copied this into C11, we would get C11 = [=$B$5 + $B$4*B10-$B$3*B10^2], with the values for x copied incorrectly (A10 becomes B10), because they are still relatively addressed. Making the address of column A absolute is achieved with $A10, B10 = [=$B$5 + $B$4*$A10-$B$3*$A10^2]. If we now copy this formula into another cell, only the row number 10 changes, e.g., into 12 when copied into any column in row 12.

Making a cell reference absolute can also be achieved by pressing the function key F4 several times. Key words for the EXCEL help: Absolute, relative, and mixed cell references.

There is a more elegant way to copy a formula down a column: clicking onto the "fill handle" (the bottom right corner) of the cell that contains the formula. Then, the cell contents are immediately continued down to the 169th row, i.e., for all cells for which there is an entry in the neighboring column, here column A, until the first empty neighbor cell is encountered.

Questions

Questions concerning Fig. 2.2 (S):

From A9 to A10, x increases by $dx = 0.1$. Why is the increase 15.7 for the next jump from −7.9 to + 8.0?[2]

Where is dx defined?[3]

Why is the name c_ in E7 provided with an underscore?[4]

Having gotten this far, we have programmed our first function. Changing the values of x_V, y_V, and $c\_$ in B1, B2, B3, the function values in column B adapt immediately. We may now proceed to the section "Graphical representation" to see the resulting curve, but should come back to learn about naming cells.

Naming cells

In the range B1:B6 of Fig. 2.2 (S), parameters are defined that are accessed in various parts of the worksheet. To call them like variables in mathematical equations, we provide the cells with the names written to the left of them. This is done by activating the range A1:B6 comprising names and values and clicking through (EXCEL 2019):

FORMULAS/DEFINED NAMES/CREATE FROM SELECTION.

A prompt appears, "CREATE NAMES FROM VALUES IN THE ☑LEFT COLUMN?". The answer is YES, that the agent has correctly detected, and we confirm this by clicking OK. For more about the NAME MANAGER, see Sect. 2.3.5. We can now refer to these cells by their names, anywhere in the spreadsheet and, indeed, throughout the whole book.

When writing, e.g., [=a] into a cell somewhere in the spreadsheet and pressing ENTER, this cell immediately gets the numerical value corresponding to a, in our

---

[2] Rows 11 to 168 are hidden. The jump is over 159 advances of dx; -7.9 + 15.9 = 8.0.

[3] The value for dx is set in cell B6, which is given the name in A6.

[4] The name c is protected for EXCEL-internal use. A name c in a cell intended to become an identifier is automatically extended by an underscore.

example, 6.2. When the content of cell B5 is changed, the value of all cells with [=a] changes as well.

We can even name cell ranges, e.g., A9:A169 with the name *x* and B9:B169 with *yP*. When activating A8:B169 and proceeding as above, the Name Manager prompts us: "CREATE NAMES FROM VALUES IN THE ☑ TOP ROW?", and we confirm this by clicking OK.

We may now write $= a + b*x + c\_*x\hat{}2$ into E9, more elegant and clearer than the formula in B7, with absolute and relative cell addressing, and copy down to D 169 to get the same values as for *y*P. For *x*, the formula in a cell takes the value in the same row in column A. The name manager has changed *c* into *c*_ because the letter *c* conflicts with a protected name in EXCEL.

2.3.2 Graphical Representation of a Function

After having created the function table of a parabola so beautifully, we would like to visualize the curve. To do so, we set the cursor into an empty cell, away from the filled cells, and click in the INSERT tab on INSERT/ CHARTS/, and on SCATTER within the CHARTS section. A blank chart is inserted.

Upon our activation of DESIGN/SELECT DATA (see Fig. 2.5b), a SELECT DATA SOURCE window opens. We click ADD, and a window as shown in Fig. 2.3 opens. For SERIES NAME, we click on cell B8 of Fig. 2.2, and for SERIES X VALUES, we activate A9:A169 and hit return. For SERIES Y VALUES, we activate B9:B169 and hit return. The empty chart changes to that shown in Fig. 2.4.

As the legend "yP" has been taken from a cell in the spreadsheet, it will adapt immediately when the cell entry is changed. If a legend in a chart is identical to that in the spreadsheet indicating the data, it helps in keeping the overview.

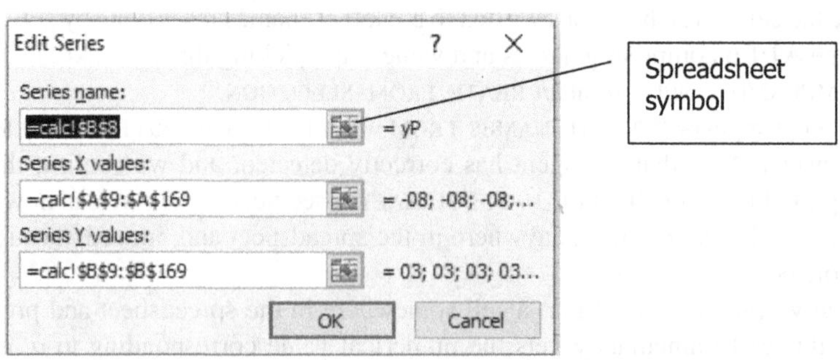

Fig. 2.3 Insertion of a data series into a chart; the series name is best taken from the worksheet by activating the relevant range, *not by entering it as text*

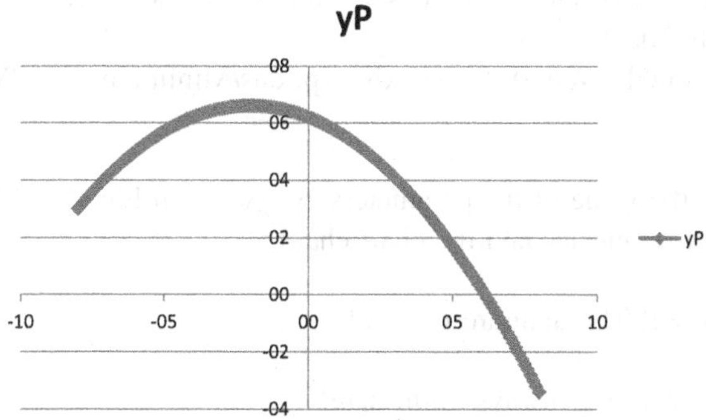

Fig. 2.4 Scatter plot after completing Fig. 2.3

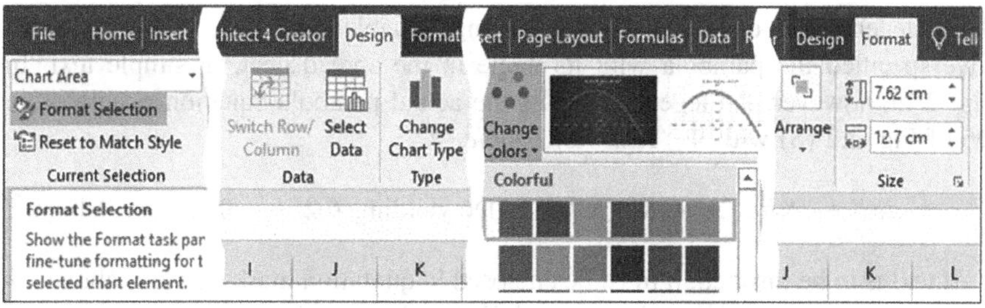

Fig. 2.5 Tabs, which are important for diagrams, after INSERT/CHARTS/SCATTER/WITH ONLY MARKERS in the Start menu (EXCEL 2019, EXCEL 2010 similar) or after activating an existing chart, **a** (left) FORMAT/CURRENT SELECTION, to the left of the start bar, to format an element of the diagram, **b** (center left) DESIGN tab / DATA group /SELECT DATA to select data to be entered into the diagram, **c** (center right) DESIGN tab /CHART STYLES, **d** (right) FORMAT/SIZE, appears after activating a diagram, to the right of the start bar, to specify the size of the diagram

Formatting the chart

We reshape the diagram according to our taste, e.g., as in Fig. 2.1a. After clicking on the diagram and HOME/FORMAT, the following components change:

- Size (7 cm high, 8 cm wide) (in the start bar to the far right, see Fig. 2.5d).

Also, after clicking on the relevant element of the diagram or selecting it from the leftmost register in the FORMAT tab (Fig. 2.5a, at first, only CHART AREA appears, but after opening the list by clicking on ▼, all items of the diagram appear), we choose before clicking FORMAT SELECTION:

- Format/Chart Area/Border /No line
- Format/Plot Area/Border Color/Solid line/Color/Black and Width/1pt

- Format/Data Series/Marker Options / Built in/Diamond, Size 4 and Marker Fill/Solid Fill/ Black
- Format/Horizontal (Value) Axis/Axis Options/Minimum /-8, /Maximum /8, /Major Unit/4.

▶ **Task** Change the value of the parameters x_V, y_V, $c\_$ in Fig. 2.2 (S) and observe how the spreadsheet entries and the chart change!

▶ **Alac** That's cool! The diagram is alive!

▶ **Tim** Once created, it's always up-to-date!

2.3.3 Smart Legends in Figures

Smart legends are created by concatenating variables and text. In Fig. 2.4, we have specified the parabola with its name in the spreadsheet, in simple text. In Fig. 2.1a, however, the legend contains the actual parabola equation generated in D6 of Fig. 2.2 (S) with Ψ *"Text" &variable.*

= *a*&" + "&*b*&"*x* + "&*c*_&"*x*" in D6 yielding 6.2 + −0.4*x* + −0.1*x*

If text is to be inserted, it must be enclosed in quotation marks, e.g., as above, "x + ". The concatenation operator is "&". When the parameters *a*, *b*, or *c*_ change, the text in D6 and the legend in the chart will immediately follow.

▶ EXCEL Ψ *"Text"& Variable* concatenation with &

▶ `Python` Ψ *„Text " + Str(Variable)* concatenation with +

Often, float values have to be rounded. When, e.g., B3 = [= 1/3], then only 0.33 is displayed in the cell. However, if B3 is inserted into a legend, 0.333333333 appears. Setting ROUND(B3, 2), 0.33 is returned: [= "c_ = "&ROUND(B3; 2)] results in the cell content [c_ = 0.33].

2.3.4 Scroll Bars

Cell contents are changed with a slider (a scroll bar)

We can play even more impressively with the curves when we change their parameters with sliders. To introduce sliders, we select DEVELOPER/CONTROLS/INSERT/ACTIVEX CONTROLS to get the tabs in Fig. 2.6a.

If the tab does not appear in your toolbar, you have to supplement the toolbar by making a tick ☑ in FILE\OPTIONS\CUSTOMIZE RIBBON in the box before DEVELOPER.

We need the slider to be an ACTIVEX CONTROL ELEMENT, and we click on the icon for the slider in the top line on the far right, with the DESIGN MODE turned on,

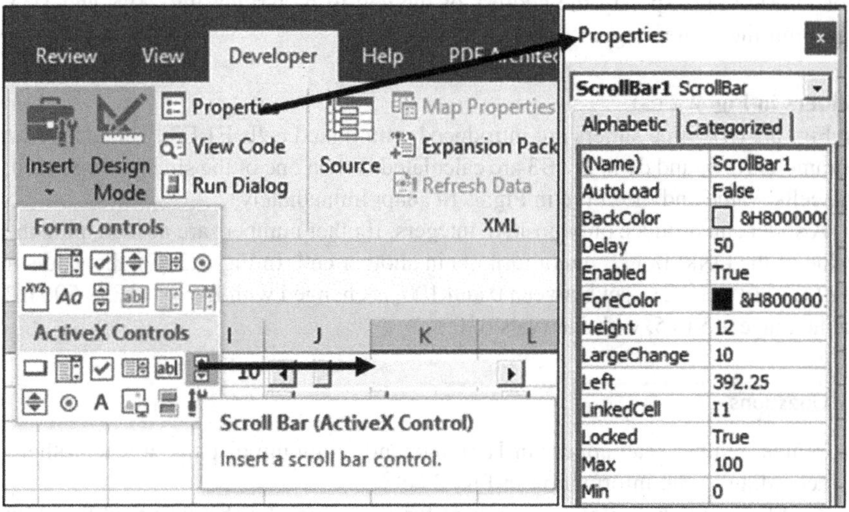

Fig. 2.6 (EXCEL 2019, EXCEL 2010 similar) **a** (left) Tab after going through DEVELOPER/CONTROLS/INSERT/ACTIVEX CONTROLS. A slider is listed in the upper row on the right, at the foot of the arrow. **b** (center) In J1:L1, a slider has been installed by clicking on the control element and pulling it up in said spreadsheet area. **c** (right) Menu for defining the properties of the slider (SCROLLBAR). It appears after we click on PROPERTIES in **a**; important parameters: LINKEDCELL, MINimum and MAXimum values

and then pull up a rectangle with the mouse at the desired place in the spreadsheet. In Fig. 2.6b, this is done in cells J1:L1. Now, with the DESIGN MODE still on, we can configure the slider. In Fig. 2.6c, the PROPERTIES list of the activated scrollbar is shown. We specify that it is cell I1 (LINKEDCELL) into which the number is to be written, and that the number should be between 0 (MIN) and 100 (MAX). We then turn the DESIGN MODE off by clicking on this icon again (see Fig. 2.6a) and move the slider's thumb with the mouse. Immediately, a number appears in I1.

DESIGN MODE is activated and deactivated by clicking on the icon. When activated, existing control elements can be modified, or new ones added. When deactivated, the control elements can be operated.

When we grab the thumb (the rectangular bar ▯ in the middle of the slider) with the cursor and move it, the output in the linked cell changes. SMALLCHANGE specifies the jumps (to the left or the right) of the numbers when we click on the (left ◁ or right ▷) edge of the slider. LARGECHANGE sets the jumps' size when we click *within the slider bar* left or right of the thumb. Try it out!

Conversely, if we change the contents of a LINKEDCELL, the new value is entered into the slider's memory, and its thumb will move.

We can use a slider (SCROLL BAR) to enter integers between 0 and 32,767 (= $2^{15} - 1$) into a cell (LINKEDCELL). Settings are specified in the PROPERTIES group:

For LINKEDCELL, specify the address of the cell to be written into. MIN and MAX can limit the value range.

Sliders in Fig. 2.2 (S)
In Fig. 2.2 (S), three sliders are introduced with linked cells E1:E3, wherefrom the parameters *a*, *b*, and *dx* in B1:B3 are calculated. When one of the sliders is operated, the cells' value and the curve in Fig. 2.1a adapt immediately.

A slider can produce only positive integers. If other numbers are needed, plug the value of the LINKEDCELL into a formula in another cell. In Fig. 2.2 (S), the value of the LINKEDCELL (E3), set between 0 and 100, is changed with B3 = [=(E3−50)/10] to the range −5 to 5, with intervals of 0.1.

Questions

Which numbers can appear in I1 (range and minimum distance to each other), according to the information in Fig. 2.6c?[5]

What is the range of the numbers for x_V in Fig. 2.2 (S), provided that Min = 0 and Max = 80 for the slider in F1:H1. What is the minimum step size?[6]

Connecting a VBA macro to a control element
The following will be important in later chapters: We can connect a VBA routine to the slider (an example is given in Exercise 10.5). A routine SUB SCROLLBAR1_CHANGE( ) … END SUB is executed each time we move the SCROLLBAR1 slider. To enter a code, click on VIEW CODE (Fig. 2.6b) and select the considered slider's name in the left-hand drop-down window that has popped up, then click CHANGE (or some other action to trigger the routine) in the right-hand drop-down window.

2.3.5 Summary: Cell References and Name Manager

▶ Ψ *The dollar makes it absolute.*

Cell references, relative, absolute, indirect
The following formulas in a cell, C3 = [=A4], C3 = [=$A4], C3 = [=A$4], C3 = [=$A$4], yield the same result: the value of cell A4 is written into the active cell C3. However, if these formulas are copied into another cell, they change. For example, in D4, there will be: [=B5], [=$A5], [=B$4], [=$A$4]. A $ sign before a column label or a row number has the effect that the label or number are held fixed when copied. A dollar makes a cell reference *absolute*. If the $ is missing, the label is incremented

[5] MIN = 0; MAX = 100; integers between 0 and 100 can appear in I1, distance = 1.
[6] B1 = [=(E1-40)/5] (see D1), range -8 (for E1 = 0) to + 8 (for E1 = 80) in steps of 1/5 = 0.2.

by the column or line spacing between the old and new cells; this is a *relative* cell reference.

The spreadsheet function INDIRECT(CELL) expects a cell address as an argument. It writes the contents of the cell with this address into the current cell. For example, with A4 = [=INDIRECt(A5)] ; A5 = [X7]; X7 = [3.4] the value in A4 will be 3.4.

Assigning names to cell ranges

[=a*x + b] instead of [=B$2*$A6 + B$3]

Individual cells, ranges in a column ("column vectors") or a row ("row vectors"), and rectangular ranges ("matrices") can be named and then inserted with their names into formulas and worksheet functions as arguments. Naming is done with the NAME MANAGER by going through FORMULAS/NAME MANAGER. For details on the NAME MANAGER, see below. Make extensive use of this feature! Doing so, you can write formulas and functions in mathematical language, e.g., [= A*SIN(k*x)] instead of [=A$1*SIN($A5*B$2)]! Such a formula is valid for a set of sine functions whose amplitudes A and wave number vectors k are stored in two row vectors with names "A" and "k" and where the independent variable x is stored in a column vector with the name "x".

Name Manager of Excel

As an example to demonstrate the properties of the NAME NANAGER, we calculate the electric field E_x in x-direction of a point charge at $(x_1, 0)$ in the xy-plane:

$$E_x = \frac{x - x_1}{\sqrt{(x - x_1)^2 + y^2}^{\,3}} \tag{2.5}$$

The definition of variables and constants and the calculation are distributed over two sheets, "Dist" and "E.x". The values of $r = \sqrt{(x - x_1)^2 + y^2}$, representing the planar distance to point $(x_1, 0)$ for x and y from -2 to 2, are calculated in Fig. 2.7a (S) (Sheet "Dist").

The x values are in B3 to F3, the y values in A4 to A8. The value for x_1 is specified in B1. The calculation can be performed with mixed cell references, as in cell B5:

$$\mathrm{B5} = \left[= \mathrm{SQRT}\left((\mathrm{B\$3} - \mathrm{Dist!\$B\$1})\char`^2 + \mathrm{A\$5}\char`^2\right)\right] \tag{2.6}$$

The formula is, however, more intuitive when cell names are used, as in D6, displayed in D2:

$$\mathrm{D6} = \left[= \mathrm{SQRT}\left((\mathrm{x} - \mathrm{x.1})\char`^2 + \mathrm{y}\char`^2\right)\right] \tag{2.7}$$

To achieve this, we have to designate, e.g., range B3:F3 with the name *x*, already present in cell G3. We activate B3:G3 and follow the menu FORMULAS/DEFINED NAMES/CREATE FROM SELECTION. To "activate" the range means

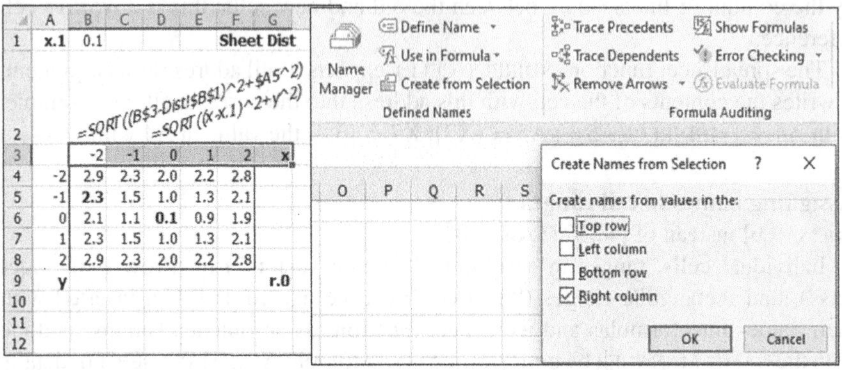

Fig. 2.7 (S) **a** (left) Sheet "Dist"; the values in the matrix B4:F8 are calculated with the values in the horizontal vector x = B3:F3 and the vertical vector y = A4:A8, together with the value of x_1 in cell $B$1, using mixed cell references in B5 and variable names in D6. **b** (right) After activating B3:G3 in **a**, the dialogue box of FORMULAS/ DEFINED NAMES suggests providing B3:F3 with the name in G3 (☑ RIGHT COLUMN)

that the cells are marked with the left mouse button pressed. A window like the one shown in Fig. 2.7b (S) pops up. The assistant has already recognized a potential name in the immediate neighborhood of the activated range, namely, in the right-most column of the activated range. This name corresponds to our intention, and we click OK.

The matrix range B4:F8 is named by activating it and selecting FORMULAS/DEFINED NAMES/DEFINE NAME. A window marked NEW NAME pops up with the REFERS TO field already filled in, because a cell range was activated before the selection. We have to fill in the NAME field, in our case, with "r.0".

The FORMULAS/DEFINED NAMES/NAME MANAGER window, displayed in Fig. 2.8 (S), gives us an overview of all of the named ranges.

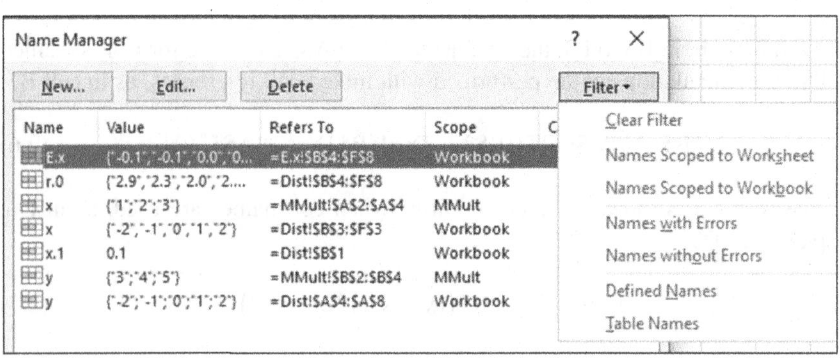

Fig. 2.8 (S) The NAME MANAGER lists all names and ranges ("refers to") and the scopes for which they are valid

	A	B	C	D	E	F	G
1					Sheet E.x		
2				=(x-x.1)/r.0^3			
3		###	###	###	###	###	#VALUE!
4		-0.1	-0.1	0.0	0.1	0.1	#VALUE!
5		-0.2	-0.3	-0.1	0.4	0.2	#VALUE!
6		-0.2	-0.8	-100	1.2	0.3	#VALUE!
7		-0.2	-0.3	-0.1	0.4	0.2	#VALUE!
8		-0.1	-0.1	0.0	0.1	0.1	#VALUE!
9							

	I	J	K	L	M	N	O
14							
15			{=(x-x.1)/r.0^3}		as matrix formula		
16	-0.1	-0.1	0.0	0.1	0.1	#N/A	#N/A
17	-0.2	-0.3	-0.1	0.4	0.2	#N/A	#N/A
18	-0.2	-0.8	-100.0	1.2	0.3	#N/A	#N/A
19	-0.2	-0.3	-0.1	0.4	0.2	#N/A	#N/A
20	-0.1	-0.1	0.0	0.1	0.1	#N/A	#N/A
21	#N/A	#N/A	#N/A	#N/A	#N/A	#N/A	#N/A
22							

Fig. 2.9 **a** (left, S) Sheet "E.x". The x component of the electric field of a point charge at (x_1, 0) is calculated. In row 3 and column G, the index is outside of the permitted range. **b** (right, S) Alternative calculation in an arbitrary position with a matrix formula

The names are valid in the entire workbook when they are first defined. The names "x" and "y" appear twice. When they were first defined in the "Dist" spreadsheet, they were valid throughout the workbook, as shown in the SCOPE column. When they are defined a second time in another sheet, here, "MMult", their scope is limited to this spreadsheet, and the previous definitions of x and y do not apply here, although they still do in the rest of the workbook.

Calculating an electric field

In sheet "E.x" [Fig. 2.9 (S)], the x-component E_x of the electric field is calculated:

$$E_x(x, y) = \frac{x - x_1}{r_0^3}$$

▶ **Tim** Which definitions for x, x_1, and r_0 are valid in sheet "E.x"?

▶ **Mag** You can find this out by trial and error or by using FILTER in the NAME MANAGER.[7]

Referring to names in a cell

The ranges (vectors x and y, matrix r_0, constant x_1) defined in the two sheets can be called in each sheet cell-wise, however, they cannot be so within the complete range of the sheet, but only within a range that matches the range in the sheet in which the name is defined, for example, the matrix r_0 only within the range B4:F8. Outside of this range, errors are reported (see Fig. 2.9a). The horizontal vector x may, for the current definition of its coordinates, only be called in columns B to F, the vertical vector y only in rows 4 to 8. The constant x_1 designates only a single cell and may be called in the whole file without restriction.

[7] The named ranges in sheet "Dist" are valid throughout the whole workbook, except for "MMult", so they are also so in "E.x".

Referring to names in a matrix function

The matrix r_0 can be called within any range of a spreadsheet when entering the formulas as a matrix function, e.g., in I16:M20 in Fig. 2.9b (S). To do so, activate the desired range, write the desired formula into the formula bar, in our case [=(x-x.1)/r.0^3], and finish with the "magic chord"! The matrix formula is enclosed in curly brackets: {=(x − x.1)/r.0^3}.

▶ Magic chord for completing matrix functions: Ψ *CTRL* + *SHIFT* + *ENTER*.

2.3.6 What Have We Learned so Far, and How to Proceed Further?

▶ **Alac** That's all really super easy. Worksheet calculation seems to be a children's game. That wasn't clear to me until now.

▶ **Tim** Well. Might we have acquired knowledge in only a narrow section for particular tasks?

▶ **Mag** We have traveled quickly across a wide area on a narrow path. This is actually a fast track to success, at least for the tasks we intend to tackle.

▶ **Tim** Is it not better to learn thoroughly so that one does not become lost when the tasks are set a little differently?

▶ **Alac** You can always tackle modified tasks through trial and error.

▶ **Mag** Yes, trial and error is a possibility. You should do this anyway with all of the programming constructs with which you are not already familiar. Nevertheless, you should also go to a bookstore or into the internet and find books that instruct you in EXCEL. Browse along the learning path you have just gotten to know. It will not take long for you to figure out which of the books explains the computational procedures in a way that you can understand. *You should buy that book*!

The same advice holds for Python, which you will get to know in the next section.

2.3.7 Python Program

Table 2.1 presents a `Python` program corresponding to Fig. 2.2 (S) for completeness. The reader may study it after having gone through Exercise 2.4.

The program consists of a list of simple assignments of type x_V = -2, with the name x_V being called the identifier, or simple formulas assigned to an identifier,

Table 2.1 Specifications corresponding to Fig. 2.2 (S)

```
xV=-2                      #Coordinates of vertex
yV=6.6
c=-0.1                     #Curvature
b=-2*c*xV                  #Coefficients of standard form
a=yV+c*xV**2
dx=0.1                     #Horizontal distance between x
x=np.arange(-8,8+dx,dx)     #Array of x values
yP=a+b*x+c*x**2             #Array of y values
print('a ={:5.2f}'.format(a),'; b ={:5.2f}'.format(b))
```

```
a = 6.20 ; b =-0.40
```

Table 2.2 Calculating the coordinates (for Fig. 2.1b) of the focus and the directrix

```
xF=xV                       #Coordinates of the focus
yF=yV+1/4/c
yD=yV-1/4/c                 #y Coordinate of directrix
```

e.g., $b = -2*c*x_V$ recurring to variables specified earlier. The calculated values of a and b (printed into the second cell) are the same as in Fig. 2.2 (S).

The term 'a = {:5.2f}' indicates that the printout starts with the text string 'a = ' and that the value of the variable *a* in format(a) is to be printed right aligned (:) in float format (f) with length 5 and 2 decimal places (5.2).

The coordinates of the focus and the directrix are calculated in Table 2.2 with the same formulas as in Fig. 2.2 (S). A plot corresponding to Fig. 2.1b is produced with the standard program `FigStd` explained in Sect. 2.4.5.

2.4 Python and NumPy Basics

We will learn how to work with program cells in the `Jupyter` notebook, and get acquainted with arrays in `numpy`. For more detailed information about `Python` specifics, the reader is referred to *Stewart, J. (2014). Python for Scientists*. Cambridge: Cambridge University Press. https://doi.org/10.1017/CBO9781107447875, Chap. 3 (*A Short Python Tutorial*), Chap. 4 (*NumPy*), and Chap. 5 (*Two-Dimensional Graphics*) or other textbooks treating the same subjects.

2.4.1 Basic Exercise

Table 2.3 displays a simple `Python` program developed in the `Jupyter` notebook.

Table 2.3 Cell structure of `Python` in the `Jupyter` notebook

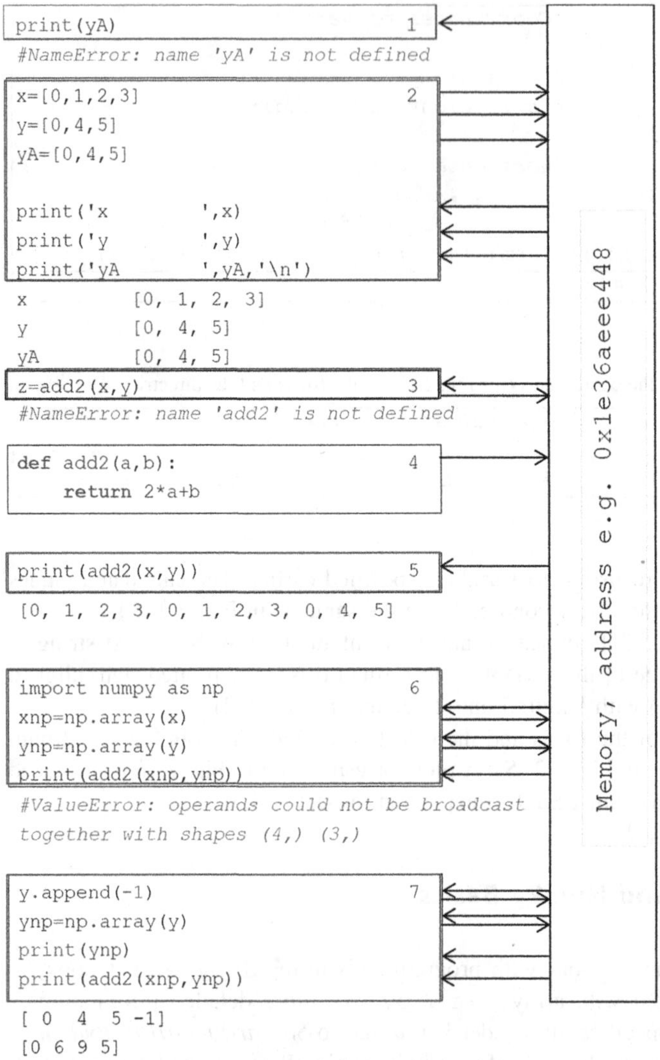

print(yA) 1
#NameError: name 'yA' is not defined
x=[0,1,2,3] 2
y=[0,4,5]
yA=[0,4,5]
print('x ',x)
print('y ',y)
print('yA ',yA,'\n')
x [0, 1, 2, 3]
y [0, 4, 5]
yA [0, 4, 5]
z=add2(x,y) 3
#NameError: name 'add2' is not defined
def add2(a,b): 4
return 2*a+b
print(add2(x,y)) 5
[0, 1, 2, 3, 0, 1, 2, 3, 0, 4, 5]
import numpy as np 6
xnp=np.array(x)
ynp=np.array(y)
print(add2(xnp,ynp))
#ValueError: operands could not be broadcast together with shapes (4,) (3,)
y.append(-1) 7
ynp=np.array(y)
print(ynp)
print(add2(xnp,ynp))
[0 4 5 -1]
[0 6 9 5]
Memory, address e.g. 0x1e36aeee448

We can start running the program with any cell. However, if referring to variables or functions, we have to define them ahead of time. Starting Cell 1 before Cell 2, or Cell 3 before Cell 4, leads to a *#NameError*.

We are now running the program cell by cell in sequence.

Cell 1: The print statement tries to get y_A from the memory. As y_A is not yet defined, a *#NameError* message is returned.

Cell 2: Three lists, characterized by square brackets, are defined, and their contents are printed out. When we now run Cell 1 again, no error message appears, but the contents of y_A are printed out.

Cell 3: The statement tries to assign a value to variable z by calling a function *add2* that is not yet defined, so that a name error occurs.

Cell 4: The function *add2* is defined. Now, running Cell 3 does not result in an error message.

Cell 5: Function *add2* is applied to variables x and y, and the result is printed out. When applied to lists, the formula $2*a + b$ is not an algebraic function; list x is concatenated twice and then list y once to yield the list $[x, x, y]$.

Cell 6: The `numpy` library is imported under the abbreviation `np`. Two `numpy` arrays are created from the lists x and y. The function *add2* is applied to x_{np} and y_{np}, yielding an error message. When applied to arrays, it is the algebraic operations addition and multiplication that are performed element-wise. As the two arrays have different size, this does not work, and a *#ValueError* message results: "*Operands could not be broadcast together with shapes (4,) and (3,)*".

Cell 7: List y is extended with a new element to have the same length as *xnp*. The function *add2* can now be applied, with the result now being an array of the same length and a linear combination of the two initial arrays. The formula $2*a + b$ is now interpreted as an algebraic operation and performed element-wise on the arrays.

Copy

Parts of the memory may get various names, e.g., the statement `AddTwo` = `add2` assigns the additional identifier `AddTwo` to the function *add2* defined in Cell 4 of Table 2.3, so that this function may also be called by `AddTwo`. Such assignments are different from making a copy, as is demonstrated in Table 2.4. The statement YA = Y[3:5] produces only a name (an identifier) for part of list Y, whereas with `YC` = `np.copy(Y[3:5])`, a new object with its own memory space is created. Y_C is an object of its own and is not affected by subsequent changes of Y.

Table 2.4 Identifier of a subarray vs. copying a subarray into a new object; only one print statement is reported in the first cell; the other print statements are similar. The results of all print statements are reported in the second and the fourth cells

```
Y=np.array([0,1,2,3,4,5,6,7,8,9])
YA=Y[3:5]
YC=np.copy(Y[3:5])
print("Y     ", Y)
```

```
Y     [0 1 2 3 4 5 6 7 8 9]
YA    [3 4]
YC    [3 4]

```

```
Y[3]=8              #Another program line
```

```
Y     [0 1 2 8 4 5 6 7 8 9]
YA    [8 4]
YC    [3 4]              #YC is independent of Y
```

2.4.2 Data Structures

Data types

The following data types are available in Python:

- `Int` (integer, unlimited size),
- `float` (8 Bytes),
- `bool` (boolean),
- `string` (text),
- `complex` (only in `numpy`).

Lists

Lists contain one or more items. They:

- may contain items of different type: `z = [1, 3.14, 'abc'. 2 + 4]`,
- are mutable, e.g., can be changed by `append` or `delete`; `x.append('new')` → `[x, 'new']`,
- can be concatenated by +: `[x] + [y]` → `[x, y]`,
- can be concatenated by *: `3*[x]` → `[x, x, x]`,
- can be sliced, i.e., subarrays can be addressed with new identifiers: `x2 = [3:-1], from the fourth (indices 0,1,2,3,...) to the penultimate element of` x.

Append, *add*, and *multiply* result in longer lists. The new identifier x2 points to a sublist of x from the 3rd to the penultimate (-1) element. The method `append`, as well as the operators + and *, are demonstrated in Table 2.3 (*def add2*). Lists can be multidimensional, but in our exercises, only one- and two-dimensional lists are used.

Sets {}

Sets contain unordered collections of unique elements to which standard mathematical set operations can be applied. These are *intersection*, *union*, *difference*, and *symmetric difference*. Examples are given in Exercise 8.4. Sets do not record element position and, as a consequence, do not support indexing, slicing, or other sequence-like behavior.

Dictionaries {}

Dictionaries comprise pairs of an identifier (data type string) and an object (arbitrary data type). The objects are addressed by their identifiers:

-
```
lis={"second": 1.234, "first": 5.678}
print(lis['first']) → 5.678
```

We do not use dictionaries in this book.

2.4.3 Python Libraries

`Python` libraries are collections of pre-compiled functions. They are open-source, supported by a community of programmers, who are always there to answer questions, e.g., on *stackoverflow.com*. We shall sometimes directly refer to such advice.

Numpy
`Numpy` is a fast and efficient array-processing package designed for numerical computing that provides functionalities comparable to MATLAB. We are using it as our standard. It is usually imported under the name *np*: `import numpy as np`, see Cell 6 in Table 2.3.

Numpy.random
Random number routines produce pseudo-random numbers. We `import Numpy.random as npr` and make use of two functions:

- `npr.rand(N)` generates an array of N random numbers between 0 and 1, (Exercises 8.2 and 8.3).
- `npr.randn(N)` generates an array of N normally distributed random numbers (Exercise 8.5).

Matplotlib
Matplotlib is a package for designing a variety of charts or even arrays of charts. We `import matplotlib.pyplot as plt`. We restrict ourselves to producing simple scatterplots with a user-defined function (Sect. 2.4.5) that is able to display nearly all results of our exercises, similar to EXCEL charts. In order to plot arrows, we introduce a second user-defined function (Sect. 3.2.7).

Scipy
`Scipy` is designed for scientific computing and is especially suited for machine learning. We need only certain functions for optimization [`minimize` (Exercise 10.6), `fsolve` (Exercise 10.5), `curve_fit` (Exercise 10.7)], linear algebra [`solve` (Exercise 5.9)], and statistics (`Chisquare` from `Scipy.stats`).

Pandas
`Pandas` mimics spreadsheet calculation within `Python`, enabling input to and output from EXCEL files and text files. We use it only occasionally, e.g., in Exercise 4.8, to make the reader aware that such things exist and are useful.

Table 2.5 Data types in numpy demonstrated with arrays built with lists from Cell 2 of Table 2.3

```
import numpy as np
xInt =      np.array(x,dtype=int)
yFloat =    np.array(y,dtype=float)
yAfloat =   np.array(yA,dtype=float)
xBool =     np.array(x,dtype=bool)
xComplex =  np.array(x,dtype=complex)
xStr =      np.array(x,dtype=str)
```

```
xInt      [   0    1    2    3]       shape (4,)
yFloat    [ 0.00 4.00 5.00 -1.00] shape (4,)
yAfloat   [ 0.00 4.00 5.00]           shape (3,)
xBool         [    0    1    1    1]
xComplex      [0.+0.j 1.+0.j 2.+0.j 3.+0.j]
xStr          ['0' '1' '2' '3']
xInt*yFloat [ 0.   4.  10.  -3.]
xInt*yAfloat  ValueError: operands could not be broadcast
                          together with shapes (4,) (3,)
```

2.4.4 Numpy Constructions

Numpy is usually imported under the name *np*: `import numpy as np`.

Ndarrays

`Ndarrays` are, in general, *n*-dimensional arrays. One- and two-dimensional ones have analogies in spreadsheets. For scientific computing, they have an advantage over spreadsheets when the data becomes large.

They:

- are immutable, with size and datatype specified when the object is introduced,
- are operated element-wise in algebraic operations: $x + y \rightarrow [x + y]$,
- can be arguments in mathematical functions.

In Table 2.5, one-dimensional arrays of various data types are built with the function `np.array(.)` expecting a list as an argument, here taken from Table 2.3. Two numerical arrays can be multiplied element-wise when they have the same shape, e.g., x_{Int} and y_{float}, but not x_{Int} and $y_{A\text{Float}}$.

Question

How do the entries of *xBool* in Table 2.5 arise?[8]

[8] Numerical 0 becomes False, $\neq 0$ becomes True. False is output as 0, True as 1.

Ab-initio constructors

The following five functions construct an array from the specifications in the argument list.

- `np.ones(N)` 1D-array of *N* ones
- `np.zeros(N)` 1D-array of *N* zeros
- `np.linspace (start, stop, number of steps, endpoint = True)`

  `np.linspace(1.5, 3.5, 3)` → [1.5 2.5 3.5], True is default.

  `np.linspace (1.5,3.5,2, endpoint = False)` → [1.5 2.5].

- `np.logspace`

  `np.logspace(1, 3, 3, base = 10)` → [10. 100. 1000.]

- `np.arange (start, stop, step)`

  `np.arange (0.0, 4.5, 1.5)` → [0.0 1.5 3.0].

In `np.arange`, the stop value is not included. This is favorable when stacking np.aranges together:

- `np.hstack([np.arange(1,4,1),np.arange(4,10,2)])`
  → [1 2 3 4 6 8]

We can start the second `np.arange` with the stop of the first `np.arange` and have the stop/start value only once.

The following two functions construct arrays of the same shape as *Array* in the argument:

- `np.ones_like(Array)` 1D-array of ones
- `np.zeros_like(Array)` 1D-array of zeros.

From the `numpy.random` library, imported as `npr`, we shall use:

- `npr.rand(N)` generating an array of *N* random numbers between 0 and 1.
- `npr.randn(N)` generating an array of *N* standard normally distributed random numbers.

Identifiers to elements of 2-dimensional arrays

In Table 2.6, we construct 1-dimensional arrays with `np.linspace` and stack them together to form a 2-dimensional array consisting of two rows. Furthermore, identifiers to single rows, elements, columns, and 2D subarrays are specified.

Table 2.6 Constructing a 2D array with two 1D arrays obtained with `linspace`; identifiers to single rows, elements, columns, and 2D subarrays; only one print statement reported in Cell 1

```
X=np.linspace(0,3,4)
Y=np.linspace(10,13,4)
Lis=np.array([X,Y])
print("np.shape(Lis)    ", np.shape(Lis))
```

```
np.shape(Lis)    (2, 4)       #2 Rows with 4 entries each
Lis
 [[ 0.00  1.00  2.00  3.00]
 [ 10.00  11.00  12.00  13.00]]

Lis[0]           [ 0.00  1.00  2.00  3.00]
Lis[0][2]        2.0

Lis[:,1]         [ 1.00  11.00]
Lis[0][2:3]      [ 2.00]
Lis[1][1:3]      [ 11.00  12.00]

Lis[0:1][1:3]
 [[ 1.00  2.00]
 [ 11.00  12.00]]
```

▶ Recommendation: Use this type of presentation, printing out the shape and content of matrices, to get an overview of your data structure!

Lis[0][2] means: from the first row (index 0), take the third element (index 2). Alternative interpretation: take the matrix element from row = 0, column = 2.

Lis[1] [1:3] means: from the second row (index 1), take the second (index 1) to the third (index 2, index 3 exclusive) elements.

Lis[0][1:-1] means: from the first row (index 0), take the second (index 1) to the penultimate (index -1) elements.

Lis[:,1] means: take the second column (index 1).

Information on arrays

- `np.size`(array) returns the total number of elements.
- `np.shape`(array)returns the shape, for 2-dimensional arrays in the form (r,c), number r, c of rows and columns, respectively.

Stacking

- `np.hstack ([list of arrays])` concatenates the 1D-arrays horizontally into a long 1D-array, is defined more generally for multidimensional arrays
- `np.stack([list of 1D-arrays])` stacks the arrays (all of the same size by necessity) in the list as rows into a 2D-array, is defined more generally for multidimensional arrays.

Functions

- `np.sin, np.cos, np.tan, np.atan2`
- `np.power`
- `np.dot.`

The function `np.dot(A, B)` performs a matrix multiplication of the two matrices A and B. The same can be achieved with the operator@: A@B. Matrix multiplication is dealt with in Exercise 2.5 in detail.

`Atan2` is *arcus tangens* calculating the angle from the x-axis given the point (x, y) . The order of arguments is `atan2(y;x)`, different from EXCEL:

`np.arctan2(`y, x`) (Python)` = ATAN2(X; Y) (EXCEL)

`np.arctan2(3, 2)` = 0.980.98 = ATAN2(2; 3).

`np.arctan2(2, 3)` = 0.590.59 = ATAN2(3; 2).

`np.arctan2(4, 1)` = 1.331.33 = ATAN2(1; 4).

2.4.5 Standard Plot Program

We create charts by using the library `matplotlib.pyplot` imported under the name `plt`. As we deal almost exclusively with scatter diagrams for functions of the form $y = f(x)$, we do not need the full versatility of `matplotlib`. To profit from this restriction, we have devised a function to plot a standard figure, `FigStd`, in Table 2.7, where all necessary style information is coded within the function

Table 2.7 User-defined function for creating a scatter plot; this cell is run in all programs at the beginning

```
import numpy as np
import matplotlib.pyplot as plt
np.set_printoptions(precision=2, threshold=10,edgeitems=3,
        formatter={'float': '{: 7.2f}'.format})

def FigStd(xlabel, xmin, xmax, dx,
              ylabel, ymin, ymax,dy,
              xlength=4,ylength=4):
    plt.figure(figsize=(xlength, ylength))
    plt.axis([xmin, xmax, ymin, ymax])
    plt.rcParams.update({'font.size': 10,
                     'font.style':'italic'})
    plt.xlabel(xlabel)
    plt.xticks(np.arange(xmin, xmax+dx, dx))
    plt.ylabel(ylabel)
    plt.yticks(np.arange(ymin, ymax+dy, dy))
    plt.plot([xmin,xmax],[0,0],'k-',lw=1) #x Axis through 0
    plt.plot([0,0],[ymin,ymax],'k-',lw=1)
```

Fig. 2.10 Parabola (the same as in Fig. 2.1b) specified in Tables 2.1 and 2.2 and plotted with Table 2.8

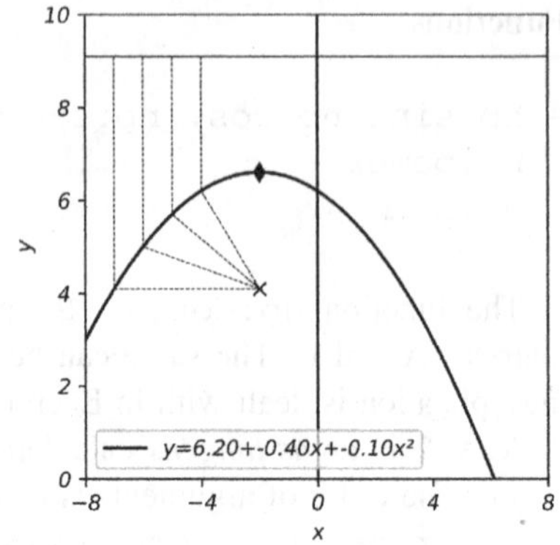

Table 2.8 Applying our standard function for scatter plots

```
FigStd('x',-8,8,4,'y',0,10,2)
lblP="y=%4.2f+%4.2fx+%4.2fx²"%(a,b,c) #Label for parabola
plt.plot(x,yP,'k-',label=lblP)     #'k-'  Black full line
plt.plot(xV,yV,'kd')               #'kd'  Black diamond
plt.plot([-8,8],[yD,yD],'k-',lw=0.5)
                                   #'lw'  Line width
plt.plot(xF,yF,'kx')
for i in range(-7,-3,1):
    plt.plot([i,i,xF],[yD,a+b*i+c*i**2,yF],
        'k--',lw=0.5)              #'k--' Black dashed line
plt.legend()#Plots the labels for the curves within the fig
```

body, and the information concerning the *x*-axis and the *y*-axis has to be specified in positional arguments. The axes' lengths in the figure have default values (`xlength = 4, ylength = 4`) that can optionally be specified otherwise.

The plot in Fig. 2.10, corresponding to Fig. 2.1b, is produced with the program in Table 2.8 (continuation of Table 2.1 and Table 2.2).

In the instruction `[yD, a + b*i + c*i**2,yF],` the values are created implicitly within the list.

Extensions

In Sect. 3.4.7, we introduce two extensions of our standard figure: a secondary *y*-axis, a logarithmic scaling of the *y*-axis with the statement `plt.yscale(value = "log")`.

Table 2.9 Formatted output with %

```
a1,a2,a3=1.2345,3.4567,-4.5678
label="y=%4.2f*x**%4.2f    +%6.1f"%(a1,a2,a3)
print(label)
```

```
y=1.23*x**3.46    +   -4.6
```

2.4.6 Formatted Output

The example in Table 2.9 shows how a formatted output can be achieved. There is a format string with three text variables "y = $*x**$ + $". The $ here is a proxy; it starts with a "%" and is followed by a format, e.g., 4.2f for a floating number 4 characters long and to be displayed with 2 decimal places. Following the string is a % sign and the name of the variables as a tuple, here %(a1, a2, a3) with the three entries replacing the three % in the format string.

2.5 Matrix Calculations in Excel and Python

We define row and column vectors and 2-dimensional matrices, and explain operations on them, parallel in EXCEL and `Python`. We apply broadcasting in EXCEL and `Python` to adapt the shape of operands so as to fit with each other in the intended operation. Finally, we get to know and apply the matrix operations transposition and inversion of linear algebra.

2.5.1 Data Structure and Nomenclature

Data structure

The data structure in scientific computing is based on arrays. `Python`, especially its `numpy` library, is designed for vectorized code, so that the programmer is forced to work with it from the first line of a code. This is its core business, described in all introductory textbooks, so we do not need to explain it here in detail.

Spreadsheet software, in contrast, is primarily designed for business calculations. In order to make it suitable for scientific computing, we have to create said vectorized data structure. This is achieved by identifying cell ranges with names so that most of the operations on vectors and matrices known from `Python` programs can also be applied in spreadsheet calculations. This is the essence of our approach to making spreadsheet calculations suitable for elementary scientific computing.

The treatment in this book is restricted to 1-dimensional arrays called vectors and 2-dimensional arrays called matrices. Vectors are of two types, row and column. We designate the shape of vectors and matrices with a tuple (*r*, *c*), where *r* and *c* are the number of rows and columns, respectively. In `Python`, a row vector is regarded as a 1-dimensional array, e.g., of shape (2,), whereas a column vector is represented as a 2-dimensional array with only one column, e.g., of shape (3, 1).

Nomenclature

We not only strive to use the same notation of variables in `Python` and EXCEL, but also to use a similar one in mathematical equations in the text. For this exercise, we have chosen the following nomenclature:

$\underline{U}, \underline{V}$ column vectors, shape (3, 1) 3 rows, 1 column
$\underline{W}, \underline{X}$ row vectors, shape (2,) 2 entries in the array
$\underline{\underline{L}}, \underline{\underline{M}}$ matrices, shape (3, 2) 3 rows, 2 columns
$\underline{\underline{N}}, \underline{\underline{O}}$ square matrices, shape (2, 2).

In the text, vectors are characterized by an underline, matrices by a double underline. Additionally, we apply subscripts and underlines to make the variable names more similar to the usual mathematical notation.

2.5.2 Operations on Arrays

Element-wise operation

Functions with scalar arguments, e.g., *cos(x),* are applied element-wise on vectors and matrices, resulting in an output of the same shape as the argument. The same holds for the algebraic operations addition and multiplication on arrays of the same shape. Function names and operations are usually the same in EXCEL and `Python`, with one important exception: *arcus tangens,* as already mentioned in Sect. 2.4.4.

Broadcasting

Broadcasting is a technical term for an essential tool in `Python`, but the underlying operations are also available in EXCEL. If two operands of an operation are of different but compatible shapes, they can be broadcast together. For a row vector of size *n* and a column vector of size *m*, this is done by repeating the row vector vertically and the column vector horizontally so as to get the same shape as an (*n, m*)-matrix. When such vectors are, e.g., multiplied, the result is a matrix of shape (*n, m*), with the elements being the element-wise product of the broadcast matrices.

Also, other operations, such as the multiplication of two vectors or the exponentiation of a vector with another vector, are performed as if each of the operands were repeated, so that matrices with a common shape are obtained that are then processed element-wise.

Linear Algebra

Functions applied in linear algebra comprise *determinant*, *inverse*, and *matrix multiplication.* Their names are listed in Table 2.10.

Table 2.10 Functions of linear algebra

EXCEL	Python
TRANSPOSE(<u>M</u>)	`np.transpose(M)`
INDEX(<u>M</u>; *r*; *c*)	<u>M</u>`[r, c]` or <u>M</u>`[r][c]`

Indexing starts in EXCEL with 1, and in Python with 0 in square brackets:

EXCEL: INDEX(<u>W</u>;1)	Python <u>W</u>`[0]`
MDETERM(<u>N</u>)	`np.linalg.det(`<u>N</u>`)`
MINVERSE(<u>N</u>)	`npl.linalg.inv(`<u>N</u>`)`
MMULT($\underline{M}_1$, $\underline{M}_2$)	$\underline{M}_1$ `@` $\underline{M}_2$ or `np.dot(`$\underline{M}_1$`,`$\underline{M}_2$`)`

In `Python`, a matrix is constructed as a list of row vectors. So, indexing <u>M</u>`[2][0]` points to the first element *[0]* in the third row <u>M</u>`[2]`. The abbreviation stands for `numpy.linalg` that is imported under that name.

Spreadsheet first
We start our overview of matrix calculations with spreadsheet constructs because the data structure there is immediately visible on the screen.

2.5.3 Matrices in Spreadsheets

Row and column vectors, matrices
In the upper half of Fig. 2.11 (S), two column vectors <u>U</u>, <u>V</u> of shape (3,1), two row vectors <u>W</u>, <u>X</u> of shape (2,), and two matrices <u>L</u>, <u>M</u> of shape (3,2) are defined, using the notation for shapes in `Python`. In the lower half of the figure, elements of the defined entities are singled out, with spreadsheet functions INDEX(VECTOR; INDEX) and INDEX(MATRIX; ROW INDEX; COLUMN INDEX). In the last two rows, the scalar product of the vectors *W* and *X* is calculated in Cartesian and polar coordinates.

The length l of a vector is determined with the equation $l = \sqrt{\sum x_i^2}$, obtained by nesting two functions: SQRT(SUMSQ(<u>W</u>). A 2-dimensional vector can be interpreted as a straight line in the *xy*-plane directed from (0,0) to (*x*,*y*). Its angle to the horizontal is determined with the *arcus tangens* function, = ATAN2(x,y), which, for vector <u>W</u>, reads as = ATAN2(INDEX(*W*;1);INDEX(*W*;2).

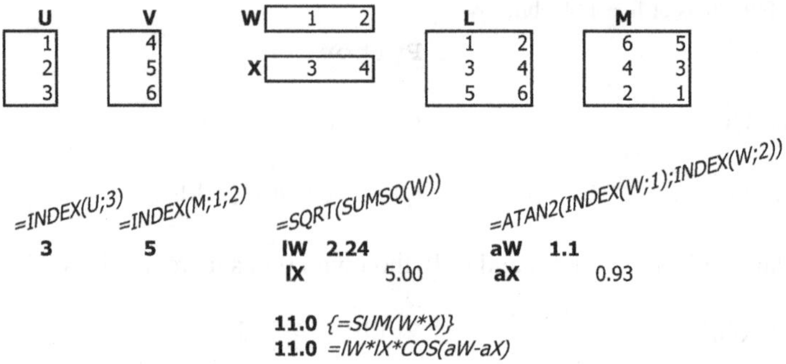

Fig. 2.11 (S) Column vectors *U*, *V*, row vectors *W*, *X*, and matrices *L*, *M* are defined as named ranges, and some operations are applied to them in the lower half of the figure

Broadcasting
Addition and multiplication of a row vector and a column vector, and of two matrices, is demonstrated in Fig. 2.12 (S).

In the upper half of Fig. 2.12 (S), two vectors are added with the operator + or multiplied with the operator *. When the operations are performed on two vectors of the same type, row or column, they must be of the same size. The operations are then performed element-wise, and the result is a vector of the same shape. Activating a larger range leads to a repetition of the 1-dimensional result [see *U**V* in Fig. 2.12 (S)]. Algebraic operations on matrices of the same shape result in a matrix of the same shape (see *L***M*).

All operations in Fig. 2.12 (S) are of the spreadsheet matrix type, as indicated by the fact that they are enclosed in curly brackets. To recall: You have to activate a range of size suitable for the result, enter the formula, and finish with the magic chord: *Ψ Ctrl, Shift, Enter*.

Multiplying or adding a column vector (*r*,) (e.g., *V*) and a row vector (,*c*) (e.g., *W*) results in a matrix of shape (*r*,*c*) (see the results for *W***V* and *V* + *W*). The operations are performed as if each of the operands is repeated, so that matrices with a common shape are obtained that are then multiplied or added element-wise.

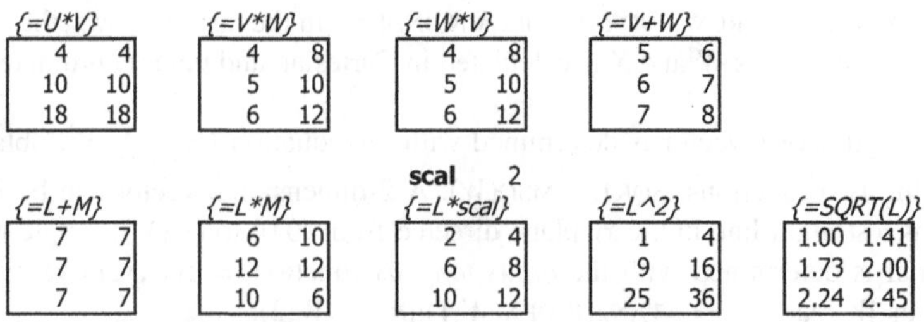

{=U*V}
4 4
10 10
18 18
{=V*W}
4 8
5 10
6 12
{=W*V}
4 8
5 10
6 12
{=V+W}
5 6
6 7
7 8
scal 2
{=L+M}
7 7
7 7
7 7
{=L*M}
6 10
12 12
10 6
{=L*scal}
2 4
6 8
10 12
{=L^2}
1 4
9 16
25 36
{=SQRT(L)}
1.00 1.41
1.73 2.00
2.24 2.45

Fig. 2.12 (S) Arithmetic operations on vectors and matrices defined in Fig. 2.11 (S)

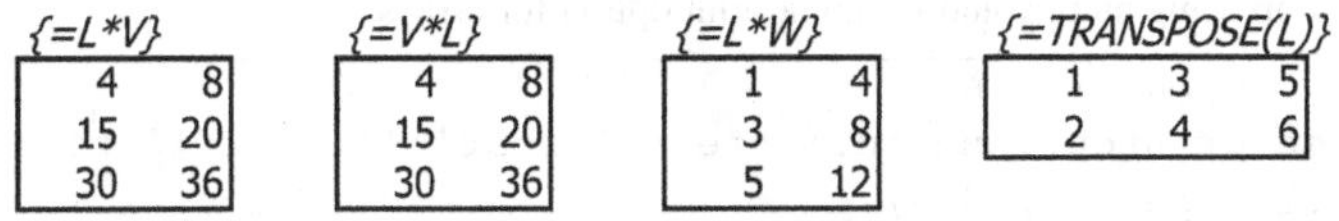

Fig. 2.13 (S) Operations on vectors and matrices with different shapes, defined in Fig. 2.11 (S) after broadcasting

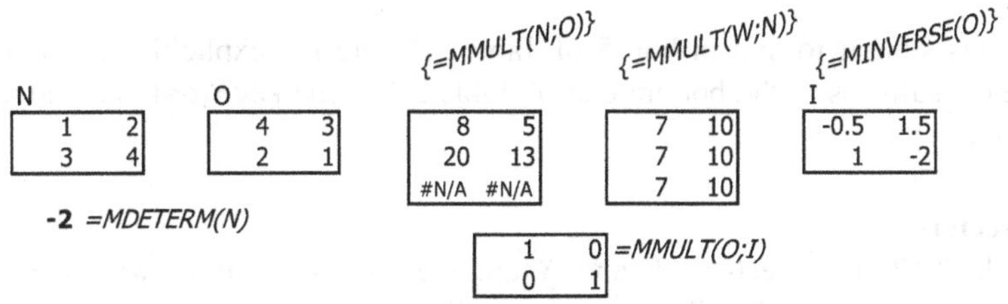

Fig. 2.14 (S) Mathematical matrix operations on the square matrices *N* and *O*

The same holds for the multiplication of a vector or a matrix with a scalar (see $\underline{L}$*scal*). Functions applied to vectors or matrices are also performed element-wise and yield a range of the same shape as the argument (see {=$\underline{L}$^2} or {=SQRT($\underline{L}$)}).

In Fig. 2.13 (S), operations combining a vector and a matrix are performed. Before an element-wise operation, the vectors are repeated so as to obtain the same shape as the matrix that also becomes the result's shape.

A matrix is transposed with the function {=TRANSPOSE(Matrix)}. In Fig. 2.13 (S), this is applied to $\underline{L}$.

Operations of Linear Algebra

Operations and functions on mathematical matrices are shown in Fig. 2.14 (S). On square matrices, the determinant can be obtained with MDETERM and its inverse matrix with MINVERSE.

Two matrices of suitable shapes, (n_1,n_2) and (n_2,n_3), can be multiplied with MMULT (*Matrix1, Matrix2*) ; the result is of shape (n_1, n_3). Multiplying the original matrix *O* with its inverse *I* yields the unit matrix.

2.5.4 Matrices in Python

Before operating with `numpy`, we have to import this library (see line 1 in Table 2.11, where we also set the `printoptions` for arrays that are to be printed, here, in the float format 0.2f with 2 decimal places and one blank between the printed numbers. The "0" in 0.2f indicates that the number of characters is not fixed but adapted to the current number. Additional separating blanks are obtained by inserting them between ' and {, or after {: as in line 3. In the following, print

Table 2.11 Importing numpy and setting a print option for arrays

```
1 import numpy as np
2 np.set_printoptions(formatter={'float':'{: 0.2f}'.format})
3 x=np.linspace(0,10,4)
4 print('x \n',x) #\n Makes a line feed
```

```
x
[ 0.00 3.33 6.67 10.00]
```

statements similar to that in line 5 of Table 2.11 are not explicitly reported, but only the results, as in the bottom cell of Table 2.11. The key word "\n" induces a line feed.

Row vectors

In Table 2.12, row vectors $\underline{W}$ and $\underline{X}$ are created as an array with the same elements as in Fig. 2.11.The length of $\underline{W}$, calculated correspondingly with `np.sqrt(np.sum(W*W))`, is the same. The scalar product of two vectors can be calculated with the function `np.dot` or with the operator @ for matrix multiplication: `W @ W`.

Column vectors

Row vectors can be transformed into column vectors by transposition, most simply with the function `np.transpose()`, but also with the extension `.transpose(1,0)` fixed to the identifier of a vector (or, more generally, of a matrix) (see Table 2.13). The original vector $\underline{U_R}$ is not defined as a row vector, but as a matrix with only one row, characterized by enclosing the list within double square brackets, [[…]]. This (1, 3)-matrix can be transposed with `np.transpose(UR)` or `UR.transpose(1, 0)` to become a (3,1)-matrix equivalent to a column vector. The transpose operation can also be applied to more-dimensional matrices.

In said table, we also demonstrate the element-wise multiplication of two column vectors $\underline{U}$, $\underline{V}$ and the broadcasting that occurs when a row vector ($\underline{W}$ from Table 2.12) and a column vector $\underline{V}$ are multiplied. An instruction [1,2,3]*[1,2] results in an error message: "Operands could not be broadcast together". The operation * is commutative: $\underline{W}*\underline{V} = \underline{V}*\underline{W}$.

Table 2.12 Creating row vectors in `Python` and determining their scalar product; print statements are not reported in Cell 1, and only their results are reported in Cell 2

`W = np.array([1, 2])`	`W, np.shape(W) [1 2]`
`X = np.array([3, 4])`	`(2,)`
`lW = np.sqrt(np.sum(W*W))`	`lW 2.24`
`ld = np.dot(W, W)`	`ld 5`
	`1 W @ W 5`

Table 2.13 Creating column vectors in Python, multiplication of two column vectors and multiplying column vector $\underline{V}$ with row vector $\underline{W}$ (from Table 2.12). The entries in the lower cells are obtained by print statements in the top cell that are not reported there, e.g., `print("np.shape(UR)", np.shape(UR))`

```
UR = np.array([[1, 2, 3]])
U1 = np.transpose(UR)
U  = UR.transpose(1, 0)
V  = np.array([[4, 5, 6]]).transpose(1,0)
```

```
np.shape(UR)   (1, 3)
UR
 [[1 2 3]]

np.shape(U1)   (3, 1)
U1
 [[1]
  [2]
  [3]]

np.shape(U)    (3, 1)
U
 [[1]
  [2]
  [3]]

U[0][0]        [1]
```

```
V
 [[4]
  [5]
  [6]]

U*V
 [[ 4]
  [10]
  [18]]

V*W
 [[ 4  8]
  [ 5 10]
  [ 6 12]]

np.sum(U*V)   32
```

Matrices

In Table 2.14, we see how 2-dimensional matrices are created with the function `np.array( ...)` . The argument of `np.array` is a list of 1-dimensional lists all of the same size. Algebraic operations such as $\underline{L}*\underline{M}$ on matrices with equal shape are again performed element-wise.

In Table 2.15, multiplication of $\underline{L}$ [shape (3,2)] with $\underline{V}$ [shape (3,1)] and $\underline{W}$ [shape (2,)] is demonstrated. These are some further examples of broadcasting. The effect of matrix transposition, `np.transpose(L)` , is also reported. Shape (3,2) indicates 3 rows with 2 elements each, and may also be interpreted as a matrix with 3 rows and 2 columns.

Linear Algebra

In Table 2.16, we learn about linear algebra operations on square matrices (shape (*n*, *n*)) using the functions `inv` and `det` of the `linalg` sublibrary of `numpy` that is often imported as *np*. `Np.linalg.det` returns the determinant and `np.linalg.inv()` the inverse of a square matrix.

Table 2.14 Creating 2-dimensional matrices in Python and performing algebraic operations on them; the entries in the bottom cells are obtained by print statements in the top cell that are not reported there, e.g., `print("L\n",L)`

```
1  L = np.array([[1, 2], [3, 4], [5, 6]])
2  M = np.array([[6, 5], [4, 3], [2, 1]])
```

<pre>np.shape(L) (3, 2)
 	
L
 [[1 2]
  [3 4]
  [5 6]]
 	
M
[[6 5]
 [4 3]
 [2 1]]</pre>
	<pre>L*M
	 [[6 10]
	  [12 12]
	  [10 6]]
	
	L**2
	 [[1 4]
	  [9 16]
	  [25 36]]
	
	np.sqrt(L)
	 [[1.00 1.41]
	  [1.73 2.00]
	  [2.24 2.45]]</pre>

Table 2.15 Product of a 2-dimensional matrix ($\underline{L}$ defined in Table 2.14) with 1-dimensional vectors $\underline{V}$, $\underline{W}$ defined in Tables 2.13 and 2.12, respectively; the entries are obtained with, e.g., `print("L*V\n", L*V)`

<pre>L
 [[1 2]
  [3 4]
  [5 6]]
 	
V
 [[4]
  [5]
  [6]]
 	
W
 [1 2]</pre>
	<pre>L*V
	 [[4 8]
	  [15 20]
	  [30 36]]
	
	L*W
	 [[1 4]
	  [3 8]
	  [5 12]]
	
	np.transpose(L)
	 [[1 3 5]
	  [2 4 6]]</pre>

The determinant of matrix $\underline{N}$ is $1{\cdot}4 - 2{\cdot}3 = -2$. `Python` deviates from that value in the 16th decimal. This is because the calculations are performed in the binary system.

The operator @ stands for matrix multiplication. The matrix product of a matrix with its inverse yields the unit matrix (see lines 18 to 20 in Table 2.16). The difference to the algebraic multiplication with the operator *, operating element-wise, is demonstrated in Table 2.17, $\underline{L}*\underline{W}$ vs. $\underline{L}@\underline{W}$.

Table 2.16 Operations of linear algebra on square matrices

```
N=np.array([[1, 2], [3, 4]])
O=np.array([[4, 3], [2, 1]])
I=np.linalg.inv(N)
```

N [[1 2] [3 4]]	np.linalg.det(N) -2.0000000000000004
O [[4 3] [2 1]]	N @ I *#Mathematical matrix multiplication* [[1.00 0.00] [0.00 1.00]]
I [[-2.00 1.00] [1.50 -0.50]]	

Table 2.17 Array multiplication with * versus mathematical matrix multiplication with @

L [[1 2] [3 4] [5 6]]	L*W [[1 4] [3 8] [5 12]]
W [1 2]	L @ W [5 11 17]

Questions concerning Tables 2.15, 2.16, and 2.17

What are the shapes of *L*, *W*, *L*W*, `np.transpose(L)`? [9]
What is the shape and the type of *N@I*?[10]
What are the shapes of *L*, *W*, *L@W*?[11]

2.6 Four Parabolas and Their Upper Envelope

Four parabolas are represented in a figure, together with their upper envelope, first produced by a spreadsheet calculation and then by an analogous `Python` program intended to serve as a basic exercise in `Python` programming and developed step by step in great detail.

---

[9] Shape(*L*) = (3, 2), shape(*W*) = (2,), shape(*L*W*) = (3, 2), shape (np.transpose(L)) = (2, 3).
[10] *N@I* is a diagonal matrix, shape = (2, 2).
[11] Shape(*L*) = (3, 2), shape(*W*) = (2,), shape(*L@W*) = (3,).

2.6.1 Graphical Representation

Figure 2.15 shows four parabolas with their vertices marked with diamonds and their upper envelope always running on the, at the respective position, top parabola.

The spreadsheet for producing this figure is given in Fig. 2.16 (S). It is briefly described in the following section on the basis of Exercise 2.2. The corresponding

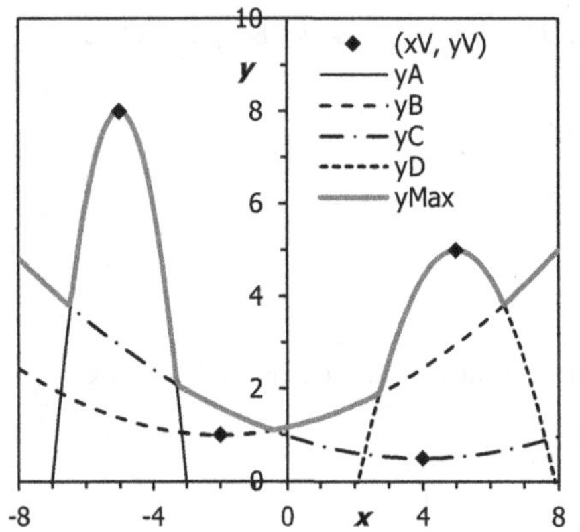

	J	K	L	M
17	-10.0	2.4	4.8	-96.4
18	-8.8	2.4	4.7	-94.8
19	-7.7	2.3	4.7	-93.3
177	-330.0	5.0	1.0	-0.4

Fig. 2.15 a (left) Four parabolas for the parameters specified in Fig. 2.16 (S). **b** (S, right) Coordinates for the four parabolas obtained with a matrix formula, J17:M177 = [{=a + b*x + c_*x^2}]

	A	B	C	D	E	F	G	H
1	**xV**	-5	-2	4	5			
2	**yV**	8	1	0.5	5		(xV, yV)	
3	**c_**	-2	0.04	0.03	-0.6			
4	**b**	-20	0.16	-0.24	6			
5	**a**	-42	1.16	0.98	-10			
6	**dx**	0.1						
7	*=A9+dx*	*=B$5+B$4*$A9+B$3*$A9^2*	*=C$5+C$4*$A10+C$3*$A10^2*	*=D$5+D$4*$A11+D$3*$A11^2*	*=E$5+E$4*$A10+E$3*$A10^2*	*=MAX(B10:E10)*	*=a+b*x+c_*x^2*	
8	**x**	**yA**	**yB**	**yC**	**yD**	**yMax**		
9	-8.0	**-10.0**	2.4	4.8	-96.4	4.8		
10	**-7.9**	-8.8	**2.4**	4.7	**-94.8**	**4.7**		
11	-7.8	-7.7	2.3	**4.7**	-93.3	4.7		
169	8.0	-330.0	5.0	1.0	-0.4	5.0		
170								

Fig. 2.16 (S) Four parabolas defined by the parameters in row vectors in rows 1 to 5; the formula reported in G7 is a better alternative for the formulas in B7:E7

Python program is developed in detail in Sect. 2.6.4, serving as the basic exercise in Python programming.

2.6.2 Data Structure and Nomenclature

x_V, y_V	arrays of the coordinates of the vertices of 4 parabolas
a, b, c	arrays of the coefficients of 4 parabolas $y = a + bx + cx^2$
x	array of x values (here, 161 from -8 to 8), separated by dx
y_A, y_B, y_C, y_C	four parabolas defined over x
G	matrix comprising y_A, y_B, y_C, and y_D as columns
y_{Max}	maximum of y_A, y_B, y_C, and y_D at the x values.

2.6.3 Spreadsheet Calculation

Figure 2.16 (S) shows a spreadsheet tabulating four parabolas for the values of x in A9:A169 and the parameters in rows 1 to 5, valid for the y-values of the parabolas in the respective column.

Question

What are the formulas for b in B4:E4 and a in B5:E4 of Fig. 2.16 (S)?[12]

The task is to write a formula into cell B9 with relative and absolute addresses so that it can be copied into the whole range B9:E169 to produce the values of the four parabolas y_A, y_B, y_C, y_D with the parameters *c_*, *b*, and *a* in range B3:E5. Parameter vector *c_* is specified directly by entering the desired values into B3:E3. Parameter vectors *b* and *a* are obtained from the coordinates of x_V and y_V of the vertex with the same formula as in Fig. 2.2 (S), realizing Eq. 2.3. In column F, the maximum y_{Max} of the four parabolas is built, giving their upper envelope. The result is shown in Fig. 2.15a.

The column vector x is defined in A9:A169 in 161 steps of dx = 0.1. The values of the four parabolas are then generated by typing the equation reported in B7 into B9 and copying right and down to E169, the range spanned by x and the row vectors *a, b, c_*. The resulting curves are shown in Fig. 2.15a, together with the vertices (x_V, y_V) and the curve y_{Max} calculated in column F.

Naming cell ranges

Row and column ranges can be provided with names (see Sect. 2.3.5). Activating A1:E5 and going through FORMULAS/CREATE FROM SELECTION/CREATE NAMES

[12] $b = -c \cdot x_V$; $a = y_V + c \cdot x_V^2$, see Fig. 2.2 (S) or Table 2.1.

Table 2.18 Specification of the coefficients of 4 parabolas and composition of their labels

```
import numpy as np
```

```
 #Specify position (xV,yv) of vertices!
xV=np.array([-5.0,-2.0,4.0,5.0])
yV=np.array([ 8.0, 1.0,0.5,5.0])
 #Define coefficients of y=a+bx+cx³
c=np.array([-2.0,0.04,0.03,-0.6])
b=-2*c*xV
a=yV+c*xV**2
```

```
 #Compose labels for the figure
lbl_1=str(a[0])+"+"+str(b[0])+"·x+"+str(c[0])+'*x²'
lbl_2=str(a[1])+"+"+str(b[1])+"·x+"+str(c[1])+'*x²'
lbl_3=str(a[2])+"+"+str(b[2])+"·x+"+str(c[2])+'*x²'
lbl_4=str(a[3])+"+"+str(b[3])+"·x+"+str(c[3])+'*x²'
print(lbl_1)
```

```
-42.0+-20.0·x+-2.0*x²
```

FROM VALUES IN THE/LEFT COLUMN does the job. Proceeding correspondingly with A8:A169 gives us a column vector named *x*. The formula can now be B9 = [=a + b*x + c_*x^2] instead of B9 = [B$5 + …], and remains the same when copied into the whole range B9:E169.

Application as a matrix formula

The formula [=a + b*x + c_*x^2] can be applied as a matrix formula within any range of suitable size. In Fig. 2.15b, the range J17:M177 has been activated, the formula entered and the process finished with the magic chord: Ψ *Str* + *Alt* + *Enter*. The same numbers appear as in B9:E169 of Fig. 2.16 (S).

2.6.4 Python Program

Now, how to do all this in `Python`?

In the first cell of Table 2.18, we import the `numpy` library. We run the cell so that the features and functions of `numpy` are available in the following cells.

Questions

What are the formulas to get a and b from the coordinates of the vertex?[13]

What are the shapes of c, b, and a in Table 2.18?[14]

[13] $a = y_v + c \cdot x_v^2$ and $b = -2 \cdot c \cdot x_v$.

[14] Shape(c) = (4,), c with 4 explicit entries; b and a have the same shape, because they are constructed with c.

Table 2.19 Creation of a column vector x

```
xS=np.linspace(-8,8,161) #Is a vector = 1D array
xR=np.array([xS])  #Is a matrix = 2D array with one row
x=xR.transpose(1,0)          #Creates a column vector

np.set_printoptions(edgeitems=2,
    formatter={'float': '{: 6.1f}'.format})
print('xS   ',xS)
print('xR   ',xR,'\n')
print("     x\n",x,'\n')
```

```
xS  [  -8.0   -7.9 ...    7.9    8.0]
xR [[  -8.0   -7.9 ...    7.9    8.0]]
```

```
     x
 [[  -8.0]
  [  -7.9]
  ...
  [   7.9]
  [   8.0]]
```

The next cell contains that part of the main program that reproduces the definitions and the data structure of the spreadsheet solution. The parameters x_V, y_V and a are specified exactly as before in numerical row vectors with four elements. The parameters b and c are obtained with the formulas mentioned above (in Sect. 2.2.1), also as row vectors.

In the third cell, the labels of the four curves in Fig. 2.15 are composed. Remember: the concatenation operator in `Python` is + (contrary to & in EXCEL), and numerical values have to be converted explicitly into a string, e.g., `str(a[0])`. They are used as the legend in the chart.

In Table 2.19, the column vector x, representing the independent variable, is constructed. We could achieve that in one statement:

`x = np.array([np.linspace(0.0,1.0,11)]).transpose(1,0)`

but use the three lines in the first cell instead, so as to make the construction clearer. First, a simple list x_S with 161 items equally spaced between −8 and 8, including 8 as the last element, is generated. In the output (bottom) cell, it shows up as a 1-dimensional row vector. To get a column vector, we first have to make it a matrix (2-dimensional array) x_R by defining an `np.array` with just one row showing up in the output cell as a list within double square brackets [[…]]. This matrix is transposed, with `x = xR.transpose(1,0)`, indicating that the axes 0 and 1 are interchanged, now yielding a matrix x with just one column. This is the desired column vector.

In Table 2.20, a two-dimensional array named G is constructed with `G = a + b*x + c*x**2`. It has the same shape and the same content as the range B9:E169 in Fig. 2.16 (S). The four columns of array G get additional names y_A to y_D, as in the spreadsheet. Remember: `Python` is zero-addressing; first index is 0.

In column F of Fig. 2.16 (S), we have calculated the maximum of the four parabolas for each value of x. The same is achieved in `Python` with the statement

Table 2.20 Creating the y values of all 4 parabolas; G[:,0] indicates first column

```
G=a+b*x+c*x**2
yA=G[:,0] #1st col.
yB=G[:,1] #2nd col.
yC,yD=G[:,2],G[:,3]
```

```
print(G,'\n')
print(yA)
yMax=G.max(axis=1)
 #Max. across the columns for every x
print(yMax)
```

```
G
[[ -10.0     2.4     4.8   -96.4]
 [   -8.8    2.4     4.7   -94.8]
 ...
 [-324.8     4.9     1.0    -0.0]
 [-330.0     5.0     1.0    -0.4]]

yA [ -10.0    -8.8 ... -324.8 -330.0]      #First column

yMax [4.82   4.74 ... 4.92   5.0]          #For every x
```

yMax = G.max(axis =1). G is a two-dimensional array, and we have to specify the axis along which the maximum value has to be found. Axis = 1 indicates that it is across columns for every entry in a row. The resulting curve is shown in Fig. 2.17 as a bold gray line.

Figure 2.17 is similar to Fig. 2.15 obtained with EXCEL, but now created by the Python program *FigStd* described in Sect. 2.4.5.

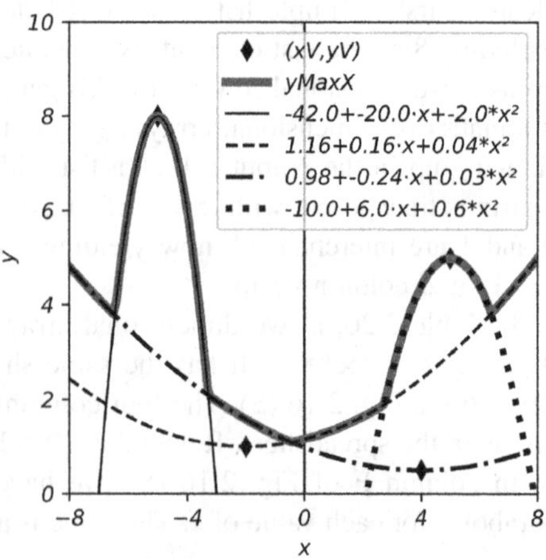

Fig. 2.17 Similar to Fig. 2.15, but produced with the Python program in Table 2.21

Table 2.21 Program for plotting Fig. 2.17

```
1  import matplotlib.pyplot as plt
2  FigStd('x',-8,8,4,'y',0,10,2)
3  plt.plot(xV,yV,'kd')
4                           #'kd' means points as black diamonds
5  plt.plot(x,yMax,color='gray',lw=4,label='yMax')
6  plt.plot(x,yA,'k-',  lw=1,   label=lbl_1)  #'k-' Black Line
7  plt.plot(x,yB,'k--',lw=1.5,label=lbl_2)  #-- Dashed line
8  plt.plot(x,yC,'k-.',lw=2,   label=lbl_3)  #-. Dash-dotted
9  plt.plot(x,yD,'k:',  lw=3,   label=lbl_4)  #:  Dotted
10 plt.legend()
11 plt.savefig('PhEx2-2 parabolas.png',dpi=1200)
```

Plotting four parabolas

The program for plotting Fig. 2.17 is given in Table 2.21, with the function `FigStd` called in the second line. It produces a figure of size 4 cm × 4 cm with scaling and labeling of the axes as specified in the arguments. The figure is made with the entries *label, minimum, maximum,* and *distance* between the ticks, each for the *x*-axis and the *y*-axis.

We add five curves with the statement `plt.plot()`. The plot statements all get the pre-syllable `plt`, the short form under which the `matplotlib` library has been imported. The *x*-values and the *y*-values of the curve to be plotted are positional arguments and have to be the two first arguments of `plt.plot`. The third argument specifies the color and style of the curve. The keyword '*ks*' in line 2 specifies that the points (x_{Ext}, y_{Ext}) are to be marked with black ('k') squares ('s'). The next plot statements contain more keyword arguments, *lw* for linewidth, and *label* for the curve's label to be displayed in the legend. Additionally, in the last plot statement, the curve's color is explicitly specified with `color='gray'`. For more information on plot styles and options, consult `Python Help!`

The legend with all labels is plotted by calling the function `plt.legend()`. Generally, you can specify its position in the figure. The default is `'loc'` = `0`, indicating that the program should choose an optimum position. Specifying `'loc'` = 2 would place the legend at the top-left part of the figure. Various style specifications can also be incorporated with keyword arguments, e.g., `fontsize` = `10`. With the last statement, the figure is saved under the indicated name and with the indicated resolution.

Question

The statement `plt.legend` will not plot the legend. Why?[15]

[15] The statement plt.legend() calls a function, and function identifiers have to be supplemented with parentheses.

Table 2.22 Extrema of matrix G along different axes

```
yMax=G.max(axis=1)           #Across columns, for every x

yMaxCol=G.max(axis=0)        #Down the rows, for every column
iMaxCol=G.argmax(axis=0) #Index of Max in the row
xMaxCol=x[iMaxCol]           #Corresponding x value

yMinCol=G.min(axis=0)
xMinCol=x[G.argmin(axis=0)] #Two statements in one
np.set_printoptions(precision=2)
```

`xMaxCol`	`yMax     [4.82 4.75 ... 4.92 5.   ]`
`[[-5.]`	`#1 Max. for every column:`
`[ 8.]`	`yMaxCol          [ 8.   5.   4.82  5.]`
`[-8.]`	`xMaxCol[:,0]  [-5.   8. -8.    5.]`
`[ 5.]]`	`#1 Min. for every column:`
	`yMinCol          [-330.  1.  0.5  -96.4]`
	`xMinCol[:,0]  [   8. -2.   4.    -8. ]`

2.6.5 Extrema Along Different Axes

In Table 2.22, we determine the extrema of the 2-dimensional matrix G, defined in Table 2.20 and containing the y values of the parabolas. The variable *yMax* contains the maxima for axis 1, i.e., across the columns (index 1), for every entry in the rows, i.e., for the same value of x. This is the upper envelope of the four parabolas.

Next, we will find the individual extrema (minima and maxima) of the four parabolas. To do so, we have to build the maxima y_{MaxCol} and minima y_{MinCol} for every single curve. This is done down the rows (index 0), for `axis` = 0. The arrays y_{MaxCol} and y_{MinCol} contain 4 entries each. The indices i_{MaxCol} at which the maxima occur in the columns of G are found with `iMax = G.argmax(axis = 0);` $x_{\mathrm{MaxCol}} = x[i_{\mathrm{MaxCol}}]$ gives the corresponding x values. For the coordinates y_{minCol} and x_{minCol} of the minima, we proceed accordingly with `argmin`.

In the printout cell, we realize that x_{Max} is a column vector, i.e., a matrix with only one column. The slice `xMaxCol[:,0]` extracts the column within the matrix and displays it as a row.

Questions

Are the values for $(x_{\mathrm{Min}}, y_{\mathrm{Min}})$ and $(x_{\mathrm{Max}}, y_{\mathrm{Max}})$ consistent with the prespecified values of the vertices (x_V, y_V)? Compare with Fig. 2.17![16]

[16] The four vertices show up as the 1st and 4th point in $(x_{\mathrm{Max}}, y_{\mathrm{Max}})$ and the 2nd and 3rd point in $(x_{\mathrm{Min}}, y_{\mathrm{Min}})$. The other points are at the boundaries of the x range.

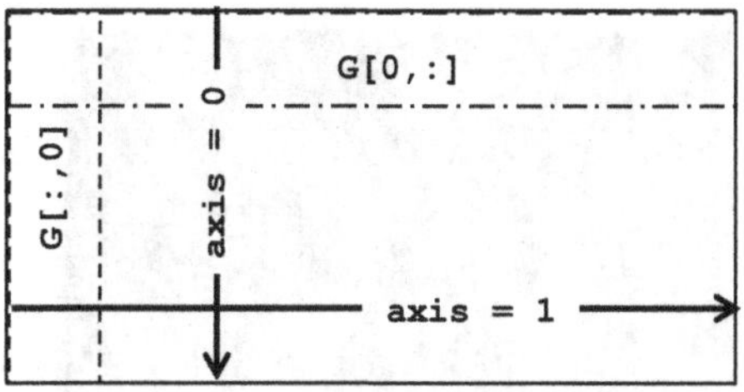

Fig. 2.18 Illustration to demonstrate indexing of matrices

The indexing of rows, columns, and axes seems not to be very intuitive. Therefore, we illustrate it in Fig. 2.18. G[0,:] addresses the first row: axis = 0 means *across* the rows yielding a number of values corresponding to the number of columns. G[:,0] addresses the first column, axis = 1 means *across* the columns yielding a number of values corresponding to the number of rows , in our example for every value of x.

2.7 Sum of Four Cosine Functions

We sum up four cosine functions with different angular frequencies ω_i. When the frequencies are multiples (overtones) of a fundamental frequency, such sums mimic the time signals of sound. Beats are generated when the frequencies are equally spaced within a small frequency range. The formula for the addition of cosines is illustrated by setting each two frequencies as equal, with the result being described by the broom rule Ψ "C*os plus cos yields mean times half the difference*".

2.7.1 Sound and a Cosine Identity

Vibrations of a string

▸ **Mag** Take a look at the microphone signal in Fig. 2.19a! How would you describe the signal?

▸ **Alac** Well, a peak is repeated periodically, and in between, there is a lot of fidgeting.

▸ **Mag** Yes and no. Yes, there is a fundamental frequency of repetition, and no, sound is not fidgeting; it is composed of harmonics.

Figure 2.19a is the record of the sound of a guitar string. What is the fundamental frequency in this case?

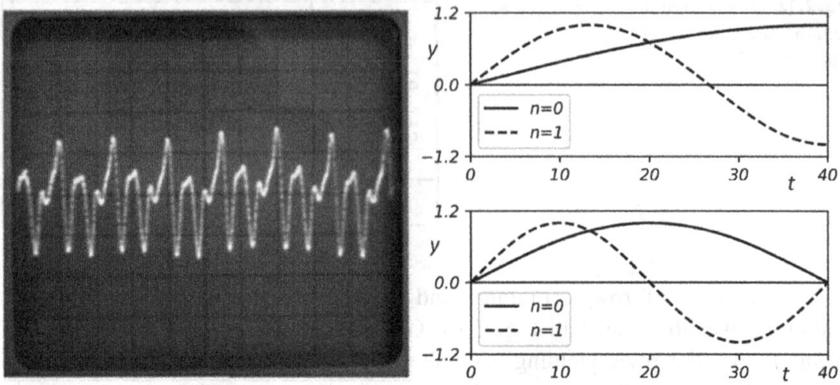

Fig. 2.19 **a** (left) Oscilloscope image of a microphone signal of a vibrating guitar string as a function of time (courtesy of Norbert Renner, University of Duisburg-Essen), time unit of the grid = 5 ms., **b** (right) Fundamental) and first harmonic oscillations (first overtones) of a reed of length l fixed on one end (top), $l = (\lambda_n/4)\cdot(2n + 1)$ and a string of length l fixed on both ends (bottom), $l = (\lambda_n/2)\cdot(n + 1)$; n indicates the number of internal nodes

▶ **Tim** The period is one-and-a-half grid distance, 7.5 ms, corresponding to a frequency of 133 Hz. But what are harmonics?

▶ **Mag** Let's consider the vibrations of a guitar string. How does a string vibrate?

▶ **Tim** It is sinusoidally excited, as in the bottom part of Fig. 2.19b.

▶ **Mag** Yes.

Figure 2.19b suggests that there are only discrete values of the wavelength with which a string can vibrate. You can determine them by considering the boundary conditions.

▶ **Tim** The sine must go through zero, where the string is clamped.

▶ **Alac** In between, it may also go through zero.

▶ **Mag** Exactly. The zero-crossings are called nodes. The boundary conditions require that the string length l be a multiple of half the wavelength, $l = (n + 1) \cdot \lambda_n/2$, $n = 0, 1, 2, \ldots$ The possible frequencies are multiples of the fundamental frequency $c/(2l)$), where c is the velocity of sound on the string. The corresponding vibrations are called harmonics. The mode with $n = 0$ is called the fundamental mode. Now, let me repeat my question: How do you now describe the microphone signal?

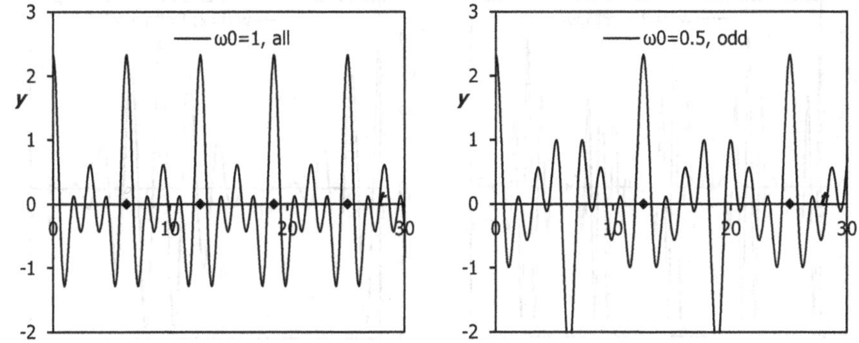

Fig. 2.20 **a** (left) Sum of a fundamental tone with $\omega_0 = 1$ and three overtones at 2 x, 3 x, 4 x ω_0. The time unit is 1 s when the values for ω are given in 1/s. **b** (right): Sum of a fundamental tone with $\omega_0 = 0.5$ and three overtones with 3 x, 5 x, 7 x ω_0. Time unit = 1 s

► **Alac** I guess it is the sum of the harmonics.

► **Mag** Right, it is the sum of the allowed vibrations with individual amplitudes.

In the case of a string clamped on both ends, the frequencies are multiples (1, 2, 3, ...) of the fundamental frequency, $f_n = f_0 \cdot (n + 1)$, or correspondingly for the circular frequencies ω_n. In cases in which one end of the vibrating medium is free, such as, for example, in a saxophone reed clamped on one side, the frequencies are odd multiples (1, 3, 5, ...) of the fundamental frequency, $f_n = f_0 \cdot (2n + 1)$. Examples are given in Fig. 2.20. There, you see the sum of four cosines with a fundamental frequency $\omega_0 = 1$ and multiples $2\omega_0$, $3\omega_0$, $4\omega_0$ (Fig. 2.20a) or multiples $3\omega_0$, $5\omega_0$, $7\omega_0$ (Fig. 2.20b).

Task in this exercise

► **Tim** Now, are we going to emulate the microphone signal?

► **Mag** Yes, but in the more general context of summing up four cosine functions whose frequencies satisfy certain conditions.

In this exercise, we shall use the form $y = A \cdot \cos(\omega \cdot t + \phi)$ with the parameters amplitude A, circular frequency ω, and zero phase ϕ. We set up a calculation model with four cosines, with their circular frequencies specified by a fundamental frequency and either three multiples thereof (to get harmonics) or three more frequencies at a distance of $d\omega$ (to get beats or demonstrate the addition theorem of cosines).

Beats and the uncertainty relation

Beats are shown in Fig. 2.21**a**. The four frequencies are again equidistant, but packed together in a small frequency range $\Delta\omega$. As control parameters for the calculation, to be systematically varied later, we take the lowest frequency ω_1 and the width of the frequency range $\Delta\omega$, i.e., the difference between the highest and the lowest

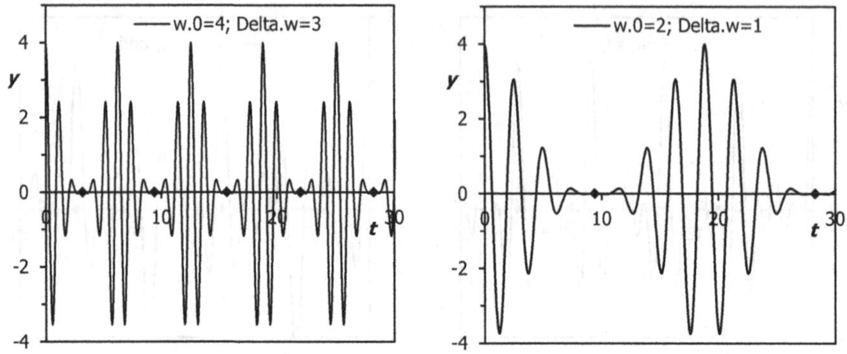

Fig. 2.21 **a** (left) Beats arising from four cosine functions; the initial angular frequency $\omega_0 = 4$, the width in the frequency range $\Delta\omega = 3$, **b** (right) another beat with initial angular frequency $\omega_0 = 2$, and width in the frequency range $\Delta\omega = 1$

frequencies. As a result, the signal clusters together into wave packets. In Fig. 2.21, they are separated by black markings. We shall determine the width of such wave packets based on an uncertainty rule.

▶ **Mag** The amplitudes of the cosine components of a beat are not freely chosen, but rather derived from binomial coefficients. How do you get such coefficients?

▶ **Alac** Well, with Pascal's triangle. Why so precise rules for the amplitudes? Can't we choose them as we like?

▶ **Mag** With said choice of amplitudes, we get a clear picture of the beat in the time domain, with the envelope being similar to a bell curve. Initially, you are to change only the lowest frequency ω_0 and the spectral width $\Delta\omega$ and observe the function's behavior in the time domain. Later, you may select the amplitudes at will and see whether the observed regularity is preserved.

▶ **Task** You are to insert points into the diagrams at the position of the nodes of the oscillations. To do so, you may create formulas that specify the marker points' coordinates when the width of a packet Δt and an initial time offset t_0 are specified. In a spreadsheet, it would be best for you to use sliders that change the coordinates so that the points in the diagram lie precisely on the nodes of the beat.

▶ **Mag** Have you figured out the rule for the width of a wave packet?

▶ **Tim** The time interval Δt for a packet is proportional to the inverse of the spectral width $\Delta t = 3\pi/\Delta\omega = 1.5/\Delta f$. The width is certainly smaller, because the signal goes to zero well before the marker points.

▶ **Mag** With that statement, you have found an uncertainty relation: $\Delta f \cdot \Delta t = 3/2 \geq 1$.

▶ **Tim** Heisenberg's uncertainty relation?

▶ **Mag** Yes, it is related to the commutation relation of time t and energy $E = hf$, $\Delta E \cdot \Delta t = \frac{3}{2}h \geq$ h.

Sum of cosines

▶ **Mag** How do you write $cos(x) + cos(y)$ as a product of two trigonometric functions?

▶ **Alac** Is it necessary to know such things? You can look it up in handbooks.

▶ **Tim** I've memorized it: "the mean times half the difference."

▶ Ψ *Cos plus Cos yields the mean, times half the difference.*

▶ **Mag** This broom rule is a useful mnemonic if you can reconstruct the full form from it:

$$cos(x) + \cos(y) = 2 \cdot \cos\left(\frac{x+y}{2}\right) \cos\left(\frac{x-y}{2}\right) \tag{2.8}$$

▶ **Alac** I'm certainly never going to forget such a crazy saying. The first cosine in the product has the average of the two primary cosines, the second half the difference.

▶ **Mag** Exactly, and if $x = 0$ and $y = 0$, then $1 + 1 = 2$ must be the result. This explains the pre-factor on the right-hand side of Eq. 2.8. We can test the sum formula for two cosines with our calculation model by making each two of our four frequencies equal so that only two different frequencies remain and setting all amplitudes to 0.5. The result of such a calculation can be seen as the thick gray curve in Fig. 2.22.

The black dotted curve corresponds to the function.

$$2\cos\left(\left(w_0 + \frac{dw}{2}\right)t\right) \text{ (the mean frequency)}$$

and the dashed curve to.

$$2 \cdot \cos\left(\frac{dw}{2} \cdot t\right) \text{ (half the difference of the frequencies).}$$

We see that the cosine with half the difference frequency (dashed line) envelopes the sum. The curve oscillates with the mean frequency (dotted line). The product of the two curves experiences a phase shift π after each zero crossing of the envelope.

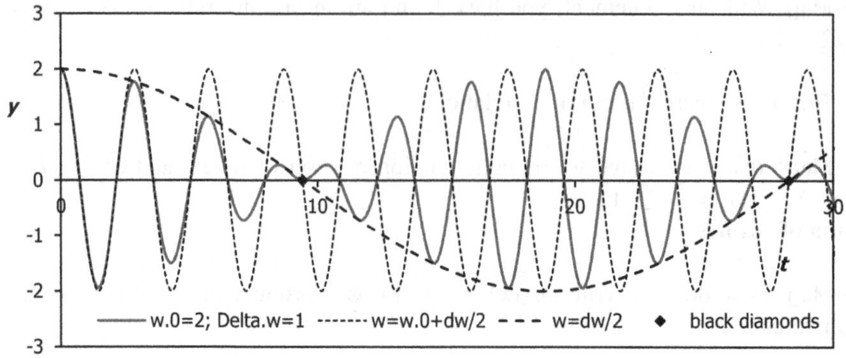

Fig. 2.22 Thick gray curve: the sum of two cosine functions with the frequencies ω_1 and $(\omega_1 + \Delta\omega)$; black dotted curve: cosine function with the mean frequency; black solid curve: envelope with half the difference of the frequencies

2.7.2 Data Structure and Nomenclature

dt	distance between adjacent points of time
t	vector of 801 equidistant points of time
c_1, c_2, c_3, c_4	cosine functions at t
c	matrix $[c_1, c_2, c_3, c_4]$
A	array of 4 amplitudes
ω	array of 4 circular frequencies
ω_0	lowest frequency
$\Delta\omega$	width of the frequency range of a beat
dω	distance between frequencies
ϕ	array of 4 phase shifts.
sumC	sum $c_1 + c_2 + c_3 + c_4$

2.7.3 Spreadsheet Layout

From the very beginning: a clear layout!

We set up a worksheet calculation to sum up four cosine functions, with their amplitudes A, angular frequencies ω, and zero phases ϕ to be chosen freely. The functions are to be calculated from $t = 0$ to $t = 32$ s for 801 sampling points. The basic Γ-structure of a suitable spreadsheet set-up can be seen in Fig. 2.23 (S). The names c_1, c_2, c_3, c_4 of the cosine functions are written in the spreadsheet with a dot separating the letter and the number, c.1, etc., because C1, etc., are cell addresses.

Overtones

The problem of generating overtones (higher harmonics) is a special case of the general task. The functions are to be calculated within a matrix range of, in our

	A	B	C	D	E	F	G	H	I	J
1		=w.0	=2*w.0	=3*w.0	=4*w.0		, all			
2	**A**	0.23	0.75	0.63	0.72					
3	**w**	**1.00**	**2.00**	**3.00**	**4.00**		**w.0**	1.00	ω_0	
4	**phi**	0	0	0	0		ω0=1, all	=J3&"="&w.0&G1		
5	0.04									
6	=A8+$A$5	=A*COS(w*t+phi)	=A*COS(w*t+phi)	=A*COS(w*t+phi)	=A*COS(w*t+phi)	=SUM(B9:E9)				
7	**t**	**c.1**	**c.2**	**c.3**	**c.4**	**sumC**				
8	0.00	0.23	0.75	0.63	0.72	2.33		Write formula into B8!		
9	**0.04**	**0.23**	0.75	0.63	0.71	**2.31**		Dragg into B8:E808!		
808	32.00	0.19	0.29	-0.11	-0.50	-0.13				

Fig. 2.23 (S) Four cosine functions c_1 to c_4 are calculated in the matrix range B8:E808 (below Γ), the cells of which always contain the same formula. The four functions are summed up in column F. The formulas in the columns are displayed in row 6 in oblique orientation. The time t (the independent variable) is defined as column vector A8:A808 (left of Γ) with the name t. The amplitudes A, the angular frequencies ω, and the zero phases φ are defined as row vectors B2:E2, B3:E3, B4:E4, respectively (above Γ). Here, the angular frequencies are multiples of the fundamental frequency ω_0

particular case, width 4 (number of functions) and height 801 (number of data points) . We write the desired values for amplitudes A, circular frequencies ω, and zero phase ϕ in rows above the matrix range and the independent variable t in a column to the left of the matrix range.

You are to organize the worksheet calculation such that you need to write a formula into only one cell that is then copied to the entire calculation area (here B8:E808) for the four functions by dragging down and to the right.

Beats

The parameters for beats are specified in H2:I4 of Fig. 2.24 (S), namely, the first frequency ω_1 and the frequency range $\Delta\omega$, from which $d\omega$, the distance between neighboring frequencies, is obtained. The amplitudes A are binomial coefficients (up to a factor of 2) and can be determined using Pascal's triangle.

Cosine identity

Figure 2.25 (S) presents a spreadsheet layout for obtaining the cosine identity shown in Fig. 2.22, the same as for beats. Here, however, $c_1 = c_3$; $c_2 = c_4$, and all amplitudes

	A	B	C	D	E	F	G	H	I	J
1		=w.0	=B3+dw	=C3+dw	=D3+dw				=Delta.w/3	
2	**A**	0.5	1.5	1.5	0.5			**Delta.w**	1.00	
3	**w**	**2.00**	**2.33**	**2.67**	**3.00**			**dw**	**0.33**	
4	**phi**	0	0	0	0			**w.0**	2	
5	0.04						w.0=2; Delta.w=1			

Fig. 2.24 (S) Parameters for a beat; ω ("omega") is coded as w

	A	B	C	D	E	F	G	H	I	J	K	L	M	N
1		=w.0	=B3+dw	=w.0	=D3+dw						Mean value			
2	A	0.5	0.5	0.5	0.5			Delta.w	1.00			half the difference		
3	w	2.00	2.33	2.00	2.33			dw	0.33					
4	phi	0	0	0	0			w.0	2		w=w.0+dw/2			
5	0.04							w.0=2; Delta.w=1				w=dw/2		
6	=A8+$A$5	=A*COS(w*t+phi)	=A*COS(w*t+phi)	=A*COS(w*t+phi)	=A*COS(w*t+phi)	=SUM(B9:E9)					=$F$8*COS((w.0+dw/2)*t)	=$F$8*COS(dw/2*t)		
7	t	c.1	c.2	c.3	c.4	SumC								
8	0	0.50	0.50	0.50	0.50	2.00					2.00	2.00		
9	0.04	0.50	0.50	0.50	0.50	1.99					1.99	2.00		
808	32	0.20	0.37	0.20	0.37	1.14					1.95	1.16		

Fig. 2.25 (S) Parameter set for a sum of two cosine functions; $c_1 = c_3$; $c_2 = c_4$; H5 displays the legend for one of the curves in Fig. 2.22, while there are formulas in K6:L6 for the angular frequencies of the fast oscillation and the envelope according to the sum formula, Eq. 2.8, applied in K8:K808 and L8:L808; the amplitudes are taken from cell $F$8. K4 and L5 contain the legend for the other two curves in Fig. 2.22

are equal (= 0.5). The legends for the fast oscillation (with medium frequency) and the envelope (with half the difference frequency) are assembled in K1:L5.

2.7.4 Python Program

Program flow in Jupyter

The program flow in the cell structure of `Jupyter` is shown in Fig. 2.26. The parameters for five subtasks are specified in five program cells. The identifiers of the parameters are the same in all five cells shown in Table 2.26 and Table 2.27. The main program gets the parameter values from the cell that has been run immediately before the main program is started. To keep track of the program flow, labels that contain all relevant parameters are created in all subtasks and displayed in the main program as legends in figures.

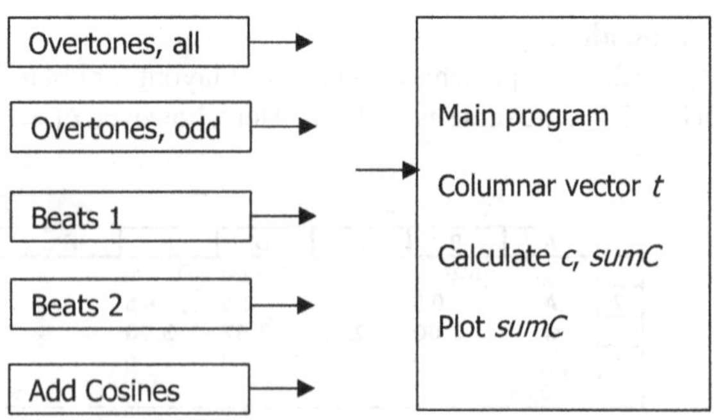

Fig. 2.26 Program flow: the parameters for five subtasks are specified in five cells and alternatively fed into the main program

Table 2.23 Main program for calculating 4 cosine functions and plotting their sum *sumC* ;y_{Min}, y_{Max}, and *FigName* are specified in Tables 2.25–2.27

```
t0=np.linspace(0,32,801)
t1=np.array([t0])
t=t1.transpose(1,0)
c=A*np.cos(w*t+phi)
sumC=np.sum(c,axis=1)

FigStd('t',0,30,10,'y',ymin,ymax,dy)
plt.plot(t,sumC,'k-',label=lbl_0)
plt.legend(loc=0,fontsize=11)    #loc = 0, best free place
plt.savefig(FigName)             #Store figure as a file!
```

Main program

The main program for calculating the four cosine functions and plotting their sum is given in Table 2.23. We assume that we have imported the two libraries `numpy` and `matplotlib.pyplot` under the shortcuts `np` and `plt`, as well as the function *FigStd* described in Sect. 2.4.5. This shall be done in all future programs and shall not be reported explicitly. The coefficients *A*, ω, ϕ, as well as the label *lbl_0* for the curve and the name *FigName* of the file name under which the figure is saved, have to be specified ahead of time. Figures 2.20, 2.21, and 2.22 all have different scaling of the *y* axis, so that the corresponding axis parameters y_{min}, y_{max}, d*y* must also be specified in the subtask cells before calling *FigStd*.

This is done in the program cells for the different situations: harmonics (Table 2.25), beats (Table 2.26), and the addition of cosines (Table 2.27).

In order to get the same data structure as in the spreadsheet solution, we have to construct *t* as a column vector. We first define our discrete time points according to column A in Fig. 2.23 (S). This is done with `t0 = np.linspace(0, 32, 801)` specifying that the range from 0 to 32 is scanned with 801 equidistant points. The endpoint 32 is included by default. If we do not want that, we have to include "`endpoint = False`" as an entry behind the three positional arguments: `np.linspace (0, 32, 801, endpoint = False)`, but we won't do that here.

The variable t_0 is now a row vector. To make it a column vector, we first transform it into a two-dimensional array by including the array t_0 within square brackets as the argument for `t1 = np.array()`, with one row only (shape = (1, 801); axis 0 has one element, a list with 801 elements. We then transpose the array between the two axes 1 and 0 by `t1.transpose(1,0)`. The shape and the first and last elements of the three vectors are reported in the second cell of Table 2.24. The column vector *t* has the same shape and contains the same numbers as the variable *t* in A8:A808 of Fig. 2.23 (S). The same holds for the matrix *c*

Table 2.24 Second cell: data structure of the variables of the main program, printed with instructions similar to those in the first cell. Third cell: matrix of the *y* values of four cosines

```
print("t0,   shape:", np.shape(t0))
print(t0)
```

t0, shape: (801,) [0.00 0.04 ... 31.96 32.00] t1, shape: (1, 801) [[0.00 0.04 ... 31.96 32.00]] t, shape: (801, 1) [[0.00] [0.04] ... [31.96] [32.00]]	c, shape: (801, 4) [[0.23 0.75 0.63 0.72] [0.23 0.75 0.63 0.71] ... [0.20 0.35 -0.04 -0.41] [0.19 0.29 -0.11 -0.50]] sumC,shape: (801,) [2.33 2.31 ... 0.10 - 0.13]

calculated as `c = A*np.cos(w*t + phi)` and for *sumC* reported in the third cell of Table 2.24.

Question

Which parameters relevant in the main program in Table 2.23 have to be specified in the sub-programs executed immediately before the main program?[17]

To calculate the four cosines, we apply the same formula, an operation on three row vectors and one column vector as in range B8:E808 in the spreadsheet (Fig. 2.23 (S)).

EXCEL: [RANGE] = [=A*COS(W*T + PHI)]

Python:`c = A*np.cos(w*t + phi)`

Both operations yield a 2D array, EXCEL by dragging into a 2D range, Python by automatic broadcasting.

You should have noticed that our spreadsheet layout has been translated line-by-line into Python. The striking similarity between the two platforms when using vector notation is the reason why mathematical-physical calculations can be sensibly performed in EXCEL. Training such calculations is a good introductory exercise for computational physics.

There is a difference between the spreadsheet and the Python solution presented here:

[17] (1) Labels that contain information on the characteristic frequency and frequency range. (2) Scalings of the y-axis. (3) Name under which the resulting figure is stored. (4) Parameters A, w, phi.

- In EXCEL, the four column ranges get the names c.1, c.2, c.3, c.4, with which they can be called.
- In `Python`, all four curves are stored in the columns of the two-dimensional list *c* and can be called by c[:,i] with $i = 0, 1, 2, 3$.

Summing up the four cosines is done in EXCEL with the formula in F9 = [=Sum(B9:E9)] of Fig. 2.23 (S) and in `Python` with `np.sum(c,axis = 1)`. The resulting variable *sumC* is a 1D vector. It is the sum across the columns (axis = 1) for a specific value of *t*, and has the rows' length. Attention: The choice of axes does not seem intuitive, but becomes apparent when *sumC* is plotted versus *t*; both have to be of the same length. For an overview, see Fig. 2.18.

Same data structure in Excel and Python

In spreadsheets, the formulas behind the values in the cells are usually hidden. To report them in our figures, we have copied their text in italic into neighboring cells. Contrastingly, in `Python`, the formulas are evident, but the data are hidden. To make them visible, we print them out in a well-structured manner (see Table 2.24). A comparison with Fig. 2.23 (S) shows that we have reached our goal to implement the same data structure in EXCEL and `Python`.

Harmonics

In Table 2.25, we specify the harmonics' parameters in the same way as in the spreadsheet, in the first cell for *all* multiples (of $\omega_0 = 1$) and in the second cell for *odd* multiples of the lowest frequency $\omega_0 = 0.5$. The values for amplitude *A*, circular frequency ω, and zero phase ϕ are specified as vectors, in `Python` realized as lists (characterized by square [] brackets). From these parameters, we compose a label *lbl_0*, later to be reported in the figure showing the result of our calculation. Furthermore, we specify in *FigName* the name under which the corresponding chart is to be stored. In lines 5 and 11, we specify the scale of the *y*-axis according to Fig. 2.20.

Table 2.25 Specifications of the parameters for harmonics

```
1  A=[0.23,0.75,0.63,0.72]
2  w=[1,2,3,4]                            #Overtones, all
3  phi=[0,0,0,0]
4  lbl_0="ω0="+str(w[0])+", all"
5  ymin,ymax,dy=-2,3,1                    #Scaling the y-axis
6  FigName="PhExI 7-4 Overtones all"      #File name for later use
```

```
7  A=[0.23,0.75,0.63,0.72]
8  w=[0.5,1.5,2.5,3.5]                    #Overtones, odd
9  phi=[0,0,0,0]
10 lbl_0="ω0="+str(w[0])+" (odd)"
11 ymin,ymax,dy=-3,3,1
12 FigName="PhExI 7-4 Overtones odd"      #In plt.savefig(FigName)
```

Table 2.26 Specifications of the parameters for beats

```
#Beats 1
A=[0.5,1.5,1.5,0.5]
Delta_w=1
dw=Delta_w/3
w1=2
w=[w1,w1+dw,w1+2*dw,w1+3*dw]
phi=[0,0,0,0]
lbl_0=("ω1="+str(w1)+"; Δω="+str(Delta_w)) #Label for curve
ymin,ymax,dy=-4,4,2
FigName="PhExI 2-7-4 Beats1"        #File name in savefig()
```

```
#Beats 2
A=[0.5,1.5,1.5,0.5]
Delta_w=3
dw=Delta_w/3
w1=4
w=[w1,w1+dw,w1+2*dw,w1+3*dw]
phi=[0,0,0,0]
lbl_0=("ω1="+str(w1)+ "; Δω="+str(Delta_w))
ymin,ymax,dy=-4,5,2
FigName="PhExI 2-7-4 Beats2"        #File name
```

Table 2.27 Specification of the parameters for the addition of cosines

```
#Addition of cosines
A=[0.5,0.5,0.5,0.5]
Delta_w=1
dw=Delta_w/3
w1=2
w=[w1,w1+dw,w1,w1+dw]
phi=[0,0,0,0]
lbl_0=("ω1="+str(w1)+"; dω="+str(np.round(dw,2)))
ymin,ymax,dy=-3,3,1                   #Scaling of y axis
FigName="PhExI 2-7-4 Sum Cos"         #File name
```

Beats and the addition of cosines

The parameters for the other situations are set in Table 2.26 (beats) and Table 2.27 (addition theorem of cosines).

2.7.5 Producing Labels (as Strings) in Excel and Python

When producing labels, the following differences between EXCEL and `Python` have to be taken into account:

EXCEL (G4 in Fig. 2.23 (S)) A1 = [="w0″ & w.0 & ", all"]

The concatenation operator is & (ampersand), and numeric values are automatically converted into a string.

Python (Table 2.25):lbl_0 = "ω0" + str(w[0]) + ", all"

The concatenation operator is + ; numeric values have to be converted explicitly into a string. In both applications, text is enclosed in quotation marks.

When numbers x have to be rounded to n decimal places, we can use ROUND(X;N) in EXCEL and `np.round(x,n)` in `numpy`.

2.8 Questions

Cell references

1. What does the broom rule Ψ *The dollar makes it absolute* tell us?

2. What formula must be written in cell B5 of Fig. 2.27 (S), with absolute and relative references, so that copying this formula into the range B5:E205 creates four sine functions?
3. (`Python`) Specify arrays A, ω, t, so that an instruction C = `A*np.cos(`ω`*`t`)` generates the four cosines C_a, C_b, C_c, C_d of Fig. 2.27(S) in one matrix C. How do you replace the # in F_{sum} = `np.#(`C`, axis` = `#)` to get the sum of the four cosines?

	A	B	C	D	E	F	G	H
1		1	2	3	4	*Amplitude*		
2		4	3	2	1	*Angular frequency*		
3								
4		**Ca**	**Cb**	**Cc**	**Cd**			
5	0						*=A*COS(w*t)*	
6	1							
7	2							
205	200							

Fig. 2.27 (S) Γ structure of a spreadsheet for displaying four cosine functions

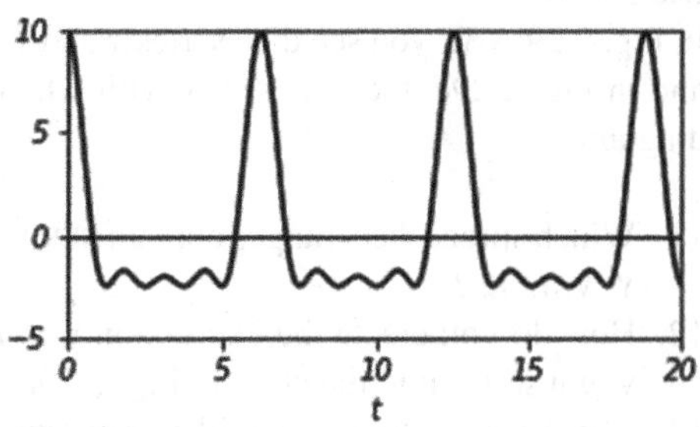

Fig. 2.28 The sum of the four cosines specified in Fig. 2.27 (S)

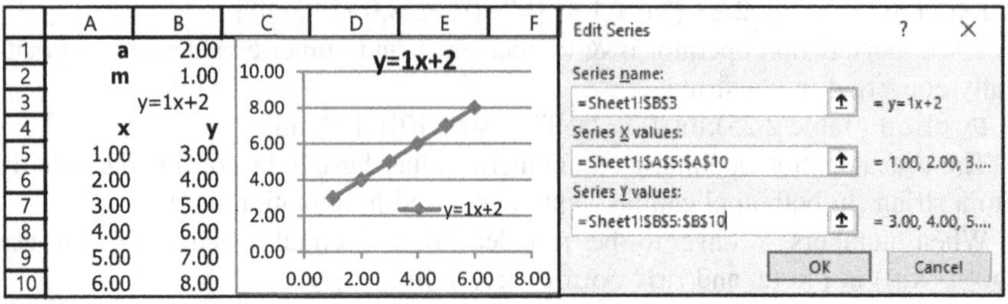

Fig. 2.29 **a** (left, S) A straight line is defined in the spreadsheet and displayed in the diagram. **b** (middle) Standard chart of the data in **a**. **c** (right) The DESIGN/SELECT DATA/EDIT dialog box, which is used to insert the data series *x*, *y* from **a** into the diagram

4. (`Python`) Write a program that realizes the spreadsheet calculation in Fig. 2.27 (S), using an instruction `C=A*np.cos(w*t)`! Calculate the sum of the four cosines and plot them as shown in Fig. 2.28, with a statement `plt.plot(t,Ctot,'k-')`!

Spreadsheet function Indirect

5. Look at the entries in eight cells of a table: A1 = 5; B2 = K; E1 = 5; K5 = 7; W1 = "E"&1; W2 = INDIRECT(W1); W3 = B2&A1; W4 = INDIRECT(W3). Which numbers appear in cells W2 and W4?

Arrays in Python

6. Array *x* is specified as `x = np.arange(-8,8 + dx,dx)` with $dx = 1$. How many elements does *x* comprise and what are its first and last elements?
7. What are the elements of `np.linspace(0,3,4)`?
8. What is the shape of arrays `U = np.array([1,2,3])`, `UR = np.array([[1,2,3]])` and `UT = UR.transpose(1,0)`?
9. What is the shape of `np.array([[1,2],[2,3],[3,4]])`?
10. Let `V = np.array([[1,2,3]]` and `W = np.array([3,2,1])`. What is the shape of `U = V * W` and `V.transpose(1,0)`?

Diagrams

In Fig. 2.29a (S), you see data series *x* and *y*, in Fig. 2.29b, the corresponding chart, and in Fig. 2.29c, the dialog box with which the data series was inserted into this diagram.

11. Which spreadsheet ranges contain SERIES NAME, SERIES X VALUES, and SERIES Y VALUES?
12. How do you create the expression $y = 1x + 2$ in the spreadsheet, and how do you insert it into the chart in Fig. 2.29b as a legend?
13. (`Python`) Below, you find a program for plotting a chart similar to the one in Fig. 2.29a, additionally with labels 'x' for the horizontal and 'y' for the vertical axis. Fill in the missing entries! The graph should be a straight line with diamonds, all black.

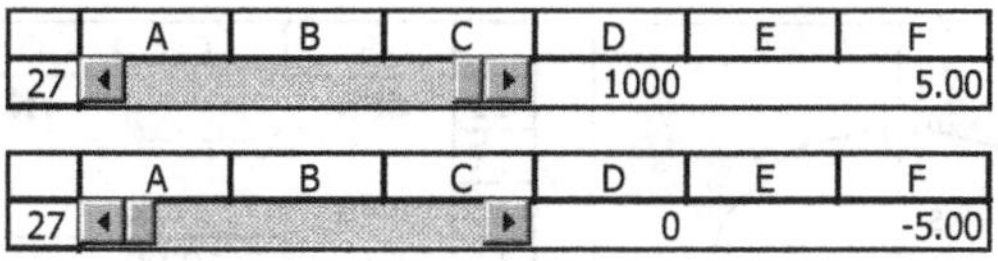

Fig. 2.30 Two settings of a slider

```
FigStd( … )
x=[ … ]
y=[ … ]
plt.plot(x,y, … )
```

14. (`Python`) How do you produce the string $y = 1x + 2$ when $a = 2.0023$ and $m = 1.001$ are specified?
15. (`Python`) How do you produce a string *3.0*exp(t/-30.0)* when $A = 3.001$ and $t_A = -30.0$ are specified?

Sliders (scroll bars)

In Fig. 2.30, you can see two settings of a slider.

16. Which is the LINKED CELL?
17. What are the minimum and maximum values of the slider?
18. The formula in F27 accesses cell D27. How does it look like?
19. The formula = (A5-500)/100 is used to generate decimal numbers between -5 and 5 recurring to a slider. What is the LINKED CELL, and what are MIN and MAX of this slider? What is the distance between two decimal numbers.

Polar coordinates

20. The coordinates of a circle are best given in polar coordinates with the angle ϕ and the radius r. How do you get the cartesian coordinates x and y needed for an xy diagram?

The figure in Fig. 2.31a is generated by the spreadsheet organization in Fig. 2.31b (S). Cells C11 and E11 have the names shown to their left. The column area B14:B26 gets the name *phi*.

21. How big are the numbers dPhi and r_K?

Apply names for cell ranges, if defined, in the answers to the next three questions!

22. What formulas are in cells B15 and B26?
23. Which formula is in column C below x?

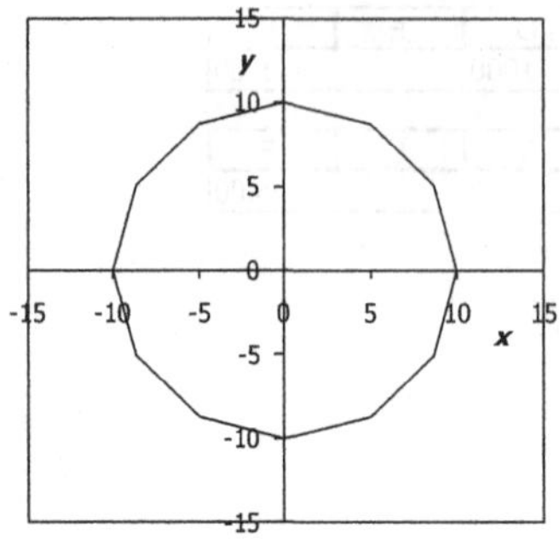

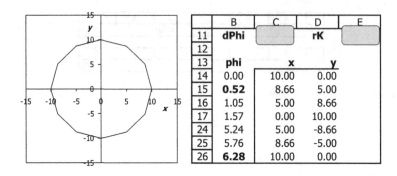

	B	C	D	E
11	**dPhi**		**rK**	
12				
13	**phi**	**x**	**y**	
14	0.00	10.00	0.00	
15	**0.52**	8.66	5.00	
16	1.05	5.00	8.66	
17	1.57	0.00	10.00	
24	5.24	5.00	-8.66	
25	5.76	8.66	-5.00	
26	**6.28**	10.00	0.00	

Fig. 2.31 **a** (left) Representation of a circle with 12 line segments; **b** (right, S) Coordinates for the circle in **a**

24. Which formula is in column D below *y*?
25. (`Python`) Complete the following program by replacing # to get a figure similar to Fig. 2.31a:

```
rK=10
phi = np.arange(0,#,dPhi)
x=rK*#
y=rK*#
FigStd(#,#,#,#,#,#,#,#)
```

Cosine functions

Figures 2.32a and b display two cosine functions.

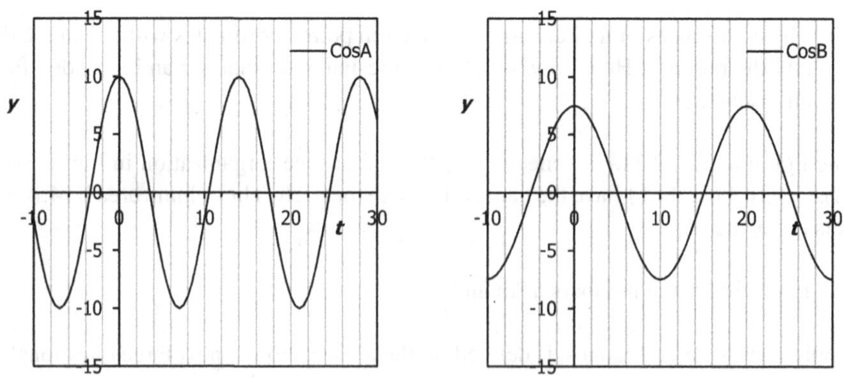

Fig. 2.32 **a** (left) Cosine function Cos_A. **b** (right) Cosine function Cos_B

26. What are the amplitudes and cycle times of the functions Cos_{A} and Cos_{B} shown in Fig. 2.32?
27. What are the angular frequencies of the two functions shown in Fig. 2.32?
28. What are the overtones to the fundamental with the frequency $f = 100$ Hz?
29. How do you interpret the broom rule: Ψ *Cos plus Cos = mean value times half the difference*?
30. A second cosine function is added to a cosine function with $f = 100$ Hz. What frequency must the second cosine function have to produce a beat of 1 Hz?

Formula Networks and Linked Diagrams

3

In this chapter, we practice clearly-structured calculations with formulas. Our aim for spreadsheets:—formulas in cells should be similar to mathematical formulas, a feat that is achieved by naming variables, calling them by their name, and clearly separating independent and dependent variables. As in every chapter, parallel solutions in `Python` are presented. For both EXCEL and `Python`, we aim to document intermediate results step by step from top to bottom and accompany them with charts. This way, the results are checked during the implementation of the formulas, and the calculation is easy to understand, even weeks later. With these rules in mind, we treat image construction for lenses, the Doppler effect, and exponential growth.

3.1 Introduction: Well-Structured Sheets and Programs

Solutions of Exercises 3.2 (Excel), 3.3.1 (Python), 3.3.5 (Python), and 3.4 (Excel) can be found at the internet address: go.sn.pub/McoItP.

Physical tasks with networks of formulas
Many tasks in secondary education, or in physics courses at colleges, or even in regard to minor subjects at universities, are based on simple formulas used to solve practical tasks. In this chapter, we solve such tasks through spreadsheet calculations and `Python` programs:

- image constructions for optical lenses,
- the Doppler effect with a general formula and for an observer off a race track, and
- exponential diode characteristics.

D. Mergel, *Physics with Excel and Python*,
https://doi.org/10.1007/978-3-030-82325-2_3

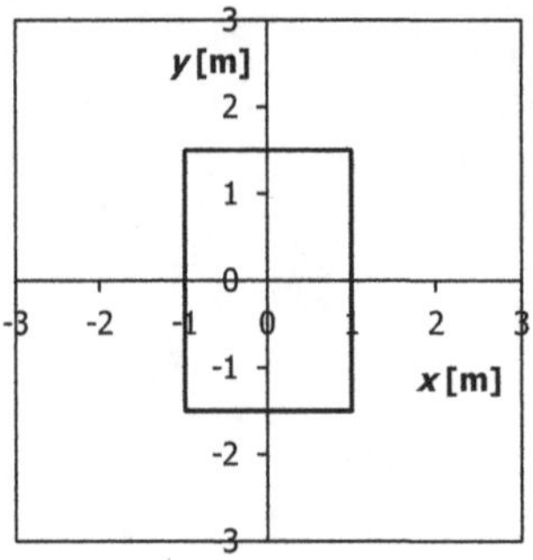

	A	B	C	D	E	F	G
1	**Designation**	**Name**	**Value**	**Unit**		**x**	**y**
2	Width	**w**	2	m		-1	-1.5
3	Height	**h**	3	m		-1	1.5
4						1	1.5
5	Area	**A**	**6**	m²	=h*w	1	-1.5
6						-1	-1.5

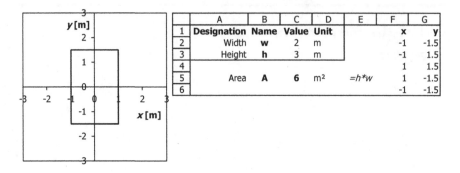

Fig. 3.1 **a** (left) A rectangle around the origin of the coordinate system **b** (right, S) Calculation of the area of a rectangle; Independent variables in B2:C3; coordinates *x*, *y* for the representation in **a** are calculated in columns F and G. **b** (right) Spreadsheet solution; coordinates for the rectangle shown in **a**, symmetrical to the origin of the coordinate system

For this purpose, a system of formulas has to be built using the results of other formulas. The spreadsheet set-ups and the `Python` programs should reflect the line of thought and be easily traceable, even weeks later.

Solve step by step and write it down in mathematical language!
The student shall learn to keep an overview by developing the solution step by step and displaying the results graphically, e.g., as optical ray construction. The diagrams should adapt automatically to any change in the parameters of the task.

Illustration: Draw a rectangle!
We consider a simple task: Specifying the width w and height h of a rectangle, calculating its area A, and drawing the rectangle as a chain of straight lines with vertices (x, y), with its center at the origin of the coordinate system (see Fig. 3.1a).

Python program
In `Python`, every object has to have an identifier (a name) so that, from the outset, the implementation is similar to mathematical formulas. For Fig. 3.1a, we apply Table 3.1.

Formula network in spreadsheets
In spreadsheet calculations, we have to pay special attention in order to achieve our goal. It is necessary to call all variables by names such that the cell formulas can be written like mathematical formulas. It is often useful to present the calculation in four columns: designation of the quantity, variable name as an identifier for cells, numerical value, and physical unit. An example is given in Fig. 3.1b. All independent variables are in a block at the top left of the sheet, with names in B2:B5 and units in D2:D5.

If one of the independent variables is changed, the entire calculation and all diagrams should follow, without corrections having to be made somewhere in the

Table 3.1 Specifying the coordinates of a rectangle around (0, 0)

```
w=2      #[m]   Width
h=3      #[m]   Height
A=h*w    #[m²]  Area
x=[-w/2, -w/2, +w/2, +w/2, +w/2]
y=[-h/2, +h/2, +h/2, -h/2, -h/2]
```

sheet. We call a spreadsheet structure corresponding to these requirements a 'formula network'. It is often useful to change independent variables with sliders that can quickly be used to get an impression of the trend of the solution (not realized in Fig. 3.1b).

The area is calculated as $F = h * w =$ height by width. If you write "= h * b" into cell C5, C5 = [= h * w], then immediately "6" appears after you hit ENTER. In columns F and G of Fig. 3.1b, the coordinates for the graphical representation in Fig. 3.1b are calculated. The formulas are the same as for x and y in Table 3.1. Cells and cell areas must be named, e.g., F2:F6 with "x". A good formula network must not only be correct, but also clear!

Provide cells in spreadsheets with names
There are several ways to name cell areas. We already became acquainted with this in the last chapter in Exercise 2.1 and Sect. 2.3.5. You can also find out more about this in the EXCEL help under the keyword "Create a name". We prefer to use the variant FORMULAS/DEFINED NAMES/CREATE FROM SELECTION, which has the advantage that the names given in the NAME MANAGER are visible in the spreadsheet, and thus contribute to the clarity of the calculation.

Check the solution with diagrams
For the representation of the rectangle in a diagram, here, Fig. 3.1a, the coordinates x, y are calculated from the values for width w and height h. The drawings should adapt automatically to any change in the parameters of the task. Therefore, in the spreadsheet of Fig. 3.1b, columns F and G contain formulas and not just numbers. The entries for w and h control all computations in the sheet and the figure, giving the impression of a "living spreadsheet".

Experience has shown that it is especially appealing to many students to work on the tasks until the diagram "obeys on command". In the case of more complex tasks, it is also easier to see whether the formula is correct.

Question

What are the formulas for x and y in Fig. 3.1b (S)?[1]

[1] $x = +w/2$ or $x = -w/2$; $y = +h/2$ or $y = -h/2$, always correctly combined.

Mathematical functions

In this chapter, we apply four mathematical functions:

- straight lines for image constructions of geometrical optics,
- exponentials for diode characteristics,
- polar coordinates with sine and cosine.

3.2 Image Construction for Focusing and Diverging Lenses

We construct the image point that a lens generates from an object point with three characteristic rays, applying the general imaging equation valid for both a focusing lens and a diverging lens. Ψ *Lens equation with plus and minus.* We draw the bundle of rays through the lens, which actually contributes to the image point. After this exercise, the reader should be able to master the straight-line equation blindfolded.

3.2.1 Straight Line Equation

In this exercise, a straight line is defined by two points (x_1, y_1) and (x_2, y_2). For a given third coordinate x_3, a coordinate y_3 is to be calculated so that the point (x_3, y_3) lies on the straight line.

Straight line equation

Given two points (x_1, y_1) and (x_2, y_2) of a straight line, the straight-line equation is

$$y(x) = y_1 + m \cdot (x - x_1) \tag{3.1}$$

or

$$y(x) = y_2 + m \cdot (x - x_2) \tag{3.2}$$

both times with the slope

$$m = \frac{y_2 - y_1}{x_2 - x_1} \tag{3.3}$$

We call the respective points (x_1, y_1) in the first straight-line equation and (x_2, y_2) in the second the *reference points* of the straight line.

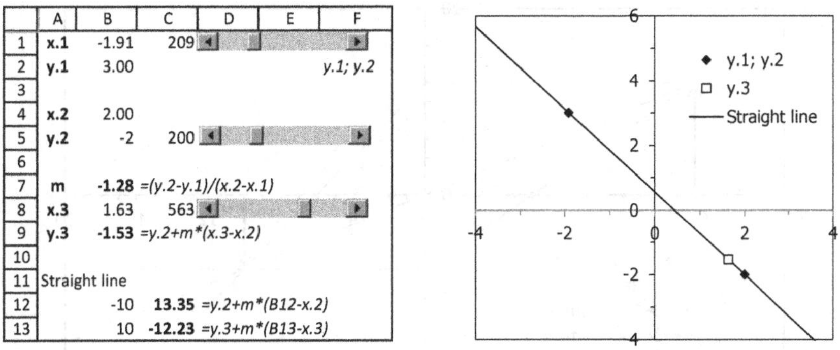

Fig. 3.2 **a** (left, S) Spreadsheet layout for the diagram in **b**; the values of x_1 and y_1 are obtained by means of sliders. **b** (right) The point "y.3" is to lie on a straight line given by the two points "y1; y2"

We demonstrate Eq. 3.1 through the spreadsheet in Fig. 3.2a. The *y*-value y_1 of the first and the *x*-value x_2 of the second defining point are directly written into cells B2 and B4, respectively. The associated values x_1 and y_2 in B1 and B5 are determined using the two sliders in D1:F1 and D5:F5. From these coordinates, the slope *m* is calculated in B7 with Eq. 3.3. The *x* value x_3 of the third point in B8 is selected with the slider in D8:F8, and the corresponding *y*-value y_3 in B9 is obtained with the straight-line equation Eq. 3.2. These three points are represented in Fig. 3.2b with diamonds.

In range B12:C13 of Fig. 3.2a (S), the straight-line coordinates are calculated for *x* values −10 to 10 extending beyond the range of the *x*-axis in Fig. 3.2b so that the straight line goes through the whole picture. The straight line is entered into the figure with SERIES X- VALUES: (B12:B13), SERIES Y- VALUES: (C12:C13).

Questions

Questions concerning Fig. 3.2a (S):

What are the linked cells for the three sliders?[2]

Which number range (the same for all) is presumably covered by the sliders?[3]

If the coordinates, to be set by sliders with MIN = 0 and MAX = 800, take on values between −4 and 4, what are the formulas in cells B1, B5, and B8, with which the coordinates are calculated from the cells linked to the sliders?[4]

[2] The linked cells are C1, C5, and C8.

[3] The number range of all sliders runs from 0 to 800.

[4] B1 = [=(C1 − 400)/100]; B5 = [=(C5 − 400)/100]; B8 = [=(C8 − 400)/100].

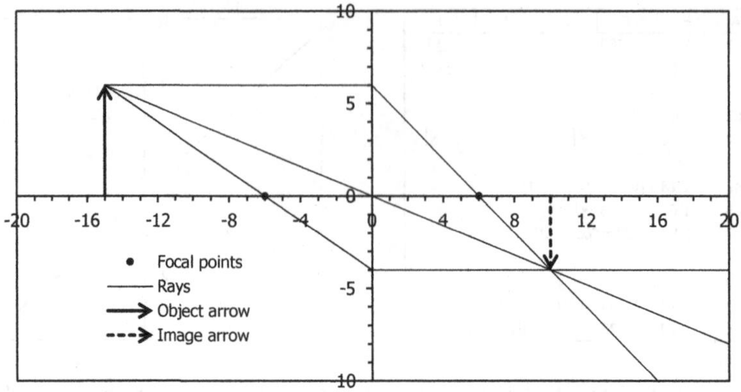

Fig. 3.3 Image construction for a focusing lens with parallel, central and focus rays

In C12 and C13, the y coordinates of the straight line are calculated with two different formulas. Why do both formulas describe the same straight line?[5]

If we have done everything correctly, the graphic presentation in Fig. 3.2a should adapt to every change of the values for y_1, x_2, x_3 by the three sliders, and the three points should lie on a straight line every time.

3.2.2 Geometrical Image Construction for a Thin Focusing Lens

Image construction by ray drawing

Figure 3.3 illustrates how the image of an object point is constructed with three characteristic rays in the xy-plane.

The x-axis represents the optical axis and also the axis of the circularly shaped lens. A thin lens is represented by its principal plane and its focal length. The shape of the lens does not play a role anymore. The principal plane is the plane $x = 0$ in which the y-axis is situated.

Question

How does the parallel ray in an image construction run?[6]

[5] The slopes are the same for both straight lines. As reference points, (x_2, y_2) has been chosen for C12 and (x_3, y_3) for C13. As both lie on the straight line, the third point also lies on the same line.

[6] The parallel ray runs from the object point parallel to the x-axis (that is, the optical axis) up to $x = 0$ and then through the image-side focus.

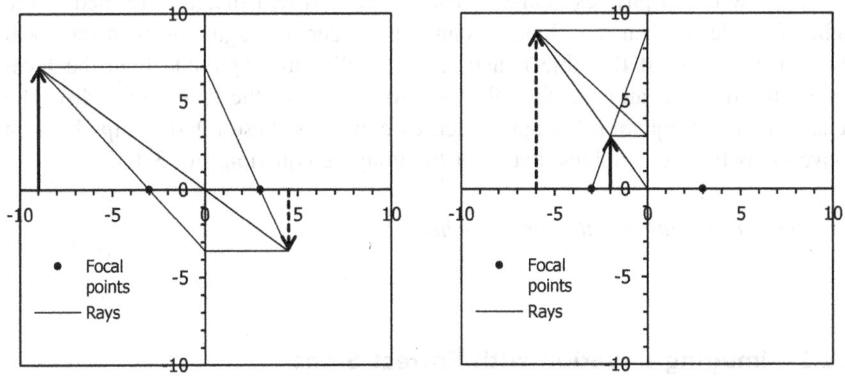

Fig. 3.4 **a** Converging lens, **a** (left) real inverted image for an object distance outside double the focal length. **b** (right) Virtual, upright image for an object distance within the focal length; corresponds to a look through a magnifying glass from the right

► **Mag** Do you remember how to determine the image point of an object point geometrically?

► **Alac** Yes, as in Fig. 3.4a. We draw two rays starting from the object point, one through the center of the lens and another parallel to the optical axis up to the principal plane and then through the focal point on the right side of the lens. The image point is where the two rays intersect.

► **Mag** So you can do it. Your construction is valid for a converging lens. However, a lens has two focal points, one on the image-side and another one on the object-side. In the image construction just described, you have exploited the fact that all rays incident parallel to the optical axis go through the *image-side focus* after having passed through the lens.

► **Tim** We have often drawn a third ray from the object point through the focal point on the left of the lens, which imagine is called the *object-side focal point.* After passing through the lens, this ray is parallel to the optical axis and then passes through the image point.

► **Mag** Yes, all three construction rays intersect at the image point, as in Fig. 3.4. That's what we want to reproduce with our exercise.

In Fig. 3.4, the image construction for a converging lens using the principal rays (parallel, central, and focal) is represented in a Cartesian coordinate system. By convention, the optical axis is the x-axis. The optical center and the lens's principal plane are respectively located in the origin of the coordinate system in the plane $x = 0$. We take it as given that the object is always to the left of the lens, i.e., *the object distance is always negative.*

We know (from physics courses) that an inverted real image is formed to the right of the lens when the object distance is bigger in magnitude than the focal length (Fig. 3.4a). If the object distance is smaller in magnitude than the focal length, the result is an erect virtual image to the left of the lens (Fig. 3.4b). The general imaging equation for optical lenses considers these relationships by sign conventions for the variables that enter the imaging equation, Eq. 3.4.

▶ Ψ *Lens equation with plus and minus!*

3.2.3 Imaging Equation with Correct Signs

▶ **Mag** Do you know the imaging equation for lenses?

▶ **Alac** Sure, I've already learned it at school:

$$\frac{1}{x_O} + \frac{1}{x_I} = \frac{1}{f} \tag{3.4}$$

where x_{O} and x_{I} are the object and the image distance, respectively, and f is the focal length.

▶ **Mag** This equation is useful only for handmade geometric constructions. For an analytical calculation, we must use a more accurate one, namely, Eq. 3.5, in which two modifications with respect to Eq. 3.4 have been introduced. Now, f_{I} is the image-side focal length, x_{I} the image distance, and x_{O} the object distance. The object distance is, in principle, negative, because the object is, by convention, placed to the left of the lens.

▶ **Alac** With the old equation, we always got the correct values for image distance and image size.

▶ **Mag** Yes, the absolute values are calculated correctly. However, no signs, plus or minus, are considered. Let's adopt the more general notation. The object distance x_o is negative if the object is to the left of the lens. The image is often upside down. This is automatically considered in Eq. 3.6, which calculates the image size y_{I} *from the object size* y_{O}.

Furthermore, Eqs. 3.5 and 3.6 are also valid for a diverging lens if a negative image-side focal length is introduced, $f_{\mathrm{I}} < 0$.

General imaging equation for lenses
For the analytical computation of images of lenses, the imaging equation has to be written with signs:

$$-\frac{1}{x_O}+\frac{1}{x_I}=\frac{1}{f_I} \tag{3.5}$$

The x-axis is the optical axis. The principal plane of the lens is in the plane $x = 0$; the object distance x_O is negative. The image-side focal length f_I is *positive* for converging lenses and *negative* for diverging lenses. The image distance x_I may result positive or negative. The imaging scale is

$$\frac{y_I}{y_O}=\frac{x_I}{x_O} \tag{3.6}$$

with y_O being the object size and y_I the image size that can be positive or negative.

Converging lens
In Fig. 3.4, you see the usual image construction for a focusing lens ($f_I > 0$) employing parallel, center, and focus rays.

Diverging lens
For the image construction of a diverging lens, you can use the same spreadsheet calculation or `Python` program as for a converging lens. You only have to enter a negative image-side focal length f_I. Figure 3.5 shows two examples (obtained from Table 3.3).

3.2.4 Beam Through a Converging Lens that Really Contributes to the Image

▶ **Mag** In your geometric ray constructions, you did not draw the cross-section of the lens. What can you say about the lens?

▶ **Alac** In all cases, the center is thicker than the edge; otherwise, it would not be a converging lens.

▶ **Tim** The principal plane of the lens is located in the plane $x = 0$. The center of the lens is at the origin of the coordinate system.

▶ **Mag** How big should the diameter of the lens be, e.g., in Fig. 3.4?

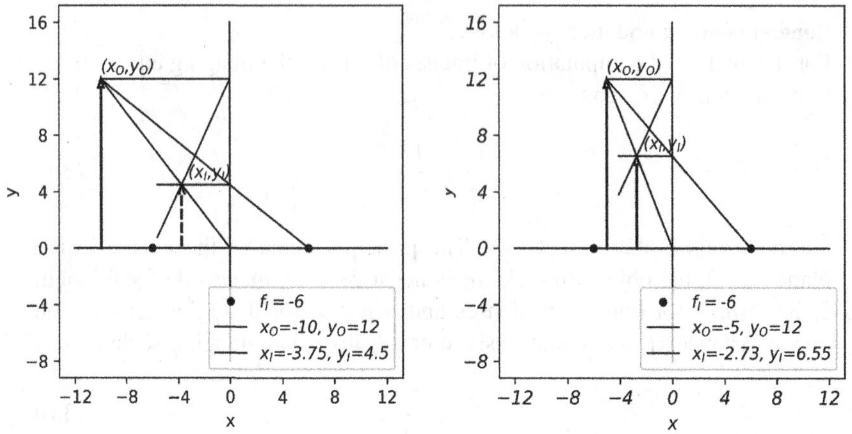

Fig. 3.5 Imaging with a diverging lens, the geometric construction being the same as for a converging lens, but with negative focal length ($f_I < 0$) (drawings obtained with the `Python` program in Table 3.4). **a** (left) upright image for an object distance larger than the focal length. **b** (right) As with **a**, but for an object distance smaller than the focal length

Table 3.2 Specifications for an image construction with a converging lens, resulting in a figure similar to Fig. 3.3

```
 #Converging lens
fI=6.0
xO=-15
yO=6
xMin, xMax, Dx = -20, 20, 5  #Scaling of figure axes
yMin, yMax, Dy = -10, 10, 2.5
FigName='Converging lens'    #File name in plt.savefig()
```

Table 3.3 Specifications for image constructions with a diverging lens, resulting in Fig. 3.5

```
 #Diverging lens, Object beyond focal length
fI=-6
xO=-10
yO=12
xMin, xMax, Dx = -12, 12, 4
yMin, yMax, Dy = -8, 16, 4
FigName='Diverging lens, outside' #In plt.savefig()
```

```
 #Diverging lens, Object within focal length
fI=-6
xO=-5
yO=12
xMin, xMax, Dx = -12, 12, 4
yMin, yMax, Dy = -8, 16, 4
FigName='Diverging lens, inside'  #In plt.savefig()
```

▶ **Alac** I would draw the lens from $y = -7$ to $y = 7$ so that the three constructing rays pass through the lens. The diameter would be roughly the same as those of the lenses used for lecture experiments.

▶ **Mag** Be cautious; think of cameras! In that case, the lens diameter is much smaller than, for example, the elephant you are photographing.

▶ **Tim** That's right. But does it mean that the construction rays do not go through the lens?

▶ **Mag** They don't, indeed. They exist only in thought and on paper. Which rays actually do contribute to the image point for a camera lens?

▶ **Alac** Only the central ray, or perhaps other rays that really do pass through the lens.

▶ **Mag** Yes, the image point is formed by a bundle of rays through the lens, as we will draw now. The lens itself has not shown up in the figures presented so far. The size of the lens is irrelevant to the image construction; only the principal plane and the focal length are needed to construct the image point.

Figure 3.6 shows the ray construction of the image point, together with the cross-section of the lens and eleven rays going from the object point through positions in the lens's full span and finally focusing in the image point. The parallel ray and the focus ray run outside the lens. They do not exist in physical reality.

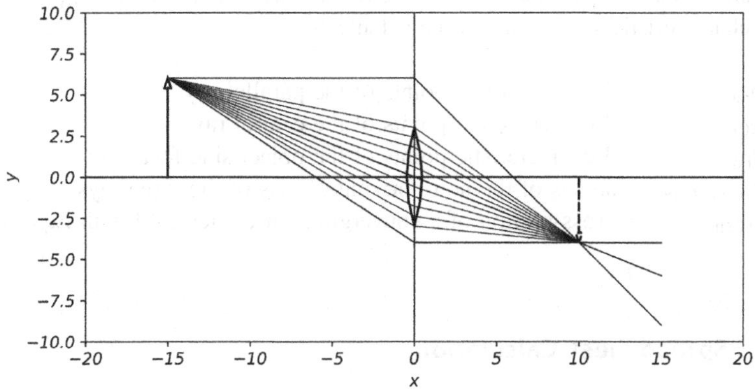

Fig. 3.6 (By program in Tables 3.4 and 3.5) Light beam contributing to the image formation in Fig. 3.3

Cross-section of the lens

We are going to add the cross-section of the lens to the drawings. To do so, we need to construct segments of a circle. We make use of the circle equation:

$$x^2 + y^2 = r^2 \tag{3.7}$$

The parameters are the coordinates x_0 of the center point on the x-axis, the radius r_K of the circle determining the lens's curvature, and the lens diameter $D_L = 2r_L$ (in front view). With $x_0 = \sqrt{r_{Lens}^2 - r_L^2}$, we calculate the distance of the center of the sphere, limiting the surface of the lens to the origin of the drawing. With

$$x = \sqrt{r_{Lens}^2 - y^2} - x_0, \tag{3.8}$$

we get the x-coordinate for a given y-coordinate on the surface of the lens.

3.2.5 Data Structure and Nomenclature

f_I	image-side focal length
(x_O, y_O)	coordinates of the object point
(x_I, y_I)	coordinates of the image point, to be calculated with the imaging equation.

Three rays for the geometrical construction of the image, without prior knowledge of the coordinates of the image point, are defined by f_I and the object point from which characteristic slopes have to be calculated.

x_{Par}, y_{Par}	3 characteristic points of the parallel ray
x_{Cen}, y_{Cen}	3 characteristic points of the central ray
x_{Foc}, y_{Foc}	3 characteristic points of the object-side focal ray
$m_{Par}, m_{Cen}, m_{Foc}$	slopes of the non-horizontal parts of the three rays
x_{Lens}, y_{Lens}	cross-section of a converging lens, calculated with Eq. 3.8.

3.2.6 Spreadsheet Calculation

Imaging equations

We calculate the ray path coordinates for imaging with a converging lens, and therewith set up an image construction that should adapt automatically whenever the parameters are changed. The coordinates of the three constructing rays' defining points are shown in Fig. 3.7 (S).

	A	B	C	D	E	F	G	H	I	J	K	L	M
1	**Specifications**									**Rays**			
2	Image-side focal length	**fI**	6.0								*Central ray*		
3	Object distance	**xO**	-15.0				slope			-15	=xO	6	=yO
4	Object height	**yO**	6.0							0	=0	0	=0
5	**Imaging equation**						**mCen**	-0.4		20		-8	=mCen*J5
6	Image distance	**xI**	10.0	=(1/fI+1/xO)^-1				=yO/xO			*Parallel ray*		
7	Image height	**yI**	-4.00	=yO*xI/xO						-15	=xO	6	=yO
8										0	=0	6	=yO
9	**Focal points**	6	=fI	0			**mPar**	-1		20		-14	=yO+mPar*J9
10		-6	=-fI	0				=-yO/fI			*Focal ray*		
11	**Object arrow**	-15	=xO	0						-15	=xO	6	=yO
12		-15	=xO	6	=yO					-6	=-fI	0	=0
13	**Image arrow**	10	=xI	0	=0		**mFoc**	-0.67		0		-4	=yO+mFoc*-xO
14		10	=xI	-4	=yI			=yO/(xO+fI)		20		-4	=L13

Fig. 3.7 (S) Imaging equation for a converging lens; the quantities related to the object and the image are designated with the indices O and I. $x = 20$ in column J indicates the right border of Fig. 3.3

The five parameters (focal length, object distance, and height, as well as the image distance and height that are dependent on them) are provided with the names in column B, with which they are entered into the image equations (B6:C7) and the coordinates of the construction rays (J:M).

The coordinates of the focal, central, and parallel rays are introduced as data series into the diagrams of Fig. 3.4a, b. We may enter the column range J3:J14 as SERIES X VALUES and L3:L14 as SERIES Y VALUES to get three separate straight lines, because empty rows separate their coordinates. The designations *Central ray*, etc., are in column K, not in column J or column L.

► *Ψ Empty rows separate curves.*

Note that, in most cells, there are formulas. So, you cannot simply copy the numbers from the spreadsheets displayed in this text. If you have done the implementation correctly, images such as those in Fig. 3.4a, b should result, automatically adapting whenever you change the parameters of focal length, object distance, and object height.

► **Alac** A fascinating experience!

The parameters in Fig. 3.7 (S) are for a converging lens ($f_I = 6.0$) and an object distance ($x_O = -15$) beyond the focal length (Fig. 3.3). With $f_I = 3.0$, $x_O = -9$, and $y_O = 7$, Fig. 3.4a results. With $f_I = 3.0$, $x_O = -2$ (within the focal length), and $y_O = 3$, Fig. 3.4b results.

For a diverging lens (Fig. 3.5), we choose $f_I = -6$ and $x_O = -10$ (**a**) or $x_O = -5$ (**b**).

```
Sub Bundle()
xO = Range("C3")
yO = Range("C4")
xI = Range("C6")
yI = Range("C7")
nR = 11  'Number of rays
RL = 3   'Lens radius
Dy = 2 * RL / (nR - 1)
r2 = 3   'First row for output
c2 = 10  'Column J
For i = 0 To nR - 1
    Cells(r2, c2) = xO
    Cells(r2, c2 + 1) = yO: r2 = r2 + 1
    Cells(r2, c2) = 0
    Cells(r2, c2 + 1) = -RL + i * Dy: r2 = r2 + 1
    Cells(r2, c2) = xI
    Cells(r2, c2 + 1) = yI: r2 = r2 + 1
    r2 = r2 + 1
Next i
End Sub
```

Fig. 3.8 (P) Drawing rays from the object point through the lens to the image point; the coordinates of the object point and the image point are read from column C of a spreadsheet, e.g., Fig. 3.7 (S). *Dy* is the distance between rays at $x = 0$

Ray bundle through the lens
We use a VBA subroutine[7] as in Fig. 3.8 (P) to draw the ray bundle that physically contributes to the image. The rays run from the object point (x_O, y_O) to a point in the lens, and finally to the image point (x_I, y_I). The coordinates of the points are in C3:C7 of a spreadsheet; they are read in the first lines of the sub-routine. The parameters of the lens are specified within the subroutine. The defining points of the rays are written into columns 6 (c_2, F) and 7 ($c_2 + 1$, G) of the spreadsheet.

3.2.7 Python Program

Specifications for three different types of images
The Python program is organized into four cells. The first three cells contain the specifications and a filename to store the resulting figure, each for

- a converging lens (Table 3.2, resulting in a figure similar to Fig. 3.3),
- a diverging lens with the object outside the focal length (Table 3.3 top, resulting in Fig. 3.5a),
- a diverging lens with the object inside the focal length (Table 3.3 bottom, resulting in Fig. 3.5b).

The fourth cell, represented in Tables 3.4 and 3.5, draws arrows representing the image and the object, together with the image construction, with rays using the specifications of one of the three initial cells that were run earlier.

The specifications in the three cells comprise not only focal length and the coordinates of the object point, but also parameters for axis scaling of the image and the name of the file wherein the image is to be stored. When the program in one of these cells is run, the resulting parameters are valid for the following image construction in Fig. 3.4. So, each of the three situations can be the basis of an image.

[7] VBA macros are introduced in Chap. 4.

Table 3.4 Drawing object and image arrows, function *ArrowP* presented at the end of this section

```
FigStd('x',xMin,xMax,Dx,'y',yMin,yMax,Dy,xlength=8)
plt.plot((-fI,fI),(0,0),'ko',
            markersize=4,label="$f_I=$"+str(fI))
ArrowP((xO,0),(xO,yO),lw=1.5)          #Object
lbl_1=r'$x_O$='+str(xO)+r', $y_O$='+str(yO)
        #'$x_O$=' becomes x_O=
plt.text(xO,yO+0.5,r"($x_O$,$y_O$)",fontsize=10)
```
```
#Calculated image
xI=1.0/(1.0/fI+1.0/xO)
yI=yO*xI/xO
Arrow((xI,0),(xI,yI),lw=1.5,ls='--') #Image
lbl_2=(r'$x_I$='+str(round(xI,2))
  +r', $y_I$='+str(round(yI,2)))      #$x_I$ as x_I in legend
plt.text(xI+0.5,yI+0.5,"($x_I$,$y_I$)",fontsize=10)
```

Table 3.5 Continuation of Table 3.4; setting up the image construction with the parameters specified in other cells without explicitly referring to the image point

```
 #Parallel ray
xPar=[xO,0,1.5*xI]
mPar=-yO/fI                         #Slope in image space
yPar=[yO,yO,yO+mPar*xPar[2]]
plt.plot(xPar,yPar,ls='-',color='k',
           lw=1,label=lbl_1)
```
```
 #Central ray
xCen=[xO,0,1.5*xI]
mCen=yO/xO                          #Slope in whole space
yCen=[yO,0,mCen*xCen[2]]
plt.plot(xCen,yCen,ls='-',
           color='k',lw=1.,label=lbl_2)
```
```
 #Ray through object-side focus
xFoc=[xO,-fI,0,1.5*xI]
mFoc=-yO/(xO+fI)                    #Slope in object space
yg0=yO+mFoc*xO
yFoc=[yO,0,yg0,yg0]
plt.plot(xFoc,yFoc,ls='-', color='k',lw=1.)
```
```
plt.legend(loc=4,fontsize=10)    #loc= 4 ,"Lower right"
plt.axis('scaled')
plt.savefig(FigName)
```

General program for drawing the image construction

The main program is in Table 3.4; it performs the image construction according to the specifications in the cell executed earlier. Arrows representing the object and the image are drawn. The coordinates of the object follow directly from the specifications

Table 3.6 User-defined function for drawing an arrow from point P_0 to point P_1 in the xy-plane (construct explained in Chap. 4)

```
def Arrow(P0,P1,c="k",ls='-',lw=1,hw=0.4):
    (x0,y0)=P0
    (x1,y1)=P1
    #c has to be given as c="k", not c='k'
    plt.arrow(x0,y0,x1-x0,y1-y0,
            length_includes_head=True,
            head_width=hw,fill=False,
            linestyle=ls, color=c,linewidth=lw)
```

in Table 3.2 or Table 3.3, whereas those of the image have to be calculated with the image equation (lines 7 and 8).

The program reproduced in Tables 3.4 and 3.5 calculates with the values obtained in one of the three cells in Tables 3.2 and 3.3. The ray constructions in Fig. 3.1 are obtained with the specifications in Table 3.5. The arrows are drawn with a function reproduced in Table 3.6 at the end of this section.

The three characteristic rays are drawn with the Python program in Table 3.5.

Plotting an arrow in Python
A user-defined function for drawing an arrow from point P_0 to point P_1 is shown in Table 3.6.

The arrow is plotted from P_0 to P_1, with color c, linewidth lw, and headwidth hw entered as keyword arguments. If the function call does not specify the values of the keywords, the default values specified in the header are taken. The coordinates of the two points have to be translated into the expected entries of the function `plt.arrow` of the `MatPlotLib` library.

Cross-section of the lens
The program for drawing the cross-section of a lens is given in Table 3.7.

The arrays y and x specify only one-quarter of the cross-section (see the values in the bottom cell). They run from (2.0, 0.0) up to (0.0, 6.0). For a complete cross-section, we need three more curves. They are obtained with the help of the functions

`np.flipud` ("flip up down") reversing the order of the elements in an array, and

`np.hstack` concatenating arrays to one long array, x_{Lens} resp. y_{Lens}.

The concatenated arrays x_{Lens} and y_{Lens} are used for the drawing performed in Table 3.8, line 21. The curve begins at (0.0, 6.0), runs to (2.0,0.0), continues to (0.0, −6.0), then to (−2.0, 0.0), and closes the cross-section by running to (0.0, 6.0).

Table 3.7 Coordinates of the cross-section of a lens with radius r_L of the disk and radius r_O of curvature of the surface of the lens

```
rL=6
rO = 10
xO=np.sqrt(rO**2-rL**2)
y=np.linspace(0,rL,3)
x=np.sqrt(rO**2-y**2)-xO
xf=np.flipud(x)                          #First becomes last
yf=np.flipud(y)
FigStd('x',-20,20,5,'y',-10,10,2.5,xlength=8)
xLens=np.hstack([ xf, x,-xf,-x])   #One long array
yLens=np.hstack([ yf,-y,-yf, y])
```

```
x   [ 2.00  1.54  0.00]
y   [ 0.00  3.00  6.00]
```

Bundle of rays through the lens

The coordinates of the $n_R = 11$ rays going through the lens are calculated in Table 3.8 with pre-specified coordinates of the object and the image point, e.g., in Table 3.4. The rays are drawn in a for-loop, starting with a y-coordinate at the bottom of the lens and increasing it by $\Delta y = 2 \cdot r_L/(n_R - 1)$. The cross-section of the lens is drawn with the plot-statement in line 21, using the coordinates x_{Lens}, y_{Lens} calculated in Table 3.7. The complete drawing, i.e., image construction, lens, and ray bundle, is shown in Fig. 3.6.

Table 3.8 Drawing a bundle of rays from the object point through the lens to the image point

```
11   #Bundle of rays through the lens
12   #Object point and image point are known.
13  nR=11                    #Number of rays
14  rL=3                     #Diameter of lens
15  x=np.zeros(3)
16  y=np.zeros(3)
17   (x[0], y[0])=(xO, yO) #Object point
18   (x[2], y[2])=(xI, yI) #Image point
19  x[1]=0
20  FigStd('x',-20,20,5,'y',-10,10,2.5,xlength=8)
21  plt.plot(xLens,yLens,'k')
22  Arrow((xO,0),(xO,yO),lw=1.5)           #Object
23  Arrow((xI,0),(xI,yI),lw=1.5,ls='--') #Image
24  Dy=2*rL/(nR-1)
25  for i in range(nR):    #Bundle of rays
26      y[1]=-rL+i*Dy      #Position in lens
27      plt.plot(x,y,'k-',lw=0.5)
```

3.3 Doppler Effect

When a sound source (a "sender") and a receiver move relative to air, the receiver perceives a frequency that is different from the transmitted one. We set up a formula for all cases of movements of the two agents on a straight line. We determine the frequency trajectory recorded at a receiver off the sender's track.

3.3.1 A Formula for All Cases

When a sound source (in the following, designated as sender *S*) and a receiver *R* are approaching or moving away from one another on a straight line, the receiver perceives a frequency different from that emitted. In the following, we develop a formula for the cases when sender and receiver move on the same straight line.

Formula for intuitive use (Doppler)
The relationship between frequencies f (frequency ratio) and speed v is given by the following formula:

$$\frac{f_R}{f_S} = \frac{c \pm v_R}{c \mp v_S} \tag{3.9}$$

The letters f, c, and v denote the frequency, the speed of sound, and the speed (≥ 0) of the agents relative to air. The upper sign in the formula is valid when the agents are approaching each other and the lower sign when they are moving apart. Note, as a mnemonic, that, above the fraction bars, there are quantities with index *R* and, below the fraction bars, quantities with index *S*, for both sides of the equation.

Ψ *Doppler effect with plus and minus*

It is best to consider which signs are to be used for every individual case. An example: S → R →; the sender moves towards the receiver, and the frequency increases ($/(c - v_S)$); the receiver moves away from the sender ($c - v_R$), and the frequency decreases; thus $f_R/f_S = (c - v_R)/(c - v_S)$.

Questions

How may Eq. 3.9 be simplified when the receiver is stationary and the sender is approaching him?[8]

[8] $f_R/f_S = c/(c - v_S)$, the received frequency becomes higher.

What frequency does the receiver hear when he travels at the same speed as the sender, (a) in front of and (b) behind the sender?[9]

Sender overtakes receiver
Let us apply Eq. 3.9 to the situation in which the sender and receiver both move in the same direction, and the sender overtakes the receiver.

Before overtaking, the sender is approaching the receiver, thereby increasing the received frequency (minus sign in the denominator). The receiver is moving away from the sender, also reducing the received frequency (minus sign in the numerator):

$$\frac{f_R}{f_S} = \frac{c - |v_R|}{c - |v_S|} \tag{3.10}$$

After overtaking, it is the other way around: The sender is moving away, the receiver is approaching, and a plus sign must be inserted in both the numerator and denominator:

$$\frac{f_R}{f_S} = \frac{c + |v_R|}{c + |v_S|} \tag{3.11}$$

Remember: $|v_R|$ und $|v_S|$ are speeds (amount of the velocities).

Motion on a straight line, analytical formula (Doppler)
We formulate a general formula in which the signs are automatically correct:

$$\frac{f_R}{f_S} = \frac{c - v_R \cdot sgn(x_R - x_S)}{c - v_S \cdot sgn(x_R - x_S)} \tag{3.12}$$

with x_S, x_R being the positions of the sender and the receiver, respectively, on the x-axis. The mathematical function *sgn* ("signum") is available as a spreadsheet function SIGN and as `np.sign` in `numpy`. To be able to better compare the formula with the previous calculations, we rewrite it with the speeds (amounts):

$$\frac{f_R}{f_S} = \frac{c - |v_R| \cdot sgn(v_R) \cdot sgn(x_R - x_S)}{c - |v_S| \cdot sgn(v_S) \cdot sgn(x_R - x_S)} \tag{3.13}$$

▶ **Tim** I could never develop a formula like that. I would always set the wrong sign or change the correct order.

▶ **Mag** Nor could I. I've been toying around with this, checking whether the outcome corresponds to the intuitive formula Eq. 3.9 in ten different situations.

---

[9] (a) $f_R/f_S = (c - v_R)/(c - v_R) = 1$; (b) $f_R/f_S = (c + v_R)/(c + v_R) = 1$; the received frequency is, in both cases, equal to the sent frequency.

	A	B	C	D	E	F	G	H	I	J	K	L	M
1								=(c.s-vR*s.vR*xRel)/(c.s-v.S*s.vS*xRel)					
2								=SIGN(x.R-x.S)					
3						fE/fS		s.vR	s.vS	x.S	x.R	xRel	
4	cs	340 m/s		S-->	R*	1.05	=(cs)/(cs-vS)	0	1	-1	1	1	1.05
5	vS	17 m/s		R*	S-->	0.95	=(cs)/(cs+vS)	0	1	1	-1	-1	0.95
6	vR	10 m/s		S*	R-->	0.97	=(cs-vR)/(cs)	1	0	-1	1	1	0.97
7				R-->	S*	1.03	=(cs+vR)/(cs)	1	0	1	-1	-1	1.03
8				S-->	R-->	1.02	=(cs-vR)/(cs-vS)	1	1	-1	1	1	1.02
9				R-->	S-->	0.98	=(cs+vR)/(cs+vS)	1	1	1	-1	-1	0.98
10				S-->	<--R	1.08	=(cs+vR)/(cs-vS)	-1	1	-1	1	1	1.08
11				<--R	S-->	0.92	=(cs-vR)/(cs+vS)	-1	1	1	-1	-1	0.92
12				<--S	R-->	0.92	=(cs-vR)/(cs+vS)	1	-1	-1	1	1	0.92
13				R-->	<--S	1.08	=(cs+vR)/(cs-vS)	1	-1	1	-1	-1	1.08

Fig. 3.9 (S) Frequency ratio for ten cases, in column F, calculated with Eq. 3.9 and with individual considerations for each case, in column M, calculated with the general formula Eq. 3.13

▶ **Alac** Proof by trial and error? You can't do that in math!

▶ **Mag** With trial and error, proof is not possible, but serious mistakes can be uncovered. You can later rigorously prove the formula.

Check the analytical formula in a spreadsheet

In Fig. 3.9 (S), ten situations are listed (D:E) in which sender and receiver move to the left or the right or one of them is at rest, and it is also distinguished as to whether the sender is to the left or the right of the receiver. In column F, the frequency ratio is calculated according to Eq. 3.9, "for intuitive use", considering which signs are to be used for each formula. In column M, the analytical formula Eq. 3.13 is applied relating to the values 1, 0, or −1 indicating the direction of the movements (s.vR, s.vS) and the relative position (x.S, x.R), with respect to the zero of the straight line, of receiver and sender. The quantity x_{Rel} is defined as $sign(x_R - x_S)$.

Checking in Python

In the `Python` program of Table 3.9, a function *Doppler* is defined realizing Eq. 3.12, taking the velocities v_S and v_R (with correct sign + or −), respectively, of the sender and the receiver, together with a keyword argument *pos* as input and returning the frequency ratio f_R/f_S. The string argument *pos* specifies the relative position of sender and receiver, 'SR' indicating that the receiver is to the right and 'RS' to the left of the sender. The same parameters as in Fig. 3.9 (S) are specified successively in a list *var1* that is passed to *Doppler* expecting three positional arguments; so, the list *var1* has to be unwrapped (*var1 in line 11). The results of *Doppler* are the same as in Fig. 3.9 (S).

Unwrapping a list with *

The function Doppler defined in Table 3.9 expects three positional arguments. The parameters of our lists are, however, specified in one object, the list *var1*. Calling *Doppler(var1)* results in an error message: `TypeError: Doppler() missing 1 required positional argument: 'vR'`. In the call of *Doppler* in line 11, list *var1* has, therefore, been unwrapped `(*var1)` so that the elements of *var1* are transferred and not the list as one object.

Table 3.9 A general formula for calculating the Doppler shift for sender and receiver moving on the same straight line

```
def Doppler(vS,vR,pos = 'SR'):
    if pos == 'SR': sgnX=1
    if pos == 'RS': sgnX=-1
    fR=c-vR*sgnX
    fS=c-vS*sgnX
    return fR/fS
c=340.0
vS=17.0
vR=10.0
var1=[vS,0.0,'SR']
print(var1,'{:5.2f}'.format(Doppler(*var1)))
```

[17.0, 0.0, 'SR']	1.05	[17.0, 10.0, 'RS']	0.98
[17.0, 0.0, 'RS']	0.95	[17.0, -10.0, 'SR']	1.08
[0.0, 10.0, 'SR']	0.97	[17.0, -10.0, 'RS']	0.92
[0.0, 10.0, 'RS']	1.03	[-17.0, 10.0, 'SR']	0.92
[17.0, 10.0, 'SR']	1.02	[-17.0, 10.0, 'RS']	1.08

3.3.2 A Sound Source Passes a Remote Receiver

Figure 3.10a illustrates the ride of a car on a straight road, on the line $y = 0$ from $x = -100$ m to $x = 100$ m. At a distance of $y_R = 30$ m off the road, a receiver is at position (0, 30). The car constantly sends out a tone of 200 Hz. Which frequency does the receiver perceive when the car passes by?

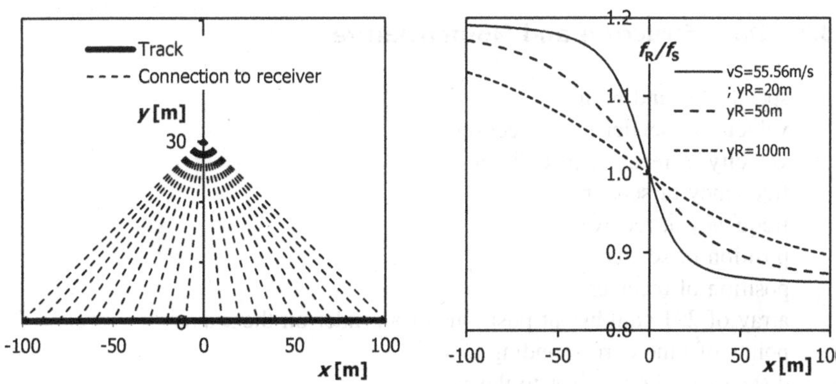

Fig. 3.10 **a** (left) An observer (receiver) at point (0; 30) hears a car (sender) passing on the x-axis (in the figure compressed). The velocity along the current connection line (dashed) determines the perceived frequency. **b** (right) The perceived frequency, relative to the frequency of the source, when a sound source passes the receiver at different distances y_R at speed $v_S = 55.56$ m/s

Velocity on line connecting sender and receiver
The velocity along the current connection line (dashed) determines the perceived frequency. The quantities v_R (receiver) and v_S (sender) in Eq. 3.9 are the velocity components on the current connecting line between sender and receiver, i.e., on the dashed lines in Fig. 3.10a.

The velocity must have the correct sign. The upper signs in Eq. 3.9 apply when the car is approaching, and the lower signs when it is driving away. For the situation described above, the sender's velocity component in the direction of the stationary receiver is to be determined by differentiating the distance with respect to time. This yields a negative velocity when the car is left of $x = 0$ and a positive velocity when is it right of $x = 0$, so that the frequency ratio

$$\frac{f_R}{f_S} = \frac{1}{1 + \frac{v_S}{c}} \tag{3.14}$$

is valid for the whole track.

Frequency curves for the distances $y_R = 20$, 50, and 100 m are displayed in Fig. 3.10b. The closer the receiver is to the track, the stronger the frequency varies with the position of the car. When the sender is at the receiver's height, the received frequency is identical to the sent frequency.

The interesting thing about the computation model in this exercise is that the time-dependent distance of the sender to the receiver is to be calculated (with Pythagoras) at each interval boundary. Then, the velocity component on the connecting line is obtained through numerical differentiation. In this way, more complicated geometries can also be treated, e.g., when the transmitter is moving at varying speed on a circular path, or if transmitter and receiver are moving on different paths.

3.3.3 Data Structure and Nomenclature

c	speed of sound in air
v_S	velocity of sender in x direction
v_R	velocity of receiver in x direction
f_S	frequency of sender
f_R	frequency at receiver
x_S	position of sender
x_R	position of receiver
x	array of 201 equidistant positions of a sender on the x-axis
$t(x)$	points of time corresponding to x
y_R	distance of the receiver to the x-axis
$dist$	current distance sender-receiver, $dist = \sqrt{x^2 + y_R^2}$.
v	relative speed along the line connecting sender and receiver

	A	B	C	D	E	F	G	H	I
1			200.00 km/h						
2	Velocity of sender	**vS**	55.56 m/s						
3	Frequency of sender	**fS**	200.00 Hz			*=B2&"="&ROUND(vS;2)&"m/s; "*			
4	Speed of sound	**c_**	340.00 m/s				*&B5&"="&yR&"m"*		
5	Distance of receiver	**yR**	20.00 m			**vS=55.56m/s; yR=20m**			
7		*=B9+1*	*=C9+(x-B9)/vS*	*=SQRT(x^2+yR^2)*	***=(dist-D9)/(t-C9)***	*=fS/(1+v/cS)*	*=f/fS*	*=(x+B9)/2*	
8		**x**	**t**	**dist**	**v**	**f**	**fNorm**	**xC**	
9		-100	0	101.98					
10		**-99**	**0.02**	101.00	**-54.47**	238.15	1.19	**-99.50**	
209		100	3.60	101.98	54.47	172.38	0.86	99.50	

Fig. 3.11 (S) A sound source moves on the x-axis past an observer at a distance d_R. The x position in column B is the independent variable. From that, the time (column C), the distance source-observer (column D), the velocity of the sender in the direction of the connecting line (E), and the observed frequency (F, G) are calculated with Eq. 3.14. Attention: The time in column C depends on the speed of the source! In column H, x_C is the center of the intervals

3.3.4 Spreadsheet Calculation "Remote Receiver"

A possible calculation model is shown in Fig. 3.11 (S), where we have chosen the x-coordinate of the car on the track as the independent variable, and time, distance, and velocity along the connecting line as dependent variables. You could just as easily choose time as the independent variable. The x in the formula C10 = [= C9 + (x - B9)/vS] refers to B10, the entry of the column vector x in the same row.

▶ **Mag** The frequency ratio is calculated numerically in columns F and G. Over which local coordinates do you plot the calculated frequencies? Perhaps over x in column B?

▶ **Tim** Well, since you asked it in that way, it probably isn't. I remember this much: We take the centers of the considered distance intervals because the speeds were calculated with the (t, x) coordinates of the interval boundaries.

▶ **Mag** Right! The centers of the intervals are calculated as x_C in column H.

3.3.5 Python Program "Remote Receiver"

A Python program that solves the task is given in Table 3.10, the main program in the upper cell, the function *FreqLine* for realizing Eq. 3.14 for the complete frequency curve in the lower cell. List slicing is used to calculate the interval centers x_C in the main program, and the velocities v when differentiating the distance *dist* in *FreqLine*.

Table 3.10 **a** (top) Specifications of the situation in which a receiver is at rest at a distance y_R off the sender's track. **b** (bottom) Function for calculating the frequency curve when y_R is given

```
vS=55.56               #Speed of sender
fS=200                 #Frequency of sender
c=340                  #Speed of sound

x=np.linspace(-100.0,100.0,201)
 #Center xC of path segments
xC=(x[1:]+x[:-1])/2

 #Time segments Dt
Dt=(x[1:]-x[:-1])/vS
 #Time t
t=np.cumsum(Dt)    #Integrates over dt
```

```
def FreqLine(yR):
    dist=np.sqrt(x**2+yR**2)
    v=((dist[1:]-dist[:-1])/Dt)
    fRS=1/(1+v/c)
    return fRS         #Frequency ratio
```

Questions

concerning Table 3.10:

What are the arguments and the global parameters accessed in *FreqLine*?[10]

The local variable x_C over which the received frequency is to be plotted is calculated in the main program. Would it be more consistent to calculate x_C in *FreqLine* and return it together with the normalized frequency (`def FreqLine(x, yR) .... return xC,fRes`)?[11]

Question

concerning Table 3.11

What does the keyword argument *lw*=1.5 in the plot function specify?[12]

What does the label in line 8 look like?[13]

How do we specify that the figure is saved with a resolution of 1200 dpi?[14]

The plot program in Table 3.11 calculates and displays the frequency curves for the three distances $y_R = 20$, 50, and 100 m.

---

[10] *FreqLine*, argument: yR, global parameters *x*, *dist*, *c*, *Dt*.

[11] Discuss!

[12] Line width *lw* = 1.5 point.

[13] The label is "yR = 20".

[14] `plt.savefig ('Doppler off, multiple.png', dpi=1200)`, `dpi` is a key word argument.

Table 3.11 Plotting several frequency trajectories with the parameters specified in Table 3.10

```
FigStd('x',-100,100,25,'f/fQ',0.8,1.2,0.1)
plt.plot([-100, 100],[1,1],ls='-',color='k',lw=1)
                         #Horizontal through y=1
yR=20                    #Distance to track
fRS=FreqLine(yR)
plt.plot(xC,fRS,'k-', lw=1.5,label='yR='+str(yR))
yR=50; fRS=FreqLine(yR)
plt.plot(xC,fRS,'k--',lw=1.5,label='yR='+str(yR))
yR=100; fRS=FreqLine(yR)
plt.plot(xC,fRS,'k-.',lw=1.5,label='yR='+str(yR))
plt.legend(loc=0,fontsize=10)
#plt.savefig('Doppler off, multiple.png')
```

3.4 Exponentials

For exponential functions, apply Ψ *Plus 1 yields times e;* plus 1 in the argument yields times e in the value. An exponential function $A \cdot \exp(-t/t_0)$ is best drawn by hand (Really? Yes, also by hand!), beginning with its tangent at $t = 0$. Diode characteristics seemingly exhibit a “kink voltage” that depends on the scaling of the y-axis; more generally, exponential functions seem to explode.

3.4.1 Explosive Character of Exponentials

Rice grains on a chessboard
In a classic bet from far eastern literature, a clever chess player agreed with his king that his winnings would be paid by placing a rice grain on the first square of a chessboard and then doubling the number of grains repetitively on each of the following 63 squares.

In Fig. 3.12a (S), the doubling of the number of rice grains is simulated with a spreadsheet calculation. In A3:A66, the 64 squares of the checkerboard are numbered from $n = 0$ to 63. In B3, a rice grain is placed on the first field ($n = 0$, B3 = [1]). In the following cells, the preceding number is doubled until the huge number $2^{63} = 9.22 \times 10^{18}$ is reached in B66 ($n = 63$), on the 64th square.

▶ *Ψ Plus (in the argument) yields times (in the result) for exponentials.*
For $y = 2^x$, plus 1 (in the argument) yield times 2 (in the result).

	A	B	C	D	E
1	=A3+1	=B3*2	=2^n		=2^D3
2	n	y	y =2^n		**2^62.4**
3	0	1	1	62.4	**6.1E+18**
4	**1**	**2**	**2**		
5	2	4	4		
6	3	8	8		
7	4	16	16		
8	5	32	32		
9	6	64	64		
66	63	9.22E+18	9.22E+18		
67	64	1.84E+19	1.84E+19		
1025	1022	4.5E+307	4.5E+307		
1026	1023	9.0E+307	9.0E+307		
1027	1024	#NUM!	#NUM!		

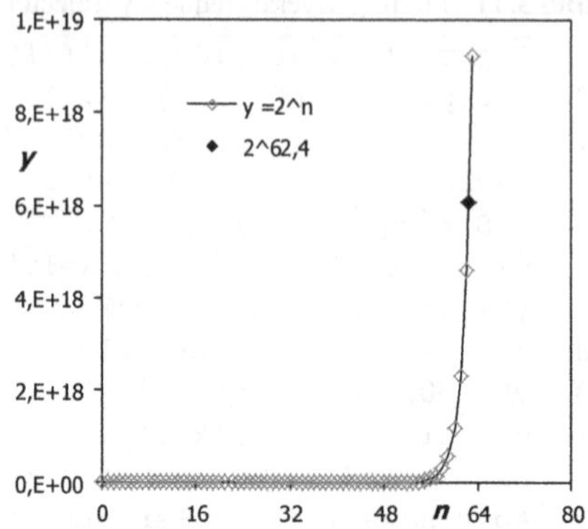

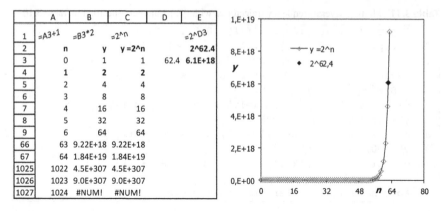

Fig. 3.12 **a** (left, S) Powers of 2, $y = 2^n$, obtained by repetitive multiplication by 2 (column B) and by potentiation (C: E). The formula in E1 refers to E3 **b** (right) Graphical representation of the powers of 2 of **a**

The values are graphically displayed in Fig. 3.12b. You can see that the explosion by a factor 10^{19} takes place on the last few squares. On squares 0–62, there are 2^{63}–1 grains, on all squares together, $2^{64}-1$ grains.

Profit

Question

What would be the share of the internationally traded rice in 2016 if the loser of the game (a wealthy medieval Sultan) had been able to deliver?[15]

What is the value of 10E3 in EXCEL and `10e3 in Python`? Be careful![16]

▶ **Mag** Can the winner satisfy his hunger with his win?

▶ **Alac** Perhaps once.

▶ **Tim** I've heard that exponential growth means explosion. So, maybe the winner can live well on his heap of rice for a week.

▶ **Mag** World rice production in 2015/16 was 470 million metric tons, but only about 5% were traded on the world market. Unlike wheat, rice is consumed by more than 95% of the population in the cultivating countries.

[15] For 64 fields, the winner would have received 2^{64}–1 = 18 × 10^{18} grains, corresponding to about 10^{18} g = 10^{12} tons of rice. This is the 2000-fold amount of the rice harvest 2015/16 of 470 million tons.

[16] 10E3 = 10 × 10^3 = 10^4 = 10,000, the same with `10e3 in Python`.

▶ **Tim** I've counted. One kilogram of rice contains about $40{,}000 = 4 \times 10^4$ grains. According to the rule of the game, more than 10,000 times the volume of one year's world trade of rice should pile up on the chessboard. Incredible!

▶ **Alac** Crazy! A disaster! This cannot be true! What's the catch?

▶ **Mag** There is no catch. The catastrophe results from the rule Ψ *plus 1 yields times 2* governing the exponential 2^{n}.

The power function $y = 2^{\mathrm{n}}$ is calculated in column C of Fig. 3.12a (S). It is entered as the worksheet formula [= 2^N] and yields the same results as the repetitive multiplication by 2 in column B. The argument of the power function does not have to be an integer. In E3, the value for the power in D3 (= 62.4) is calculated and inserted as a filled diamond in Fig. 3.12b.

Maximum float number

We continue doubling the preceding number in the worksheet beyond $n = 63$ until the application can no longer store the resulting number. As of row 66 in Fig. 3.12a (S), the numbers are represented in exponential form, $9.22\text{E+}18^{+18} = 9.22 \times 10^{18}$. The number 2^{1024} can no longer be calculated in EXCEL2019; see the error message #NUM! in row 1027.

In `Python 3`, the `int` format can store arbitrarily large numbers. However, if the number is to be calculated as float, the same limit holds as in Excel: 2**1023 gives 8.988e+307 and 2**1024 returns the `OverflowError: int too large to convert to float`.

3.4.2 General Exponential Function

The power function can be generalized to $y = a^{\mathrm{x}}$, where a and x are real numbers. If a is Euler's number $e = 2.718$, then the power function becomes the known exponential function with the formula [= EXP(...)] (EXCEL) or `np.exp()` (`Python`).

Exponential function with characteristic length

The normalized exponential function is usually written in mathematical textbooks as follows:

$$f(x) = |a| \cdot exp(ax) = |a| \cdot e^{ax} \tag{3.15}$$

The letter e denotes Euler's number $e = \sum_0^\infty \frac{1}{n!} = 2.718$. Normalized means that the integral from 0 to $+\infty$ (for $a < 0$) or from 0 to $-\infty$ (for $a > 0$) is 1. The integral is dimensionless because the product of the units of dx and a is 1.

It is, however, often physically more sensible to write the exponential with a characteristic x value x_0:

$$f(x) = \left|\frac{1}{x_0}\right| \cdot \exp\left(\frac{x}{x_0}\right) = \left|\frac{1}{x_0}\right| \cdot e^{\frac{x}{x_0}} \tag{3.16}$$

Thus, the unit of x_0 is equal to the unit of x, e.g., a length or a time, and has an intuitive meaning: the tangent at $x = 0$ intersects the y-axis at the amplitude $|1/x_0|$, and the x-axis at the characteristic length x_0.

When the function is specified with an amplitude $\frac{A}{|x_0|}$, its integral from 0 to, respectively, $+\infty$ (for $a < 0$) or $-\infty$ (for $a > 0$), is A.

3.4.3 Representation in a Diagram

First, the tangent at $t = 0$!
Two exponential functions, together with their tangents at $t = 0$, are shown in Fig. 3.13a. Here, the independent variable is time t.

▶ **Mag** Do you now know how to draw an exponential function $A_e \cdot exp(t/t_0)$ by hand on a piece of paper?

▶ **Alac** Sure! First, mark the amplitude A_e on the vertical axis and the characteristic time t_0 on the horizontal and pass a straight line through the two points. The exponential curve approaches the tangent at $t = 0$ and the horizontal axis for $t \to \infty$ or $t \to -\infty$, depending on the characteristic parameter's sign.

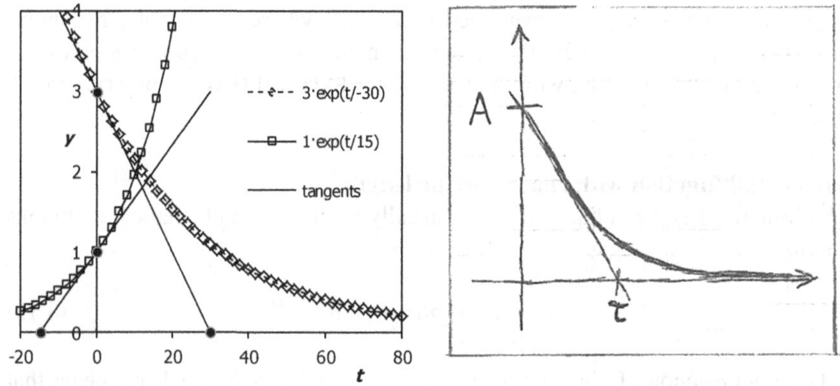

Fig. 3.13 **a** (left) Two exponential functions $A_e \cdot exp(t/t_0)$ with their tangents at the intersections with the y-axis. **b** (right) How to draw an exponential function by hand: First, the tangent as a straight line with its intersections with the x- and y-axes!

▶ **Mag** Correct, just as in Fig. 3.13b! Keep in mind:

▶ Ψ *Plus one in x, times e in y*
Ψ *Expo with kink and straight line*

▶ **Tim** "Plus one" and "Straight line" are clear, but why "kink"?

▶ **Mag** You have seen this in Fig. 3.12**b**; it will be explained in Sect. 3.4.4. An essential feature of an exponential is an explosion on a suitably scaled *y*-axis.

3.4.4 Diode Characteristics *I(U)*

The current I through a semiconductor diode depends exponentially on the applied voltage U. The $I(U)$ characteristics of a semiconductor diode are described by an exponential function passing through zero:

$$I = I_s \cdot \left(\exp\left(\frac{U}{U_T} \right) - 1 \right) \tag{3.17}$$

This function has two parameters: the strength of the reverse current I_s also called the saturation current, and the thermal voltage U_T, which is given by $k_B T/e$, with k_B being the Boltzmann constant, e the elementary charge, and T the absolute temperature. At room temperature, $U_T = 25$ mV. The current through a diode is zero when the applied voltage is zero.

We will represent such a function for $I_s = 1 \times 10^{-14}$ A and $U_T = 0.025$ V in various plots (see Fig. 3.14).

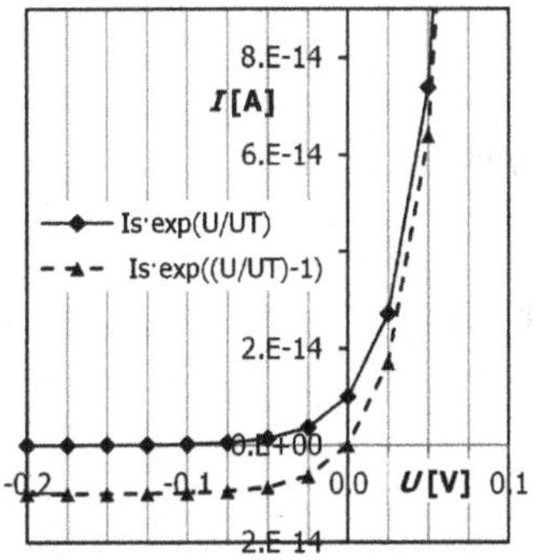

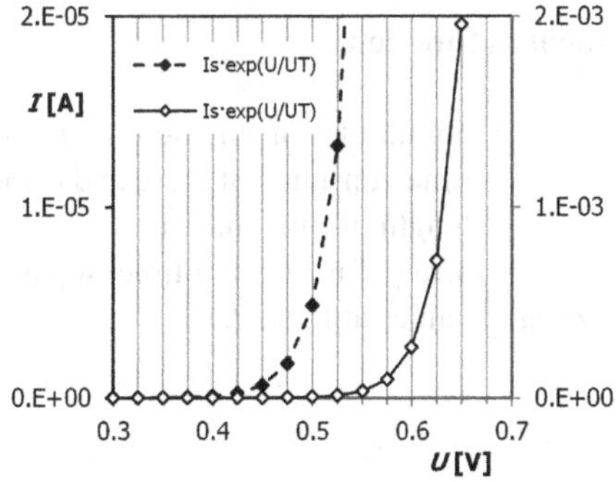

Fig. 3.14 **a** (left) Diode characteristics and associated exponential function, representation for small currents. **b** (right) Twice the same exponential as in **a**, but with different scaling of the I axes; left y-axis for the left curve, right y-axis for the right curve; the vertical grid lines have a distance $U_T = 25$ mV

Questions

Which of the curves in Fig. 3.14a, b are exponential functions? Which are shifted on the I-axis? Which are shifted on the U-axis?[17]

At which U-values in Fig. 3.14b are you most likely to find the "kink points", in electronic engineers' jargon?[18]

Also shown in Fig. 3.14a is an exponential function without the term -1 in the parentheses of Eq. 3.17. This function intersects the I-axis at I_S. It grows by a factor of $exp(1) \approx 2.7$ whenever U progresses by dU. The curves in Fig. 3.14a appear to be exponential, the way they are usually represented.

For exponential functions to the base *e,* the broom rule is Ψ *Plus 1 yields times e*. In the concrete case, this means that, if U advances by U_T (the distance between the vertical gridlines in Fig. 3.14), *exp* increases by a factor of about 2.7.

In Fig. 3.14b, the same data as in Fig. 3.14a are shown twice, only with different scaling of the two I-axes: I_{Max} is now at 2.5×10^{-5} A (left vertical axis) or 2.5×10^{-3} A (right vertical axis); the curves seem to exhibit a kink at about 0.5 and 0.6 V. In electrical engineering, this voltage is called the "kink voltage", "knee voltage" or "cut voltage".

The increase of the U-value by a factor of 2.7 when progressing from $U = 0.475$ to $U = 0.500$ V looks like a steep rise. It increases, however, by the same factor when it progresses from $U = 0$ to $U = 0.025$ V but then appears like the familiar soft curvature of the exponential. The position of the seeming kink on the U-axis is a function of the scaling of the I-axis.

In Fig. 3.15a, the I-axis is logarithmic. In this representation, the diode characteristics appear as a straight line, except for the points below $U = 0.05$, because, there, the term -1 is quantitatively significant. A kink is nowhere to be seen.

3.4.5 Data Structure and Nomenclature

Exponential function

A, B	amplitudes of functions exp_A and exp_B
t_A, t_B	time constants of exp_A and exp_B
d*t*	length of time interval
t	array of instants of time, separated by d*t*
exp_A, exp_B	values at times *t*.

[17] All curves represent exponential functions. The diode curve in Fig. 3.14a is shifted downwards on the I-axis by the saturation current I_S, so that it passes through zero. The curves in Fig. 3.14b have not been shifted on the U-axis.

[18] At about 0.5 and 0.6 V.

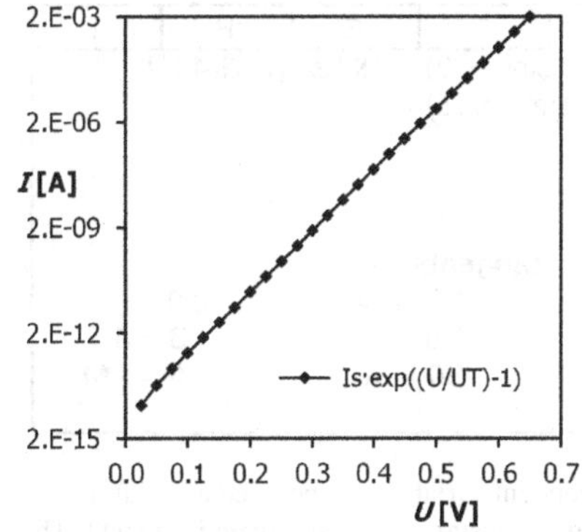

	A	B	C	D
1	**Is**		1.00E-14	
2	**UT**		2.50E-02	
3	**dU**		2.50E-02	
4				
5				
6	=A8+dU	=Is*EXP(U/UT)	=Is*(EXP(U/UT)-1)	
7	**U**	**exp**	**I**	
8	-0.200	3.35E-18	-1.00E-14	
9	**-0.175**	9.12E-18	-9.99E-15	
58	1.050	1.74E+04	1.74E+04	

Fig. 3.15 **a** (left) Diode characteristics, logarithmic scale of the I axis (semi-log plot). The curve can only be represented for $I > 0$, because the logarithm is defined only for positive values. **b** (right) Spreadsheet layout for calculating diode characteristics, I_s = saturation current, U_T = temperature voltage, $dU = U_T$

Diode characteristics

- I_s saturation current
- U_T thermal voltage, 25 mV at room temperature, characteristic parameter of the exponential
- dU interval width, here, $dU = U_T$
- U sequence of voltages, dU apart
- *exp* exponential function with I_s and U_T as amplitude and characteristic voltage
- I diode current, $I(U)$.

3.4.6 Spreadsheet Calculation

Initial slope of the exponential function

In Fig. 3.16 (S), two exponential functions are calculated for 51 points equidistant in t,

$$exp_A(t) = A \cdot exp\left(\frac{t}{t_A}\right) = A \cdot e^{\frac{t}{t_A}} \tag{3.18}$$

and exp_B correspondingly. The characteristic time, t_A or t_B, may be positive or negative. The distance dt between the points is set in B3.

► For the exponential function $y = e^x$, applies; Ψ *Plus 1 (in the argument) yields times e (in the value).*
For the power of 2, $y = 2^n$ holds: Ψ *Plus 1 yields times 2.*

	A	B	C	D	E	F	G	H	I
1	**A**	3.0	**tA**	-30.0		3·exp(t/-30)	=A&"·exp(t/"&tA&")"		
2	**B**	1.0	**tB**	15.0		1·exp(t/15)			
3	**dt**	2.0							
4	=A6+dt	=A*EXP(t/tA)		=B*EXP(t/tB)					
5	**t**	**expA**		**expB**		**tangents**			
6	-20.0	5.8		0.3		**30**	=-tA	0.0	
7	**-18.0**	5.5		0.30		0.0		**3**	=A
8	-16.0	5.1		0.34		**-30**	=tA	**6**	=2*A
56	80.0	0.2		207.13					

Fig. 3.16 (S) Two exponential functions whose independent variable *t* is specified as column vector A6:A56 and their parameters amplitude *A*, *B* and time constant t_A, t_B are given in A1:D2. The legends for the functions are compiled in F1 and F2 with the formula in G1 of type Ψ *"Text" & Variables*

The table in Fig. 3.16 (S) has a typical Γ layout. The 51 values of the independent variable time *t* are located to the left of Γ in a column vector named "*t*". The parameters *A*, B, and t_A, t_B of the curves are specified above Γ, as well as the time interval d*t*, the distance on the horizontal axis between the calculation points. The initial value of *t*, here, −20, has to be entered into cell A6. The values for the remaining 50 *t* values are determined successively from the respective predecessor. The *t*-predecessor for cell A7 is cell A6, A7 = [=A6 + d*t*]. In the formula, A6 is not provided with a dollar sign. It is a relative reference, so that the address of the addressed cell adapts during copying. Therefore, A56 = [=A55 + d*t*].

Questions

Suppose that A6 contains an initial time $t = 5$ and the length of a time segment is stored in a cell with the name d*t* and has the value 2. What values for time *t* are in cells A7 and A8?[19]

How can the coordinates of the tangent to the exponential $A_e \cdot exp(t/t_0)$ at $t = 0$ be derived from the parameters of the exponential function?[20]

What is the distance on the horizontal axis between the functions' calculation points in Fig. 3.13a created from Fig. 3.16 (S)?[21]

What is the value of $\int_0^{\infty} 3 \cdot \exp(-\frac{t}{30}) dt$?[22]

► **Task** Change the parameters *A*, t_A, and d*t* and check whether the diagram reacts accordingly! This is the case if each cell contains the correct formula. Remember:

[19] A6 = [5], A7 = [= A6 + dt] = 7; A8 = [= A7 + dt] = 9.
[20] Straight line through the points (0, *A*) und ($-t_e$, 0).
[21] The distance of adjacent points is d*t* = 2 (see Fig. 3.16 (S), B3).
[22] The definite integral is 90, based on Eq. 3.16, because $3 = \frac{1}{30} \cdot 90$.

you cannot merely transfer the numbers from the figures to your worksheet. Most cells contain a formula; only sometimes are numerical values entered directly.

Diode characteristics

A possible calculation model is presented in Fig. 3.15b (S).

The parameters I_s and U_T of the diode characteristics are set in the named cells C1:C2. In C3, the horizontal distance between neighboring sampling points is defined. Here, we have chosen $dU = U_T = 0.025$ V that is valid for room temperature. The *I-U* characteristics are illustrated in Fig. 3.14. They intersect the *I*-axis at 0, showing that no current flows without applied voltage.

Formatting the axis of a diagram

The axis of a diagram is formatted by activating it with the left mouse button and then clicking FORMAT. In the CURRENT SELECTION group to the left of the ribbon, VERTICAL (VALUE) AXIS shows up in the bar. Upon clicking FORMAT SELECTION, a window opens that allows you to set the minimum, maximum and other parameters of the axis. If the axis is to be scaled logarithmically, the appropriate box, ☑ LOGARITHMIC SCALE, must be activated with a checkmark.

3.4.7 Python Program

Initial slope

A `Python` program for drawing the exponentials of Fig. 3.13a and their slopes at $t = 0$ is given in Table 3.12.

The list t of time instants is created by `np.arange` $(-20, 80 + dt, dt)$, mimicking the construction of the time vector in the spreadsheet of Fig. 3.16 (S) with, e.g., A8 = [=A7 + d*t*]. The lower limit −20 corresponds to the value in A6. To come to 80 (in A56), we have to specify 80 + dt as the upper limit, because `np.arange` creates an interval that is open at its right end with the endpoint excluded. Exp_A and exp_B are constructed from t with the corresponding amplitudes and time constants

Table 3.12 **a** (top) Specifications of two exponential functions, equivalent to rows 1 through 3 in Fig. 3.16 (S); **b** (bottom) labels for the two functions created in lines 7 and 8

```
1  dt=2.0
2  A,tA= 3.0,-30.0          #Amplitude and time constant
3  B,tB=1.0,15.0
4  t=np.arange(-20,80+dt,dt)
5  expA=A*np.exp(t/tA)
6  expB=B*np.exp(t/tB)
7  lblA=(str(A)+'*'+'exp(t/'+str(tA)+')')
8  lblB=(str(B)+'*'+'exp(t/'+str(tB)+')')
```

```
lblA   3.0*exp(t/-30.0)
lblB   1.0*exp(t/15.0)
```

Table 3.13 Specifications for diode characteristics, the same as in Fig. 3.15b

```
Is=1e-14
UT=2.5e-2
dU=2.5e-2
U=np.arange(-0.2,1.05+dU,dU)
exp=Is*np.exp(U/UT)
I=Is*(np.exp(U/UT)-1)        #Diode characteristics
```

(A, t_A) and (*B*, t_B). The curves can be plotted with our standard function `StdFig`. In Fig. 3.13a, they are represented with open symbols. In `Python`, this is specified by `plt.plot(…., fillstyle='none', …)`.

Questions

What are the instructions for plotting the tangents in Fig. 3.13a?[23]

What are the size and last element of:

- `L1 = np.arange(−20, 80+dt, dt)` with $dt = 1$,[24]
- `L2 = np.linspace (−20, 80, 100)`,[25]
- `L3 = np.linspace (−20, 80, 101)`[26]
- `L2 = np.linspace (−20, 80, 100, endpoint=False)`.[27]

Diode characteristics

A `Python` program corresponding to the spreadsheet layout in Fig. 3.15b is shown in Table 3.13.

Subplots

A plot like Fig. 3.14b with two vertical axes cannot be achieved with our standard figure, the function *FigStd* defined in Sect. 2.4.5. We have to refer to the function `subplots` of the `pyplot` library (see Table 3.14). *I(U)* is plotted twice, with `ax1.axis([0.3, 0.7, 0, 2e-5])` for the primary (left) *y*-axis from 0 to 2e−5 and with `ax2.axis([0.3,0.7,0,2e-3])` for a second *y*-axis from 0 to 2e−5, declared with `ax2 = ax1.twinx()` as the secondary (right) *y*-axis.

Logarithmic scaling

Logarithmic scaling of an axis can be achieved with `plt.yscale(value="log")` and `plt.xscale(value="log")`.

---

[23] `plt.plot([0,tA],[A,0],'k-')`, compare F5:H8 in Fig. 3.16 (S)!

[24] L1, size 101, last element L1[−1] is 80.

[25] L2, size 100, last element 80.

[26] L3, size 101, last element 80.

[27] L4, size 100, last element 80 − 100/100 = 79.

Table 3.14 Setting up a diagram similar to Fig. 3.14b with primary and secondary axes, variables specified in Table 3.13

```
fig, ax1 = plt.subplots()
ax1.plot(U, I, 'k-')
ax1.axis([0.3, 0.7, 0, 2e-5])
for x in np.arange(0.3,0.7,dU):
    ax1.plot([x,x],[0,2e-5],'k-',lw=0.6)

ax2 = ax1.twinx()       #Secondary axis
ax2.axis([0.3,0.7,0,2e-3])
ax2.plot(U, I, 'r--')
plt.show()
```

3.5 Questions

General advice

1. To practice programming, translate the spreadsheet solutions of this chapter into `Python` and compare with the suggested programs!

Python-specific

2. Let `x=np.linspace(−100,100,101)`. What is the distance between neighboring elements? What does x[1:]−x[:−1] look like? What are the first and last elements of (x[1:] + x[:−1])/2 ?
3. Let A = 3.001 and $t_A = -30.0$. How can we produce a label "3.0*exp(t/−30)"?

Concerning Table 3.15:

4. What is the size of the list *col*?
5. What are the first two elements of *col*?
6. What are the last two elements of *col*?
7. How do you compile the four lines into one line with list comprehension?

Check your answers by programming!

Table 3.15 Producing a list x*y

```
col=[]
for x in range(1,11):
    for y in range(1,6):
        col.append(x*y)
```

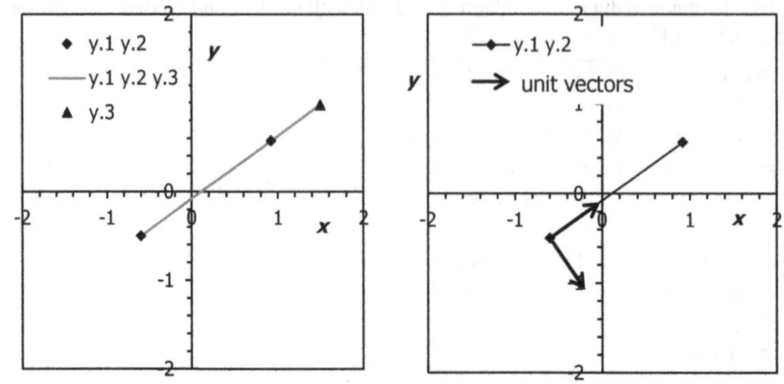

Fig. 3.17 **a** (left) The segment defined by two points (x_1, y_1) and (x_2, y_2) is extended to a third point whose x-coordinate x_3 can be selected arbitrarily. **b** (right) The unit vectors in the direction of the line and perpendicular to the line are attached to the line, defined in Fig. **a**, at the left point

Broom rules

8. Explain the broom rules:
   Ψ *Lens equation with plus and minus.*
   Ψ *Blank lines separate curves.*
   Ψ *Doppler effect with plus and minus.*

Straight-line equation

A straight line is defined by two points (x_1, y_1) and (x_2, y_2) (see Fig. 3.17a).

9. What is the equation for determining the distance between the two points?
10. Which equation must be used to find the value y_3 for a given horizontal position x_3 so that (x_3, y_3) lies on the straight line?
11. Deduce from the coordinates of the two points the direction vector (D_x, D_y) of the straight line normalized to length 1!
12. How do the coordinates of the two points result in the vector (P_x, Py), the perpendicular to the straight line and normalized to length 1?

    A spreadsheet layout for three points on a straight line is shown in Fig. 3.18 (S).

13. What is the linked cell and MIN and MAX of the slider in F4:H4?
14. What formulas are in B3 (input from E3) and D4 (input from E4)?
15. Write a `Python` program that performs the calculations of Fig. 3.18 (S)! Replace the function of the sliders with simple assignments with random functions: `E1=` ... ; `E4=` ... !

	A	B	C	D	E	F	G	H	I	J	K	M	N	P
1	*Three points on a straight line*								*Unit vectors*					
2	x.1	-0.6	y.1	-0.5						*along the straight line*				
3	**x.2**	**0.92**	y.2	0.57	92					x.1	-0.6	y.1	-0.5	
4	x.3	1.5	**y.3**	**0.98**	57					**x.p**	-0.03	**y.p**	-0.1	
5										*perpend. to the straight line*				
6	*Length of segment 1-2*				*Slope of segment 1-->2*					x.1	-0.6	y.1	-0.5	
7	l.12	2.66			m.12	0.70				**x.v**	-0.2	**y.v**	-1.07	

Fig. 3.18 (S) Spreadsheet layout used to create the coordinates for Fig. 3.17**a** and **b**, B3 and D4 depend on the outputs on the sliders. The slider in F4:H4 goes from 0 to 80

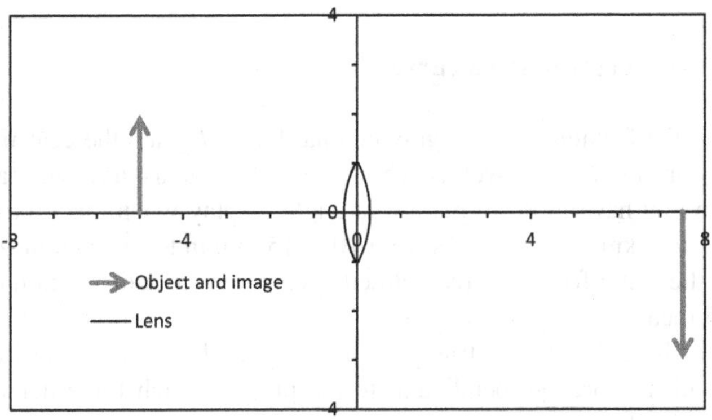

Fig. 3.19 Incomplete image construction for imaging with a converging lens

Image construction with converging and diverging lenses

16. In school, one usually learns the equation $1/f = 1/o + 1/i$ for imaging with converging lenses (o is object distance, i is image distance). How is this imaging equation modified according to DIN 1335[28] and made suitable for numerical calculation in spreadsheets and `Python` both for converging and diverging lenses?
17. How is the magnification factor defined in DIN 1335?
18. What characterizes a converging lens in the imaging equation?
19. What characterizes a diverging lens in the imaging equation?
20. Draw the rays for the image construction in Fig. 3.19!
21. What is the image-side focal length?
22. Draw the bundle of rays that contributes to the image formation!

[28] Equations 3.5 and 3.6.

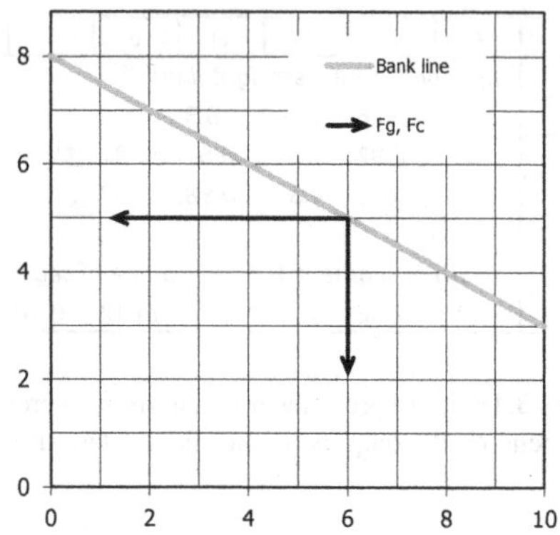

Fig. 3.20 Gravitational F_g and centrifugal F_c force when driving through a curve

Forces when driving through a curve

23. What are the formulas for the gravitational force F_g and the centrifugal force F_c on a car of mass m, traveling with speed v through a curve with radius r?
24. What do you have to do to get a true-angle display when the x-axis is scaled from -2 to 8 km and the y-axis from 10 to 15 km, in EXCEL and in `Python`?
25. How is the static friction force defined? What does a static friction coefficient $\mu = 0.5$ mean?
26. Draw in Fig. 3.20, with a triangle ruler, the resulting force, the force in the plane, and the force perpendicular to the plane! Which force determines the static friction?
27. Draw a vector in the bank line and another one perpendicular to it!
28. What are the vector equations for determining the quantities of Question 26?

Doppler effect

$$\frac{f_E}{f_Q} = \frac{c \pm v_E}{c \mp v_Q} \tag{3.19}$$

29. What do the letters in Eq. 3.19 stand for? Which signs are to be used when? Adjust the formula for the three cases in Fig. 3.21!

A car (sender S) drives along the x-axis. Its position at time t is specified in an array named x_S. A pedestrian (receiver R) moves along the y-axis towards the x-axis. Its position at time t is indicated in an array named y_R.

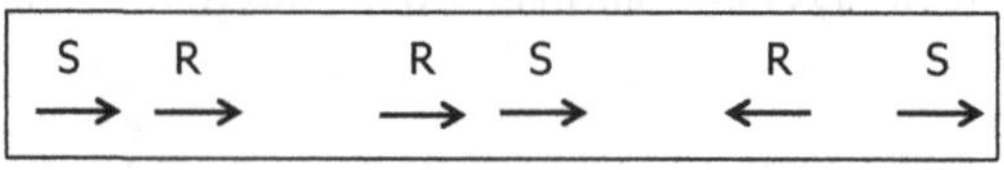

Fig. 3.21 A sound source S and a sound receiver R move on a straight line

30. What is the distance d_{SR} between car and pedestrian as a function of t?
31. What is the formula for calculating the relative velocity v_{SR} in the direction of the link line? To answer this question, you have to define vectors x_R, y_R and d_{SR}.

Exponential function

32. To draw an exponential function freehand on paper, it is useful to start with a straight line as a guide. How is this straight line determined by the exponential function parameters, amplitude A and time constant τ?
33. An exponential function increases from 1 to 2 when the argument is increased from $t = 0.0$ to 0.1 s. How much does it increase if the argument is increased from 0.0 to 0.2 s? Think binary! How much does it increase if the argument is increased from $t = 0.8$ to 1.0 s? Sketch the same exponential function twice, each time with the t-axis from 0 to 1.2, but with a y-axis scaling from 0 to 4 for the first sketch and from 0 to 16×4 for the second sketch! What do you see in the second sketch that is not obvious in the first sketch?

Macros with Visual Basic and Their Correspondences in Python

4

We practice the *basic programming structures*: loops, branches, sub-routines (FOR, IF, SUB/`def`), with particular emphasis on data exchange between spreadsheets and procedures. We will learn to obtain the sequences for EXCEL-typical spreadsheet operations by recording the associated commands with a macro recorder. With this knowledge, we will:

- draw dense-packed crystal planes,
- decode protocols of measuring instruments and compile clear summaries of the results,
- systematically modify the parameters of calculation models with log and control procedures and continuously enter the results of the calculations into another range of the spreadsheet, and
- outsource complicated formulas into user-defined spreadsheet functions in order to arrange the tables more clearly.

We will present parallel solutions in `Python` creating the drawings with the library `turtle`.

4.1 Introduction: For, If, Sub/Def

Solutions of Exercises 4.3 (Excel), 4.5 (Python), 4.7 (Excel), and 4.8 (Python) can be found at the internet address: go.sn.pub/gTtbiH.

Tim worries, Alac brags

► **Mag** This chapter will teach us to program, to let macros interact with spreadsheet calculations, and to realize parallel solutions in `Python`.

D. Mergel, *Physics with Excel and Python*,
https://doi.org/10.1007/978-3-030-82325-2_4

▶ **Tim** That sounds pretty demanding. Is it at all manageable for beginners like me?

▶ **Mag** Quite clearly: Yes. Many people have already achieved that, even students who had never before written a line of code. Visual Basic is a good-natured programming language that does not require much knowledge, at least not for the tasks that we want to tackle. This chapter will not only take away your fear of programming, but you will also find it fun to write programs that do amusing things.

▶ **Tim** Well, people who had just finished the course told me that they had to deal with routines, macros, programs, and procedures.

▶ **Mag** Don't worry, we don't make any distinction among those terms, we use them all synonymously. Our programs include both EXCEL-typical command sequences and classical algorithmic structures.

▶ **Alac** I'm not afraid of VISUAL BASIC or `Python`; after all, I already took a course about another programming language and used it to write funny programs.

▶ **Mag** That is certainly a good prerequisite for faster success. You will have an easier time than Tim. Nevertheless, don't take the tasks for granted. In our course, programming tasks are combined with physics exercises (so much for our hopes for just "funny":-)). I've often experienced instances in which mere programmers have been discouraged by their limited progress in this kind of programming and have given up.

▶ **Alac** So, I'm essentially learning more physics?

▶ **Mag** Not only that. You should certainly understand more about physics after the course than before. Nevertheless, the exercises will also familiarize you with good programming skills: to develop systematically, to document carefully, to detect and correct mistakes.

▶ **Tim** One more question. Many workplaces require programming skills in special programming languages. Wouldn't it be better for me to learn such languages from the onset?

▶ **Mag** Don't worry.' In our tasks with VISUAL BASIC and `Python`, you will have room to make a sufficient number of mistakes to learn from so that you can be confident in becoming a computer expert. The algorithmic constructions are the same in all programming languages. More important than acquiring special knowledge at an early stage is that you gain the ability to cope with "hard" programming tasks and master the rules for good programming.

How do we proceed?

This course cannot give a general introduction to programming, because that would require an excess of repetition of things that are already well described in specific textbooks. As in Chap. 2, we will do a basic exercise that you should follow step by step and convert into your own program. It contains all of the commands and constructions that we will need later, but not much more. However, this basic knowledge will enable you to find your way around in EXCEL help or `Python` internet aid and choose suitable textbooks for programming with VISUAL BASIC and `Python`.

Only a few algorithmic constructions

In our programs, we use constructions that are the same in all algorithmic languages: *Loops, logical branches, sub-routines* (FOR, IF, SUB/`def`). Special EXCEL instructions, e.g., for handling files or creating drawings, will be provided to us by the macro recorder, which records commands that the user makes when setting up spreadsheet calculations.

Three types of program will be used repeatedly in the following chapters: log procedures, formula procedures, and user-defined functions.

Rep-log and scan-log procedures

Two tools can be used sensibly for every task in EXCEL: Sliders and log procedures. With a slider, variables in a spreadsheet can easily be changed manually, and the user can see how the results change, as we have learned in the basic Exercise 2.2. A *rep-log procedure* systematically changes independent variables in the spreadsheet and continuously writes the calculation results into another spreadsheet range. The spreadsheet calculation works like a function. The equivalent in `Python` is functions that are repetitively called in loops. *Scan-log procedures* scan a range of a spreadsheet and write the data into another range in a structured manner. The program in Fig. 4.5 (P) is a simple example. In Chaps. 8 to 10, log procedures repeat stochastic experiments to illustrate statistical rules and laws.

User-defined functions in visual basic

We generally recommend performing calculations step by step in several rows or columns. If the calculation runs without errors, spreadsheet space and calculation time can be saved if you execute this calculation in a user-defined spreadsheet function, with values transferred from the spreadsheet into the program, and with values written into the spreadsheet by the function, exactly as we learned with built-in spreadsheet functions, e.g., COS(x) or SIN(x). As examples, we shall implement functions for the scalar and the vector products of three-dimensional vectors, as usual, in both Visual Basic and `Python`.

Pandas

`Pandas` ("Python Data Analysis") is a library for `Python`, based on `NumPy`. It is designed for data management and analysis and works with structured data (`DataFrame` (2-dimensional)) and time series (`Series` (1-dimensional)), thus mimicking spreadsheet calculations. We shall use it in this book only in Exercise 4.8 “Processing the protocol of a measuring device”.

4.2 Basic Exercise: For-Loops

We get to know the VISUAL-BASIC-Editor and practice reading cell contents and filling in cells. We are using For-loops to execute tasks one after the other with systematically changed parameters. This often involves incrementing a running index in the loop that indicates a cell’s position in the spreadsheet. Such loops are called *loop2i*, obeying the broom rule Ψ *Continue counting in the loop*!

4.2.1 Visual-Basic-Editor 1: Editing

In the menu ribbon (see Sect. 1.7), click on the main group DEVELOPER and then on the tab VISUAL BASIC; the “SHEET1 (CODE)” window appears (Fig. 4.1).
If the right lower rectangle (below “(General)”) is gray, double-click the SHEET1 line in the MICROSOFT EXCEL OBJECTS so that it turns white. Instructions that you write into this sheet in the VISUAL BASIC editor are executed in Sheet1.

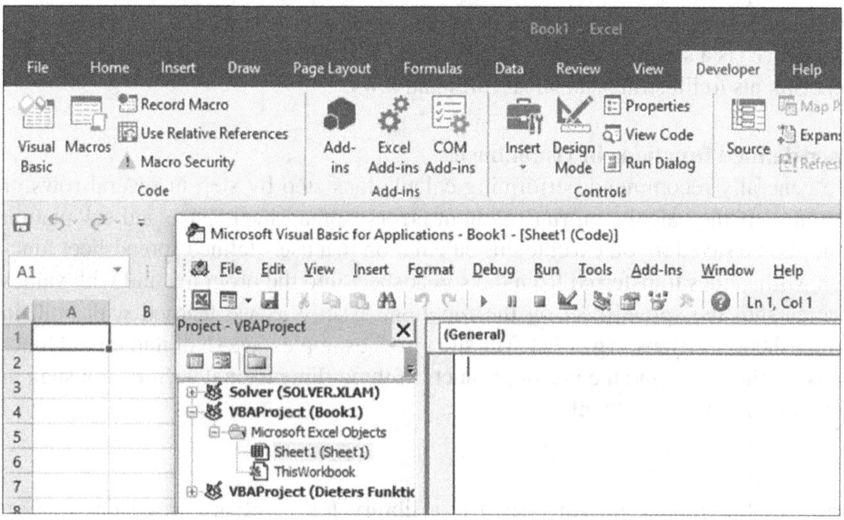

Fig. 4.1 Spreadsheet and associated VISUAL BASIC sheet after executing DEVELOPER/VISUAL BASIC (EXCEL 2019)

4.2.2 Programming

Enter values and formulas into cells

Write SUB *Name* into the first line of the white area! For "Name", enter your own name. The editor adds one line END SUB. You can now write statements between SUB Name and END SUB. In Fig. 4.2b, SUB *Annegret* has been created with four instructions.

Cell B1 in a spreadsheet can be addressed in two ways, with CELLS(1,2) or with RANGE("B1"). CELLS(*r*,*c*) addresses the cell in the *r*th row and *c*th column. Cell A1 (= CELLS(1,1)) is filled with a text in quotation marks, here, with "Annegret". Of course, you should write your own name.

Individual cells can also be addressed with instructions of the type [RANGE("A1") = …], as in lines 3 and 5 of the macro in Fig. 4.2b. In loops, the addressing with CELLS(*r*,*c*) can be used more effectively, if, e.g., the row index *r* or the column index *c* is to be scanned systematically.

▶ **Task** First, write only this one line [CELLS(1,1) = "Annegret"] into the VBA sheet and start the program by pressing the start button ▶ (high-lighted in Fig. 4.2b). For the procedure to be executed, the pointer I must be somewhere in the procedure. Then, insert the other lines of SUB ANNEGRET one after the other, execute the macro after each line and observe what happens in the spreadsheet:

- a number is written into Cell B1, here, 12.25.
- CELLS(2,1) = CELLS(1,2) means that cell A2 (CELLS(2,1)) is being filled in with the contents of cell B1 (CELLS(1,2)). The content of B1 is transferred once to A2 by the program, and A2 remains unchanged, even if the content of B1 is changed later.
- The text " = A1" is written into cell B2, interpreted in the spreadsheet as a formula. The corresponding cell is filled in with the contents of cell A1, B2 =

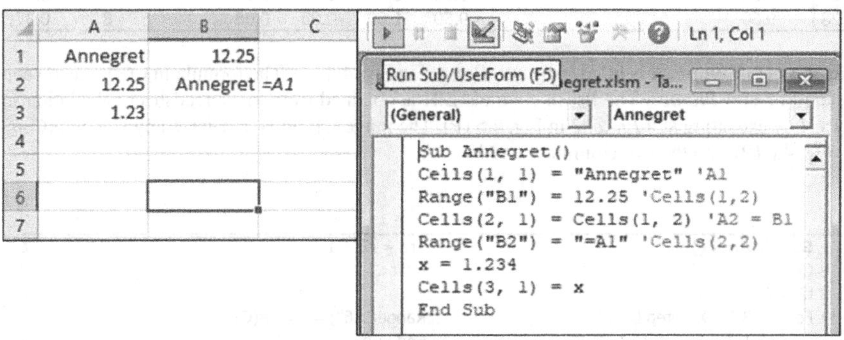

Fig. 4.2 SUB *Annegret* in **b** (right) writes into range A1:B3 in **a** (left). The text after the apostrophe' is interpreted by the VBA interpreter as a comment and not as program code

[=A1]. If the contents of A1 in the spreadsheet are now changed, the new value also appears in cell B2.

▶ **Task** Delete all entries in the spreadsheet again and then run through the program step by step by placing the cursor I in the program and repeatedly pressing the function key F8 (also obtained and explained in the VBA developer, Fig. 4.9, by DEBUG/ STEP INTO). Step by step, the previously deleted entries will appear again in the spreadsheet.

▶ **FOR loops** The macros in Fig. 4.4 (P), Fig. 4.5 (P), and Fig. 4.6 (P), wherein FOR loops are used, fill in the spreadsheet in Fig. 4.3 (S).

SUB *Protoc1* in Fig. 4.4 (P) fills in column A. Line 2: the text "x" is written into A1. In the FOR loop, the variable x is incremented from 3 to 9.5 in steps of 0.25 and written into cells in A. The variable x takes on 3 as the first value and is then incremented by 0.25 each time the loop is traversed until the value 9.5 is reached. The variable of the FOR loop, here, x, is called the *loop index.* In line 9, a formula is entered into C6, normal text enclosed in quotation marks but starting with an equal sign.

	A	B	C	D	E	F	G	H	I	J	K
1	x		9.50		**x**	**Cos(x)**	**Sin(x)**	**Tan(x)**			
2	3.00		-1.00	*=COS(C1)*	3.00	-0.99	0.14	-0.14		3	9.50
3	3.25		-0.08	*=SIN(C1)*	3.25	-0.99	-0.11	0.11		4	
4	3.50		0.08	*=TAN(C1)*	3.50	-0.94	-0.35	0.37		5	x
5	3.75				3.75	-0.82	-0.57	0.70		6	Cos(x)
6	4.00		0.99	*=COS(C1)^2*	4.00	-0.65	-0.76	1.16		7	Sin(x)
7	4.25				4.25	-0.45	-0.89	2.01		8	Tan(x)
8	4.50				4.50	-0.21	-0.98	4.64			
23	8.25				8.25	-0.39	0.92	-2.39		3	0.08
24	8.50				8.50	-0.60	0.80	-1.33		4	=TAN(C1)
25	8.75				8.75	-0.78	0.62	-0.80		5	3.50
26	9.00				9.00	-0.91	0.41	-0.45		6	-0.94
27	9.25				9.25	-0.98	0.17	-0.18		7	-0.35
28	9.50				9.50	-1.00	-0.08	0.08		8	0.37

Fig. 4.3 (S) Column A is filled in by SUB *Protoc1* in Fig. 4.4 (P). The formula in C6 has also been entered by SUB *Protoc1*. The list in columns E:H is obtained from the spreadsheet calculation in C1:C4 by executing SUB *Protoc2* in Fig. 4.5 (P). The range C1:H4 is transferred by SUB *ScanCopy* in Fig. 4.6 (P) into the two columns J and K

```
1 Sub Protoc1()
2 Cells(1, 1) = "x"
3 r2 = 2
4 For x = 3 To 9.5 Step 0.25
5   Cells(r2, 1) = x
6     r2 = r2 + 1
7   Next x
8
9 Range("C6") = "=Cos(C1)^2"
10 End Sub
```

Fig. 4.4 (P) SUB *Protoc1* fills in A in Fig. 4.3 (S) and a formula into C6. Syntax for calling a cell: CELLS (ROW, COLUMN)

```
1 Sub Protoc2()
2 r2 = 2     'Running index for writing to cells
3 Cells(1, 6) = "Cos(x)"
4 For x = 3 To 9.5 Step 0.25
5  Cells(1, 3) = x   'Column C
6  Cells(r2, 5) = x  'Column E
7   Cells(r2, 6) = Cells(2, 3)  '6 = column F
8   Cells(r2, 7) = Cells(3, 3)  '7 = column G
9   Cells(r2, 8) = Cells(4, 3)  '8 = column H
10  r2 = r2 + 1
11 Next x
12 End Sub
```

Fig. 4.5 (P) SUB *Protoc2* changes the value in cell C1 (line 5) and writes the function values from C2:C4 consecutively into the columns F (6th) to H (8th) of the spreadsheet in Fig. 4.3 (S)

Loop2i structure

In Fig. 4.4 (P) , we have introduced a *running index* r_2, which specifies the row of the cell to be filled in. It is set in line 2 to 2 before the start of the loop and is incremented by 1 in line 6 at the end of each loop cycle, so that the values of x are sequentially written into lines 2 to 28. We will often use such structures, call them *Loop2i*, because they comprise two indices, and we memorize them with a broom rule Ψ.

▶ *Ψ Loop2i: Continue counting (the running index) in the loop!*

Spreadsheet calculation used as a function

In cells C2:C4 of Fig. 4.3 (S), formulas with the trigonometric functions COS, SIN, and TAN are written by hand with the argument in C1, e.g., C1 = [9.50] and C2 = [=COS(C1)]. These spreadsheet calculations are used by SUB *Protoc2* in Fig. 4.5 (P) as a function. It is a typical *rep-log procedure*, changing a parameter in a spreadsheet calculation and writing the calculation results to another range of the spreadsheet.

SUB *Protoc2* changes the value in cell C1 in the ($x =$)-loop, which is used as an argument in the functions in C2:C4, and transfers the results of the spreadsheet calculation from C2:C4 (one below the other) to F to H (side by side).

The statement in line 3 is: CELLS(1,6) = "Cos(x)". The quotation marks indicate a text in between, which is to be written as text into the cell. An instruction CELLS(1,6) = COS(x) would cause the program first to calculate the cosine of the variable x, to which one would have to have assigned a value somewhere earlier in the procedure, and then write the result, a number, into the cell. An instruction CELLS(1,6) = ' = COS(x)' would write a formula into the cell, similar to line 9 in Fig. 4.4 (P).

All cell references with fixed row and column indices can also be expressed with RANGE, e.g., RANGE("C2") instead of CELLS(2,3).

Questions

concerning SUB *Protoc2*, Fig. 4.5 (P)

```
1 Sub ScanCopy()
2 r2 = 2              'Row 2
3 c2 = 10             'Column J
4 For r = 1 To 4      'Rows 1 to 4
5   For c = 3 To 8    'columns C to H
6     Cells(r2, c2) = c
7     Cells(r2, c2 + 1) = Cells(r, c)
8     r2 = r2 + 1
9   Next c
10  r2 = r2 + 1
11 Next r
12 End Sub
```

Fig. 4.6 (P) SUB *ScanCopy* writes the contents of the range A1:F4 of Fig. 4.3 (S) consecutively into columns J and K of the same spreadsheet. CELLS (2,3) corresponds to C2 in the spreadsheet

What do the instructions in lines 3 and 5 have to be if you want to address with RANGE?[1]

Which instructions must be added to the code to write the headings in G1 and H1 of Fig. 4.3 (S)?[2]

How would you have to change the for-loop in lines 4 to 11 if you want to read the x values from column A?[3]

Nested loops

SUB *ScanCopy* in Fig. 4.6 (P) transfers the range (r = 1 to 4: c = 3 to 8), i.e., C1:H4 of the table, to columns J and K of Fig. 4.3 (S).

Range C1:H4 is read horizontally, row by row, and written consecutively vertically into column K ($c_2 + 1 = 11$) with the ($c =$) loop; line 8: The running index r_2 is incremented, indicating the next free row in the spreadsheet.

To be read, the range with the two coordinates row number r (from 1 to 4) and column number c (from 3 to 8) must be scanned with the cells being addressed with CELLS(R,C). This is done with two nested loops, an outer loop (FOR $r =$), and an inner loop (FOR $c =$) that is called within the outer loop and ends in line 9 with NEXT C. SUB *ScanCopy* also writes the index c into column J (line 6, $c_2 = 10$ from line 3).

The line index r_2 is incremented by one at the end of each of the two loops FOR $c =$ and FOR $r =$. The increment in the inner loop (FOR $r =$) causes the adjacent entries in a row of the table, e.g., C1:H1, to be written consecutively into rows 2 to 7 of J, J2:J7. The increment in the outer loop (FOR $c =$) causes a row, e.g., row 8 in the table in Fig. 4.3 (S), to be skipped.

---

1 RANGE(„F1") = "Cos(x)": RANGE(„C1") = x.

2 RANGE(„G1") = "Sin(x)": RANGE(„H1") = „Tan(x)"; do not forget the quotation marks!

3 FOR r = 2 TO 28: x = CELLS(r, 1): ... To put multiple statements on one line in VBA, separate the statements by a colon ":"!

Questions

Using the variables x and r_2 in the *loop2i* in SUB *Protoc2* in Fig. 4.5 (P), explain the broom rule: Ψ *Loop2i: Continue counting (the running index) in the loops!*[4]

It would have been easier to specify the columns in SUB *ScanCopy* in Fig. 4.6 (P) as numbers, i.e. CELLS(r_2, 11) instead of CELLS(r_2, $c_2 + 1$). Does the variant CELLS(r_2, $c_2 + 1$) offer any advantage?[5]

Why is J8:K8 in Fig. 4.3 (S) not filled in?[6]

What is the value of r_2 at the end of SUB *ScanCopy* in Fig. 4.6 (P)?[7]

How do we proceed further?

In this basic exercise, we have learned how to read content from cells into procedures and fill in cells. We have also become familiar with FOR loops and the special construction of *loop2i* with a loop index and a running index. In this chapter's next exercises, we will get to know sub-routine calls (SUB) and logical queries (IF) . FOR, SUB, IF are already the essential basic structures of programming, which we will repeat in `Python` (`def` instead of SUB) and apply in the following chapters over and over again.

4.3 Macro-Controlled Drawings with FOR, SUB, IF

We construct a macro for drawing filled circles of variable diameter at different coordinates in the spreadsheet. The required instructions are obtained by recording macros generated when a circle is inserted and formatted as a shape by hand. They are combined in a sub-routine to be called from the main program, specifying the circles' position. Similarly, we get the instructions for drawing rectangles and triangles. We are practicing the basic structures of programming: FOR, IF, SUB.

[4] The rows from $r = 1$ to 4 and the columns from $c = 3$ to 8 are scanned in the nested for-loops. The 24 scanned values are stored in successive rows. The index of these rows, r_2, must be incremented in the inner FOR loop after every entry.

[5] If the data is to be output to another range of the spreadsheet, only one parameter for the columns, namely, c_2, must be adjusted in addition to r_2.

[6] Because, in *ScanCopy* in Fig. 4.6 (P), at the end of the loop FOR $r = 1$ to 4, the index r_2 is incremented without data having previously been written into cells in that row.

[7] At the end of *ScanCopy*, the following applies: $r_2 = 2$ (initial value) $+ 4 \times 6 = 24$ (c-loop) $+ 4$ (r-loop) $= 28$ (Row 28 in Fig. 4.3 (S)).

4.3.1 Macro Recorder

We are going to record the commands that are executed when we insert an ellipse into a spreadsheet and format it. The recorded macro is converted into a subroutine that is called several times by the main program with modified coordinates.

In Fig. 4.7a, you see a decorative spiral drawn with the tools acquired in this exercise. The starting point is a macro (Fig. 4.7b), recorded when an ellipse was inserted and formatted.

What do We Learn in This Exercise?

▶ **Mag** Once you have completed this task, you can create images like the one in Fig. 4.7a.

▶ **Alac** Great, that will amaze my friends!

▶ **Mag** More importantly, you will master Visual Basic statements such as those in Fig. 4.7b (P).

▶ **Tim** Terribly complicated! I will never be able to keep all of that in my head at the same time.

▶ **Mag** You're not supposed to. Figure 4.7b (P) contains a series of instructions that the Macro Recorder has recorded when an ellipse has been inserted by hand.

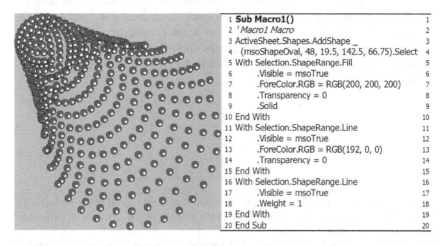

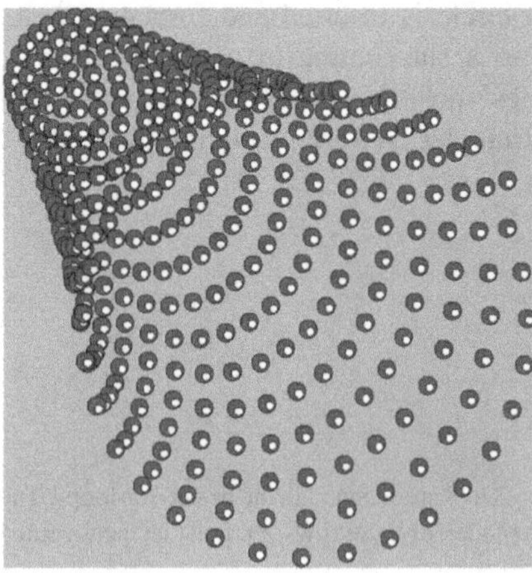

```
Sub Macro1()
'Macro1 Macro
ActiveSheet.Shapes.AddShape _
 (msoShapeOval, 48, 19.5, 142.5, 66.75).Select
With Selection.ShapeRange.Fill
       .Visible = msoTrue
       .ForeColor.RGB = RGB(200, 200, 200)
       .Transparency = 0
       .Solid
End With
With Selection.ShapeRange.Line
       .Visible = msoTrue
       .ForeColor.RGB = RGB(192, 0, 0)
       .Transparency = 0
End With
With Selection.ShapeRange.Line
       .Visible = msoTrue
       .Weight = 1
End With
End Sub
```

Fig. 4.7 **a** (left) Decorative spiral, drawn by a macro. **b** (right, P) Macro recorded by the macro recorder while an ellipse is inserted into the spreadsheet. Superfluous instructions have been deleted. If possible, do not write such code by hand! Get it using DEVELOPER/ RECORD MACRO and modify it as needed!

▶ **Alac** So, everything is done by the macro recorder?

▶ **Mag** No, you still have to modify the recorded code, introduce variables and learn the basic program constructions: loops (FOR i = … TO …), logical branches (IF THEN … ELSE …), and sub-routines. (CALL SUB(*a*, *b*, *c*, …)).

The tab DEVELOPER/RECORD MACRO

We want to apply a VBA macro to draw a series of filled circles. First, we have to get the elementary commands for drawing a circle. These can be found, in principle, in manuals for VISUAL BASIC FOR APPLICATIONS. Nevertheless, we make life easier for us and use the macro recording function. You can find it in the main register tab DEVELOPER, (see Fig. 4.8). Further explanations can be found in EXCEL help under the keyword CREATE A MACRO.

If the main tab DEVELOPER does not appear in your ribbon, you must activate it in the EXCEL options, with FILE/OPTIONS/CUSTOMIZE RIBBON/MAIN TABS/☑ DEVELOPER.

Circles, Squares, Triangles, by Hand and by Macro

After turning on the macro recording function, we draw a circle by hand (INSERT/ILLUSTRATIONS/SHAPES) and format it. For example, we select the color of the filling and the thickness and color of the border. When we have finished the drawing, we end the macro recording by clicking the STOP RECORDING button, which appears in the toolbar in place of RECORD MACRO. The recorded macro is in a project MODULE (see Fig. 4.9 under "Modules"), not in a VBA sheet connected with a spreadsheet.

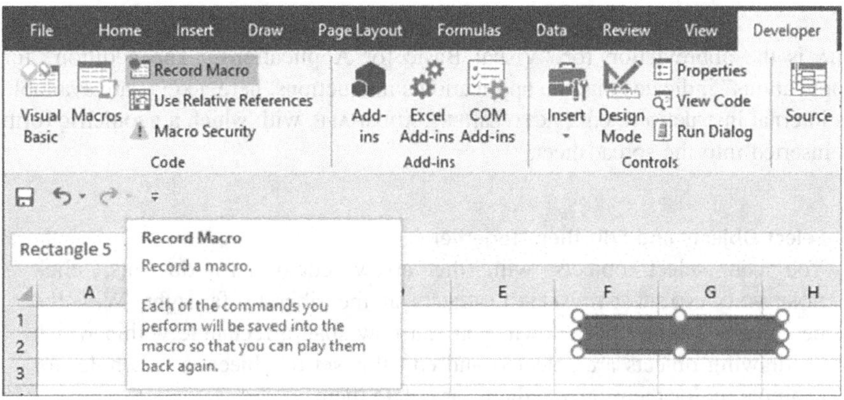

Fig. 4.8 The DEVELOPER/RECORD MACRO tab records all program code associated with the spreadsheet operations performed by the user, e.g., introducing a rectangle as in F1:H2 (INSERT/ILLUSTRATIONS/SHAPES). The VISUAL BASIC button (far left) activates the VISUAL BASIC EDITOR (see Fig. 4.1)

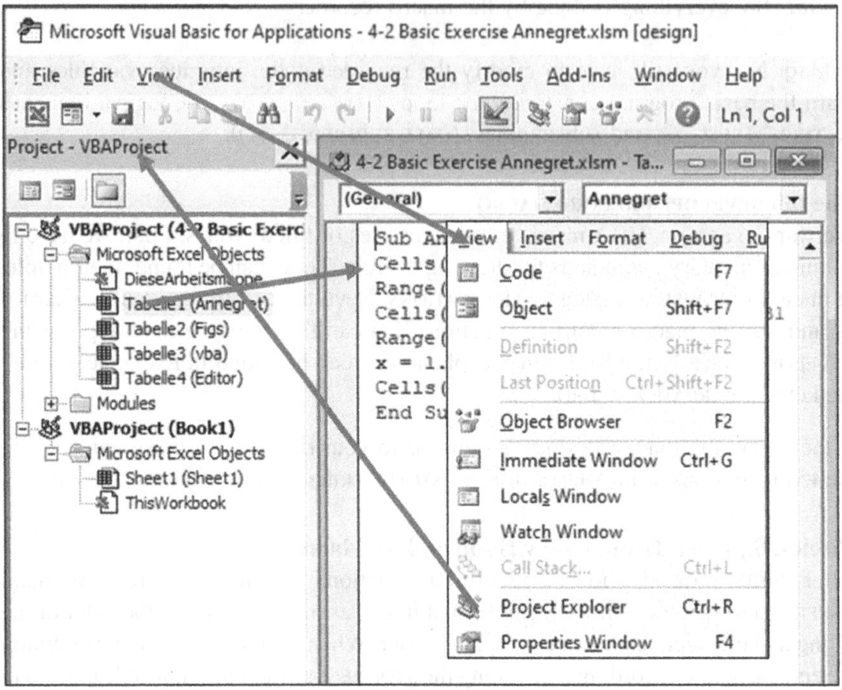

Fig. 4.9 Visual basic editor. You have to activate PROJECT EXPLORER (with VIEW/PROJECT EXPLORER) to see all open files. The recorded macro is located in Module 1 (hidden in the group MODULES) of the VBA project (4–2 Basic Exercise *Annegret*)

VBA is the abbreviation for "Visual Basic for Applications". The addition "for applications" indicates that the application's instructions, here, EXCEL, are available as internal instructions, e.g., ACTIVESHEET. ADDSHAPE, with which a geometric form is inserted into the spreadsheet.

Select Objects and Edit them Together

You can select objects with the arrow cursor. To do this, click HOME/FIND AND SELECT/↖SELECT OBJECTS at the ribbon's far-right. With the new mouse pointer held down, you can now span a rectangle within which all drawing objects are selected and edit this set of objects as a whole, for example, color them, group them, or delete them.

Within a VBA macro, use the command ACTIVESHEET.DRAWINGOBJECTS.SELECT to select all drawing objects in the active worksheet. With SELECTION.DELETE, you can delete all objects.

4.3.2 Visual-Basic Editor 2: Macro Recording, Debugging

We can review the result of our macro recording in the Visual Basic editor. This editor is activated when you click on DEVELOPER/VISUAL BASIC (far left in Fig. 4.1) or press ALT F11. A window like that in Fig. 4.9 appears when the program page of a sheet or a module is additionally double-clicked. SHEET1 has been clicked here, which already contains SUB *Annegret* from Sect. 4.3.2.

Upon clicking on the "View" tab, a menu opens up that has been placed over SUB *Annegret* in Fig. 4.9.

We click on the PROJECT EXPLORER button, and the PROJECT—VBAPROJECT sub-window appears. In this window, each worksheet (SHEET1, SHEET2, SHEET3, SHEET4) is assigned a VBA sheet in which Visual Basic code can be generated and edited. In Fig. 4.9, SHEET1 (Tabelle1(Annegret)) has been clicked, and in the editor, the macro SUB ANNEGRET from Sect. 4.2.2 has popped up.

Since we have already recorded a macro, another object MODULE1 appears under ⊞ Modules . It contains the program code SUB MACRO1, which we have transferred to Fig. 4.7b (P), with four instructions:

- Lines 3 and 4: An ellipse (MSOSHAPEOVAL) has been created. The first two numbers in the argument list are the *x* and *y* coordinates; both are measured from the upper left corner of the spreadsheet. The next two numbers in the list are the two diameters of the ellipse.
- Line 7: The area within the ellipse is colored.
- Line 13: The border of the ellipse is colored.
- Line 18: The thickness of the border of the ellipse is specified.

You can edit the macro commands in the editor like normal text. The syntax must, of course, comply with the rules of the VBA interpreter.

Debug/Step Into
Let the macro run again; best if you do it step by step! If you place the cursor in a program in the Visual Basic Editor and press the function key F8, each step of the program is executed individually (DEBUG/STEP INTO). You can then see exactly what is happening and check whether the drawing is changing as you expect. You can also change the instructions before they are executed. Going through a macro step by step is a good way to detect programming errors.

If you place the cursor on a variable name, the value of that variable will pop up.

► **Task** Change the coordinates and the size of the diameters by modifying the instructions!

► **Mag** Now, the real programming starts, with loops and sub-routines!

4.3.3 Programming Elements

Variables Instead of Numbers

In SUB MACRO2 in Fig. 4.10 (P), we have replaced the current numbers in ADDSHAPE with variables *x*, *y*, d*x* and d*y*, to which we have assigned values in lines 2 to 5. If we run this macro, one of the shapes in Fig. 4.11a (S) is created, or a similar one if other values have been chosen.

The colors in lines 9, 12, 32, and 35 are composed of red, green, and blue components (intensity between 0 and 255) via the specification RGB (red, green, blue).

```
1 Sub Macro2()
2 x = 400
3 y = 20
4 dx = 100
5 dy = 50
6 ActiveSheet.Shapes.AddShape(msoShapeOval, _
7     x, y, dx, dy).Select
8 With Selection.ShapeRange.Fill
9     .ForeColor.RGB = RGB(220, 220, 220)
10 End With
11 With Selection.ShapeRange.Line
12     .ForeColor.RGB = RGB(180, 0, 0)
13     .Weight = 1
14 End With
15 End Sub
16
17 Private Sub CommandButton1_Click()
18 Call Circles
19 End Sub

20 Sub Circles()
21 For i = 1 To 3
22     Call Disc(i * 50 + 60, i * 25)
23 Next i
24 End Sub
25
26 Sub Disc(x, y)
27 'x and y are the coordinates of the center
28 d = 50 'diameter of the circle
29 ActiveSheet.Shapes.AddShape(msoShapeOval, _
30     x - d / 2, y - d / 2, d, d).Select
31 With Selection.ShapeRange.Fill
32     .ForeColor.RGB = RGB(220, 220, 220)
33 End With
34 With Selection.ShapeRange.Line
35     .ForeColor.RGB = RGB(180, 0, 0)
36     .Weight = 1
37 End With
38 End Sub
```

Fig. 4.10 **a** (left, P) Variable names are introduced, MACRO1() from Fig. 4.7b (P) becomes *Macro2*(). The macro SUB *CommandButton1* is triggered by the command button in Fig. 4.11b. **b** (right, P) *Macro2()* is converted into a sub-routine *Disc(x,y)*, which is called repeatedly by the main program *Circles* with various values for *x* and *y*, with the result in Fig. 4.11b

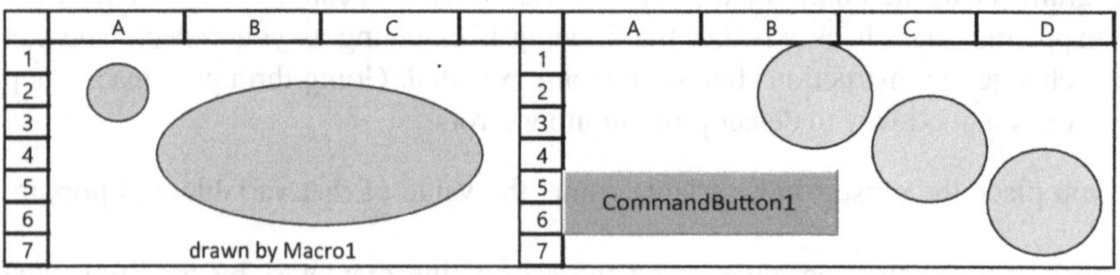

Fig. 4.11 **a** (left, S) Circle and ellipse after executing SUB *Macro2* in Fig. 4.10a (P). **b** (right, S) Result of the procedure *Circles* in Fig. 4.10b (P)

Questions

Which color is created with RGB(180, 0, 0)?[8]

Which color is created with RGB(220,220,220)?[9]

How can you tell that a circular disc, and not an elongated ellipse, is produced with SUB *Disc*?[10]

Sub-routines

We want to summarize the relevant instructions in a sub-routine "Disc", which contains, in the procedure header, the coordinates (*x, y*) of the center of the circle in the parameter list, *Disc(x, y)*. This sub-routine is called from a main program with different values for (*x, y*). Figure 4.10b (P) suggests a solution for this task, with the main program SUB *Circles* and the sub-routine SUB *Disc*. Superfluous specifications in the recorded macro have been deleted.

The diameter of the circular disc is set to $d = 50$ in SUB *Disc* (line 28). Grey is now selected as the fill color, lines 31, 32. The line width remains as before (… LINE.WEIGHT = 1). These parameters cannot be changed by the main program, because they are not in the procedure header.

When placing the circular disc in the spreadsheet, note that the center of the circle is passed via the procedure header (SUB *Disc(x,y)*), but that it is the upper left corner of the shape that must be specified in the drawing command.

The ratio of the scaling in Visual Basic to the grid scale in the spreadsheet can be seen from the following data:

- A circle with diameter 100 points has a diameter of 3.53 cm.
- 28.4 point correspond to 1 cm.
- 28.5; 28.6; 28.7 point all correspond to 1.01 cm.
- 28,8 point correspond to 1.02 cm.

Transferring parameters to sub-routines

► When parameters are passed to sub-routines, the order in the argument list in the procedure header is decisive; the names in the main program are not significant.

A procedure header in Fig. 4.10b (P) reads SUB *Disc(x,y)*. This sub-routine is called in SUB *Circles*() with CALL *Disc*(I*50 + 60, I*25). The first entry in the header in SUB *Disc* is taken over as *x*, and the second entry as *y*. We often name the variables

[8] RGB(180, 0, 0) is a strong red, intensity 180 of 255.

[9] RGB(220,220,220) is a light grey; red, green and blue are equally present.

[10] In lines 29, 30, ACTIVESHEET.SHAPES.ADDSHAPE(…,,., *d, d*), the same variable *d* is used for both diagonals of the ellipse.

in the main program the same as in the sub-routine. So, we could write $x = i*50 + 60$ and $y = i*25$ within the loop in SUB *Circles()* and then call CALL *Disc(x, y)*, with the same result as above.

If we executed CALL *Disc(y, x)*, the first entry, here, *y* from the main program, would be interpreted as *x* in the subprogram and the second entry as *y*. The row of the three circular discs would start at A8 and go down more steeply. In the main program, we could also choose completely different variable names, e.g., *a* and *b*, and then proceed with CALL *Disc(a, b)* or CALL *Disc(b, a)*.

▶ Name the variables such that you are best able to keep an overview!

Questions

concerning Fig. 4.10b (P)

Which three circle centers are passed to SUB *Disc(x,y)* in SUB *Circles*?[11]

Which argument in CELLS(*a*,*b*) stands for the row index in the spreadsheet?[12]

Main Program

A main program is characterized by the fact that it contains no parameters in the procedure header. Only main programs are executable programs. Sub-routines generally contain parameters in the header that must be assigned values by a higher-level program. Examples:

- SUB *CIRCLES()* in Fig. 4.10b (P) is a main program that the user can start.
- SUB *DISC(X,Y)* in Fig. 4.10b (P) is a sub-routine with *x* and *y* in the procedure header. It cannot run on its own, but can only be called by another procedure with specified values for the parameters *x* and *y*.

▶ **Task** Change the procedure so that, in addition to the coordinates of the center point, the diameter *d* of the circle and the thickness *w* of the boundary (SHAPE OUTLINE) are selected in the main program and are transferred to the sub-routine as parameters in the procedure header![13]

FOR loop

The main program *Circles* calls the sub-routine DISC in the loop (FOR $i =$) three times. The centres of the circular disks are set to (110, 25), (160, 50) and (210,75) for $i = 1, 2, 3$. The drawing resulting from these specifications is shown in Fig. 4.11b.

[11] *(x, y)* = (110, 25), (160, 50) and (210, 75).

[12] The first argument, *a*, stands for the row: CELLS(*row*, *columns*).

[13] SUB DISC(X,Y,D,W), LINE 36, WEIGHT = W.

A FOR loop is used in the macro *Circles*. The general syntax for a FOR loop is:

```
FOR x = xmin TO xmax STEP delta_x
    {LIST OF COMMANDS}
NEXT x
```

An example with integers:

```
r2 = 10
FOR N = -211 TO 453 STEP 12
    CELLS(R2, 2)=N
    R2 = R2 +1
NEXT N
```

When this loop is executed, the loop index *n* assumes the values −211, −199, …, 437, 449. CELLS(10,2) to CELLS(65,2) are filled in. In the argument of CELLS, the row number comes first and the column number second. Cells B10 to B65 are therefore filled in with −211, −199, …, 449.

A further example is the loop in SUB *Circles()*, in which the sub-routine *Disc* is called three times:

```
FOR I = 1 TO 3
    CALL DISC(I*50+60, I*25)
NEXT I
```

▶ **Task** Develop a macro for drawing a row of rectangles! "Develop" means that you get the instructions with RECORD MACRO and redesign the recorded macro using variables, sub-routines, and loops.

▶ **Task** Write a macro that draws a (4 × 4) array of filled circles, the colors thereof being composed of fractions of red and green, with the green fraction systematically increasing in each row and the red fraction systematically increasing in each column!

Command Button, Design Mode On/Off

In Fig. 4.11b, a command button has been inserted in "Design mode" in A5:B6 with DEVELOPER/INSERT/ACTIVEX CONTROLS/COMMAND BUTTON (visible when the mouse is over the ▭ icon).

In the PROPERTIES card, revealed by right-hand clicking on COMMAND BUTTON/PROPERTIES, COMMANDBUTTON1 (as text) is assigned both as a NAME and a CAPTION for the command button. As is usually the case with controls, with the DESIGN MODE turned on (click the DESIGN MODE button on the DEVELOPER tab, see Fig. 4.1 and Fig. 1.1 of Sect. 1.7), the PROPERTIES can be changed, e.g., name and caption. When the design mode is switched off (by clicking the DESIGN MODE button again), the control can be operated.

The procedure SUB COMMANDBUTTON1_CLICK in Fig. 4.10a (P) is assigned to the command button. In detail, proceed as follows: In the VISUAL BASIC Editor (Fig. 4.9), click on the arrow ▽ at (GENERAL). A list opens up in which SUB COMMANDBUTTON1 appears. This entry is activated by clicking on it. Next, click on the arrow ▼ next to the cell with the inscription "Annegret" (as shown in Fig. 4.9, or the name you have chosen). A list opens up in which CLICK appears, together with other commands. Click on this entry, and SUB COMMANDBUTTON1 is immediately completed to SUB COMMANDBUTTON1_CLICK. The upper line in the VBA editor will now read COMMANDBUTTON1 ▼; CLICK().

In our case, in Fig. 4.10a (P), only one procedure, SUB *Circles*, is called. We could omit SUB COMMANDBUTTON1_CLICK by naming the command button *Circles* and completing SUB *Circles()* to SUB *Circles*_CLICK(). Please note that we would have to change the *name* of the button, which is independent of its *caption.*

4.4 A Checkerboard Pattern (Excel)

We obtain the VBA commands for drawing elementary geometric shapes by recording macros and incorporate them into a sub-routine that executes the drawing in the desired layout. The shapes' position in the spreadsheet is passed to the procedure via its header or via global variables. The respective shapes, as well as the color of their borders and interiors, are selected randomly.

4.4.1 Checkerboard, Same-Colored and Multi-colored

Checkerboard with same-colored shapes

In this exercise, a checkerboard of rectangles, triangles, and circles, as shown in Fig. 4.12a, is to be drawn. The procedure for this is shown in Fig. 4.13 (P).

SUB DRAWI1 in Fig. 4.13 (P) is the main program that randomly calls one of the subroutines *Rect, Ova,* or *Tria*, ten times in each of eight rows, drawing a rectangle, ellipse, or triangle at the current position of *x* and *y*. Its core is a nested loop with two loop indices, *k* for the row and *i* for the column address within a row.

The variable *ROT* in SUB *drawi1* determines whether a rectangle, an ellipse (oval) or a triangle shall be drawn. In line 5, the variable *ROT* is randomly assigned a value 0, 1, or 2. Chance is brought in by the function RND() generating a random number between 0 and 1, which is then multiplied by 3. This real number is turned into an integer by INT (into the variable ROT). To give some examples: INT(0.75*3) = INT(2.25) = 2; INT(0.22*3) = INT(0.66) = 0; INT(0.54*3) = INT(1.53) = 1.

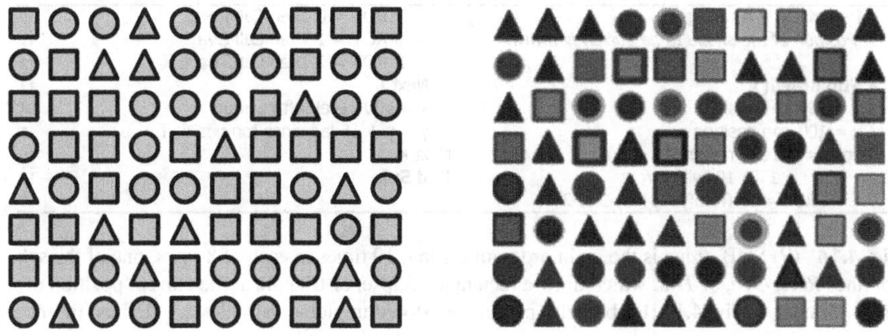

Fig. 4.12 **a** (left) A checkerboard pattern of rectangles, circles and triangles, all equally formatted, drawn with SUB *drawi1* in Fig. 4.13 (P). **b** (right) Like **a**, but with forms differently formatted, filled with different colors, and surrounded with borders of different thickness and different color, drawn with SUB *drawi* in Fig. 4.14 (P)

```
1 Sub drawi1()
2 x = 100
3 For k = 1 To 8 'next row
4    For i = 1 To 10 'within row
5       ROT = Int(Rnd() * 3) '0, 1 or 2
6       If ROT = 0 Then Call Rect(x, y)
7       If ROT = 1 Then Call Ova(x, y)
8       If ROT = 2 Then Call Tria(x, y)
9       x = x + 15
10   Next i
11   x = 100 'reset left position
12   y = y + 15 'advance top position
13 Next k
14 End Sub
```

Fig. 4.13 (P) Procedure SUB *drawi1* with which Fig. 4.12a is drawn

In lines 6 to 8, logical IF queries determine which shape is drawn. After the shape has been drawn, the *x* value is increased by 15 (line 9).

▶ **Task** First, draw only one row by omitting the loop (FOR *k* = ...)! The sub-routines *Rect(x,y)*, *Ova(x,y)* and *Tri(x,y)* should be written according to the model of *SUB Disc(x,y)* in Fig. 4.10b (P). Apart from MSOSHAPEOVAL, MSOSHAPERECTANGLE and MSOSHAPETRIANGLE have to be used.

▶ **Task** Draw the complete checkerboard pattern!

A randomly more colored checkerboard pattern

We draw 8 rows with 10 shapes each, such as shown in Fig. 4.12b. The format of the shapes, namely, the color to be filled in and the color and thickness of the border, is now determined using a random number. In addition, the positions (left, top) of the shapes are not passed to the sub-routines through the procedure header, but via global variables. The main procedure can be found in Fig. 4.14 (P), a typical sub-routine in Fig. 4.15 (P).

```
1 Private x, y As Single
2 'Position of the shape to be currently drawn
3
4 Sub drawi()
5 x = 100 'left position
6 y = 100 'top position
7 For k = 1 To 8 'next row
8     For i = 1 To 10 'within row
9         ROT = Int(Rnd() * 3) '0, 1 or 2
10            If ROT = 0 Then Call Rect
11            If ROT = 1 Then Call Ova
12            If ROT = 2 Then Call Tria
13        Next i
14        x = 100 'reset left position
15        y = y + 15 'advance top position
16    Next k
17    End Sub
18
```

Fig. 4.14 (P) SUB *drawi* is the main program calling, 10 times in each of 8 rows, one of the sub-routines *Rect*, *Ova*, or *Tria*, which draw a rectangle, ellipse, or triangle at the current position of *x* and *y*. Similar to Fig. 4.13 (P), but with parameters stored in global variables *x*, *y* defined in line 1

```
1 Sub Rect() 'Draws a rectangle of width 10 and height 10
2 ActiveSheet.Shapes.AddShape(msoShapeRectangle, x, y, 10, 10).Select
3 Call Lin(255 * Rnd(), 63 * Rnd(), 63 * Rnd(), 10) 'Rim of shape
4 'Formats rim: Red fully varied, green and blus, half intensity, strength
5 Call Interi(63 * Rnd(), 128 + 127 * Rnd(), 63 * Rnd()) 'Interior of shape
6 'Formats interior: Green always more intense than 50%
7 x = x + 15 'Advances position in the row
8 End Sub
```

Fig. 4.15 (P) SUB *Rect* draws a rectangle with a fixed size, but with randomly selected colors for the border (line 3) and interior (line 5). The instructions for coloring are executed in the *Lin* and *Interi* sub-routines in Fig. 4.16 (P)

4.4.2 Global Variables

Global variables are also valid in sub-routines. They must be declared as PRIVATE or PUBLIC before the routines (see line 1 in Fig. 4.14 (P)). Variables of type PRIVATE are only available in the module in which they are declared, those of type PUBLIC in the whole workbook. The positions (Left, Top) of the shapes are now stored in global variables *x* and *y* that can be read and modified by each sub-routine.

The data type SINGLE in line 1 of Fig. 4.14 (P) denotes a single-precision floating-point number stored in 4 bytes. Decimal numbers of the data type DOUBLE are stored in 8 bytes. You can find out more about other data types with EXCEL help in the VBA-Editor.

▶ Don't just copy the macros if you already have some programming practice! Rehearse the sequence of instructions in your mind and get the commands for drawing shapes through macro recording!

In SUB *Rect* in Fig. 4.15 (P), a square is drawn, and its interior and border are formatted with the sub-routines *Lin* and *Interi* in Fig. 4.16 (P). The position is taken from the global variables (*x*, *y*) and passed to MSOSHAPERECTANGLE through the procedure header of *AddShape*. Both side lengths are fixed to 10. The arguments

```
1 Sub Lin(r, g, b, w)
2 red, green, blue and weight w
3 With Selection.ShapeRange.Line
4     .ForeColor.RGB = RGB(r, g, b)
5     .Weight = w
6 End With
7 End Sub
8 Sub Interi(r, g, b)
9 'red, green, blue
10 With Selection.ShapeRange.Fill
11     .ForeColor.RGB = RGB(r, g, b)
12 End With
13 End Sub
14
```

Fig. 4.16 (P) SUB *Lin* and SUB *Interi* color the border and the interior of the shape (line 2), respectively, according to the variables *r* (red), *g* (green), and *b* (blue). The thickness (weight) of the border is specified in *w*, set to 10 in Fig. 4.15 (P)

in the headers of *Lin* and *Interi* are generated in SUB *Rect*(...) with RND() which returns a random number between 0 and 1.

Questions

concerning Fig. 4.15 (P):

The procedure header of SUB *Rect* is empty. How does the sub-routine know the position in the spreadsheet where the rectangle is to be drawn?[14]

Which color dominates the border of the shapes?[15]

4.5 A Checkerboard Pattern (Python)

We draw a multi-colored checkerboard pattern using the `Python` library *turtle*, setting up a similar program structure as for EXCEL in Sect. 4.4, but considering the differences in code from Visual Basic.

4.5.1 Turtle

In order to draw a set of shapes with `Python`, we use the library `turtle`. This simple plot program's illustrative idea is that of turtles running across the screen, thereby creating colorful traces.

In Table 4.1, the libraries `turtle` and `numpy.random` are imported and a turtle named *t* is created. This will be the first cell in every program; it must be run before the functions are compiled, because they resort to these libraries. Generally, several turtles can be active at the same time. We use, however, only

[14] SUB *Rect* accesses the global variables *x*, *y*.

[15] Line 5 in Fig. 4.15 (P): Green = 128 + 127*RND() is represented at least with strength 128.

Table 4.1 Importing relevant libraries; creating a screen with a title; the internet address in line 1 points to an introduction to `Turtle` graphics

```
1  #https://docs.python.org/3.3/library/turtle.html
2  import turtle
3  import numpy.random as npr
4  t=turtle.Turtle()      #Creates a turtle with name t
5  turtle.title("Checkerboard")
```

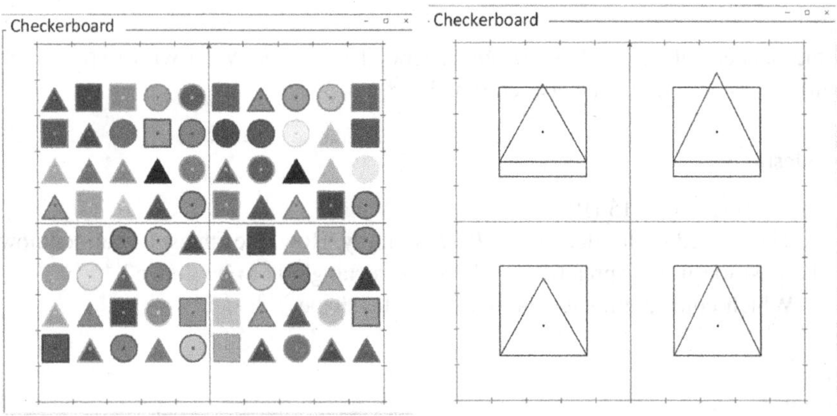

Fig. 4.17 **a** (left) Turtle screen created by the `Python` program in this section, in a frame spanned by the points (−400, −400) and (400, 400). **b** (right) Triangles of different size and at different positions in a square

one `turtle` instance called *t*. In the second cell in Table 4.1, a screen is created with a program-specific name, here, *Checkerboard*.

The standard size of the screen is 1000 pt × 800 pt spanned between (−500, −400) to (500, 400). Such a screen with a checkerboard pattern is shown in Fig. 4.17a. `Turtle` and EXCEL apply coordinate systems with different origins. In EXCEL, it is at the upper left corner of the spreadsheet, and all coordinates are positive. In `Turtle`, the origin is at the center of the screen, and the coordinates are positive or negative.

Attention: The turtle window may be below the Python window.

Different types of triangle

The centroid (x_C, y_C) of a triangle, corresponding to its center of gravity, is calculated as

$$x_C = \frac{x_1 + x_2 + x_3}{3}, \; y_C = \frac{y_1 + y_2 + y_3}{3}$$

It is marked in Fig. 4.17b with dots.

The function *drawTria*, used to draw Fig. 4.17a, draws an equilateral triangle (bottom left in Fig. 4.17b) whose centroid does not coincide with the square's center, in contrast to the equilateral triangle at top left. The bottom right triangle has its top point at the midpoint of the upper side of the square. It is shifted in the upper right figure so that its centroid coincides with the square's center.

Basic functions and measurements

The basic functions (always completed with parentheses ()), attributed to the instance of a turtle, e.g., `t.penup()` or `t.pos()`, are.

Settings:

`.pen( ... )`	specifying `speed`, `pensize`, `pencolor`, `fillcolor`
`.pu()`	pen up
`.pd()`	pen down
`.setpos(x,y)`	moving to the specified position
`.setheading(ϕ)`	setting the orientation as angle ϕ in ° with respect to the x-axis
`.rt(ϕ)`	right turn by angle ϕ in °
`.lt(ϕ)`	left turn by angle ϕ in °
`.fd(r)`	step forward by r pixels
`.bk(r)`	step backward by r pixels
`.dot(s,c)`	plots a dot with diameter s and color c at current position
`.begin_fill()`	beginning to fill in the contour
`.end_fill()`	ending to fill in the contour

Measurements:

`.pos()`	returns Cartesian coordinates of turtle
`.heading()`	returns angle of direction
`.distance(x,y)`	returns distance to point (x, y)

Question

How do you get the position of a turtle named *doro* in polar coordinates? There are two possibilities.[16]

4.5.2 Differences to Visual Basic

We will demonstrate some syntactic differences between `Python` and VISUAL BASIC by means of the function *drawSquare*, which creates a square shape.

[16] `r = doro.distance(0,0), phi = doro.heading()/180*np.pi() or x = doro.pos(0), y = doro.pos(1), r = np.sqrt(x**2 + y**2), phi = np.arctan2(y,x)`

Table 4.2 Function *drawSquare* in Python, drawing a square with center (*x*, *y*) and size *sz* that is filled if the keyword variable *fill* is set to True

```
def drawSquare(x,y,sz,fill=False):
    r=sz/2
    t.pu()                        #Pen up!
    t.setpos(x-r,y-r)             #Run to position!
    t.pd()                        #Pen down!
    if fill==True: t.begin_fill()
    for i in range(4):
        t.fd(sz)                  #Forward!
        t.lt(90)                  #Turn left!
    if fill==True: t.end_fill()   #Fill square with color!
    t.pu()
    t.setpos(x,y)
```

In Table 4.2, it is implemented in Python with four segments of equal length and a 90° turn after each segment; the features of the pen have been specified in a superordinate program. In Fig. 4.18 (P), this is realized in a Visual Basic sub-routine that is called from the main program SUB DSQ().

Positional and keyword arguments

The procedure headers in both cases, Python and VBA, contain positional arguments *x*, *y*, *sz*, and a keyword argument fill that decides whether or not the contour is filled (with the default set to False).

The first three positions in the header must contain appropriate values when the function is called. It is the position in the header that determines the variable in the procedure to which the value is assigned. In Fig. 4.18 (P), a variable *y* with value 40 is passed through the third position to the procedure, where it is assigned to a variable named *sz*.

The keyword argument *fill* is defined in the procedure header with a default value taken in the procedure, if not specified otherwise.

```
Dim col(2) As Integer

Sub dSq()
  col(0) = 100
  col(1) = 200
  col(2) = 0
  y = 40
  For a = 100 To 200 Step 50
    Call drawSquare(a, 100, y, fill:=True)
  Next a
  Call drawSquare(100, 100, y, fill:=True)
End Sub

Sub drawSquare(x, y, sz, _
  Optional fill As Boolean = False)
  ActiveSheet.Shapes.AddShape( _
      msoShapeRectangle, x, y, sz, sz).Select
  Selection.ShapeRange.fill.Visible = msoFalse
  If fill = True Then
  With Selection.ShapeRange.fill
    .Visible = msoTrue
    .ForeColor.rgb = rgb(col(0), col(1), col(2))
  End With
  End If
End Sub
```

Fig. 4.18 (P) *DrawSquare* realized in visual basic for EXCEL

Global parameters

The first line in the VISUAL BASIC program in Fig. 4.18, before the procedures, defines an array *col* of integers with three elements accessible in the whole module. It is written in the main program on lines 4 to 6 and read in *drawSquare* on line 21.

In `Python`, arrays can be declared anywhere in the program, e.g., by col = [100, 200, 0]. All variables, as well as arrays, are valid in subordinate functions unless the variable name is again declared in a function with an equal sign, e.g., *col* = [10, 100, 10], creating a new object with own memory space.

Grouping blocks of code

In `Python`, indentation has a syntactic function. A block of code is necessarily grouped by the same amount of indentation. In Table 4.2, there are two examples: the statements of the function are all indented by 4 spaces; the two statements to be executed in the for-loop are further indented by another 4 spaces.

EXCEL uses code words to state the end of a block. In Fig. 4.18 (P), the procedure (sub-routine) ends with END SUB, the IF block with END IF, and the WITH block with END WITH. The (FOR A =) block in the main program ends with NEXT A. Although indentation does not have a syntactic function in Visual Basic, we use it to maintain a better overview of the program structure.

Case-sensitivity

`Python` is case-sensitive: `True` and `False` have to be written with capital T and F; *x* and *X* are two different variables. Visual Basic is case-insensitive: an input "true" is automatically changed to "TRUE"; *x* and *X* are regarded as the same variable. When we change the case of the first letter anywhere in the program, names in other places will automatically adapt.

4.5.3 Checkerboard with Squares, Triangles, and Circles

User-defined functions for square, triangle, circle, dash

To draw a checkerboard pattern as in Fig. 4.17a, we need functions that draw, besides the rectangle already realized in Table 4.2, a triangle and a circle. They should have the same procedure header as *drawSquare*, with the center point (*x*,*y*) and the size *sz* as positional arguments and *fill* as a keyword argument.

Drawing a triangle in Table 4.3 is similar to how it is done in *drawSquare*. Drawing a circle with radius *r* is achieved with the built-in function `circle(r)` (see Table 4.4). There, the turtle runs along a circle of radius *r*, starting at its current position and with its current direction.

In the three functions mentioned above, the turtle runs along a shape, starting at its current position and with its current direction. It was, however, intended by the programmer that the turtle start running straight to the right. This is indeed

Table 4.3 Drawing a triangle with center (*x*, *y*)

```
def drawTria(x,y,sz,fill=False):
    r=sz/2
    t.pu()
    t.setpos(x-r,y-r)               #Left lower edge
    t.pd()
    if fill==True: t.begin_fill()
    for i in range(3):
        t.fd(sz)                    #Forward!
        t.lt(120)                   #Turn left!
    if fill==True: t.end_fill()
    t.pu()
    t.setpos(x,y)
```

Table 4.4 Drawing a circle with center (*x*, *y*)

```
def drawCircle(x,y,sz,fill=False):
    r=sz/2
    t.penup()
    t.setpos(x,y-r)
    t.pendown()
    if fill==True: t.begin_fill()
    t.circle(r)                 #Is a function within turtle
    if fill==True: t.end_fill()
    t.penup()
    t.setpos(x,y)
```

assured in our current main program, but nevertheless, not making this intention explicit is considered a big mistake in Software Engineering.

Questions

The turtle named *t* in Tables 4.2, 4.3, and 4.4 is a global instance accessed within the functions. This is possible because you have only one turtle running. What do you do if several turtles are on the field?[17]

In Table 4.4, it is implicitly assumed by the programmer that the turtle is heading straight to the right at start, a big programming mistake. How do you avoid this bug?[18]

[17] The turtle name has to be an argument, e.g. *t*, `def drawSquare(t, x, y, sz, fill = false)`, so that it is no longer regarded as a global instance in the functions.

[18] Introduce `t.setheading(0)` before the turtle starts running. If necessary, store the original direction at the beginning, e.g., `phi0 = t.heading()` and reset it at the end with `t.setheading(phi0)`.

Table 4.5 Drawing a dash

```
def dash(ds):
    t.rt(90)        #Right turn
    t.fd(ds)
    t.rt(180)       #180° to the right
    t.fd(2*ds)
    t.rt(180)
    t.fd(ds)        #Back to zero
    t.lt(90)        #90° to the left, original direction
```

Table 4.6 Drawing the checkerboard pattern with a nested loop; when lines 3 and 20 are activated, the turtle runs faster

```
1   turtle.clearscreen()
2   t=turtle.Turtle()
3   #turtle.tracer(0, 0)
4    #Draw checkerboard pattern
5   for rn in range(-280,285,80):
6       for c in range(-360,365,80):
7           r=npr.rand()
8           g=npr.rand()
9           b=npr.rand()
10          tup=(r,g,b)      #Red, green, blue
11          tup2=(g,b,r)     #g=Red, b=green, r=blue
12          t.pen(pencolor=tup2, fillcolor=tup,
13                                pensize=4, speed=0)
14          if r<0.33:
15              drawTria(c,rn,60,fill=True)
16          elif r<0.67:
17              drawSquare(c,rn,60,fill=True)
18          else: drawCircle(c,rn,60,fill=True)
19          t.dot()
20  #turtle.update()
```

The function *dash* in Table 4.5 draws a dash at the current position perpendicular to the current turtle heading, of extension d*s* to both sides.

Main program

The main program has three parts. The first part is shown in Table 4.1, importing the necessary libraries and creating a screen with the title *Checkerboard*. Then, in Table 4.6, the checkerboard pattern is drawn with a nested loop over 10x positions and 8 y positions, with randomly choosing one of our three shapes.

Before we call the *draw** functions, the pen specifications have to be set in the main program, `pencolor` and `fillcolor` in our program, by `rgb` (red, green, blue) in standard mode with values between 0 and 1, randomly chosen with `npr.rand()`.

Table 4.7 Drawing a frame with dashes around the checkerboard

```
21  #Draw frame with dashes
22  t.pen(pencolor="black", pensize=1, speed=10)
23  t.penup()
24  t.setpos(-400,-400)        #Left bottom corner
25  t.pendown()
26  for k in range(4):         #4 straight lines
27      for i in range(10):    #10 segments with dashes
28          dash(6)
29          t.fd(80)           #Forward!
30      dash(6)
31      t.lt(90)               #Left turn 90°!
32  #turtle.update()
```

Table 4.8 Drawing the axes of the coordinate system

```
33  #Axes of the coordinate system
34  t.pu(), t.setpos(-400,0), t.pd()
35  t.fd(800)                           #Forward!
36  t.pu(), t.setpos(0,-400), t.pd()
37  t.setheading(90)                    #Direction 90° to x axis
38  t.fd(800)
39  #turtle.update()
```

The keyword variable pencolor expects a tuple with 3 elements to specify the color. The individual variables *r* (red), *g* (green), *b* (blue) are set in lines 7 to 9 and assembled into two different tuples, *tup* for `fillcolor` and *tup2* for `pencolor`.

Frame around the figure in Turtle

The main program is continued in Tables 4.7 and 4.8, through drawing of a frame around the checkerboard and the axes of the coordinate system, respectively.

Questions

What type of triangle is specified in *drawTria*, Table 4.3: equilateral, or acute-angled?[19]

How do you draw a triangle touching a square at two neighboring corners and the center of the opposite side (see Fig. 4.17b, bottom right)?[20]

---

[19] The function *drawTriangle* draws a triangle with all angles equal to 60° so that it becomes equilateral.

[20] Let the `turtle` run with `setpos(..)` along `(x-sz/2,y-sz/2)`, `(x + xz/2, y-sz/2)`, `(0, y + sz/2)`, back to `(x-sz/2,y-sz/2)!`

What is the direction of the turtle after having been guided by the code snippet in Table 4.7?[21]

How do we speed up Python's turtle function?
The answer is found with an internet search (2020):

https://stackoverflow.com/questions/16119991/how-to-speed-up-pythons-turtle-function-and-stop-it-freezing-at-the-end

with the answers:

- (1) Set `turtle.speed()` to fastest.
- (2) Use the `turtle.mainloop()` functionality to do work without screen refreshes.
- (3) Disable screen refreshing with `turtle.tracer(0, 0)`, then, at the end, do `turtle.update()`

We trigger variant (3) when we activate lines 8 and 25 in Table 4.6. The instruction `turtle.tracer(0,0)` eliminates the millisecond delays that occur when the screen is updated after every turtle change. The screen is refreshed with the complete picture by `turtle.update()` (Table 4.8).

4.6 Drawing Densely-Packed Atomic Layers; Crystal Physics

We draw two different stackings of three planes with densely packed spheres that correspond to the cubic face-centered (*fcc*) or hexagonal dense-packed (*hdp*) crystal structure.

4.6.1 Program Structure and Geometry

In this task, a top view of a close-packed plane of atoms is to be drawn with a program based on a nested For-loop. The subordinate loop draws a horizontal row of discs, representing the atoms, touching each other. This is achieved by shifting a new disk to the right by the diameter d relative to the previous disc. In a higher-level loop, the rows are to be shifted one after the other in the plane so that the circular disks touch each other in a hexagonal arrangement, as shown in Fig. 4.19.

In a second development step, the main program resulting from the first step is to be converted into a procedure that draws a plane and to which the coordinates

[21] The turtle turns 90 + 180 + 180 − 90 = 360° = 0° toward its original direction.

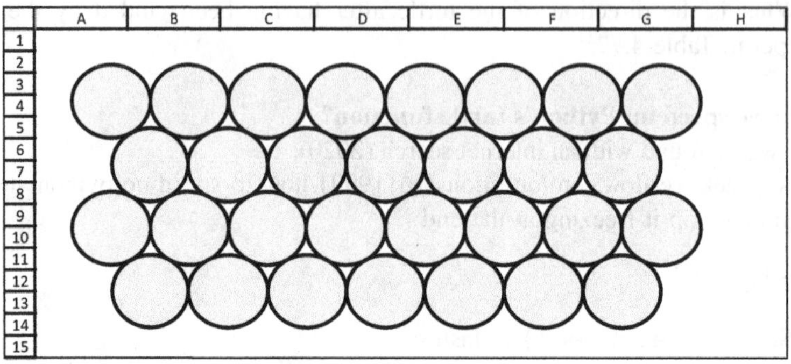

Fig. 4.19 (S) Hexagonally packed plane, drawn with SUB DIEB from Fig. 4.22 (P), here, however, in half-size

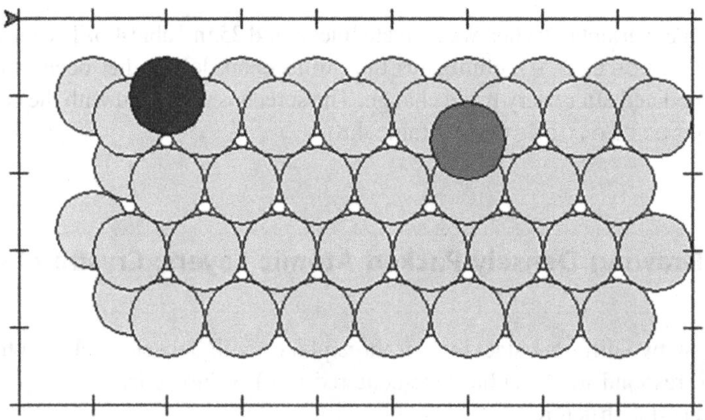

Fig. 4.20 Two densely packed planes and only two "atoms" in the third plane; dark grey "atom" with dotted border top left: position as in the hexagonal close packing (hcp); middle grey "atom" with black border: position as in the face-centered cubic (fcc) shape; the drawing is obtained with Tables 4.11 and 4.12

of the first disc are passed. An extended main program puts a second plane onto the gaps in the first plane, and ultimately places two discs, one at a position typical of the face-centered cubic (*fcc*) structure and the other one characteristic of hexagonal densest packing (*hdp*), resulting in Fig. 4.20.

Questions

How many neighbors does a sphere in a close-packed plane have?[22]

[22] Six nearest neighbours.

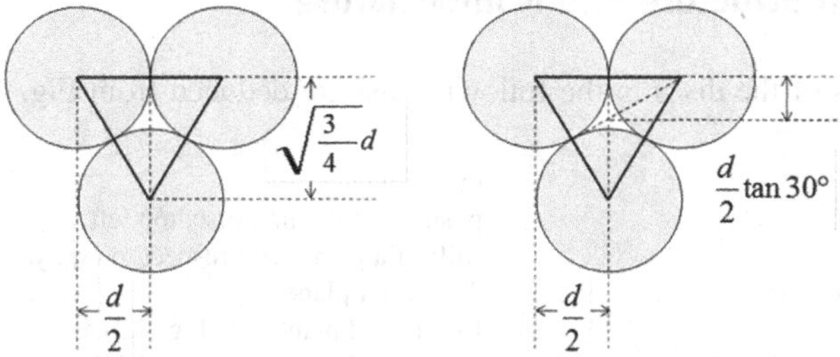

Fig. 4.21 **a** (left) Geometry of a hexagonal packing in a plane, displacement of the second row of atoms with respect to the first row of atoms. **b** (right) Position of an atom in the second plane

How many neighbors does a sphere in a stack of close-packed planes have?[23]

The drawings in Fig. 4.21 indicate the coordinates of the centers of the circular disc in the planes. Figure 4.21a gives the position of a disc in a row relative to the previous row, and Fig. 4.21b that of a disc in a plane relative to the previous plane.

Three stacked planes

The first row of circular disks starts at (x_0, y_0). The second row is offset from the first row in the x- and y-directions by distances indicated in the geometric construction in Fig. 4.21a. A second plane is to be placed over the first one, with the circular disks lying over the gaps in the first plane (see Fig. 4.21b), resulting in the assembly of light grey discs in Fig. 4.20.

For the third plane, there are two possibilities:

- It lies exactly above the first plane, as in the crystal structure of hexagonal close packing (*hcp*). For the drawing in Fig. 4.20, only one circular disk is placed in the correct position (dark grey, top left in the picture).
- It lies above the still visible gaps in the first plane, as in the cubic face-centered (*fcc*) crystal structure. In Fig. 4.20, this is done only for one circular disk, drawn in middle grey with a black border. The displacement of the third plane in the x- and y-directions is, for *fcc*, twice as large as the displacement of the second plane, both with respect to the first plane.

[23] 12 nearest neighbours, 6 of them in its own plane, 3 below and 3 above.

4.6.2 Data Structure and Nomenclature

The positions of the discs in the following list are deduced from Fig. 4.21:

d	disc diameter
(x_0, y_0)	position of the first disc, top left
(del$X$, del$Y$)	shift of a plane with respect to (x_0, y_0)
del$X = 0$, del$Y = 0$	for the 1st plane
del$X = d/2$	for the 2nd plane (see Fig. 4.21b)
del$Y = -d/2 \cdot tan(30/180 \cdot \pi)$	for the 2nd plane (see Fig. 4.21b)

$x = x_0 + \text{del}X + \text{d}x + \text{i} \cdot d$ position of disc i.

$\text{d}x = \text{del}X + 0$	1st and 3rd rows (see Fig. 4.21a)
$= \text{del}X + d/2$	2nd and 4th rows (see Fig. 4.21a)
$y = y_0 + \text{d}y$	y position of disc i (see Fig. 4.21a)
$\text{d}y = \text{del}Y + d_{34} \cdot n$	nth row (see Fig. 4.21a and Eq. 4.1)

with:

$$\text{d}_{34} = d \cdot \sqrt{3/4} \tag{4.1}$$

read from Fig. 4.21a.

4.6.3 Excel

A row of discs
We draw four rows of circular disks so that the disks touch each other, using four consecutive FOR-loops (see SUB *DiEb* Fig. 4.22 (P)).

The sub-routine SUB *Disc(x,y)*, already reported in Fig. 4.10b, is called with individual positions x, y passed via the procedure header. Contrary to the figure, the disc's diameter is set in the main procedure to $d = 100$. In each row, the x-position of the following circle is shifted to the right by a circle's diameter. The y-position is always the same for a row.

Stacking planes
To draw several planes, one on top of the others, we convert the main program *DiEb* described above into the sub-routine SUB *Plane(delX, delY)* (repeated in Fig. 4.23 (P)) to which the coordinates of the top left disc are transferred. The displacements *delX* and *delY* can be determined with the help of Fig. 4.21b.

```
Sub DiEb()
x0 = 100
y0 = 100
d = 100 'diameter of the disc
'1st row of discs
For i = 0 To 7
    Call Disc(x0 + i * d, y0)
Next i
'2nd row of discs
dx = d / 2 'shift with respect to 1st row
dy = d * Sqr(3 / 4)
For i = 0 To 6
    Call Disc(x0 + dx + i * d, y0 + dy)
Next i

'3rd row of discs
dx = 0 'shift with respect to 1st row
dy = d * Sqr(3 / 4) * 2
For i = 0 To 7
    Call Disc(x0 + dx + i * d, y0 + dy)
Next i
'4th row of discs
dx = d / 2 'shift with respect to 1st row
dy = d * Sqr(3 / 4) * 3
For i = 0 To 6
    Call Disc(x0 + dx + i * d, y0 + dy)
Next i
End Sub
```

Fig. 4.22 (P) SUB *Dieb* for drawing a plane; the four rows of atoms are drawn using four loops, result shown in Fig. 4.19 (S)

```
Sub Plane(delX, delY)
x0 = 100 'Offset to left upper corner
y0 = 100 'of the worksheet
d = 100 'Diameter of the disc
dx = delx
dy = dely
For i = 0 To 7
    Call Disc(x0 + dx + i * d, y0 + dy)
Next i
dx = d / 2 + delx
dy = d * Sqr(3 / 4) + dely
For i = 0 To 6
    Call Disc(x0 + dx + i * d, y0 + dy)
Next i

dx = 0 + delx
dy = d * Sqr(3 / 4) * 2 + dely
For i = 0 To 7
    Call Disc(x0 + dx + i * d, y0 + dy)
Next i
dx = d / 2 + delx
dy = d * Sqr(3 / 4) * 3 + dely
For i = 0 To 6
    Call Disc(x0 + dx + i * d, y0 + dy)
Next i
End Sub
```

Fig. 4.23 (P) SUB *DiEb* is converted into a sub-routine *Plane* to which the initial coordinates are transferred by a higher-level program

In the main program SUB *hcp_fcc* in Fig. 4.24 (P), SUB *Plane* is called twice, for the initial layer with del$X = 0$ and del$Y = 0$ and for the second layer with its

```
Sub hcp_fcc()
x0 = 100
y0 = 100
Call Plane(0, 0)
delx = 100 / 2
dely = 100 / 2 * Tan(30 _
            / 180 * 3.14159265)
Call Plane(delx, dely)
hdp

dy = 0
dx = 100 * 1
Call Disc(x0 + dx, y0 + dy)
'fcc
dy = 100 * Sqr(3 / 4) - 100 / 2 * Tan(30 _
            / 180 * 3.14159265)
dx = 100 * 5
Call Disc(x0 + dx, y0 + dy)
End Sub
```

Fig. 4.24 (P) Main program, which calls SUB *Plane* twice, places two atoms on top (with SUB *Disc* from Fig. 4.10b (P)), and thus draws a picture similar to Fig. 4.20

discs on the gaps of the first layer. There are two possibilities for the third plane, *hcp* or *fcc*, represented with one disc each in lines 12 and 17.

Questions

In Fig. 4.23 (P), the variables in the header are called *delX* and *delY*. In the body of the procedure, the formulas refer to different names, *delx* and *dely*. Will this discrepancy lead to error messages?[24]

How do you have to change SUB *Disc* in Fig. 4.10b (P) so that *d* becomes a global variable?[25]

Grouping and copying shapes in various picture formats

To group the shapes into an integrated picture, activate the white arrow SELECT OBJECTS in the FIND&SELECT tab (far right in the HOME tab of the EXCEL ribbon, Fig. 1.1 in Sect. 1.7), drag the selection rectangle around the shapes, then click FORMAT/GROUP. You can now copy the group as one graphic into other applications, e.g., into a Word file or a PowerPoint file.

► **Task** Group your drawing into an image of type *png*, *tif*, or some other image format, and copy this image to another area of the spreadsheet or to another application, e.g., to a PowerPoint file! To do so, select the object, click COPY, move the cursor to another location in the table, click PASTE/PASTE SPECIAL, and select the desired format.

4.6.4 Python

In the first cell of Table 4.9, the relevant libraries are imported. In the next cell, a screen with the name "Crystal planes" is created, and global parameters are specified, with the disc diameter *d* being set to 50 and the position of the first disc in the upper left corner at $(x_0, y_0) = (-175, 175)$. In the third cell, the first row of discs is drawn, comprising eight discs drawn from left to right, using the function *drawCircle* in Table 4.4. When developing the program, you should check this snippet of code and see whether a row of gray discs is really plotted from left to right.

Comments concerning Table 4.9:

- Line 1, link for an introduction into the basic features of turtle,
- Line 5, the screen is cleared. This is important when developing a program, and you have to improve the code and repeat it again and again until it runs error-free.

[24] No, Visual Basic is case-insensitive, *contrary to* `Python`.

[25] *Skip line 28 "d = 50", insert* DIM d AS INTEGER as the first program line, and specify $d = \ldots$ somewhere in the program before SUB *Disc* is called for the first time.

Table 4.9 Importing relevant libraries; creating a screen and setting global parameters; drawing 1st row of discs, function *disc* from Table 4.4

```
#https://docs.python.org/3.3/library/turtle.html
import turtle
import numpy as np
import numpy.random as npr
turtle.clearscreen()
t=turtle.Turtle()            #Create turtle with name t!
turtle.title("Crystal planes")
tup=(0.9,0.9,0.9)            #Light grey
t.pen(pencolor="black", fillcolor=tup, pensize=1, speed=10)

x0=-175
y0=175
d=50                         #Disc diameter
 #1st row of discs
dx=0
dy=0
for i in range(8):
    drawCircle(x0+dx+i*d,y0+dy,d,fill=True)
print(t.pos())               #Current position
```

Table 4.10 Function for drawing a close-packed plane; del*X*, del*Y* position of top left disc with respect to (x_0, y_0), *drawCircle* from Table 4.4

```
def Plane(delX,delY):
    #1st row of discs
    dx=0+delX
    dy=0+delY
    for i in range(8):
        drawCircle(x0+dx+i*d,y0+dy,d,fill=True)
    #2nd row of discs
    dx=d/2+delX
    dy=-d*np.sqrt(3/4)+delY
    for i in range(7):
        drawCircle(x0+dx+i*d,y0+dy,d,fill=True)
    #3rd row of discs
    dx=0+delX
    dy=-d*np.sqrt(3/4)*2+delY
    for i in range(8):
        drawCircle(x0+dx+i*d,y0+dy,d,fill=True)
    #4th row of discs
    dx=d/2+delX
    dy=-d*np.sqrt(3/4)*3+delY
    for i in range(7):
        drawCircle(x0+dx+i*d,y0+dy,d,fill=True)
```

Table 4.11 Plotting the first and second planes

```
22  Plane(0,0)
23  delX=50/2
24  delY=-50/2*np.tan(30/180*np.pi)
25  Plane(delX,delY)
```

Questions

concerning Table 4.9

What is the position of the turtle after having drawn the first row?[26]

What are the pen color and fill color of the turtle?[27]

What is the global parameter accessed in *drawCircle*?[28]

To produce a figure like that in Fig. 4.20, two planes and, additionally, two discs have to be placed at appropriate positions. The function for drawing a close-packed plane is shown in Table 4.10. Its arguments are the shifts *delX* and *delY* of the first disc's position with respect to (x_0, y_0). The four rows of discs are drawn with loops calling *drawCircle* (from Table 4.4), always with the same variables, but with dx and dy set for each row individually, according to Eq. (4.1).

The first two planes are drawn with the program in Table 4.11. Two atoms in the third plane representative of the hexagonal-dense packed and face-centered cubic structures, respectively, are drawn in Table 4.12. For *hdp*, the position is just on top of a disc in the first plane, whereas for *fcc*, x and y get a double shift from a disc in the first plane, whereby the x position becomes identical to the neighboring disc in the first plane and the y position is in the center of a gap in the second plane.

Question

In line 29 of Table 4.12, dx = 50*1 is specified; the number 50 is precisely the disc diameter. Nevertheless, what is bad about this instruction, and likewise about the instructions in lines 36 and 37, and lines 23 and 24 in Table 4.11?[29]

[26] The position of the turtle after the first row is x = -175 + 7·50 = +175; y = 175, as can be deduced from lines 18 and 19 in Tab. 4.9.

[27] `pencolor` = "black", `fillcolor` = tup, tup = (0.9, 0.9, 0.9), a light gray.

[28] The disc diameter d is a global parameter accessed in `drawCircle`. It is also used in line 18 to specify the shift of the position of the current disc with respect to the preceding one.

[29] When the value of the global parameter d is changed, these instructions do not follow. Set dx = d*1 in lines 47, 48, 53, 60, 61 instead!

Table 4.12 Drawing an atom of the third plane, either *hdp* or *fcc*

```
 #hdp
tup=(0.3,0.3,0.3)          #Dark grey
t.pen(pencolor="black", fillcolor=tup,pensize=1, speed=0)
dx=50*1
dy=0
drawCircle(x0+dx,y0+dy,d,fill=True)

 #fcc
tup=(0.6,0.6,0.6)          #Middle grey
t.pen(pencolor="black", fillcolor=tup,pensize=1, speed=0)
dx=50*5
dy=-50/2*np.tan(30/180*np.pi)*2
drawCircle(x0+dx,y0+dy,d,fill=True)
```

4.7 Text Processing

Operations on strings (texts) are practiced by the example of swapping letters within words. We will use the functions *Len*, *Split*, *Join* in both Visual Basic and `Python`. The functions LEFT, RIGHT, and MID are used in VISUAL BASIC to cut out pieces of strings, achieved in `Python` through list slicing.

4.7.1 Cutting and Joining Strings

Swirling characters

▶ **Text** It is siad taht a txet can be undesrtood eevn if you lvaee olny the begiinnng and the end lteter in ecah wrod in plcae but exahcnge middle lettsre. Do you beileve taht or is it nonnesse?

▶ **Task** Write a pogrram that fsirt rades a text from a sersadehept. This text is then to be bokren down into wsodr. The idnuvidial words are tnorsfarmed so that the frsit and last letrets raiemn in pcela, but the iennr ltrtees are ramlondy swapped.

We are going to write a program that exchanges letters in an originally correctly written text. Such a program has already swirled the first two paragraphs of this description. The number of letter swappings in the above text: 1 (Text), 2 (Task).

Plain text of the first two paragraphs

▶ **Text** It is said that a text can be understood even if you leave only the beginning and the end letter in each word in place but exchange middle letters. Do you believe that, or is it nonsense?

▶ **Task** Write a program that first reads a text from a spreadsheet. This text is then to be broken down into words. The individual words are transformed so that the first and last letters remain in place, but the inner letters are randomly swapped.

VBA instructions for text processing

We need the following VBA instructions that affect character strings:

LEN(...), SPLIT(...), JOIN(...) , LEFT(...), RIGHT(.) and MID(...).

An overview is given in the box. For more information, refer to VBA help! There, you can find out, for example, about the MID function:

Returns a Variant (String) containing a specified number of characters from a string.

▶ **Syntax** MID (STRING, START [, LENGTH]).

STRING	Required. String expression from which characters are returned.
START	Required; Long. Character position in STRING at which the part to be taken begins.
LENGTH	Optional; Variant (Long). Number of characters to return.

Text processing in VBA and Python

VISUAL BASIC	Python
SPLIT(STRING, [SEPARATOR...)	`String.split("separator")`

Splits a string expression into words and stores them in an array. Unless otherwise specified, a space is interpreted as a separator between words.

LEN(STRING)	`len(String)`

Determines the length of a given string.

LEFT(STRING, LENGTH)	`String[:Length]`

Cuts out a piece of length *Length* from a character string starting from the left.

RIGHT(STRING, LENGTH)	`String[-Length:]`

Cuts out a piece of length *Length* from a character string starting from the right.

MID(STRING, START, LENGTH) `String[Start: Stop]`

Cuts out a piece of length LENGTH or (`Stop - Start`) from a character string starting from position START and to the right of it. Attention: VISUAL BASIC starts indexing from 1, `Python` from 0.

JOIN(ARRAY[, DELIMITER) `'Delimiter'.join(Array)`

Returns a string created by joining several substrings contained in an array. If '`Delimiter`' is omitted, the space character (" ") is used.

Questions

Let the variable *Tx* contain the text "Cutting out". With which instructions do you get the first, last, and second characters of *Tx*? How do you copy the string "ing" from *Tx* to a new variable *Wd*? In Visual Basic,[30] in Python?[31]

Consider the string *Sente* = "This is. Our goal." How do you separate the string into the two sentence fragments terminated by full stops? Use the Split command! How do you get a new string JS = "This is." including a full stop? EXCEL?[32] `Python`?[33] Compare!

4.7.2 Data Structure and Program Flow

Text	text to be processed, words separated by spaces
Words	list of the words in *Text*

For every *Word* in *Words*:

Exchange two letters	with function *ExchLett(Word)*
Split *Word*	→ *Letters* list of letters in *Word*
Join *Letters*	→ *NewWord*
Join sequence of *NewWord*	→ *NewSentence*

[30] LEFT(*Tx*, 1), RIGHT(*Tx*, 1), MID(*Tx*,2, 1), *Wd* = MID(*Tx*, 5, 3); the first character has index 1.

[31] Tx[0], Tx[-1], Tx[1], Wd = Tx[4:7]; the first character has index 0.

[32] SINGSENT = SPLIT(SENTE, "."): DIM NEWSENT(1) AS STRING: NEWSENT(0) = SINGSENT(0): NEWSENT(1) = ".": JS = (NEWSENT, ""), 5 statements; the colon is the separator between statements in a line.

[33] `SingSent = Sente.split("."); JS = ''.join ([SingSent[0], "."]),` 2 statements; the semicolon is the separator between statements in a line.

In the VISUAL BASIC program in Sect. 4.7.3, we use one array of words containing the original words at the beginning and the new words at the end. In Python in Sect. 4.7.6, we are using two arrays: Words, with original data remaining unchanged, and WordsNew obtained consecutively by appending one scrambled word after the other.

4.7.3 Excel

In the following, first, a complete program (main program *Scramble* and sub-routine *XWord*) that solves the task is introduced. Do not copy it, but rather continue reading! Then, the program is developed step by step in test macros to follow the effect of the individual instructions.

Please be aware that, in the following continuous text, according to our spelling convention, the words in SMALL CAPS are VBA internal terms, while the words in *italics* are invented by the programmer.

Split (Sentence)

The main program *Scramble()* (Fig. 4.25 (P)) reads a text from cell A1 of the spreadsheet (*Text* = CELLS(1,1)), splits it into words (*Words* = SPLIT(*Text*), line 3) and passes the words one by one to the sub-routine *XWord(Word)* (line 7) which exchanges two letters. If two letters are to be exchanged for a second time, line 8, now commented out, has to be activated. The instruction SPLIT specifies the variable *Words* automatically as an array. For the data type *array*, see Sect. 4.7.5.

The sub-routine *XWord (Word)* (Fig. 4.26 (P)) splits the word transferred via the header into individual letters (lines 18–22), randomly exchanges two inner letters (lines 23–28), puts together the new word in lines 30 to 34, and returns the modified word to the higher-level procedure from which the sub-routine was called.

```
1 Sub Scramble()
2 Text = Cells(1, 1)
3 Words = Split(Text)
4 For IS = 0 To UBound(Words)
5   'Cells(IS + 1, 2) = Words(IS)
6   'activate for Scramble_test
7     Call XWord(Words(IS)) 'Letters are exchanged
8     'Call XWord(Words(IS)) '... a second time
9 Next IS
10 newText = Join(Words, " ")
11 Cells(7, 1) = newText
12 End Sub
```

Fig. 4.25 (P) SUB *Scramble* reads a text from cell A1, splits it into words stored in the array *Words*, and passes the words one by one to the sub-routine *XWord*. The words returned by *XWord* (in the variable *Word* in the header) are assembled into a new sentence in line 10 and output to cell A7 (CELLS(7,1)) of the spreadsheet. Line 8, now a comment, has to be activated when two letters are to be exchanged for a second time

```
13 Sub XWord(Word)
14 Dim Letter(20) As String
15 Debug.Print (Word)
16 lWord = Len(Word) '#Letters in the word
17 If lWord >= 4 Then
18    For n = 1 To lWord
19        'Letters are singled out.
20        Letter(n) = Mid(Word, n, 1)
21        'Cells(n, 3) = Letter(n)
22    Next n
23    n1 = Int(Rnd() * (lWord - 2)) + 2
24    n2 = Int(Rnd() * (lWord - 2)) + 2
25    'Random positions 2 to lWord-1
26    L0 = Letter(n1)
27    Letter(n1) = Letter(n2)
28    Letter(n2) = L0
29
30    For n = 1 To lWord
31        Word = Join(Letter, "")
32    'Cells(n, 4) = Letter(n)
33    'Output to spreadsheet in Xword_test
34    Next n
35 End If
36 End Sub
```

Fig. 4.26 (P) SUB *XWord* detects the length of the transferred word (line 16), splits it into letters (lines 18–22), swaps two inner letters (lines 23–28), and reassembles the letters into the modified word (line 31). Commented lines 21 and 32 have to be activated to obtain SUB *XWord_Test*, mentioned in Sect. 4.7.4

Join (Sentence)

The main program SUB *Scramble* in Fig. 4.25 (P) reassembles the modified words into a text (*newText* = JOIN(*Words*, " "), line 10) and writes it into cell A7 (Cells(7,1) = *newText*). The second entry " " in JOIN causes a space to be inserted after each element of the array *Words*.

VBA terms and user-defined variable names

In the procedures *Scramble* and *XWord*, there are terms that VBA assigns a *precisely defined meaning* to:

- CELLS(r, c): cell in the row *r* and column *c* of the current spreadsheet,
- FOR ... TO ...; DO WHILE ... LOOP; ON ERROR GOTO;
- The functions SPLIT(...); JOIN(...); INT(...); RND().

Such terms are printed in the text in SMALL CAPS.

There are also eleven *variable names*, which the programmer has invented himself/herself:

- *Text, Words, lSent, newText, lWord, Letter*, n_1, n_2, L_0, *x*, *n*.

He might as well have taken eleven letters:

- *a, b, c, d, e, f, g, h, i, j, k*,

or eleven combinations of letters and numbers:

- *a1, a2, a3, a4, b5, b6, b7, b8,* × *1,* × *2,* × *3.*

All combinations of letters and numbers are allowed as variable names, but the first character must be a letter. Such names are *italicized* in the text. To keep the program clear, you should choose variable names that easily convey their meaning in the program.

Attention: *l* (small el) and *I* (capital i) can easily be confused! Variable names in different places of the program then look the same but designate two different variables. So, it is better to use a capital *L*: *LWord* instead of *lWord*.

Questions

What should lines 2 and 11 of Fig. 4.25 (P) be if the statement is formulated with RANGE instead of CELLS?[34]

Which variable names in Fig. 4.25 (P) did the programmer come up with himself/herself?[35]

Which variable names in Fig. 4.26 (P) did the programmer come up with himself/herself?[36]

Do lines 23 and 24 of Fig. 4.26 (P) guarantee that two letters are always exchanged?[37]

4.7.4 Programming Step by Step

Step by step, we develop a program that performs the text swirling described in Sect. 4.7.1. It interacts with the spreadsheet, i.e., reads from cells and fills in the spreadsheet cells as shown in Fig. 4.27 (S).

	A	B	C	D
1	A sentence is to be decomposed.	A	s	s
2		sentence	e	e
3		is	n	n
4		to	t	t
5		be	e	n
6		decomposed.	n	e
7	A sentnece is to be decomposed.		c	c
8			e	e

Fig. 4.27 (S) A1 contains the sentence to be processed. B, C, D, and cell A7 are filled in by the program. The individual words are in B, the individual letters of the second word are in C, those of the swirled word in D

[34] *Text* = RANGE("A1"); RANGE("A7") = *newText*.

[35] *Text, Words, lS, newText*.

[36] *Word, Letter, lWord, n, n_1, n_2, Lo*.

[37] No, n_1 and n_2 can be identical, so that there is no effective exchange.

We modify the procedures SUB *Scramble* in Fig. 4.25 (P) by activating the still out-commented line 5 (filling in column B of Fig. 4.27 (S)) and SUB *XWord* by activating lines 21 (filling in column C) and 32 (filling in column D).

The sub-routine *XWord_Test(Word)* is the same as *XWord*(*Word*) in Fig. 4.26 (P), however, with the lines 21, 22, and 33 to 35 activated that now output intermediate results into the spreadsheet.

Question

concerning Fig. 4.26 (P):

Which program lines guarantee that the first and last letter of a word are not displaced? From which range of the spreadsheet is the text to be processed read? Into which ranges of the spreadsheet are the individual words of the text, the individual letters of a word, the swirled letters, and the modified text written?[38]

After checking these macros to see if they do what we want them to do, we transform them into a procedure that reads a sentence from cell A1 and outputs the changed sentence in A2, like SUB *Scramble* in Fig. 4.25 (P). We now only swap letters from the word's interior, i.e., leave the first and last letters as they are.

▶ **Task** Do this exercise with other texts as well, and surprise your friends with playful letters!

4.7.5 VBA Constructs

The data type Array in VBA

Arrays are declared in VBA as follows:

DIM *Variable name(shape)* AS *data type*

For example, [DIM *Fel*(2) AS DOUBLE] defines an array with three cells (to be addressed with 0, 1, 2), where each cell can contain a real number of type DOUBLE. DIM AR(2,3) AS INTEGER defines a two-dimensional array of integers of shape 3 rows × 4 columns.

4.7.6 Python

The basic functions and methods for text processing, namely, splitting a text into words, a word into letters, and, the other way around, joining letters to form a new

[38] Text read from A1 (CELLS(1,1)), words written into B, letters of the selected word into C, swirled letters into D (CELLS(R,4)), modified text into A7. Compare with Fig. 4.27 (S)!

Table 4.13 Basic functions for text processing

```
1 Text="""A sentence is decomposed."""
2 Words=Text.split()
3 Letters=list(Words[1])
4 L=Letters[4]
5 NewWord=':'.join(Letters[1:-1]) #Concatenate with ":"
```

6 print(Words)	['A', 'sentence', 'is', 'decomposed.']
7 print(Letters)	['s', 'e', 'n', 't', 'e', 'n', 'c', 'e']
8 print(L)	e
9 print(NewWord)	e:n:t:e:n:c

word, are presented in Table 4.13. The `join()` method creates a new string from "precursor strings", e.g., a list of letters. The letters in the new word are separated by a colon as specified in a prefix to `join` (see lines 5).

These basic functions are again applied in Table 4.14 to split a longer text into words, pass each individual word to the function *ExchLett* (reported in Table 4.15), and join the scrambled words into a new text, with blanks as separators. The content of the variable *Text* starts with three quotation marks """, indicating that the following text covering several lines up to the next three quotation marks """ is a string. To enter multi-line strings, use triple codes to start and end them![39]

The function *ExchLett* in the third cell of Table 4.15 uses the function `random.sample` from the `random` library to choose two different internal letters and then exchange them. This function was found with a search in *stack-overflow.com*. Its syntax and mode of action can be deduced from the second cell, which presents the result of line 23.

In lines 30 to 32, the temporary variable L_0 is introduced in order to swap two variables. The code in line 34 does the same, but without the use of any temporary variable. In `Python`, a backslash (\) indicates that the instruction line is continued. Statements can also be split up after a comma.

[39] https://stackoverflow.com/questions/10660435/pythonic-way-to-create-a-long-multi-line-string.

Table 4.14 Rewriting scrambled words in a text **a** (left) program; **b** (right) result

```
Text="""It is said that you can read a text also if you
leave only the beginning and the end letter in each word as
they are and swirl two middle letters. Do you believe that
or is it nonsense?"""

Words=Text.split()
WordsNew=[]
for i in range(len(Words)):
    WN= ExchLett(Words[i])
    WordsNew.append(WN)
sentNew=' '.join(WordsNew)        #Concatenate with blank
print(sentNew, "\n")
```

```
It is siad taht you can raed a txet aslo if you lvaee olny the
beginning and the end lteter in ecah wrod as tehy are and
swril two mildde ltteers. Do you belveie taht or is it
noesnnse?
```

Table 4.15 Swapping internal letters

```
#https://stackoverflow.com/questions/9755538/
#how-do-i-create-a-list-of-random-numbers-without-
duplicates

import random
random.sample(range(0,10), 10)
```

```
   [7, 2, 5, 9, 0, 6, 3, 4, 8, 1]
```

```
def ExchLett(Word):
    #Split a word into letters
    Letters=list(Word)
    #Exchange two internal letters of a word
    if len(Letters)>=4:
        n=random.sample(range(1,len(Letters)-1), 2)
        L0=Letters[n[0]]
        Letters[n[0]]=Letters[n[1]]
        Letters[n[1]]=L0      #Join letters into a word!
    return ''.join(Letters)
```

```
Letters[n[0]],Letters[n[1]=\
Letters[n[1]],Letters[n[0]]
```

Table 4.16 Scrambling a text passage

```
1  Task="""Write a program that first reads a text from a
   spreadsheet. This text is then to be broken down into
   words. The individual words are transformed so that the
   first and last letters remain in their place, but the inner
   letters are randomly swapped."""
2  Words=Task.split()
3  WordsNew=[]
4  for i in range(len(Words)):
5      WN= ExchLett(Words[i])
6      WordsNew.append(WN)
7  sentNew=' '.join(WordsNew)
8  print(sentNew)
```

```
Wrtie a porgram taht frist rdaes a txet form a sprdaesheet.
Tihs txet is tehn to be broekn dwon itno worsd. The indaviduil
wodrs are transofrmed so taht the fisrt and lsat lettres
remian in thier palce, but the inner lettres are rdnaomly
seappwd.
```

Questions

Write lines 30 to 32 of Table 4.16, swapping two variables using the temporary variable `L0`, as one statement![40]

How can lines 5 and 6 of Table 4.15 be merged into one statement?[41]

How can you introduce an additional line of code into Table 4.16 to achieve two letter swappings in a word?[42]

Table 4.16 processes the second text passage, now performing two letter exchanges.

4.8 Processing the Protocol of a Measuring Device

Continuous text is decomposed and rearranged in tables. Knowledge acquired by processing texts in the previous exercise is applied to separate text and numbers in reports created by measuring instruments, with the aim to recognize code words and rearrange the essential results in tables.—The `Python` program is based on the library `Pandas`.—A piece of advice: With

[40] `Letters[n[0]], Letters[n[1]] = Letters[n[1]], Letters[n[0]]`.

[41] WN = EXCHLETT(EXCHLETT(WORDS[I])).

[42] Introduce `WN = ExchLett(WN)` after line 5!

the knowledge gained in this exercise, you may earn a small bit of extra income from a side job in scientific projects!

4.8.1 Protocol of a Measuring Device

Many measuring devices output a plain-text file as a protocol containing both text and numbers. As an example, we will use the output of chemical analysis with RBS (Rutherford Back Scattering) concerning the composition of four Nb-doped TiO_2 layers on a silicon substrate (spreadsheet in Fig. 4.28 (S)). Every layer contains different fractions of Ti (titanium), O (oxygen), Nb (niobium), and Ar (argon). We will convert this information into a table, as in Fig. 4.29 (S).

	A	B	C	D
1	T1.lay	T2.lay	T3.lay	T4.lay
2	!----------------	!----------------	!----------------	!----------------
3	d=0.20E18	d=0.25E18	d=0.30E18	d=0.35E18
4				
5	Ti#,1	Ti#,1	Ti#,1	Ti#,1
6	O#,2.5	O#,2.4	O#,2.3	O#,2.2
7	Nb#,0.03	Nb#,0.05	Nb#,0.07	!----------------
8	Ar#,0.01	!----------------	Ar#,0.008	s=
9	!----------------	s=	!----------------	Si#,1
10	s=	Si#,1	s=	
11	Si#,1		Si#,1	

Fig. 4.28 (S) Protocol of RBS measurements, transferred into an EXCEL spreadsheet; row 1 = names of four different samples; parameters: d = number of atoms per cm^2; Ti, O, Nb, Ar = elements found in the layer with their indices in the chemical formula

	A	B	C	D	E	F
1						
2	**SampNam**	**NAtoms**	**Ti**	**O**	**Nb**	**Ar**
3	**T1**	2,00E+17	1	2,5	0,03	0,010
4	**T2**	2,50E+17	1	2,4	0,05	
5	**T3**	3,00E+17	1	2,3	0,07	0,008
6	**T4**	3,50E+17	1	2,2		
7						

Fig. 4.29 (S) The data from Fig. 4.28 (S) have been written into this table in a spreadsheet with the name "TabLay". Each sample has its own row. The first row is left blank in order to insert an index for the next free row later

```
36 For r = 3 To r2
37     cl = Cells(r, sample) 'e.g.: d=0.20E18
38     Le = Len(cl) 'length of the string
39     cll = Left(cl, 2) 'take first two letters
40     If cll = "d=" Then NAtoms = Right(cl, Le - 2)
47 Next r
```

Fig. 4.30 (P) Cuts off the first two letters of a data string (line 39) and checks whether this is the code word "d ="

```
Sub DecodeRBS()
'Original data in sheet "RBS-data"
'Table with sample characteristics in "TabLay"
Dim cl, cll As String
'Write headers!
r3 = 2
Sheets("TabLay").Select
Cells(r3, 1) = "SampNam"
Cells(r3, 2) = "NAtoms"
Cells(r3, 3) = "Ti"
Cells(r3, 4) = "O"
Cells(r3, 5) = "Nb"
Cells(r3, 6) = "Ar"
r3 = r3 + 1 'next free row for output
For sample = 1 To 4
    Sheets("RBS-data").Select 'original data
    'Get sample name!
    SampNam = Cells(1, sample) 'sample name
    Le = Len(SampNam) 'length of the name
    SampNam = Left(SampNam, Le - 4)
                    ' ".lay" is removed
    'Identify range with information on sample!
    For r1 = 3 To 30 'scans rows 3 to 30
        cl = Cells(r1, sample) 'Content of cell
        cll = Left(cl, 5)
        If cll = "!----" Then r2 = r1 - 1
        'r2 = last row of information on the sample
    Next r1
```

Fig. 4.31 (P) Complete program for rearranging the raw data from Fig. 4.28 (S) into a table as in Fig. 4.29 (S); continued in Fig. 4.32 (P)

4.8.2 Detection of Code Words

The information in a column of Fig. 4.28 (S) is to be decoded and stored in a row of Fig. 4.29 (S). The main task is to identify certain code words that indicate the physical or technical quantity to which the following numbers refer. An extract from the complete Visual Basic decoding program (Fig. 4.32 (P)) can be found in Fig. 4.30 (P).

The data set for a layer is read line by line. The data strings "d = 0.20E18" and "Ti#,1" contain information about the number of atoms per cm^2 and the titanium content in the sample, respectively. When the program processes the file, it is not clear from the outset what type of data string is currently involved. For decoding, therefore, the first parts of the data line are separated (line 39 in Fig. 4.30 (P)), and it is queried as to whether this part is "d =" (line 40). If this is the case, the following string is interpreted as a number and written into the corresponding variable; for "d =", this is *NAtoms* (line 40 in Fig. 4.30 (P)).

In `Python`, the task is tackled with the library `Pandas`, which mimics spreadsheet calculation.

4.8.3 Data Structure and Nomenclature

wb	name in `Pandas` of the workbook containing the data
sh	name in `Pandas` of the worksheet within the workbook
T1, T2, T3, T4	identifiers of samples

„!----"	code word to indicate the end of useful information
"d="	two-character code word for atomic coverage, atoms per cm^2
"O#,"	three-character code word for oxygen
"Ti#,", "Nb#,", "Ar#,"	four-character code words for titanium, niobium, and argon
r_2	last row of useful information
r_3	current row in the spreadsheet for output
NAtoms	number of atoms per cm^2
O, Ti, Nb, Ar	fraction of the respective element

4.8.4 Excel

The complete decoding procedure can be found in Fig. 4.31 (P) and Fig. 4.32 (P). SUB *DecodeRBS* writes headings into the spreadsheet "TabLay" (lines 7 to 13), successively reads rows 3 through 30 from the spreadsheet "RBSData" (line 23, index r1), determines the last row of useful data (line 26), decodes the useful data (FOR loop in rows 36 through 47), and finally writes the decoded data row by row (index r_3) into the spreadsheet "TabLay" (rows 49 through 55).

The loops are of type *loop2i*, with the loop index r and the running index r_3, indicating the next free row in the table and set to 2 at the beginning (line 6) and incremented in line 56 after the extracted values for the parameters have been entered into the output table within the For-loop with the index r scanning the input table.

Since the code words have different lengths ("d =" has two letters, "Ti#," has four letters), in lines 39, 41, and 43, two, three, and four letters are, one after the other, cut off from the beginning of the data string and it is checked as to whether they correspond to one of the code words. The part of the data string following

```
29 'Remove old information!
30 NAtoms = Empty
31 O = Empty
32 Ti = Empty
33 Nb = Empty
34 Ar = Empty
35 'Decode information in rows 3 to r2!
36 For r = 3 To r2
37    cl = Cells(r, sample) 'e.g.: d=0.20E18
38    Le = Len(cl) 'length of the string
39    cll = Left(cl, 2) 'take first two letters
40    If cll = "d=" Then NAtoms = Right(cl, Le - 2)
41    cll = Left(cl, 3) 'take first three letters
42       If cll = "O#," Then O = Right(cl, Le - 3)
43    cll = Left(cl, 4) 'take first four letters
44       If cll = "Ti#," Then Ti = Right(cl, Le - 4)
45       If cll = "Nb#," Then Nb = Right(cl, Le - 4)
46       If cll = "Ar#," Then Ar = Right(cl, Le - 4)
47 Next r
48 'Write decoded data into a different sheet!
49 Sheets("TabLay").Select
50 Cells(r3, 1) = SampNam
51 Cells(r3, 2) = NAtoms
52 Cells(r3, 3) = Ti
53 Cells(r3, 4) = O
54 Cells(r3, 5) = Nb
55 Cells(r3, 6) = Ar
56 r3 = r3 + 1 'next free row
57 Next sample
58 End Sub
```

Fig. 4.32 (P) Continuation of Fig. 4.31 (P)

the cut-off contains a number separated with RIGHT(..) and then assigned to the corresponding variable (e.g., *NAtoms*, *Ti*, ...). This separation is easy because all numbers have the same format.

Questions

Which code word is queried in Fig. 4.30 (P)?[43]

Which code words in Fig. 4.32 (P) have length 4?[44]

Module in VBA

Since the program SUB *DecodeRBS* refers, with SHEETS("RBS-data").SELECT and SHEETS("TabLay").SELECT, to two different spreadsheets, it must be operated in a module. If it stands in the VBA sheet associated with a spreadsheet, it operates only in that spreadsheet with instructions like CELLS(*r*,*c*) = .

VBA keyword empty

In lines 36 to 47 of Fig. 4.32 (P), the variables O, Ti, Nb and Ar are only filled in if the assigned code words occur in the original protocol. If a code word does not occur in the current data line, the content of the associated variable is not overwritten, and the old value persists. To prevent this from happening, we have entered the assignments in lines 30 to 34, e.g., Nb = EMPTY. EMPTY is a VBA keyword, to ensure that the variables do not contain any value when a new sample is processed.

Sample T4 does not contain any Nb; cell E6 in Fig. 4.29 (S) thus remains empty. If we had not set *Nb* = EMPTY before decoding, then *Nb* would still contain the value 0.07 of the previous sample and would have been incorrectly entered into the table of results. This would be a grave error in content!

The table is continued the next time the macro is called

► **Tim** The presented program sorts the data of exactly four samples into another table. What if a new set of samples comes in and the table is to be continued? Can the program remember the value of the index r_3 for the next free row and use it for the next call?

► **Mag** No, the program forgets the values of the variables when it finishes its execution.

► **Alac** Then, I will simply adapt the code before each new call. In line 14, the next free row in the table of Fig. 4.29 (S) is entered, $r_3 = 7$, and for the last FOR loop index in line 15, I will enter the current number of samples.

---

43 „d = " is queried.

44 The code words „Ti#,", „Nb#," and „Ar#," have the length 4.

▶ **Mag** That's a practical idea, and it works. But it can also be done more elegantly with the following two pieces of program.
You can use the first row in Fig. 4.29 (S), still empty, e.g., cell A1, to store the number of the first free row after the previous entries, and cell D1 to specify the number of new samples. This information is then read with r_3 = RANGE("A1") in line 14 and ... TO RANGE("D1") in line 15. Cell A1 is now overwritten with the value of r_3 by a new instruction at the end of SUB *DecodeRBS*. The number of new samples must be entered manually in D1 when a new data series is to be decoded, or the programmer can devise a query to automatically determine the number of samples in the raw data of Fig. 4.28 (S).

DO ... LOOP UNTIL
The next free row can also be determined by querying the current cell content in a loop to see if the current cell is empty:

```
DO
  r3 = r3 + 1
LOOP UNTIL CELLS(r3, 1) = EMPTY
```

Similarly, the number of NOT EMPTYs in a new raw data file can be obtained.

For more information on the instructions DO ... LOOP, DO WHILE ... LOOP, and DO ... LOOP UNTIl, see EXCEL help!

4.8.5 Python

`Pandas` ("Python Data Analysis") is a library for `Python`, based on `NumPy`. It is designed for data management and analysis and works with structured data (`DataFrame` (2-dimensional)) and time series (`Series` (1-dimensional)), thus mimicking spreadsheet calculations. We use it only in this exercise to read data from an EXCEL book, decode the measurement protocol, and write the results into an EXCEL sheet, in the same form as in Fig. 4.29 (S).

R1C0 in Pandas
In Table 4.17, the EXCEL workbook 'RBS_data.xlsx' (Fig. 4.28 (S)) is opened, so that its data are available in `Pandas`. Without further specification, the workbook must be in the same directory as the Python program. The data in *Sheet1* are entered into a two-dimensional matrix *sh*. This matrix can be addressed in A1 or R1C0 reference style (see second cell). In A1 style, a column is addressed by a letter and a row by a number starting at 1; in R1C0 style, both are addressed by numbers, *with columns being numbered starting at 0.*

With the statement in line 5, the contents in the cells will be copied directly. If a cell contains a formula, this formula will be transferred. If it is desired that all

Table 4.17 **a** (top cell) opening an EXCEL workbook; **b** (bottom cell) addressing cells and ranges and reading their content

```
import numpy as np
import pandas as pd
import openpyxl #A Python library to read/write Excel 2010

wb=openpyxl.load_workbook('RBS_data.xlsx')
sh=wb['Sheet1']
```

```
Sh:               <Worksheet "Sheet1">
sh['A1'].value:   T1.lay
sh[1][0].value:   T1.lay
```

```
for r in range(1,4):
    print(sh[r][0].value)
```

```
range(1,6):
    T1.lay
    !--------------------------
    d=0.20E18
    None
    Ti#,1
```

formulas be evaluated and only the results transferred, the opening has to include a keyword argument data_only = True:

$$\texttt{wb} = \texttt{openpyxl.load\_workbook('RBS\_data.xlsx', data\_only} = \texttt{True).}$$

In Table 4.18, the sample name is extracted from the first entry in $c = 0$ (column A), reported in the second cell. The content of the bottom cell is produced by the print statements shown explicitly in lines 5 to 9. In contrast to this, print statements

Table 4.18 Extracting the sample name from the first entry in column *c*; print statements resulting in the output cell (bottom cell) are explicitly reported in lines 5 to 9

```
1  c=0
2  SN=list(sh[1][c].value)[:-4] #Delete ".lay"
3  SampNam=''.join(list(SN))
4
5  print("sh[1][c]                 ",sh[1][c])
6  print("sh[1][c].value           ",sh[1][c].value)
7  print("list(sh[1][c].value)",list(sh[1][c].value))
8  print("SN                       ",SN)
9  print("SampNam                  ",SampNam)
```

```
sh[1][c]             <Cell 'Sheet1'.A1>
sh[1][c].value       T1.lay
list(sh[1][c].value) ['T', '1', '.', 'l', 'a', 'y']
SN                   ['T', '1']
SampNam              T1
```

Table 4.19 Getting the last row of useful information; line 5 contains an error

```
1 def LastRow(c):
2     for r1 in range(2,15):
3         cl=sh[r1][c]          #Cell address
4         if cl.value!=None:
5            cl1=''.join(list(cl.value)[:5])
6            #Indentation 4 spaces!
7             if cl1=='!----': r2=r1-1
8     return r2
```

are usually omitted in our tables reporting Python programs; only the results are usually reported in an output cell.

Questions

What are the values of `sh[1][0].value` and `list(sh[1][0].value)[0]`?[45]

Line 5 of Table 4.19 contains a bug. Which one?[46]

In Table 4.19, column *c* is scanned for the occurrence of "!—", the code word signaling the end of the information on the first layer.

Table 4.20 reports the function for decoding the string in the cell in column *c*, row r_2. The Python code mimics the VBA procedure in Fig. 4.32 (P).

In Table 4.21, a data frame *out* is created reproducing the EXCEL sheet in Fig. 4.29 (S). A For loop runs over *sample* with sample data, determining the last row r_2 with useful information and decodes the range rows 3 to r_2 of the column. The print statement in line 11 produces the output in the lower cell.

Output from Pandas to an Excel file

Writing our results into an EXCEL file requires some care. The simple statement in Table 4.22a creates a new file *RBS_data3.xlsx* and writes our frame *out* into a sheet with the name *Sheet3*. If a file with said name already exists, it is overwritten. The keyword arguments `header = False, index = False` cause row 1 and column 1 of Table 4.21 (bottom cell) not to be output.

To write the data into an already existing file, we have to open that file and specify a writer (lines 8 and 9 in Table 4.22).

[45] `sh[1][0].value -> T1.lay, list(sh[1][0].value)[0] -> T.`

[46] Indentation with respect to the `if` line is only 3 spaces; it must be 4 spaces.

Table 4.20 Function for decoding a string

```
def decode(c,r2):                  #Column c, row r2
    nAtoms=None
    O=None
    Ti=None
    Nb=None
    Ar=None
    #Decode information in rows 3 to r2!
    #First index in array starts with 0
    for r in range(2,r2+1):
        cl=sh[r][c]                #Cell address
        if cl.value!=None:         # != means "not equal"
            cl1=''.join(list(cl.value)[:2])
            if cl1=='d=':
                            nAtoms=''.join(list(cl.value)[2:])
            cl1=''.join(list(cl.value)[:3])
            if cl1=='O#,':   O =''.join(list(cl.value)[3:])
            cl1=''.join(list(cl.value)[:4])
            if cl1=='Ti#,': Ti=''.join(list(cl.value)[4:])
            if cl1=='Nb#,': Nb=''.join(list(cl.value)[4:])
            if cl1=='Ar#,': Ar=''.join(list(cl.value)[4:])
    nAtoms=float(nAtoms)
    return(nAtoms,Ti,O,Nb,Ar)
```

4.9 User-Defined Functions

We code functions in VISUAL BASIC that can be applied in spreadsheets just like built-in functions. As examples, we realize the vector operations scalar product and cross-product with three-component vectors as arguments and with, respectively, a scalar and a vector as the return variable.

4.9.1 User-Defined Functions as Add-In

Functions in modules

We often call built-in functions in cells, e.g., trigonometric functions with formulas like B5 = [=B$1*cos(B$2*$A5]). We can also create functions ourselves and apply them in the same way. As an example, we implement a function *CosSq(x),* calculating the square of a cosine. This has to be done in a module that we create with INSERT/MODULE in the project explorer (see Fig. 4.33a).

The function is implemented in the corresponding Visual Basic sheet MODULE1. The qualifier is FUNCTION, not SUB, as for procedures. For a function, a value must be

Table 4.21 Data frame reproducing the structure of the EXCEL sheet in Fig. 4.29 (S); *NaN* stands for "Not a number"

```
out = pd.DataFrame(index=range(1,8), s=list('ABCDEF'))
title=['SampNam','NAtoms','Ti','O','Nb', 'Ar']
out.iloc[1]=title
 #iloc is integer position from 0 to length-1 of the axis
for sample in range(4):
    SampNam=sh[1][sample].value
    SampNam=SampNam[:-4]       #Remove .lay
    r2=LastRow(sample)
    nAtoms,Ti,O,Nb,Ar=decode(sample,r2)
    result=[SampNam,nAtoms,Ti,O,Nb,Ar]
    out.iloc[sample+2]=result
print(out)
out.to_excel("output.xlsx",sheet_name="Sheet2",
  header=False,index=False)    #A…F and 1…7 not transferred
```

```
#NaN means "Not a number"
           A          B    C    D     E      F
1        NaN        NaN  NaN  NaN   NaN    NaN
2    SampNam     NAtoms   Ti    O    Nb     Ar
3         T1      2e+17    1  2.5  0.03   0.01
4         T2    2.5e+17    1  2.4  0.05   None
5         T3      3e+17    1  2.3  0.07  0.008
6         T4    3.5e+17    1  2.2  None   None
7        NaN        NaN  NaN  NaN   NaN    NaN
```

Table 4.22 **a** (top cell) Creates a new file and writes the data into the specified sheet; **b** (bottom cell) Opens an existing file and adds a new sheet

```
out.to_excel('RBS_data3.xlsx',sheet_name="Sheet3",
    header=False,index=False)
 #Creates a new workbook with one sheet "Sheet3".
 #Overwrites 'RBS_data3.xlsx' if it exists already.
```

```
"""https://stackoverflow.com/questions/20219254/
    how-to-write-to-an-existing-excel-file-
    without-overwriting-data-using-pandas"""
wb2 = openpyxl.load_workbook('RBS_data2.xlsx')
writer = pd.ExcelWriter('RBS_data2.xlsx',
                                    engine='openpyxl')
writer.book = wb2 #Necessary for not deleting other sheets
out.to_excel(writer, "data3",header=False,index=False)
writer.save()
```

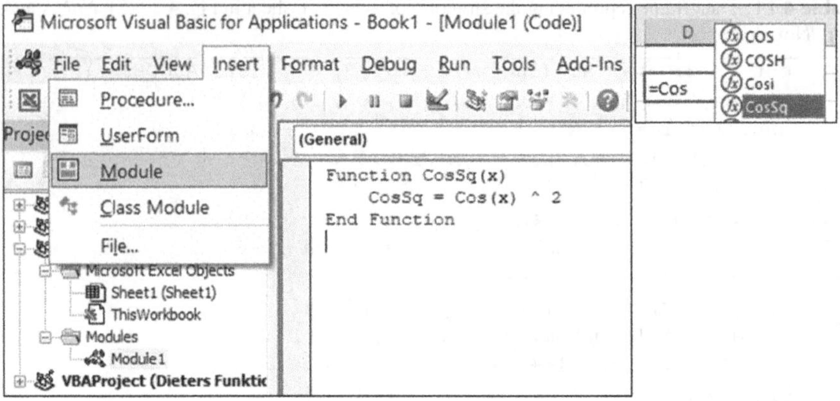

Fig. 4.33 **a** (left) PROJECT-EXPLORER window; the user-defined spreadsheet function *CosSq* is in MODULE1. **b** (right) *CosSq* pops up after typing "=Cos" among the other functions starting with *cos*

assigned to the function identifier within the function's body, in our case $cosSq = \cos(x)\hat{}2$. When now "=Cos" is written into a cell, a list pops up with all functions starting with *Cos*, including our *CosSq* (see Fig. 4.33b).

If you want to use your functions in every EXCEL file, they must be saved as an ADD-IN. To do so, create an EXCEL file, enter your function codes into VBA modules, and finish with: SAVE AS/EXCEL ADD-IN. This add-in must be activated in the EXCEL options. Upon selecting FILE/OPTIONS/ADD-INS/ a list appears with an entry "☐ Dieter's Functions" that must be included by ticking the box ☑. In the VBA editor, VBA PROJECT *Dieters Funktionen*.xlam now also appears in the project explorer under PROJECT–VBA PROJECT (see Fig. 4.33a, bottom line).

4.9.2 Scalar Product and Vector Product

We are developing functions for the scalar and vector products of two three-dimensional vectors stored in cell ranges that are entered as arguments in the functions. Let's consider two three-dimensional vectors:

$$\underline{r}_1 = (x_1, y_1, z_1) \text{ and } \underline{r}_2 = (x_2, y_2, z_2)$$

Their scalar product is defined as.

$$\underline{r}_1 \cdot \underline{r}_2 = x_1 \cdot x_2 + y_1 \cdot y_2 + z_1 \cdot z_2 \tag{4.2}$$

Their vector product (or cross product) is defined as.

$$\underline{r}_1 \times \underline{r}_2 = (y_1 \cdot z_2 - y_2 \cdot z_1, z_1 \cdot x_2 - z_2 \cdot x_1, x_1 \cdot y_2 - x_2 \cdot y_1) \tag{4.3}$$

```
1 Function Scl(r1 As Range, r2 As Range)
2 Scl = r1(1) * r2(1) + r1(2) * r2(2) + r1(3) * r2(3)
3 End Function
4
5 Function Crs(r1 As Range, r2 As Range)
6 Dim cs(2)
7 cs(0) = r1(2) * r2(3) - r1(3) * r2(2)
8 cs(1) = r1(3) * r2(1) - r1(1) * r2(3)
9 cs(2) = r1(1) * r2(2) - r1(2) * r2(1)
10 Crs = cs
11 End Function
12 Function Crsm(r1 As Range, r2 As Range)
13 Dim cs(2, 2)
14 cs(0, 0) = r1(2) * r2(3) - r1(3) * r2(2)
15 cs(1, 0) = r1(3) * r2(1) - r1(1) * r2(3)
16 cs(2, 0) = r1(1) * r2(2) - r1(2) * r2(1)
17 cs(0, 1) = cs(1, 0)
18 cs(0, 2) = cs(2, 0)
19 Crsm = cs
20 End Function
21
22
```

Fig. 4.34 (P) User-defined functions for the scalar product *Scl* and the vector product *Crs* of two three-dimensional vectors r_1 and $r_{2;}$; the function *Crsm* can process and output both column and row vectors

The output of the scalar product is one number returned into the cell with the corresponding formula; that of the vector product is a set of three components to be entered into a row range or a column range.

These two products are calculated with the two user-defined spreadsheet functions *Scl* and *Crs* in Fig. 4.34 (P).

Questions

How many components does the array *cs*(2) in Fig. 4.34 (P) have?[47]

What are the differences between the arrays named *cs* in the functions *Crs* and *Crsm* in Fig. 4.34 (P)?[48]

Scalar product

The scalar product is easy to program. It can be calculated in one code line (line 2 in Fig. 4.34 (P)). Two three-dimensional cell ranges must be entered as arguments. These can both be column ranges or both row ranges or one column range and one row range (see Fig. 4.35 (S)). Consequently, the variables in the function header are declared as RANGE.

Vector product

The result of a vector product is again a vector. In Fig. 4.36 (S), two column vectors $\underline{a}$ and $\underline{b}$ are defined in range A2:B4. The row vectors $\underline{c}$ and $\underline{d}$ in range B6:D7 contain the same coefficients as $\underline{a}$ and $\underline{b}$. In column D, the cross-product $\underline{a}$ x $\underline{b}$ is calculated with spreadsheet formulas.

[47] The array DIM *cs*(2) has the three components *cs*(0), *cs*(1), *cs*(2).

[48] In *Crs*, a one-dimensional array (type *cs*(2)) is written, in *Crsm*, a two-dimensional array (type *cs*(2,2)).

	A	B	C	D	E	F	G	H	I	J	K	L
1	a	b							12	=Scl(A2:A4;B2:B4)		
2	1	2		c_	1	2	3		12	=Scl(A2:A4;E3:G3)		
3	2	2		d	2	2	2		12	=Scl(a;b)		
4	3	2							12	=Scl(c_;d)		
5									12	=Scl(a;d)		

Fig. 4.35 (S) Contains the results of the user-defined spreadsheet function *Scl*, which calculates the scalar product of two three-dimensional vectors

	A	B	C	D	E	F	G	H	I	J	K	L	M	N	O
1	a	b		a x b			=crs(a;b)	=crs(c_;d)	=crs(c_;b)			=crsm(a;b)	=crsm(c_;d)	=crsm(c_;b)	
2	3,0	-2,0		-10,0	=A3*B4-B3*A4		-10,0	-10,0	-10,0			-10,0	-10,0	-10,0	
3	2,0	8,0		-20,0	=A4*B2-B4*A2		-10,0	-10,0	-10,0			-20,0	-20,0	-20,0	
4	2,5	5,0		28,0	=A2*B3-B2*A3		-10,0	-10,0	-10,0			28,0	28,0	28,0	
5															
6	c_	3,0	2,0	2,5			-10,0	-20,0	28,0	=crs(a;b)		-10,0	-20,0	28,0	=crsm(a;b)
7	d	-2,0	8,0	5,0			-10,0	-20,0	28,0	=crs(c_;d)		-10,0	-20,0	28,0	=crsm(c_;d)
8							-10,0	-20,0	28,0	=crs(a;d)		-10,0	-20,0	28,0	=crsm(a;d)

Fig. 4.36 (S) The vector product $\underline{a}$ x $\underline{b}$ is calculated in column D using spreadsheet formulas; the range G2:I8 contains results of the user-defined function *crs*, which can only output row vectors (wrong results in G2:I4); the range L2:N8 contains results of the user-defined spreadsheet function *crsm*, which can accept and output row and column vectors

In columns G to I, the function *Crs* is used. As you can see from the results, this function accepts row and column vectors as input, but only returns correct values if they are output as row vectors (see G6:J8); the results in G2:I4 are wrong.

In columns L to N, the function *Crsm* from Fig. 4.34 (P) is applied, which can output the result either as a row vector (e.g., L6:N6) or as a column vector (e.g., L2:L4). This is because, in this function, a 3×3 matrix is written into the range declared with DIM *cs*(2,2), of which only one row or one column is output if only one row range or one column range is activated.

The functions *Crs* and *Crsm* must be called as matrix functions. In Fig. 4.36 (S), for example, the area G6:I6 was activated, the formula entered according to J6 and closed with the magic chord Ψ *(Ctl + Shift) + Enter*. In L2:L4, a column area was activated, and in L6:N6, a row area, so that in each case, vectors with three components are returned by *crsm*.

Questions

Why is *cs*(2,2) in FUNCTION *Crsm* sufficient as an array for a 3 x 3 matrix?[49]

49 DIM *cs*(2,2) is a (0, 1, 2) x (0, 1, 2)-Matrix. The indices begin at 0.

Table 4.23 Two column vectors *a*, *b* and two row vectors *c*, *d* are specified

```
a=np.array([[1,2,3]]).transpose(1,0)
b=np.array([[2,2,2]]).transpose(1,0)
c=np.array([1,2,3])
d=np.array([2,2,2])
```

a	b		
[[1] [2] [3]]	[[2] [2] [2]]	c d	[1 2 3] [2 2 2]

Table 4.24 Scalar product, line 8 deactivated

```
5 def Scl(r1,r2):
6     o=r1[0]*r2[0]+r1[1]*r2[1]+r1[2]*r2[2]
7     out=o
8     #if type(o) == np.ndarray: out = o[0]
9     return out
```

Scl(a,b)	[12]
Scl(c,d)	12
Scl(a,d)	[12]

How can this function be used to output row and column vectors to a spreadsheet?[50]

4.9.3 Python

Scalar product

The specifications in Table 4.23 for column vectors *a*, *b*, and row vectors *c*, *d* are the same as in Fig. 4.35 (S).

The function *Scl* (for "scalar product") as reported in Table 4.24 corresponds literally to the Visual Basic function of the same name (Fig. 4.34 (P)). Its output is, however, only a scalar if two row vectors are multiplied.

Column vectors are two-dimensional arrays; a scalar product with one of them is broadcast into a one-dimensional array (see lines 10 and 12 of Table 4.24). If a scalar is always desired, line 8 has to be activated by removing the # character.

Vector product or cross product

A function *Crsm* for calculating the cross-product of two three-dimensional vectors is reported in Table 4.25.

[50] Because internally a 3×3-matrix is created, see explanations for Fig. 4.36 (S)!

Table 4.25 Cross product of two three-dimensional vectors, output optionally transposed in line 16 to become a column vector

```
1 def Crsm(r1, r2, C):
2     cs=np.empty(3)
3     cs[0]=r1[1]*r2[2]-r1[2]*r2[1]
4     cs[1]=r1[2]*r2[0]-r1[0]*r2[2]
5     cs[2]=r1[0]*r2[1]-r1[1]*r2[0]
6     if C==True:cs=np.array([cs]).transpose(1,0)
7     return cs
```

Table 4.26 Cross-product of row and column vectors

```
a=np.array([[3,2,2.5]]).transpose(1,0)
b=np.array([[-2,8,5]]).transpose(1,0)
c=np.array([3,2,2.5])
d=np.array([-2,8,5])
```

Crsm(a,b) [-2.00 4.00 -2.00] Crsm(c,d) [-2.00 4.00 -2.00]	Crsm(c,b,True) [[-2.00] [4.00] [-2.00]]

In Table 4.26, again, two column vectors *a*, *b* and two row vectors *c*, *d* are specified. Their pairwise cross-product, obtained with *Crsm*, is reported in the lower cells of the table. If only the vectors are transferred to the function, row vectors are returned (bottom left cell), whereas a column vector is returned if the optional parameter *C* is assigned "True" (bottom right cell).

4.10 Questions and Tasks

Densely packed planes

1. How many neighbors does a sphere have in a closely-packed plane?
2. How many neighbors does a sphere have in a stack of closely-packed planes?

Program-controlled drawings

3. Write a macro that writes the numbers 1 to 20 in a diagonal of a table, e.g., in cells A1, B2, etc.
4. What does the broom rule Ψ *Empty lines separate curves* mean?

Record macro

The diagram in Fig. 4.37 (S) has been created with the macro recorder switched on. The program code can be found in Fig. 4.38 (P). The diagram has been formatted

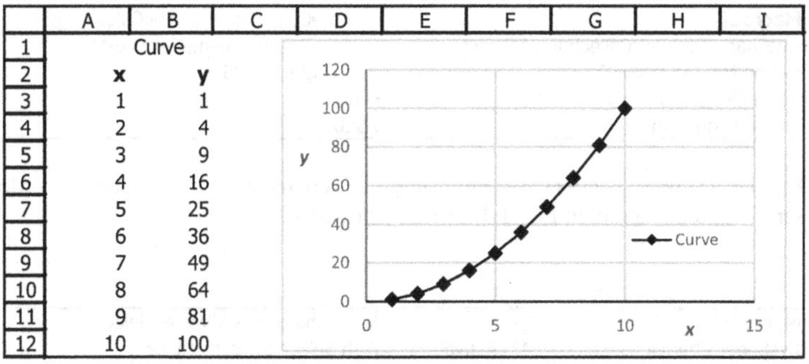

Fig. 4.37 (S) Diagram of the data in columns A and B

```
1 Sub Macro1()
2     ActiveSheet.Shapes.AddChart2(240, xlXYScatterLines).Select
3     ActiveChart.SeriesCollection.NewSeries
4     ActiveChart.FullSeriesCollection(1).Name = "=Diagram!$B$1"
5     ActiveChart.FullSeriesCollection(1).XValues = "=Diagram!$A$3:$A$12"
6     ActiveChart.FullSeriesCollection(1).Values = "=Diagram!$B$3:$B$12"
7 End Sub
```

Fig. 4.38 (P) Instructions recorded by the macro recorder when the diagram in Fig. 4.37 (S) was created

with the programs in Fig. 4.38 (P) and Fig. 4.40 (P). Your task is to analyze the VISUAL BASIC programs and redraw the diagram with our standard `FigStd` and `plt` (`matplotlib.pyplot`) commands in `Python`.

Plotting a diagram

5. Of what type is the diagram in Fig. 4.37 (S) (LINE, BAR, or SCATTER)?
6. What is the equation for *y*?
7. How do you create the arrays *x* and *y* in `Python`?
8. With SUB MACRO1 in Fig. 4.38 (P), retrace how the chart was created and interpret the program lines 3 to 6!
9. (`Python`) What does a header in `FigStd` (`numpy` and `matplotlib`) look like when leading to a diagram like that in Fig. 4.37 (S)?
10. SUB MACRO2 in Fig. 4.39 (P) has recorded the instructions executed to format the data series in the diagram of Fig. 4.37 (S). Interpret the formatting instructions! How do you implement them in the plot command of `plt.plot(?)` of the `plotlib` library?
11. SUB MACRO3 in Fig. 4.40 (P) has recorded the commands executed to format the *x*-axis in the diagram of Fig. 4.37 (S). Interpret the instructions that follow the two WITH SELECTION commands!

```
Sub Macro2()
  ActiveChart.FullSeriesCollection(1).Select
  Selection.MarkerStyle = 2
  Selection.MarkerSize = 7
  With Selection.Format.Line
    .ForeColor.ObjectThemeColor _
          = msoThemeColorText1
    .Weight = 1.25
  End With
End Sub
```

Fig. 4.39 (P) SUB MAKRO2 contains the commands that were recorded by the macro recorder when the data series for the diagram in Fig. 4.37 (S) was formatted

```
Sub Macro3()
  ActiveSheet.ChartObjects("Chart 1").Activate
  ActiveChart.Axes(xlCategory).Select
  With Selection.Format.Line
     .ForeColor.ObjectThemeColor \
                        '= msoThemeColorText1
     .Weight = 1#
  ActiveChart.Axes(xlValue).Select
  With Selection.Format.Line
    .ForeColor.ObjectThemeColor \
              '= msoThemeColorBackground1
    .ForeColor.Brightness = -0.5
  End With
End Sub
```

Fig. 4.40 (P) SUB MACRO3 contains commands recorded by the macro recorder when the diagram in Fig. 4.37 (S) was formatted

12. Write a Python program that produces a similar diagram with FigStd (numpy and matplotlib) ! It should include formatting the data series in the function header and changing the thickness of the *x*-axis within FigStd.

Text processing in VISUAL BASIC and Python

The following three questions refer to both Visual Basic for EXCEL and Python. The variable *Tx* contains the text "We are cutting."

13. With which instructions do you get the first, the last, and the 4th letters of *Tx*?
14. How do you copy the fragment "re cu" from *Tx* to a new variable *Wd*?
15. Of what type are the variables *A, B, C,* and *D* in the commands *A* = *Split(B)* and *C* = *Join(D)*?

Ψ Loop2i; continue counting in the loop!

16. Write a macro SUB *XY1*() that writes all products $x \cdot y$ from $x = 1$ to 10 and from $y = 1$ to 5 successively into a spreadsheet, with *x* and *y* being integers!
17. Do the same in another macro SUB *XY2()* for *x* and *y* being half-integers (1, 3/2, 2, 5/2, …)!
18. Do the same as in SUB *XY2()* in a new macro SUB *XY3()*, but insert a blank line after every third entry into the spreadsheet!
19. Create a similar Python program using the .append method and nested loops running over arrays *x* and *y* created by np.linspace! For the equivalent of inserting an empty row into a spreadsheet, append 'None' to the list!
20. Make a hand-drawn sketch of straight-line segments in the *xy* plane, with the *x* values in line 1 and the *y* values in line 2 of Fig. 4.41 (S)!

	A	B	C	D	E	F	G	H	I	J	K	L
1	**x**	0,5	1,0		1,5	2,0		2,0	1,5		1,0	0,5
2	**y**	1,5	2,0		2,0	1,5		1,0	0,5		0,5	1,0

Fig. 4.41 (S) Coordinates for straight-line segments in the *xy* plane

Rep-log procedure

In Fig. 4.42a, a circle is represented, calculated in the spreadsheet in Fig. 4.43 (S) in columns G:I. The coordinates of the centers of these circles are taken from the table in Fig. 4.41 (S). Figure 4.42b shows four circles whose coordinates have been obtained from the table in Fig. 4.43 (S) with a rep-log procedure that systematically changes the center point and the radius.

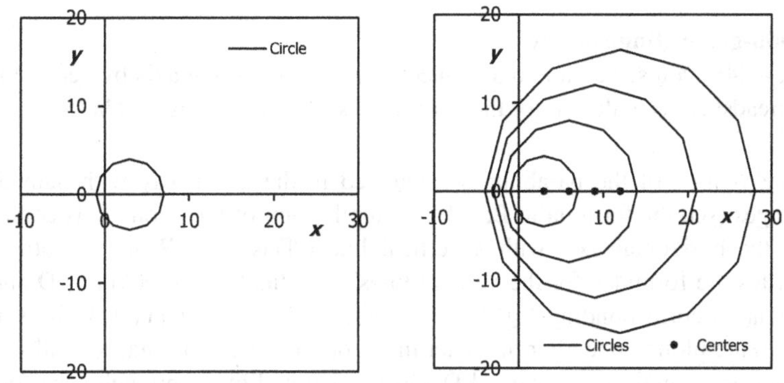

Fig. 4.42 **a** (left) A circle with $r_0 = 4$ and $x_0 = 3$. **b** (right) The circle from **a** was enlarged three times and shifted along the x-axis; in the diagram, all four circles are represented as one data series

	A	B	C	D	E	F	G	H	I	J	K	L	M	N	O
1	**r.0**	4.00		**c.s**	4.00										
2	**x.0**	3.00		**v.F**	3.00			*Circle*			*Circles*			*Centers*	
3							=G5+dphi	=r.0*COS(phi)+x.0	=r.0*SIN(phi)						
4		30 °					**phi**	**x**	**y**		**x**	**y**		**xC**	**yC**
5	**dphi**	0.524					0.00	7.00	0.00		7.00	0.00		3	0
6							**0.52**	6.46	2.00		6.46	2.00		6	0
17							6.28	7.00	0.00		7.00	0.00			

Fig. 4.43 (S) Spreadsheet calculation for Fig. 4.42. The variable c_s contains the factors with which the radius of the circle is increased. The variable v_F is the velocity with which the circle center is shifted on the *x*-axis

21. What are the radii and the coordinates of the circles' center points in Fig. 4.42b?
22. The coordinates of the circle are to be calculated four times by changing the parameters in the table and then to be stored successively in columns Q and R (column indices 17 and 18, respectively) of the spreadsheet and graphically displayed in a diagram as one data series (to yield Fig. 4.42b). The centers' coordinates are in columns N and O (column indices 14 and 15, respectively). The circle radius should increase with $r_0 = v_S \cdot t$, and the center point should be shifted on the x-axis with $x_0 = v_F \cdot t$. Write such a log procedure in Visual Basic!
23. For the answer of Question 23 in `Python`, create empty lists Q and R and extend them with the `.append` method in a nested loop! Instead of an empty line, insert "`None`"! Finally, plot (Q, R) with a correct label and observe whether four separated circles show up!

Formula-generating routine

In Fig. 4.44b, you see a parabola connected to the horizontal axis by vertical lines. The spreadsheet calculation for the coordinates is shown in Fig. 4.44a (S).

24. The y-value of the parabola is calculated in the usual way with named cell ranges (see the formula in B5). Every tenth point of the parabola is connected to the horizontal axis with a vertical line ("Dashes"). Write a routine that writes the formulas for the vertical lines' coordinates into columns D and E!
25. Create a corresponding `Python` program with two variants. (a) The vertical lines are plotted one after the other in a loop. (b) The coordinates of all vertical lines are stored in lists named D and E separated by empty cells, and plotted as one data series.

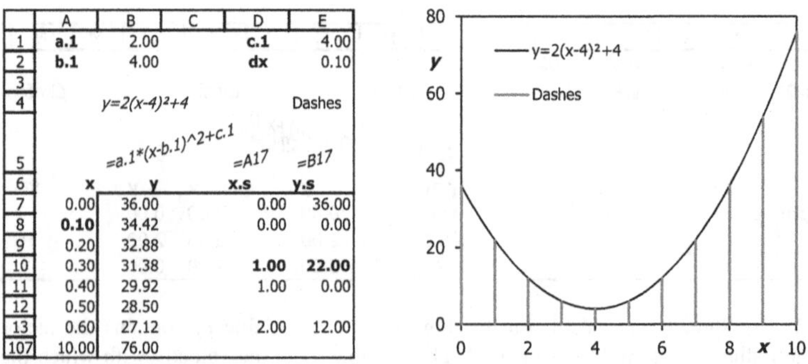

	A	B	C	D	E
1	a.1	2.00		c.1	4.00
2	b.1	4.00		dx	0.10
3					
4		y=2(x-4)²+4			Dashes
5		=a.1*(x-b.1)^2+c.1		=A17	=B17
6	x	y		x.s	y.s
7	0.00	36.00		0.00	36.00
8	0.10	34.42		0.00	0.00
9	0.20	32.88			
10	0.30	31.38		1.00	22.00
11	0.40	29.92		1.00	0.00
12	0.50	28.50			
13	0.60	27.12		2.00	12.00
107	10.00	76.00			

Fig. 4.44 **a** (left, S) Polynomial $y = a_1(x - b_1)^2 + c_1$ in columns A and B; coordinates x_s, y_s for the vertical lines of Fig. b in columns D and E. **b** (right) Display of the data from **a**

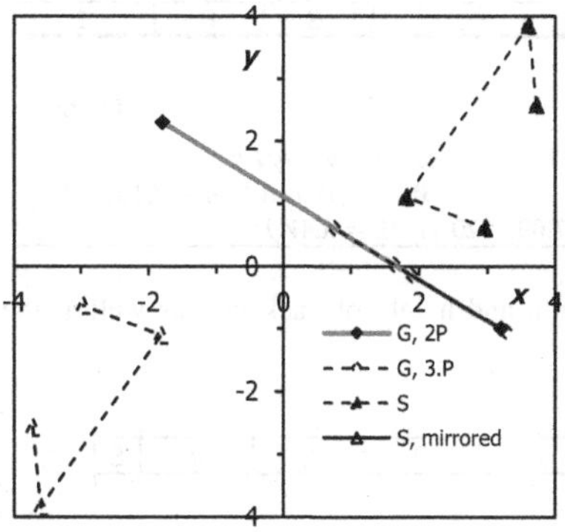

	A	B	C	D	E	F	G	H	I	J
1			G, 2P							
2		x	y							
3	1	-1.8	2.3							
4	2	3.20	-1			4				
5										
6			G, 3.P		S			S, mirrored		
7		=-4+8*RAND()	=Straight($B$3;$C$3;$B$4;$C$4;B10)		=RAND()*$F$4	=RAND()*$F$4		={Mirr(E10;F10)}		
8		x	y		xS	yS		xSm	ySm	
9		1.89	-0.14		3.72	2.58		-3.72	-2.58	
10		3.27	**-1.05**		3.60	3.85		**-3.60**	**-3.85**	
11		1.63	0.03		1.81	1.11		-1.81	-1.11	
12		0.76	0.61		2.97	0.62		-2.97	-0.62	

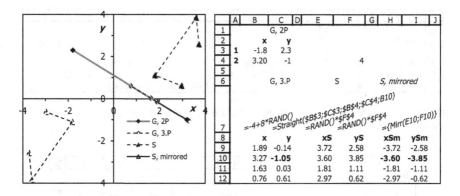

Fig. 4.45 **a** (left) Two points "G, 2P" define a straight line (18a); the points in the third quadrant are mirrored at the zero point into the first quadrant (18b). **b** (right, S) Spreadsheet calculation for **a**; the coordinates for the points on the straight line are in columns B and C; the coordinates in columns H and I are the mirror images of the coordinates in columns E and F

User-defined spreadsheet functions

In Fig. 4.45a, the results of two user-defined spreadsheet functions are displayed. One function adds additional points, "G, 3.P", onto a straight line "G, 2P" defined by two points. The other function mirrors points in the first quadrant at (0, 0) into the third quadrant. The coordinates of the points in Fig. 4.45a are calculated in the spreadsheet of Fig. 4.45b (S).

26. What are the names of the two user-defined functions reported in line 7 in Fig. 4.45b (S)?
27. Write a function (VISUAL BASIC or `Python`) of the type $y_3 = f(x_1, y_1, \ldots)$ that calculates the y-value y_3 of a third point from the coordinates of the two defining points of a straight line and the x-value x_3 of the third point!
28. Write a function (VISUAL BASIC or `Python`) of the type $(x_{sp}, y_{sp}) = f(x,y)$ that mirrors the coordinates x and y of a point at the origin of the coordinate system!

Macros

29. You want to trigger a macro whenever a slider named SCROLLBAR1 is changed. What is the name of the associated macro?[51]
30. At a mail-order company, some data from all outgoing packets are entered in the spreadsheet of Fig. 4.46 (S). Write a protocol routine (Visual Basic or `Python`) that reads some data from Fig. 4.46 (S) and enters it into a table as in Fig. 4.47 (S)! The packets have to be numbered consecutively.

[51] SUB SCROLLBAR1_CHANGE().

	A	B	C	D	E	F	G	H	I
1		*Name*	Mary B.						
2	*Running number*		46						Protoc
3									
4	Width (cm)	**w**	5	Volume (l)	**V**	1.80	*=w*l*h/1000*		
5	Length (cm)	**l**	18	Surface (m²)	**S**	0.11	*=(w*l+w*h+l*h)*2/10^4*		
6	Hight(cm)	**h**	20	Time		07.09.2020 17:21	*=NOW()*		

Fig. 4.46 (S) Table section in which the width, length, and height of packages, as well as the sender's name, are to be entered

	K	L	M	N	O	P	Q	R	S
6	12	*next free row*							
7	***Number***	***Name***	***Time***	***Width/cm***	***Length/cm***	***Height/cm***	***Volume/l***	***Surface/m²***	
8	43	Otto L.	7.9.20 17:03	17	17	14	4.05	0.15	
9	44	James L.	7.9.20 17:16	20	10	10	2.00	0.10	
10	45	Henry M.	7.9.20 17:16	18	28	8	4.03	0.17	
11	46	Mary B.	7.9.20 17:21	5	18	20	1.80	0.11	

Fig. 4.47 (S) The data from Fig. 4.46 (S) are to be reorganized in this way

Basic Mathematical Techniques

5

With the methods learned in Chaps. 2–4 (list processing, programming constructs for, if, sub/def), we practice differentiation, integration, and calculating with vectors. We get to know a new technique: solving systems of linear equations with matrix calculation.

5.1 Introduction: Calculus, Vectors, and Linear Algebra

Solutions of Exercises 5.3 (Excel), 5.5 (Python), 5.6 (Python), and 5.7 (Excel) can be found at the internet address: go.sn.pub/VYYbJL.

Straight-line segment
Straight-line segments are central for vector calculation, and also for calculus, because we approximate all curves by sequences of such elements. In the introductory Exercise 5.2, we calculate vector entities related to a straight-line segment, e.g., line vector and mid-perpendicular, as well as length, slope, and area enclosed with the x-axis.

Differentiation and integration
We get to know simple techniques with which the first and second derivatives of a function can be obtained (Exercise 5.3) and with which a function can be integrated. With integration, the area between the curve of a function and the x-axis (Sect. 5.4.1) and the length of a curve (Sect. 5.4.2) can be calculated. We pay special attention to the x-value over which the results must be plotted: at the beginning, in the middle, or at the end of the interval for which the elementary calculation was done.

D. Mergel, *Physics with Excel and Python*,
https://doi.org/10.1007/978-3-030-82325-2_5

Vectors in the plane
We discuss the addition and the scalar product of two vectors and convert planar Cartesian and planar polar coordinates one into the other (Exercise 5.5). In Exercise 5.6, the tangents to and the perpendiculars on a polynomial are drawn. As a physically relevant example, we calculate the forces acting on a car running through a banked curve (Exercise 5.7). In Exercise 5.8, we calculate a mobile's equilibrium by using the mathematical construction of a weighted average.

Systems of linear equations
Systems of linear equations (with small rank) are solved by bringing them into matrix form and forming the inverse matrix of the coefficient matrix (Exercise 5.9). With this method, we get the coefficients of a polynomial from a given set of points on the curve and determine currents in electrical networks with Kirchhoff's rules. In `Python`, we also use functions from the `numpy.linalg` library for these tasks.

Mathematical functions
In Sect. 5.10, some useful mathematical functions are listed, both in EXCEL and `numpy` notation, together with short descriptions.

5.2 Straight-Line Segment Under a Magnifying Glass

A straight-line segment is specified by its two endpoints. Its center, line vector, and perpendicular vector are obtained with vector operations. The basic pieces of calculus, d*x*, d*y*, d*A* (the area under the curve), are calculated.

5.2.1 Under a Magnifying Glass

The functions we are investigating in this textbook are all continuous. They are approximated as polylines, i.e., as sequences of straight-line segments. Therefore, all operations of calculus, such as differentiation, integration, and integration along a line, are based on such properties of segments as length, slope, line vector, and mid-perpendicular.

In this exercise, we specify a straight segment in a plane by its two endpoints $\underline{A}$, $\underline{B}$ formulated as positional vectors. The unit line vector $\underline{AB}$ of the segment and its perpendicular $\underline{AB\_p}$ to be erected at the midpoint $\underline{AB\_C}$ are calculated with matrix operations. An example is shown in Fig. 5.1.

The equation of the line passing along the segment is

$$y = y_A + m \cdot (x - x_A) \tag{5.1}$$

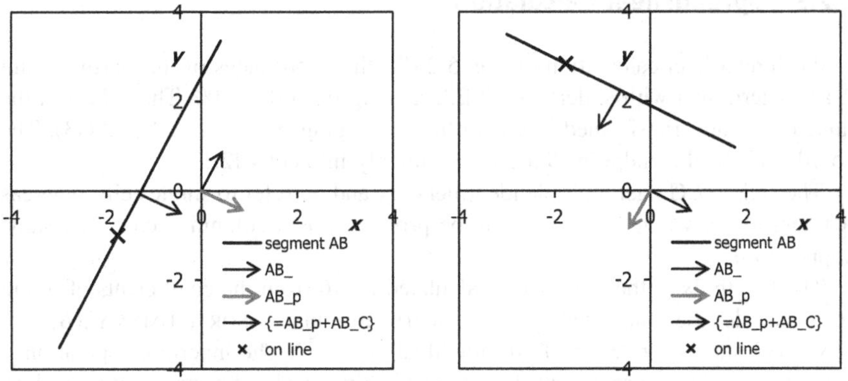

Fig. 5.1 Line vector and mid-perpendicular of a straight-line segment. The cross is drawn with a function based on the coordinates of the two endpoints, **a** (left) with the settings from Fig. 5.2 (S), **b** (right) other settings

with m being the slope of the segment and (x_A, y_A) the coordinates of point $\underline{A}$. In Fig. 5.1a, the point (y_s, x_s) on this line for $x_s = -1.76$ is represented by a cross. We use vector calculation, e.g., $\underline{\underline{P}} \cdot \underline{AB}$ to obtain the unit vector perpendicular to the segment with $\underline{\underline{P}}$, the 90° rotation matrix, and $\underline{AB}$, the unit line vector.

5.2.2 Data Structure and Nomenclature

A, B	the two endpoints of a straight segment in the plane
x_A, y_A	coordinates of point A
x_B, y_B	coordinates of point B
$dx = (x_B - x_A)$	their distance in the x-direction
$dy = (y_B - y_A)$	their distance in the y-direction
$ds = \sqrt{dx^2 + dy^2}$	length of the segment
m	slope dy/dx
$y = y_A + m \cdot (x - x_A)$	function describing the line through the segment
$\Delta A = (y_A + y_B)/2 \cdot dx$	area between the segment and the x-axis
$\underline{A} = (x_A, y_A)$	vector representation of point A
$\underline{B} = (x_B, y_B)$	vector representation of point B
$\underline{AB\_C} = (\underline{A} + \underline{B})/2$	center of the segment
$\underline{AB} = (\underline{B} - \underline{A})/ds$	unit line vector
$\underline{\underline{P}} = \begin{bmatrix} 0 & -1 \\ 1 & 0 \end{bmatrix}$	90°- rotation matrix.
$\underline{AB\_p\_} = \underline{\underline{P}} \cdot \underline{AB}$	unit vector perpendicular to the segment.

5.2.3 Spreadsheet Calculation

In the spreadsheet calculation of Fig. 5.2 (S), the coordinates of the starting point $\underline{A}$ are determined with sliders in A2:B3, ranging from 0 to 100. The values in the linked cells are transformed into coordinates ranging from −4 to 4 (G2:G3). The coordinates of the endpoint $\underline{B}$ are typed directly into cells J2:J3.

The cells G2:G3 get multiple identifiers: x_A and y_A refer to single cells, whereas $\underline{A}$ refers to the whole range and can be processed as a column vector. The same applies to $\underline{B}$.

The length ds of the segment is calculated in A6 from the coefficients of $\underline{A}$ and $\underline{B}$. It can also be calculated with the matrix formula {=SQRT(SUMXMY2($\underline{B}$;$\underline{A}$))}. SUMXMY2 stands for "Sum of all individual $(x - y)^2$". The innermost operation $\underline{A}$ – $\underline{B}$ involves two matrices, but the output is only a scalar. So, we have to enclose the formula in curly brackets and finish with the magic chord Ψ *Ctl* + *Shift* + *Return*.

The result of this calculation is shown in Fig. 5.1a. The perpendicular unit vector $\underline{AB\ p}$ is also drawn from the center $\underline{AB\_C}$ of the line to $\underline{AB\_C + AB\_p}$ (J6:J7).

In Fig. 5.3 (S), we calculate the primary segments of calculus dx, dy, and the area dA between the segment and the x-axis. Furthermore, we set up the equation

	A	B	C	D	E	F	G	H	I	J	K
1							A_			B_	
2	[slider]			55.00		**xA**	0.40	=(D2-50)/12.5	**xB**	-3.00	
3	[slider]			92.00		**yA**	3.36	=(D3-50)/12.5	**yB**	-3.50	
4	=SQRT((yA-yB)^2+(xB-xA)^2)			{=(A_-B_)/length_AB}			=MMULT(P_;AB_)		{=(A_+B_)/2}	{=AB_p+AB_C}	
5	**ds**			**AB_**			**AB_p**		**AB_C**		
6	7.66		0.00	0.44		0.00	0.90	=D7	-1.30	-0.40	
7			0.00	0.90		0.00	-0.44	=-D6	-0.07	-0.51	
8											

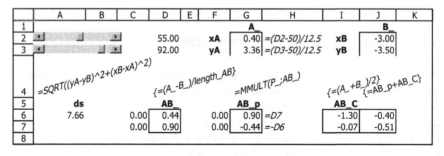

Fig. 5.2 (S) Coordinates of points $\underline{A}$ and $\underline{B}$, line vector $\underline{AB}$ and center position vector $\underline{AB\_C}$; the perpendicular vector $\underline{AB\_p}$ is obtained from $\underline{AB}$ through matrix multiplication with $\underline{P}$, the 90° rotational matrix $\underline{P}$ presented in Fig. 5.3 (S)

	L	M	N	O	P	Q	R	S	T
5	area	**dA**	0.24	=(yA+yB)/2*dx		**P_**			
6		**dx**	-3.40	=xB-xA		0	-1		
7		**dy**	-6.86	=yB-yA		1	0		
8	slope	**m**	2.02	=dy/dx					
9									
10	on line	**x**	-1.76	28	[slider]				
11		**y**	-1.00	=yA+m*(x-xA)					

Fig. 5.3 (S) Continuation of Fig. 5.2 (S). Characteristics of a straight-line segment important for calculus, and the equation for the line running along the segment

$y = y_A + m\cdot(x - x_A)$ of the line along the segment. The value of x is determined with a slider so that the cross in Fig. 5.1 runs along this line.

5.2.4 Plotting Vectors with Python Matplotlib

To plot vectors with `Python`, we have to use the function `arrow` of the library pyplot,to be imported with import `matplotlib.pyplot` as `plt`. We integrate this function into a user-defined function *ArrowP* with standard formatting parameters (see Table 5.1).

In EXCEL, arrowheads are a design feature of a line: … LINE/END ARROW TYPE.

The constructor arguments of `plt.arrow` comprise, among others, the keyword arguments:

`width`	float (default: 0.001) width of full arrow tail
`fill`	bool
`linestyle` or `ls`	{'-', '–', '-.', ':', '', (offset, on–off-seq), …}
`linewidth` or `lw`	float or None
`head_length`	float or None (default: 1.5 * `head_width`) length of arrow head
`overhang`	float (default: 0, triangular) fraction that the arrow is swept back

Some arrows are plotted in Fig. 5.4 to demonstrate the effect of the constructors. The arrow pointing upwards is drawn with our standard function *ArrowP* (overhang = 1) in Table 5.1.

5.3 Differentiation

We learn how to approximate the first and second derivatives of a function *f(x)* numerically with difference quotients between neighboring calculation

Table 5.1 User-defined function *ArrowP* for drawing an arrow from point P_0 to point P_1 in a plane; the argument list of `plt.arrow` does not contain all possible keyword arguments (similar to Tables 5.3, 5.4, 5.5 and 5.6)

```
def ArrowP(P0,P1,c="k",ls='-',lw=1,hw=0.2):
    (x0,y0)=P0 #Decomposes the foot position vector
    (x1,y1)=P1 #Decomposes the tip position vector
    print(lw,hw)
    #c has to be given as c="k", not c='k'
    plt.arrow(x0,y0,x1-x0,y1-y0,
              length_includes_head=True,
              head_width=hw,overhang=1,fill=False,
              linestyle=ls, color=c,  linewidth=lw)
```

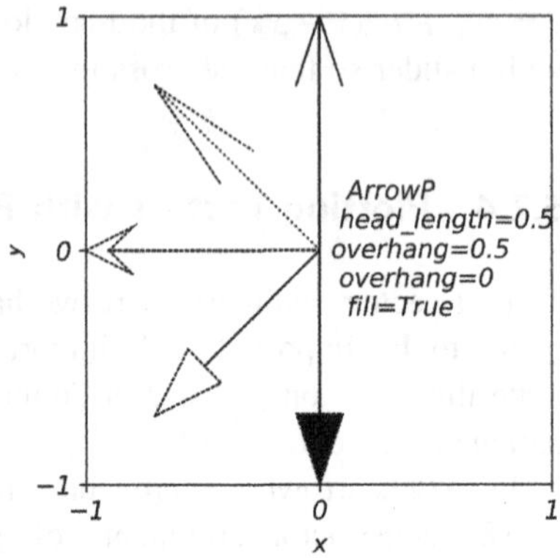

Fig. 5.4 Arrows plotted with different constructor arguments in the procedure head of `plt.arrow`

points. The first derivative must be plotted over the middle between two grid points, the second derivative over the central of three grid points.

5.3.1 First and Second Derivative

The first derivative of a curve is the slope of its tangent to the curve at the specified *x*-value. The derivative of a function *f(x)* with respect to *x* is defined as

$$\frac{df(x)}{dx} = \lim_{\Delta x \to 0} \frac{f(x + \Delta x) - f(x)}{\Delta x} \tag{5.2}$$

More specifically, Eq. 5.2 is called the right derivative.

Such an approach to the limit cannot be carried out for discrete functions; they are specified as a list of *x*-values ($x = [\ldots x_i \ldots]$) and *y*-values ($y = [\ldots y_i \ldots]$). Instead, we calculate the difference quotient of neighboring grid points i and $i + 1$:

$$\frac{df\left(\frac{x_i + x_{i+1}}{2}\right)}{dx} \approx \frac{f(x_{i+1}) - f(x_i)}{x_{i+1} - x_i} = \frac{y_{i+1} - y_i}{x_{i+1} - x_i} \tag{5.3}$$

For our discrete functions, the first derivative in the center of a segment is approximated by $\Delta y/\Delta x$, the slope of the segment between adjacent points. The values are to be plotted over the center of the interval. It is evident that the smaller Δx is, the better the accuracy of the approximation.

The second derivative of a function is the derivative of the first derivative, thus,

$$\frac{d^2 f(x)}{dx^2} = \frac{d}{dx}\left(\frac{d}{dx} f(x)\right) \tag{5.4}$$

which is equivalent to applying Eq. 5.2 twice. The second derivative describes the change of the slope, and is thus a measure of the curvature of the curve. The difference equation for grid points x_i at equal intervals Δx is

$$\begin{aligned}\frac{d^2 f(x)}{dx^2} &\approx \frac{1}{\Delta x}\left(\frac{f(x_{i+1}) - f(x_i))}{\Delta x} - \frac{f(x_i) - f(x_{i-1})}{\Delta x}\right) \\ &= \frac{f(x_{i+1}) - 2f(x_i) + f(x_{i-1})}{\Delta x^2}\end{aligned} \tag{5.5}$$

The second derivative at point x can be calculated directly without a detour via the first derivative with function values at the grid points x, $x - dx$ and $x + dx$. It must be plotted over point x (x_i in Eq. 5.5), the coordinate of the middle grid point.

Sine function

To give an example, we differentiate the sine function, knowing beforehand that its first and second derivatives are the cosine and the negative sine, respectively. So, we can check whether our numerical calculations reproduce this result. This is indeed confirmed in Fig. 5.5, with 100 calculation points in one period 2π. The shape of a cosine is clearly visible in the numerically calculated y_{1d} in Fig. 5.5a, and that of a negative sine in y_{2d} in Fig. 5.5b.

Oscillation of a mass-spring-system

Consider the oscillation $z(t)$ of a mass-spring system. The second derivative $\ddot{z}$ with respect to time is the acceleration a, which, in turn, is proportional to the restoring

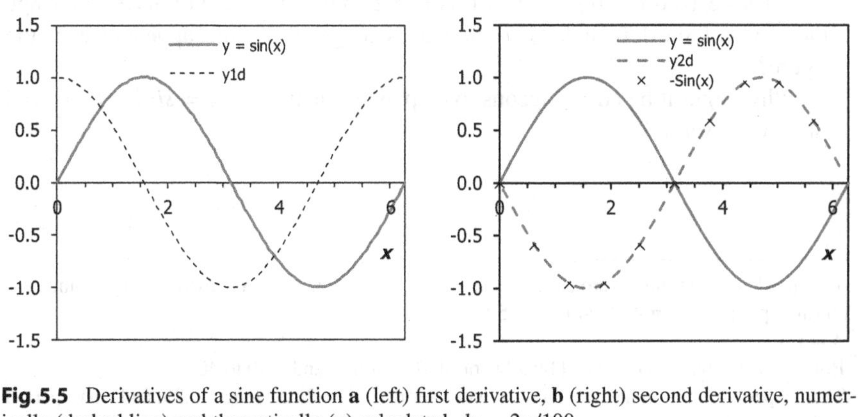

Fig. 5.5 Derivatives of a sine function **a** (left) first derivative, **b** (right) second derivative, numerically (dashed line) and theoretically (x) calculated, $dx = 2\pi/100$

force $-kz$ of the spring. The curvature of the displacement is thus proportional to the force. This leads to the simple equation of a harmonic oscillator based on Newton's law:

$$F = -k \cdot z = m \cdot a = m \cdot \ddot{z}$$

or

$$\ddot{z} = -\left(\frac{k}{m}\right) \cdot z \tag{5.6}$$

The curvature of the displacement is thus proportional but opposed to the force. A sine function is a solution to this differential equation, as can be seen in Fig. 5.5b.

Question

Let $y = \sin(x)$ in Fig. 5.5 be the displacement of an oscillator. At which positions x is the speed of the oscillator maximum, and at which positions is it zero?[1]

Composite function

We are now considering a composite function, i.e., for which the argument is not just an independent variable, but also a function. For the numerical derivative, there is no difference from a simple function. We build the two arrays x and $y = sin(f(x))$ and proceed with $\Delta y/\Delta x$.

In Fig. 5.6a, the function $y = A \cdot \sin(kx)$ with $A = 0.1$ and $k = 2\pi$ is differentiated first once and then twice. The results are again the cosine and the negative sine, however, with different amplitudes, $0.1 \cdot 2\pi$ and $-0.1 \cdot (2\pi)^2$, respectively.

Questions

What is the amplitude A of the curve $A \cdot sin(kx)$ in Fig. 5.6a?[2]

Let A be 10 times bigger than it is in Fig. 5.6a. How do you have to change the scale of the left and the right y-axes to get the same appearance in the figure?[3]

Why might it be advantageous to display a function $f(t) = sin(2\pi t)$ instead of $f(x) = sin(x)$?[4]

---

[1] The speed is maximum at zero crossings ($x = 0, \pi, 2\pi$ in Fig. 5.5, at the extrema in y_{1d}) and zero at turning points ($x = \pi/4, 3\pi/4$ in Fig. 5.5).

[2] $A = 0.1$.

[3] Both axes also have to be scaled by a factor of 10: -4 to 4 and -40 to 40.

[4] Then, a period duration has the length 1, a quarter period ($\pi/2$) the length 0.25, and a half period (π) the length 0.5, thus always at simple rational numbers that are clearly visible on the x-axis.

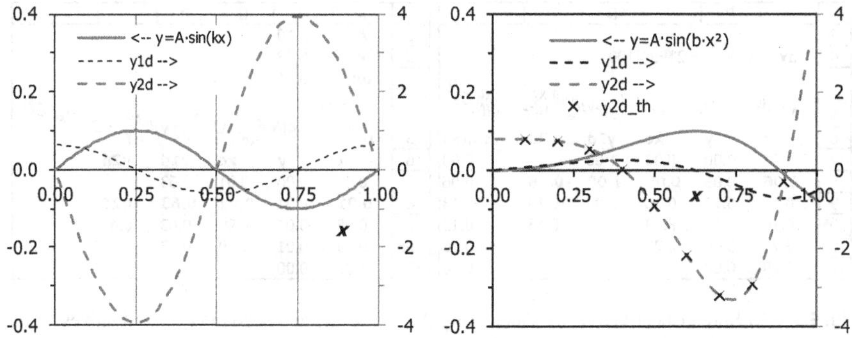

Fig. 5.6 **a** (left) $y=A \cdot sin(2\pi x)$ and its first and second derivative. **b** (right) $y = A \cdot sin(bx^2)$ with $A = 0.1$ and $b = 4$. its first and second numerical derivative, together with the analytical second derivative Eq. 5.7. For the derivatives, the right y-axis is valid.

Why might it be advantageous to divide the values of $f''(t)$, the second derivative of $sin(2\pi t)$, by $4\pi^2$?[5]

In Eq. 5.7, we have chosen a parabola as an argument for the sine:

$$y = \mathrm{A} \cdot \sin\left(bx^2\right) \tag{5.7}$$

The function $y = A \cdot \sin\left(bx^2\right)$ and its numerically obtained derivatives are displayed in Fig. 5.6b with continuous lines.

Analytical derivatives with the chain rule

The derivatives of a composite function $y = f(z), z = g(x)$ are obtained with the chain rule:

$$\frac{dy}{dx} = \frac{df}{dz} \cdot \frac{dg}{dx} \tag{5.8}$$

For $y = \sin(2\pi x)$, we get

$$y' = -\cos(2\pi x) \cdot 2\pi \text{ and } y'' = -\sin(2\pi x) \cdot (2\pi)^2 \tag{5.9}$$

and for $y = \sin\left(b \cdot x^2\right)$:

$$\begin{aligned} y' &= -\cos\left(bx^2\right) \cdot 2bx \\ y'' &= -\sin\left(bx^2\right) \cdot (2bx)^2 + \cos\left(bx^2\right) \cdot 2b \end{aligned} \tag{5.10}$$

[5] Then, we would expect an amplitude -1, which is easier to verify in the diagram.

	A	B	C	D	E	F
3						
4	**dx**	0.0628	=2*PI()/100			
5	=A7+dx	=SIN(x)	=(x+A9)/2	=(B9-y)/(A9-x)	=(y1d-D7)/(xc-C7)	=-SIN(x)
6	**x**	**y**	**xc**	**y1d**	**y2d**	**-sin(x)**
7	0.00	0.00	0.03	1.00		0.00
8	**0.06**	0.06	**0.09**	**1.00**	**-0.06**	-0.06
9	0.13	0.13	0.16	0.99	-0.13	-0.13
105	6.16	-0.13	6.19	1.00	0.13	0.13
106	6.22	-0.06	6.25	1.00		0.06
107	6.28	0.00				0.00

	A	B	C	D	E	F
2	**A**	0.10				
3	**k**	6.28				
4	**dx**	0.01				
5	=A7+dx	=A*SIN(k*x)	=(x+A9)/2	=(B9-y)/(A9-x)	=(y1d-D7)/(xc-C7)	
6	**x**	**y**	**xc**	**y1d**	**y2d**	
7	0.00	0.00	0.01	0.63		
8	**0.01**	0.01	**0.02**	**0.63**	**-0.25**	
105	0.98	-0.01	0.99	0.63	0.49	
106	0.99	-0.01	1.00	0.63		
107	1.00	0.00				

Fig. 5.7 (S) Spreadsheet layout for calculating the first (y_{1d}) and second (y_{2d}) derivatives of the sine function **a** (left) for *sin(x)* and **b** (right) for *sin(kx)*

Some points of y'' are also displayed in Fig. 5.6b with crosses. They lie on the curve obtained with numerical differentiation, indicating that our procedure is correct, especially our decision to plot the result over the middle of the three points from which the second derivative is obtained.

5.3.2 Data Structure and Nomenclature

x, y arrays specifying the points of the curve, x values dx apart,
x_C array with the x-values of the center of the segments,
y_{1d} array containing the values of the first derivative,
y_{2d} array containing the values of the second derivative.

5.3.3 Spreadsheet Layout

A spreadsheet layout for the differentiation of the sine function is given in Fig. 5.7 (S).

Questions

concerning Fig. 5.7 (S):

Interpret the formula in C5, valid for C8![6]

What are the lengths of the arrays x, x_C, y_{1d}, and y_{2d}?[7] Also, compare with Table 5.2!

[6] The spreadsheet formula in C5 calculates the x-value of the centre of the interval.

[7] len(x) = R107 – R6 = 101, len(xC) = 100, len(y1d) = 100, len(y2d) = 99 (see also Table 5.2).

Table 5.2 **a** (top) Function *deri* for determining the derivative of $y = f(x)$; **b** (bottom left) First and second derivatives of $y = sin(2\pi x)$; **c** (bottom right) Lengths of the arrays *x, y1d,* and *y2d*

```
def deri(x,y):
    dx=x[1:]-x[:-1]
    dy=y[1:]-y[:-1]
    xC=(x[1:]+x[:-1])/2          #x Center of segment
    return xC,dy/dx
```

```
x=np.linspace(0,1,101,
endpoint=True)
y=np.sin(2*np.pi*x)
x1,y1d=deri(x,y)
x2,y2d=deri(x1,y1d)
```

len(x)	101
len(y1d)	100
len(x1)	100
len(y2d)	99
len(x2)	99

The formula in D8 is D8 = [=(*B9-y*)/(A9-*x*]. Which value does EXCEL take for *x* and *y*?[8]

How do you get a selection of –sin(*x*) in column F of 10 points, as presented in Fig. 5.5b?[9]

We plot (x_c, y_{1d}). Show that the formulas in E5 (valid for E8) is correct![10]

5.3.4 Python Program

In the Python program corresponding to Fig. 5.7 (S), the derivatives are built with a function *deri*, shown in Table 5.2a, requiring *x* and *y* as input and returning x_C, the center of the intervals, and $\Delta y/\Delta x$. The Python program in Table 5.2b treats $y = \sin(2\pi x)$.

The quantities d*x*, d*y*, and x_C are obtained by slicing the arrays *x* and *y*. Their length is, by construction, one less than that of *x* and *y,* as can be verified in Table 5.2c. We no longer have to think about the values of *x* for which the derivative is a good approximation, because this is already done within the function and returned as x_C. For the second derivative, we apply *deri* twice.

Composite function

In Table 5.3, we are considering the composite function $y = \mathrm{A}\cdot\sin(bx^2)$ (Eq. 5.10).

The results for the 10 *x*-values specified in line 7 are shown in Fig. 5.6b. The numerically calculated values (marked –) and the theoretical ones (marked with x)

[8] For *x* and *y* in C8 and D8, the values in the same (8th) row are taken from the column ranges with names *x* (A8) and *y* (B8).

[9] With a VBA procedure comprising a loop of the type: FOR R = 7 TO 107, …, R2 = R2 + 1.

[10] Row *n*-1: x-dx, xc = x-dx/2, y1d.

Row *n*: x, xc = x + dx/2,y1d,y2d.

To get y_{2d} at the position *x*, we have to calculate the difference quotient for *y* at *x* + dx/2 and *x*-dx/2.

Table 5.3 First and second derivatives of $A \cdot \sin(bx^2)$

```
x=np.linspace(0,1,101,endpoint=False)
A=0.1
b=4
yf=A*np.sin(b*x**2)
x1,yf1d=deri(x,yf)
x2,yf2d=deri(x1,yf1d)
xx=np.linspace(0,1,10)
yf2d_th=(A*np.cos(b*xx**2)*2*b
        -A*np.sin(b*xx**2)*(2*b*xx)**2)
        #Theoretical second derivative
```

of the second derivative coincide, indicating that our simple numerical recipe yields sufficiently accurate derivatives.

Question

Over which arguments, type x or x_c, must the function y_{f2d-th} in Table 5.3 be plotted so that it correctly represents the second derivative of the function y?[11]

5.4 Integration

We determine the area under a sine curve and the length of a polynomial.

5.4.1 Area Under a Curve

Integral function, definite integral

The integral of a function $f(x)$ between x_1 and x_2 corresponds, in the simple cases we are dealing with, to the area limited by the curve, the x-axis, and two vertical boundaries. It can be positive or negative. The area under a curve between two adjacent interpolation points is calculated using the trapezoid rule visualized in Fig. 5.8a. The area of a trapezoid of width Δx is

$$F_{\Delta x}(x) = \frac{f(x - \Delta x) + f(x)}{2} \cdot \Delta x \tag{5.11}$$

The integral from x_1 to x_2 is the sum of all trapezoids in that region. It must be represented in the diagram over x_2, the end of the integration interval.

[11] The function y_{f2d-th} has to be plotted over xx, the equivalent of x, because it is the analytically determined second derivative of $A \cdot sin(bx^2)$ at position xx.

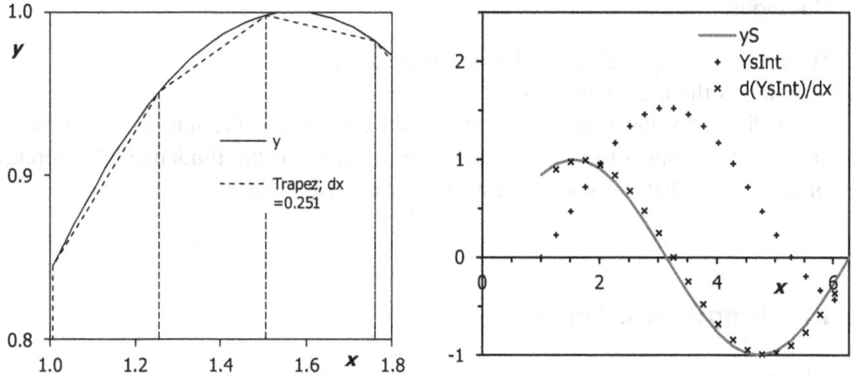

Fig. 5.8 **a** (left) Numerical calculation of an integral with the trapezoidal rule, dx = 0.251. **b** (right) Integral of the sine function and differentiation of the integral displayed over x (column A in Fig. 5.11 (S))

Figure 5.9 shows the integral function (a) of a sine function and (b) of a polynomial. The integral function of a *sine* is a *cosine*. The integral in Fig. 5.9a is theoretically $F(x) = (1 - \cos(x))$. The integral function of a polynomial of the nth order is a polynomial of the $(n + 1)$th order.

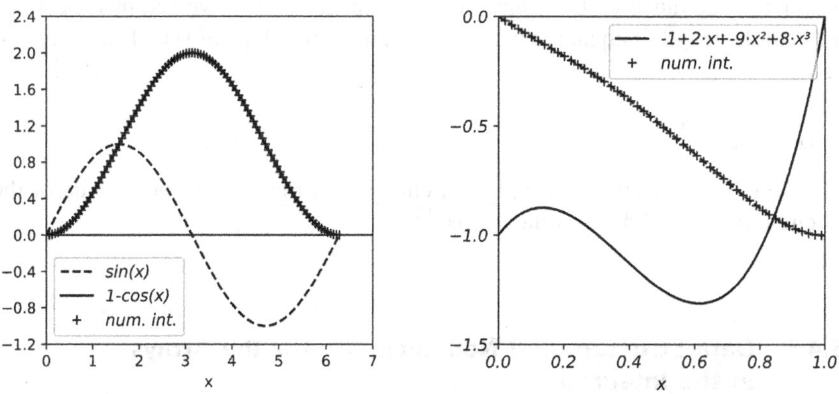

Fig. 5.9 Curves obtained with the `Python` program in Table 5.4. **a** (left) Integration of the sine function. **b** (right) Integration of a 3rd order polynomial (lower curve) (here, negative values of the integral)

Questions

What is the analytical integral function of a *sine*?[12]

What is the area under a *sine* arc?[13]

In Fig. 5.8b, the numerically obtained derivative of the integral function is shifted with respect to the original curve, contrary to the fundamental theorem of calculus. What, do you suspect, is the reason for this?[14]

5.4.2 Length of a Curve

The length Δs of a straight segment is easily calculated as

$$\Delta s = \sqrt{(\Delta x^2 + \Delta y^2)} \tag{5.12}$$

$$\mathrm{d}s = \sqrt{1 + \left(\frac{dy}{dx}\right)^2} \cdot \mathrm{d}x \tag{5.13}$$

The length of a curve may be approximated by the length of the sequence of segments used to approximate it, i.e., the sum of the lengths of all segments.

How is the function "length of a curve" correlated with the derivative of that curve? Looking at Fig. 5.10a, we may state that the slope of the length is always positive; and the bigger the *absolute value* of the derivative of the function, the bigger the derivative of the length. Looking at Eq. (5.13), we see that the slope $\mathrm{d}s/\mathrm{d}x$ of the length is equal to the absolute value of the slope of $y(x)$ if $(\mathrm{d}y/\mathrm{d}x)^2 \gg 1$.

Question

Where does the curve "Length of a circle" as a function of the y value of the circle in Fig. 5.10b cross the x-axis?[15]

5.4.3 Data Structure and Nomenclature for the Arrays in the Integration

$\mathrm{d}x$ horizontal distance between the vertices

[12] $\int \sin(x)dx = \cos(x)$.

[13] The area under a *sine* arc is 2, as can be seen form the value of the integral function at $x = \pi$ in Fig. 5.9a, based on the theorem $\int_a^b f(x)dx = F(b) - F(a)$.

[14] The derivative is wrongly plotted over the end of the intervals, not correctly over their center.

[15] The "Length of a circle" crosses the x-axis at $x = 0$ (for $y = 0$), $x = \pi$ (for $y = 0$ after having gone through a half circle, and $x = 2\pi$ (for $y = 0$, after having gone through a full circle).

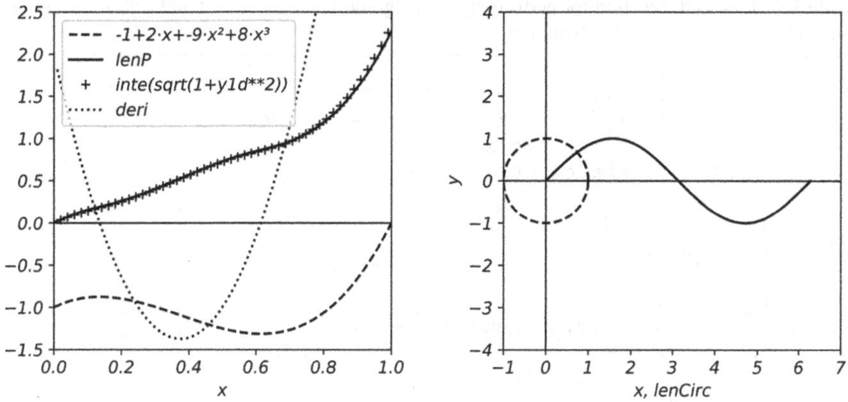

Fig. 5.10 **a** (left) Length of a polynomial, calculated with Eq. (5.12); with `def lenCurve` and a second time as *len2* in line 12 in Table 5.5. **b** (right) Length of a circle (reported on the *x* axis) as a function of the *y*-value of the circle; attention: the horizontal axis represents the dependent variable

x_P	*x* coordinates of a polynomial
y_P	*y* coordinates of the polynomial
$Y_{P_{Int}}$	integral function of the polynomial
len_P	length of y_P
x	*x* coordinates of a sine and a cosine
y_S	*y* coordinates of a sine function
y_c	*y* coordinates of a cosine function
$Y_{S_{Int}}$	integral function of the sine function
len_{Circ}	length of the circle defined by polar coordinates: $x = y_C$ and $y = y_S$.

5.4.4 Python Program

Area under a curve

A `Python` function, `def` *inte*, for performing the integration of a discretized function $y = f(x)$ with the trapezoid approximation is shown in Table 5.4. The returned array *integ* has the same length as *x*. Its first element is zero, because that was introduced by `np.zeros(len(x))`, and the first element is not overwritten. With the `Python` program in the next cells, we perform integrations of a *sine* function (middle cell) and a 3rd order polynomial (bottom cell). The results are displayed in Fig. 5.9.

Length of a polynomial

The approximation of Eq. 5.12 is implemented in the Python function `def` *lenCurve(x,y)* in Table 5.5, returning the cumulated sum of the individual elements d*s* (`np.cumsum(ds)`) as a function of *x*.

In the second cell of Table 5.5, the length of the polynomial $y_P(x)$ of Table 5.4 is calculated, to be displayed in Fig. 5.10a. First, with *lenP* = *lenCurve(x,yP)* and then (line 12) as *len2,* applying Eq. (5.13).

Table 5.4 First cell: function for performing an integration of $y = f(x)$; second cell: $f(x) = sin(x$; third cell: $f(x)$ is a 3rd order polynomial

```
1  def inte(x,y):
2      dx=x[1:]-x[:-1]
3      yC=(y[1:]+y[:-1])/2    #y Center of segment
4      integ=np.zeros(len(x))
5      integ[1:]=np.cumsum(yC*dx)
6      return integ
```

```
7  x=np.linspace(0,2*np.pi,101, endpoint=True)
8  ys=np.sin(x)
9  yc=np.cos(x)
10 YsInt=inte(x,ys)
```

```
11 x=np.linspace(0,2,100,endpoint=True)
12 a,b,c,d=-1,2,-9,8
13 yP=a+b*x+c*x**2+d*x**3
14 YpInt=inte(x,yP)
```

Table 5.5 Function returning the length of the curve $y = f(x)$; x and y_P are defined in Table 5.4

```
1  def lenCurve(x,y):
2      ds=np.zeros(len(x))
3      dx2=(x[:-1]-x[1:])**2
4      dy2=(y[:-1]-y[1:])**2
5      ds[1:]=np.sqrt(dx2+dy2)
6      return(np.cumsum(ds))
```

```
7  FigStd('x',0,1,0.2,'y',-1.5,2.5,0.5)
8  plt.plot(x,yP,'k--')
9  lenP=lenCurve(x,yP)
10 plt.plot(x,lenP,'k-')
11 xC,y1d=deri(x,yP)                  #Derivative
12 len2=inte(xC,np.sqrt(1+y1d**2)) #Length accord. to formula
13 plt.plot(x[:-1],len2,'k+')
14 plt.show
```

Questions

Express lines 3 and 4 of Table 5.4 in one instruction! An `nd.array` is to be returned![16]

Design suitable labels for the plots in lines 8, 10, and 13 of Table 5.5!

Length of a circle

In Table 5.6, we calculate the length of a circle. The full length is reported in the second cell. It is within 0.2 ‰ of the value of 2π. The length *lenCirc* (value on the x-axis) as a function of the y coordinate y_C of the circle (regarded as the independent variable) is displayed in Fig. 5.10b.

[16] `Integ = np.array([0,*np.cumsum(yC*dx)])`

Table 5.6 Calculating the length of a circle and plotting it as a function of the *y*-values of the circle

```
phi=np.linspace(0,2*np.pi,101, endpoint=True)
yS=np.sin(phi)           #x-coordinates of the circle
yC=np.cos(phi)           #y-coordinates of the circle
lenCirc=lenCurve(yC,yS)
FigStd('x, lenCirc',-1,7,1,'y',-4,4,1)
plt.plot(yC,yS,'k--')
plt.plot(lenCirc,yS,'k-')
```

```
lenCirc[-1]   6.282
(2*np.pi)     6.283
```

5.4.5 Spreadsheet Solution

Area under the curve

A spreadsheet solution corresponding to Table 5.4 for the *sine* function is given in Fig. 5.11a. The integration is performed in column C by operating on the individual cells of y and taking the constant d*x* for all $x_{i+1} - x_i$. If the horizontal distance between the vertices is not constant, we have to replace d*x* in D8 by (x-A7), and so on.

The integration in Fig. 5.11a (S) starts at $x = 1.005$ (A7). The corresponding start value for the integration, zero, was entered into C7. The following cells accumulate the areas of the trapezoids. The resulting integral Y_{sInt} is shown in Fig. 5.8b, together with y_S, this, however, for dx = 0.251/10.

Question

According to the first fundamental theorem of calculus, the derivative of the integral over *f*(*x*) should again yield the function *f*(*x*). In Fig. 5.8b, however,

	A	B	C	D	E	F	G
4	**dx**	0.251					
5	=A7+dx	=SIN(x)	=C7+(y+B7)/2*dx	=(YsInt-C7)/dx		Sub Trapez()	
6	**x**	**yS**	**YsInt**	**d(YsInt)/dx**			**Trapez**
7	1.005	0.844	0			1.005	0.000
8	**1.256**	0.951	**0.23**	**0.90**		1.005	0.844
9	1.508	0.998	0.47	0.97		1.256	0.951
10	1.759	0.982	0.72	0.99		1.256	0.000
11	2.010	0.905	0.96	0.94		1.256	0.951
12	2.262	0.771	1.17	0.84		1.508	0.998
107	26.138	0.844	0.00	0.76			

```
Sub Trapez()
r2 = 7
For r = 7 To 11 Step 1
  Cells(r2, 6) = Cells(r, 1) 'x
  Cells(r2, 7) = 0
  r2 = r2 + 1
  Cells(r2, 6) = Cells(r, 1) 'x
  Cells(r2, 7) = Cells(r, 2) 'y
  r2 = r2 + 1
  Cells(r2, 6) = Cells(r + 1, 1) 'x.next
  Cells(r2, 7) = Cells(r + 1, 2) 'y.next
  r2 = r2 + 1
Next r
End Sub
```

Fig. 5.11 **a** (left, S) Integration of $y = sin(x)$, columns F and G contain the data series for the trapezoids shown in Fig. 5.8a. **b** (right, P) VBA procedure for writing the coordinates of the trapezoids into columns F and G

d(*YsInt*)/d*x* is shifted relative to the function $y_S(x)$ to the right. Inspecting Fig. 5.11a, find out why that's the case![17]

In column D, the integral is differentiated, and the result is also displayed in Fig. 5.8b as a function of x listed in column A. In this plot, the derivative d(Y_{sInt})/dx is shifted to the right with respect to the original function. The reason for this flaw is that the derivative is plotted over x (D8 is plotted over A8), the end of the interval in which the derivative is built, instead of over the center of the interval.

Coordinates of the trapezoids by a scan-log procedure
Figure 5.11b (P) shows the program code for a procedure that writes the coordinates for the trapezoids from the table data for x and dY_{sInt}/dx in Fig. 5.11a into the columns F and G of Fig. 5.11a (S). Rows 7–12 of columns A and C are scanned with the loop index r, and three points of a trapezoid are transferred with each loop cycle to F and G, with the running index r_2 being incremented three times in each run. The result "Trapez" is shown in Fig. 5.8a.

5.5 Vectors in the Plane

Polar and Cartesian coordinates are converted one into the other. Two vectors are added, and their scalar product is built. Perpendicular bisectors are erected on line segments. Arrows representing forces are attached to application points in the xy-plane.

5.5.1 Vectors

Vectors in polar and Cartesian coordinates
Vectors have a magnitude l and a direction that, in the plane, can be determined by the angle α to the positive x-axis. Alternatively, a vector can be defined by Cartesian coordinates (V_x, V_y). The two coordinate systems can be transformed one into the other by

$$\begin{aligned} V_x &= l \cdot \cos(\alpha) \\ V_y &= l \cdot \sin(\alpha) \end{aligned} \tag{5.14}$$

[17] The derivative d(Y_{sInt})/dx is plotted versus x listed in column A. The derivative calculated for a segment is, however, to be plotted over the horizontal center of this segment.

$$l = \sqrt{V_x^2 + V_y^2}$$
$$\alpha = arcus\ tangens(V_x, V_y) \tag{5.15}$$

Attention: The order of the arguments in *arcus tangens* is different in the EXCEL and `Python` functions:

EXCEL	`Python`
ATAN2(*x*, *y*)	`np.arctan2(`*y*, *x*`)`

Vector addition
Two vectors are added by adding their Cartesian coordinates individually:

$$\underline{W} = \underline{V} + \underline{U}, \quad (W_x, W_y) = (V_x + U_x, V_y + U_y).$$

Vector addition is illustrated for forces in Fig. 5.12a with the axes F_x and F_y scaled in units of N (Newton). The vectors are represented as arrows with their bases in the origin of the coordinate system and the coordinates of their head points being the coordinates of the vector. The resulting vector $\underline{W}$ points to the corner of the parallelogram spanned by $\underline{V}$ and $\underline{U}$.

In Fig. 5.12b, the plane *xy* is displayed, scaled in units of m with the arrows representing the vectors of Fig. 5.12a attached at a point of application, here, (2, 4), after being scaled with a scalar with the physical dimension m/N (here, 0.8 m/N) to get the same physical unit as the axes, namely, m (meter). The scaling factor is chosen so that arrows of convenient length result that fit into the chart.

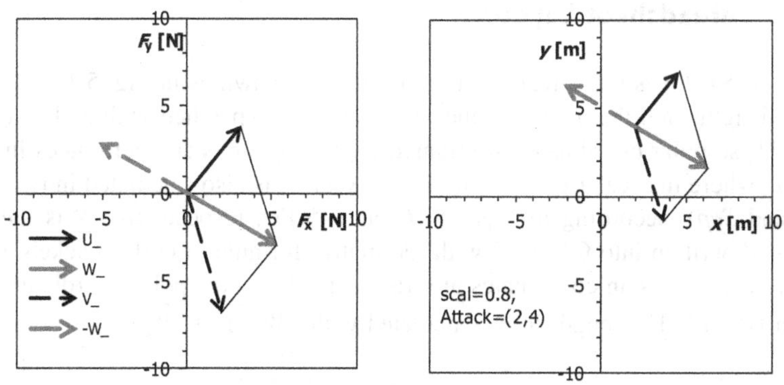

Fig. 5.12 **a** (left) Vector addition of forces in the (F_x, F_y) plane with axes in physical units N; two vectors $\underline{U}$ and $\underline{V}$ are added to produce the resulting vector $\underline{W}$. **b** (right) Vector arrows attached to a point in the *xy*-plane with axes in physical units m

Question

What is the physical unit of the scaling factor for forces in the *xy*-plane?[18]

Scalar product
The scalar product of two vectors $\underline{V}$ and $\underline{U}$ can be calculated in two ways:

- by multiplying their Cartesian coordinates and summing up the products:

$$\underline{U} \cdot \underline{V} = U_x \cdot V_x + U_y \cdot V_y \tag{5.16}$$

- or using polar coordinates with the included angle $\gamma = \alpha_U - \alpha_V$:

$$\underline{U} \cdot \underline{V} = l_U \cdot l_V \cdot \cos(\gamma) \tag{5.17}$$

5.5.2 Data Structure and Nomenclature

$\underline{U}, \underline{V}, \underline{W}$	vectors with two components
l_U, l_V, l_W	length of $\underline{U}$, $\underline{V}$, and $\underline{W}$
a_U, a_V, a_W	angles of $\underline{U}$, $\underline{V}$, and $\underline{W}$ to the x-axis
$\underline{Attack}$	point of application in the plane, vector with two components (*x* and)
$\underline{U}_{at}$, $\underline{V}_{at}$, $\underline{W}_{at}$	coordinates of heads of arrows attached to the point of application.

5.5.3 Spreadsheet Layout

Fig. 5.13 (S) shows the spreadsheet layout for the drawings in Fig. 5.12.

The lengths and the angles of the two vectors are specified with sliders (scroll bars). These polar coordinates are transformed into Cartesian coordinates in rows 7 and 8, where the scalar product of the two vectors is also calculated in two ways ($\underline{U}\cdot\underline{V}$ and *Scp*), according to Eqs. (5.16) and (5.17), respectively. $\underline{W}$ is the sum of $\underline{U} + \underline{V}$, written into C11:C12 with the matrix formula in curly brackets {=U_ + V_}. The Cartesian coordinates are transformed into the polar coordinates (l_W, α_W) in I10:I12. The length l_W is calculated with $\sqrt{W(1)^2 + W(2)^2}$.

Questions

concerning Fig. 5.13 (S):

[18] The physical unit of the scaling factor is m/N, so that the length of the arrows is in m.

	A	B	C	D	E	F	G	H	I	J
1					50		length	lU	5.00	=E1/10
2					140		angle	aU	0.87	=(E2-90)/180*PI()
3					71			lV	7.10	=E3/10
4					17			aV	-1.27	=(E4-90)/180*PI()
5										
6			U_			V_				
7	.x	0.00	3.21	=lU*COS(aU)	0.00	2.08	=lV*COS(aV)	U·V	-19.33	=SUMPRODUCT(U_;V_)
8	.y	0.00	3.83	=lU*SIN(aU)	0.00	-6.79	=lV*SIN(aV)	Scp	-19.33	=lU*lV*COS(aU-aV)
9										
10			W_			-W_		lW	6.0614	=SQRT(INDEX(W_;1)^2
11		0.00	5.29	{=U_+V_}	0.00	-5.29	{=-W_}			+INDEX(W_;2)^2)
12		0.00	-2.96		0.00	2.96		aW	-0.51	=ATAN2(C11;C12)

Fig. 5.13 (S) Specification of two vectors $\underline{U}$ and $\underline{V}$ by their length and their angle to the x-axis with sliders (rows 1–4); transformation to Cartesian coordinates (C7:F8); their scalar product (twice in H7:I8); their sum (C11:D12); and the polar coordinates of the sum (I10:I12)

What is the apparent range of the numbers generated by the slider in A2:D2?[19]

How do you get positive and negative angles between $-\pi/2$ and $\pi/2$ from an always positive output of a slider with a range of 0–180?[20]

Where are the length and angle of the vector $\underline{W} = \underline{U} + \underline{V}$ calculated?[21]

With which instruction do you get the first entry in the named range *W_*?[22]

In rows 11 and 12 of Fig. 5.13 (S), the vector sum of the two vectors is built with a matrix formula $\underline{W}\_ \{= \underline{U} + \underline{V}\}$. Addition is not possible in polar coordinates; if the vectors are specified by length and angle, they must be converted to Cartesian coordinates before addition.

Figure 5.14 (S) contains the extension of the calculation for obtaining the coordinates of arrows representing vectors applied at a point (*Attack*) in the plane. *Attack* is specified in Z4:Z5. The vector arrows in the xy-plane go from *Attack* to *Attack_* + $\underline{U} \cdot$ *scal*, and so on. All four arrows can be entered into a chart together as one series by specifying AB4:AL4 as SERIES X-VALUES and AB5:AL5 as SERIES Y-VALUES, because the respective ranges in the spreadsheet are separated by empty cells.

Ψ *Empty cells separate curves.*

---

[19] The slider in A2:D2 ranges from 0 to 180, angles in degree.

[20] With a formula as in I2 = [=(E2-90)/180*Pi()], one can get positive or negative values.

[21] The length and angle of the new vector $\underline{W}$ are calculated in I10 (l_W) and I12 (a_W).

[22] INDEX(W_;1); in EXCEL, the first entry is indexed as 1, contrary to `Python` where indexing starts with 0.

	Z	AA	AB	AC	AD	AE	AF	AG	AH	AI	AJ	AK	AL	AM
2	**scal**	**0.8**		*{=U_*scal+Attack_}*			*{=V_*scal+Attack_}*			*{=W_*scal+Attack_*			*{=-W_*scal+Attack_}*	
3	**Attack_**			U_			V_			W_			-W_	
4	2.00		2.00	4.57		2.00	3.66		2.00	6.23		2.00	-2.23	
5	4.00		4.00	7.06		4.00	-1.43		4.00	1.63		4.00	6.37	

Fig. 5.14 (S) Calculating the coordinates of arrows representing the vectors $\underline{U}$, $\underline{V}$, $\underline{W}$ in the *xy*-plane of Fig. 5.12a

5.5.4 Python Program

The `Python` program corresponding to Fig. 5.13 (S) is given in the first two cells of Table 5.7. The results of the calculation are shown in the third cell (bottom right) of that table. They correspond exactly to the values in Fig. 5.13 (S). Lines 16-19 calculate the coordinates of the vector arrows in the plane, similar to Fig. 5.14 (S). `W_.shape` = (2,) indicates that $\underline{W}$ is an array with two elements. The values for the scalar product and the polar coordinates of $\underline{W}$ are exactly the same as in Fig. 5.13 (S).

Notice! The argument order of *arcus tangens* is different in the EXCEL and `Python` functions (see Sect. 5.5.1).

The main program in Table 5.8 calls *ArrowP* (Table 5.1 in Sect. 5.2.4) four times. It is a continuation of Table 5.7 and can refer to the data, e.g., *U_at*, specified therein. The resulting diagram is shown in Fig. 5.15.

Table 5.7 **a** (top) `Python` program for defining two vectors with their length and their angle to the *x*-axis, being transformed into Cartesian coordinates and their scalar product being built; **b** (bottom left) the sum of the two vectors is built and the coordinates of all vectors are calculated when attached to a point of attack in the plane; **c** (bottom right) reports values and shapes of some variables

```
1  lU=5.0          #Length of vector U
2  aU=0.8727       #Angle to x-axis
3  U_=lU*np.array([np.cos(aU),np.sin(aU)])
4  lV=7.10
5  aV=-1.274
6  V_=lV*np.array([np.cos(aV),np.sin(aV)])
7
8  Scp=lU*lV*np.cos(aU-aV)
9  UV=U_@V_        #Dot product
```

```
10 W_=U_+V_
11 lW=np.sqrt(np.sum(W_**2))
12 aW=np.arctan2(W_[1],W_[0])
13
14 Attack_=np.array([2.00,4.00])
15 scal=0.8
16 U_at=U_*scal+Attack_
17 V_at=V_*scal+Attack_
18 W_at=W_*scal+Attack_
19 W_at_opp=-W_*scal+Attack_
```

```
UV           -19.33
Scp          -19.33

W_          5.29  -2.96
W_.shape      (2,)
lW            6.06
aW           -0.51
```

Table 5.8 Continuation of Table 5.7; the program plots four arrows and the parallelogram of forces, *Arrow*P from Table 5.1

```
FigStd('x',-10,10,5,'y',-10,10,5)
ArrowP(Attack_,U_at,lw=1.5)
ArrowP(Attack_,V_at,lw=1.5)
ArrowP(Attack_,W_at,lw=1.5)
ArrowP(Attack_,W_at_opp,lw=1.5)
plt.plot([U_at[0],W_at[0],V_at[0]],
         [U_at[1],W_at[1],V_at[1]],'k--')
                                  #Parallelogram, 3 sides
plt.axis('scaled')                #Axis lengths to scale
```

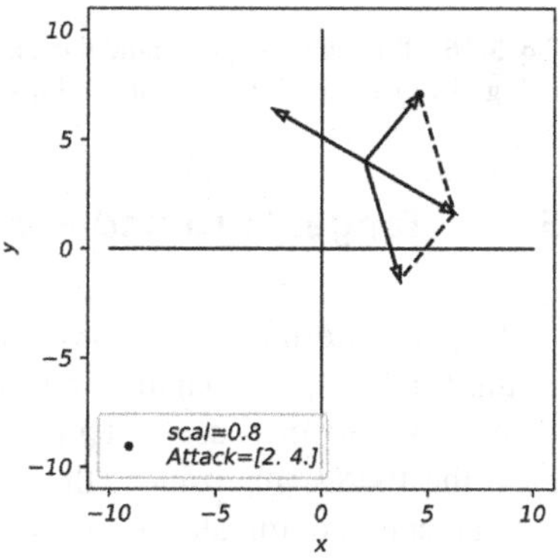

Fig. 5.15 Vector diagram corresponding to Fig. 5.12b, but drawn with the Python program in Table 5.8

Questions

How do you produce the labels shown in Fig. 5.12b (EXCEL) and Fig. 5.15 (Python)?[23]

What is the effect of the statement `plt.axis('scaled')`?[24]

---

[23] Excel: “scal = ”&scal&”; Attack = (“&Index(Attack_,1)&”,”&Index(Attack_,2)&”)”.

Python: `lbl = "scal = " + str(scal) + "\nAttack = " + str(Attack_)`.

[24] The lengths of the axes in the figure correspond to the scaling of the axes specified in the program (see Fig. 5.15).

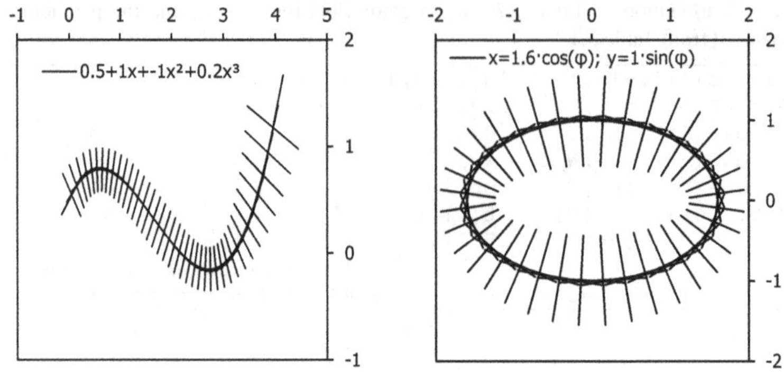

Fig. 5.16 Tangents and perpendiculars, **a** (left) for a polynomial as obtained from the spreadsheet in Fig. 5.18 (S), and **b** (right) on an ellipse constructed using polar coordinates (see legend)

5.6 Tangents to and Perpendiculars on a Curve

Tangents are fixed to a curve and perpendiculars are erected on it, taking a third-degree polynomial as an example. All calculations are performed with vectors. `Python` uses a loop to draw a multitude of segments. In contrast, in the EXCEL realization, the curve is calculated in a spreadsheet, and the segments' coordinates are generated with a scan-log procedure.

5.6.1 At/On a Polynomial and an Ellipse

We determine and display tangents to and perpendiculars on a polynomial, first with a `Python` program and then through the combination of a spreadsheet calculation and a VBA procedure. An example is shown in Fig. 5.16a. In Fig. 5.16b, the same construction is shown for an ellipse.

The coordinates of the ellipse are obtained with

$$x = a_X \cdot \cos(\phi);\ y = a_Y \cdot \sin(\phi) \tag{5.18}$$

Questions

What are the coefficients a_X and a_Y of Eq. 5.18 for the ellipse in Fig. 5.16b[25]

[25] From the legend, we infer $a_X = 1.6$ and $a_Y = 1$.

Table 5.9 Program for determining the endpoints of segments representing the tangents to and the perpendiculars on a curve

```
 #Define polynomial of 3rd order
a,b,c,d=0.5,1.0,-1,0.2
dx=0.1
x=np.arange(0,4+dx,dx)
y=a+b*x+c*x**2+d*x**3

 #Segments, their centers and lengths
xC=(x[1:]+x[:-1])/2
yC=(y[1:]+y[:-1])/2
Dx=x[1:]-x[:-1]
Dy=y[1:]-y[:-1]

scal=5        #Scaling the length of arrows
 #Tangentials
xL=xC[1:]-Dx[1:]*scal
xR=xC[1:]+Dx[1:]*scal
yL=yC[1:]-Dy[1:]*scal
yR=yC[1:]+Dy[1:]*scal
 #Perpendiculars
xpL=xC[1:]-Dy[1:]*scal
xpR=xC[1:]+Dy[1:]*scal
ypL=yC[1:]+Dx[1:]*scal
ypR=yC[1:]-Dx[1:]*scal
```

In Fig. 5.16a, the perpendiculars do not seem to be perpendicular to the curve, contrary to Fig. 5.16b. What is the reason for this?[26]

5.6.2 Data Structure and Nomenclature

a, b, c, d	coefficients of the polynomial $y = a + b \cdot x + c \cdot x^2 + d \cdot x^3$
x	x-coordinates of the vertices, dx apart
y	y values of the function
x_C, y_C	center of the segments
dx, dy	lengths of the segments in the x and y directions
x_L, y_L	left coordinates of the tangential segments
x_R, y_R	right coordinates of the tangential segments
x_pL, y_pL	left coordinates of the perpendicular segments
x_pR, y_pR	right coordinates of the perpendicular segments.

5.6.3 Python Program

A `Python` program for our task is given in Table 5.9. The segments' centers and lengths are elegantly obtained with one instruction each through slicing, as are, again, the arrays for the coordinates of the tangential and perpendicular segments.

[26] In Fig. 5.16a, the lengths of the x- and y-axes do not conform to the scaling (-1 to 5) and (-1 to 2). In Fig. 5.16b, the lengths of the axes have been adjusted to the scaling.

Table 5.10 Program for plotting the results of Table 5.9

```
FigStd('x',-1,5,1,'y',-1,2,0.5)
plt.plot(x,y)
for i in range(0,len(xC)-1):
    plt.plot([xL[i],xR[i]],
             [yL[i],yR[i]],'k-',lw=0.5)
for i in range(0,len(xpL)-1):
    plt.plot([xpL[i],xpR[i]],
             [ypL[i],ypR[i]],'k-',lw=0.5)
plt.axis('scaled') #Only effective downstream (at the end)
plt.savefig('PhEx 5.5 polynomial.png',dpi=1200)
```

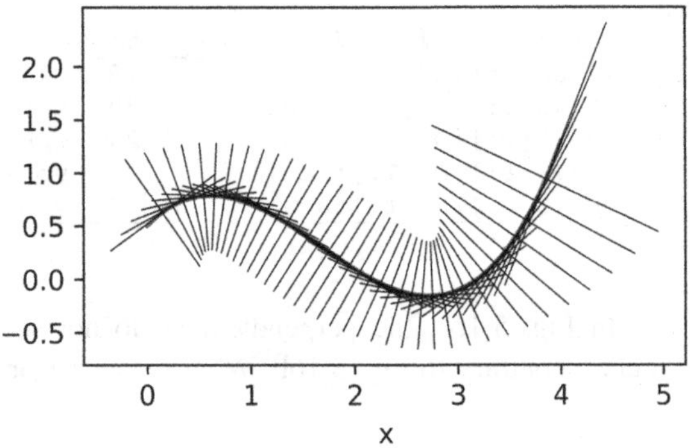

Fig. 5.17 Tangents to and perpendiculars on a polynomial; the coordinates are calculated in Table 5.9 and drawn with the program in Table 5.10

Tangents are drawn through the center (x_C, y_C) of a segment ranging from

$$(x_C - dx \cdot scal, y_C - dy \cdot scal) \text{ to } (x_c + dx \cdot scal, y_c + dy \cdot scal).$$

Scal is a scalar chosen to give a suitable length to the segments representing the tangents and perpendiculars in the figure. No explicit for-loop is necessary to determine the various coordinates, contrary to the VBA code in Sect. 5.6.4. In `Python`, this is implicitly done with slicing.

The result of the calculation in Table 5.9 is plotted by the program in Table 5.10 with the resulting chart in Fig. 5.17. Here, looping over all tangents and all perpendiculars is chosen, contrary to Fig. 5.18b (S), where all coordinates are written into one column for *x* and another one for *y*, with empty cells separating segments so that they can be entered as one series into the figure. With the statement `plt.axis('scaled')`, the axes' lengths are adapted to the axes' scaling. Here, the perpendiculars are visibly orthogonal to the curve, contrary to Fig. 5.16a. With the last statement in Table 5.10, the diagram in Fig. 5.17 is stored as a *png* file.

5.6.4 Spreadsheet Solution

In the spreadsheet of Fig. 5.18a (S), the coefficients of a polynomial of the 3rd degree are specified, and 41 points (*x*, *y*) on that curve are calculated similar to

	A	B	C	D	E
1	**a**	0.50		**cc**	-1.00
2	**b**	1.00		**d**	0.20
3			0.5+1x+-1x²+0.2x³		
4	**dx**	0.10			
5					
6		*=a+b*x+cc*x^2+d*x^3*			
7	x	y		by VBA procedure	
8	0.0	0.50		-0.15	0.36
9	0.1	0.59		0.25	0.73
10	0.2	0.66			
47	3.9	1.05		0.66	0.98
48	4.0	1.30		0.64	0.58

```
1 Sub TangVert()
2 scal = 5
3 r2 = 8
4 For r = 8 To 47
5    xA = Cells(r, 1)
6    yA = Cells(r, 2)
7    xB = Cells(r + 1, 1)
8    yB = Cells(r + 1, 2)
9    'center
10   xC = (xA + xB) / 2
11   yC = (yA + yB) / 2
12   dx = (xB - xA) * scal
13   dy = (yB - yA) * scal
14   'tangentials
15   Cells(r2, 4) = xC - dx
16   Cells(r2, 5) = yC - dy
17   r2 = r2 + 1
18   Cells(r2, 4) = xC + dx
19   Cells(r2, 5) = yC + dy
20   r2 = r2 + 2
21   'perpendiculars
22   Cells(r2, 4) = xC - dy
23   Cells(r2, 5) = yC - dx
24   r2 = r2 + 1
25   Cells(r2, 4) = xC + dy
26   Cells(r2, 5) = yC + dx
27   r2 = r2 + 2
28 Next r
29 End Sub
30
```

Fig. 5.18 **a** (S, left) Table for calculating the x and y coordinates of a 3rd order polynomial; the columns D and E from row 8 contain the outputs of the VBA procedure in (**b**). **b** (P, right) VBA procedure for calculating the coordinates of the segments representing the tangents and perpendiculars to the curve $y = f(x)$ and storing them all in columns D and E

Table 5.9. The coordinates of the segments, representing the tangents at and the perpendiculars on the curve, are calculated in a VBA procedure SUB *TangVert* in Fig. 5.18b (P).

The subroutine is of the type *scan-log* and applies the construction *loop2i* The loop index r runs down A8:A47, scanning the values in columns 1 and 2. The index r_2 specifies the row of the output in columns D and E and is incremented in lines 17, 20, 24, and 27. In each cycle, the coordinates of the tangent and those of the perpendicular are calculated and written one after the other into the same columns, always separated by blank lines so that they are plotted as isolated segments when entered as one series into a chart.

5.7 Banked Curve

We calculate the forces acting on a vehicle running through a banked curve. The gravitational and centrifugal forces are decomposed into vectors on the road and perpendicular to it. The components are combined to get the forces pressing the vehicle onto the road and pushing it perpendicular to its track on the road.

5.7.1 Cross-Section of the Road

Turns on roads are generally banked so that the surface is inclined towards the inside of the turn. The reason for this is that the car is in less danger of being pushed out of the turn if the centrifugal force $\underline{C}$ is not fully working parallel to the road's surface. In addition, the gravitational force $\underline{G}$ also has a component parallel

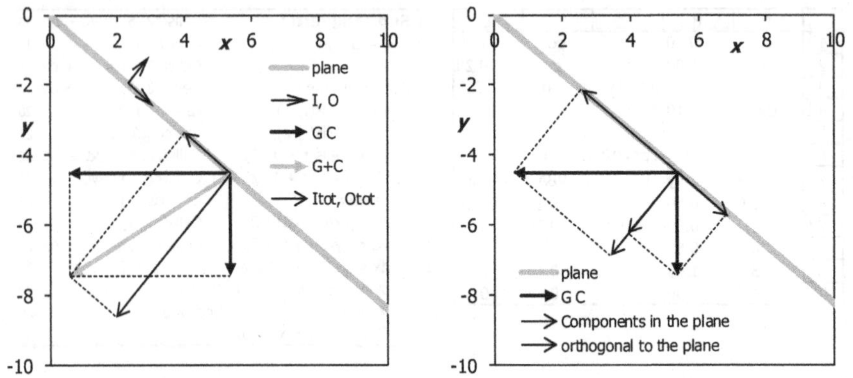

Fig. 5.19 **a** (left) The total force $\underline{G} + \underline{C}$ is decomposed into components parallel and orthogonal to the line (cross-section of the road). **b** (right) The two vectors $\underline{G}$ and $\underline{C}$ are decomposed individually into components parallel and orthogonal to the bank line. Inclination angle $\alpha = 40°$

to the surface towards the inside of the turn. We represent the situation with a straight line ("bank line") describing the road's cross-section.

We treat the task with 2-dimensional vectors, namely, unit vectors parallel and orthogonal to the bank line and vectors representing the forces acting on the car, gravitational mg and centrifugal $-mv^2/r$, where m, v, and r are the mass and velocity of the vehicle and the radius of the curve, respectively. The signs are valid for the situation of Fig. 5.19. As the mass occurs in both forces, we need not consider it explicitly when interested in comparisons between the two forces. So, we use accelerations with sizes $G = g$ and $C = v^2/r$. The components of the accelerations in the bank line and perpendicular to it are calculated with scalar products with the in-plane and out-of-plane unit vectors.

The acceleration parallel to the road drives the car out of its track. The acceleration perpendicular to the bank line presses the vehicle onto the road, thus determining the frictional force.

Questions

What are convenient physical units of $\underline{G}$ and $\underline{C}$?[27]

Is it sensible to interpret $\underline{G}$ and $\underline{C}$ as accelerations?[28]

[27] As we have defined these quantities as accelerations, their unit is N/kg.

[28] Open to debate. Pro: The mass does not play any role. But does the frictional force depend on the mass?

5.7.2 Data Structure and Nomenclature

α	inclination of the track towards the horizontal
$\underline{I} = [cos(\alpha), -sin(\alpha)]$	unit line vector
$\underline{O} = [sin(\alpha), cos(\alpha)]$	unit vector orthogonal to the bank line
$\underline{G} = [0, 9.81]$	gravitational acceleration [N/kg] = [m/s^2]
rr	radius of the curve
v	speed of the car
$\underline{C} = [-v^2/rr, 0]$	centrifugal acceleration
$\underline{Attack}$	point of attack of the forces
$\underline{G_I}$, $\underline{C_I}$	components in the plane
$I_{tot} = \underline{G_I} + \underline{C_I}$	total force in the plane
$\underline{G_O}$, $\underline{C_O}$	components orthogonal to the plane
$O_{tot} = \underline{G_O} + \underline{C_O}$	total force orthogonal to the plane.

5.7.3 Python Program

In the first cell of Table 5.11, the four basic vectors, line vector $\underline{I}$, orthogonal vector $\underline{O}$, gravitational acceleration $\underline{G}$, and centrifugal acceleration $\underline{C}$, are determined from the parameters of the exercise: inclination angle α, gravitational acceleration *g*, speed *v* of the car and radius *rr* of the curvature of the road. In the second and third cell of that table, vector operations are applied to decompose the accelerations into components parallel and orthogonal to the bank line. $\underline{G}$@$\underline{I}$ is a matrix multiplication, for one-dimensional vectors equivalent to the scalar product. The resulting values are listed in Table 5.12. They have the same values as those in the spreadsheets in Sect. 5.7.4.

In Table 5.13, the coordinates of the arrows, representing the vectors in the *xy*-plane with suitable length, are calculated. The starting point $\underline{Attack}$ of all arrows is given as multiple *PoA* of the unit line vector, starting at the origin (0, 0) of

Table 5.11 First cell: specifying line vectors $\underline{I}$ (in-plane) and $\underline{O}$ (orthogonal to plane), and gravitational $\underline{G}$ and centrifugal $\underline{C}$ acceleration

```
a=40                           #Angle of inclination, degrees
lbl_1="α="+str(a)+"°"
a*=np.pi/180                   #Angle in radian
rr=100                         #[m], Radius of curve
v=40                           #[m/s], Speed of car

I_=np.array([np.cos(a),-np.sin(a)])
O_=np.array([-I_[1],I_[0]])
G_=np.array([0,-9.81])         #Gravitational acceleration
C_=np.array([-v**2/rr,0])      #Centrifugal acceleration
```

```
GI_=G_@I_*I_
CI_=C_@I_*I_
Itot_=GI_+CI_
```

```
GO_=G_@O_*O_
CO_=C_@O_*O_
Otot_=GO_+CO_
```

Table 5.12 Numerical values of the vectors specified in Table 5.11, values for GO_ and CO_ are the same as in Fig. 5.21 (S)

I_	[0.77 -0.64]	GI_	[4.83 -4.05]
O_	[0.64 0.77]	CI_	[-9.39 7.88]
G_	[0.00 -9.81]	Itot_	[-4.56 3.83]
C_	[-16.00 0.00]	Otot_	[-11.44 -13.64]

Table 5.13 Calculating the coordinates of the arrows in Fig. 5.19; PoA is a scalar determining the point of attack on the bank line

```
PoA=7
scal=0.3
Attack_=np.array(I_*PoA)
G=G_*scal+Attack_
C=C_*scal+Attack_
GI=GI_*scal+Attack_
CI=CI_*scal+Attack_
Itot=Itot_*scal+Attack_
GO=GO_*scal+Attack_
CO=CO_*scal+Attack_
Otot=Otot_*scal+Attack_
GpC=(G_+C_)*scal+Attack_
```

Table 5.14 Plotting the arrows that represent the various accelerations in Fig. 5.19a with *ArrowP*

```
FigStd('x',0,10,2,'y',-10,0,2)
PoA=7

plt.plot([0,20*I[0]],[0,20*I[1]],c='0.7',lw=3,ls='-')
                             #Plane
At_=I_*3                     #Attack point of the unit vectors
scal=1
Arrow(At_[0],At_[1],
          At_[0]+I_[0]*scal, At_[1]+I_[1]*scal)
Arrow(*At_,*(At_+O_*scal))   #* Decomposes the array

At_=Attack_
Arrow(*At_,*G_)
Arrow(*At_,*C_)
Arrow(*At_,*GpC_)
Arrow(*At_,*Itot_)
Arrow(*At_,*Otot_)
  #Parallelogram of forces:
plt.plot([C[0],GpC[0],G[0]],[C[1],GpC[1],G[1]],'k:')
plt.plot([Itot[0],GpC[0],Otot[0]],
               [Itot[1],GpC[1],Otot[1]],'k:')
```

the bank line. The endpoints are obtained with instructions like G = G_*scal + Attack. Plotting the arrows is achieved with Table 5.14.

Question

Interpret the statement in Table 5.11: G_@I_*I_![29]

[29] G_@I_is a matrix multiplication of two vectors, equivalent to G_[0]*I_[0] + G_[1]*I_[1], yielding a scalar. G_@I_*I_is the multiplication of the vector I_ with this scalar.

5.7.4 Spreadsheet Solution

A spreadsheet solution of the banked-curve problem is shown in Fig. 5.20 (S).

We specify the line vectors I̲ and O̲ and the forces G̲ and C̲ as column vectors with two components. We have to pay attention to operator precedence when calculating the centrifugal force $-v^2/rr$. The values of the coordinates are the same as in Table 5.12, left cell.

▶ In EXCEL, the *sign* operator – has precedence over the *power* operator ^. We therefore have to enter $= -(v\text{\textasciicircum}2/rr)$. In `Python`, $-v{**}2/rr$ will accomplish the task.

In Fig. 5.21 (S), G̲ and C̲ are decomposed into components parallel and orthogonal to the banked curve's characteristic line. The scalar product is obtained with SUMPRODUCT. In Fig. 5.22 (S), the components of the corresponding arrows in the plane are calculated.

A̲t̲t̲a̲c̲k̲ is the starting point of all arrows, and G, C, GI, CI, I_{tot}, GO, CO, O_{tot} are the endpoints of the arrows representing the various accelerations in the xy-plane.

We can create a living figure by introducing sliders to vary the parameters α, rr, and v.

5.8 Weighted Average

Inclination of road	angle	40 °	I_		O_		G_		C_	
	a	0.70 radian	0.77	*=COS(a)*	0.64	*=SIN(a)*	0		-16	*=-(v^2/rr)*
Radius of curve	rr	100 m	-0.64	*=SIN(a)*	0.77	*=-COS(a)*	-9.81		0	
velocity of car	v	40 m/s					plane			
		144 km/h					0	15.32	*{=I_*20}*	
							0	-12.86		

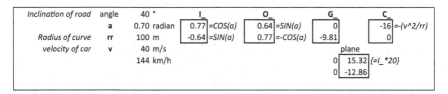

Fig. 5.20 (S) Specifying line vectors I̲ and O̲, and gravitational G̲ and centrifugal C̲ acceleration, same specifications as in Table 5.11, same values as in Table 5.12

Components in the plane			*orthogonal to the plane*		
=SUMPRODUCT(G_;I_)*I_	=SUMPRODUCT(C_;I_)*I_	=GI_+CI_	=SUMPRODUCT(G_;O_)*O_	=SUMPRODUCT(C_;O_)*O_	=GO_+CO_
GI_	CI_	Itot_	GO_	CO_	Otot_
4.83	-9.39	-4.56	-4.83	-6.61	-11.4
-4.05	7.88	3.83	-5.76	-7.88	-13.6

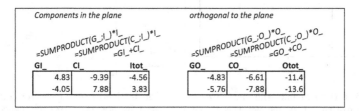

Fig. 5.21 (S) Continuation of Fig. 5.20 (S), decomposing the gravitational and centrifugal forces into components parallel and orthogonal to the bank line

poA	7	scal	0.3							
=I_*poA	{=G_*scal+Attack_}		{=GI_*scal+Attack_}		{=Itot_*scal+Attack_}			{=Otot_*scal+Attack_}		{=Attack_+(G_+C_)*scal}
Attack_	G	C	GI	CI	Itot	GO	CO	Otot		G+C
5.36	5.36	0.56	6.81	2.55	3.99	3.91	3.38	1.93		0.56
-4.50	-7.44	-4.50	-5.72	-2.14	-3.35	-6.23	-6.86	-8.59		-7.44

Fig. 5.22 (S) Continuation of Fig. 5.21 (S), calculating the coordinates of the vector arrows to be drawn in the *xy*-plane; the values of the coefficients are to be compared with the corresponding values in Table 5.12, right cell

The equilibrium of a mobile with two arms is calculated with the law of the lever. The mathematical construct is a weighted average. The calculation uses vectors for forces and arms.

5.8.1 A Mobile with Two Arms

Figure 5.23a shows a mobile with one horizontal crossbar and weights attached at its left and right ends, balanced by a counter-force applied at the center of gravity of the construction. The crossbar is supposed to be weightless. The equilibrium is calculated with the law of the lever:

$$g \cdot m_L \cdot x_L = g \cdot m_R \cdot x_R$$

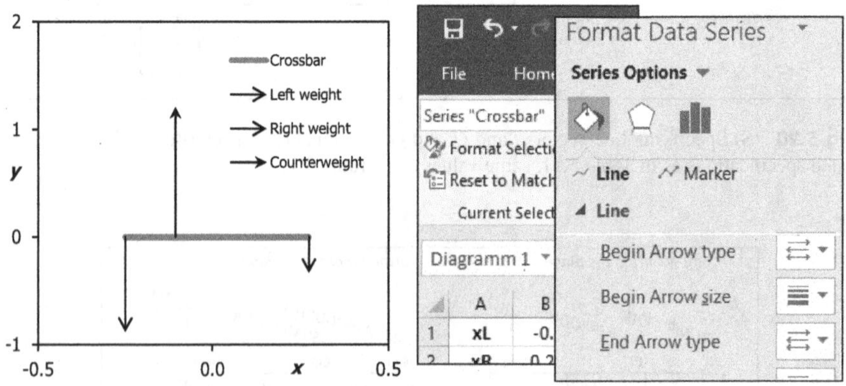

Fig. 5.23 **a** (left) Mobile with one horizontal crossbar. **b** (right) EXCEL menu for editing the series "Crossbar" in **a**; the arrowheads are introduced by FORMAT DATA SERIES/SERIES OPTIONS/FILL&LINE/END ARROW TYPE

where m_L and m_R are the masses attached at the ends of the bars and x_L and x_R their distance to the fulcrum.

We use the vector formulation for the equilibrium of torques:

$$\underline{W}_\mathrm{L} \times \underline{X}_\mathrm{L} = \underline{W}_\mathrm{R} \times \underline{X}_\mathrm{R}$$

The construction is to be done with all positions and forces specified as two-dimensional vectors. Weights and counter-force are to be represented as arrows in the xy-plane.

The x-coordinate x_C of the equilibrium point, where the counter-force has to be applied to hold the mobile in equilibrium, is calculated as a weighted average with the masses as weights:

$$x_C = \frac{m_L \cdot x_L + m_R \cdot x_R}{m_L + m_R} \tag{5.1}$$

5.8.2 Data Structure and Nomenclature

$\underline{L}$	left arm of the crossbeam
$\underline{R}$	right arm of the crossbeam
x_L, x_R	length of the arms of the crossbeam from $x = 0$
m_L, m_R	masses at the crossbeam
$\underline{w}_\mathrm{L}$	weight at the left arm
$\underline{w}_\mathrm{R}$	weight at the right arm
$\underline{CoG}$	center of gravity (x_C, 0)
x_C	horizontal position of center of gravity, from $x = 0$
$\underline{w}_\mathrm{Anti}$	upward counter force applied at $\underline{CoG}$.

All underlined entities are two-dimensional vectors represented in the programs with an underscore at the end, e.g. *L_*.

5.8.3 Python Program

In Table 5.15, the horizontal positions (distance to $x = 0$) of the end of the arms of the crossbar on the horizontal $y = 0$ and the forces are specified.

In Table 5.16, the arrows representing the weights at the crossbar's ends and the counter-force to be applied at the equilibrium point are plotted with *ArrowP* from Sect. 5.2.4. A diagram like the one in Fig. 5.23a results (lines 16–18 ignored).

Table 5.15 Specifying the vectors (arms of the crossbar, weights and counter-force) of the mobile in Fig. 5.23a

```
#Scalars
xL = -0.25
xR = 0.274
mL = 0.88
mR = 0.33
xC =(mL*xL+mR*xR)/(mL+mR)
```

```
#Vectors
L_  = np.array([xL,0])
R_  = np.array([xR,0])
CoG_  = np.array([xC,0])
wL_  = np.array([0,-mL])
wR_  = np.array([0,-mR])
wAnti_ = np.array([0,mL+mR])
```

Table 5.16 Plotting the vectors of Table 5.15 as arrows in the plane of Fig. 5.23a

```
14  FigStd('x',-0.5,0.5,0.5,'y',-1,2,1)
15  plt.plot([xL,xR],[0,0],'k-')
16  plt.plot(*(L_+wL_),'kd',fillstyle="none")
17  plt.plot(*(R_+wR_),'kd',fillstyle="none")
18  plt.plot(*(CoG_+wAnti_),'ks',fillstyle="none")
19  ArrowP(L_,L_+wL_,hw=0.03)
20  ArrowP(R_,R_+wR_,hw=0.03)
21  ArrowP(CoG_,CoG_+wAnti_,hw=0.03)
```

Questions

concerning Table 5.16:

What are the arguments transferred by `*(L_+wL_)` to `plt.plot(...)`?[30]

What do the instructions in lines 16–18 do (result not appearing in Fig. 5.23a).[31]

5.8.4 Spreadsheet Calculation

A spreadsheet calculation corresponding to the Python program in Table 5.15 is laid out in Fig. 5.24 (S). All vectors are specified as column vectors with two coordinates, one for the x and the other for the y direction. In the lower half from left to right, we have the arms $\underline{L}$ and $\underline{R}$ of the crossbar, the arrows representing the weights at the left and the right end, and the counter-force applied at the center of gravity.

[30] `L_ + wL_= [xL, -mL]` (an array), `*[xL, -mL] = xL, -mL` (the elements of an array).

[31] The instructions plot diamonds at the tips of the weights and a square at the tip of the counter-force.

	A	B	C	D	E	F	G	H	I	J	K	L	M	N	O
1	**xL**	-0.25												{=-(wL_+wR_)}	
2	**xR**	0.274		**L_**		**R_**		**CoG_**		**wL_**		**wR_**		**wAnti_**	
3	**mL**	0.88		-0.3	=xL	0.27	=xR	-0.1	=xC	0		0		0	
4	**mR**	0.33		0		0		0		-0.9	=-mL	-0.3	=-mR	1.21	
5															
6	**xC**	**-0.107**	=(mL*xL+mR*xR)/(mL+mR)												
7															
8				=L_	=R_		=L_	=L_+wL_		=R_	=R_+wR_		=CoG_	=CoG_+wAnti_	
9															
10		*x*		-0.3	0.27		-0.3	-0.3		0.27	0.27		-0.11	-0.1	
11		*y*		0	0		0	-0.9		0	-0.3		0	1.21	
12															

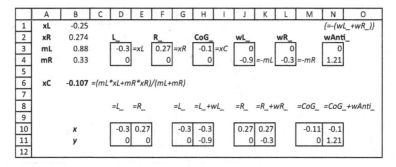

Fig. 5.24 (S) Defining the elementary vectors and calculating the anti-force

Question

How do you enter the three arrows into a figure as in Fig. 5.23a?[32]

5.9 Systems of Linear Equations

We treat two problems resulting in a system of linear equations solved with matrix inversion: calculating the standard form of a polynomial of the 3rd degree from four points and determining the currents in a network of resistances, voltage, and current sources with Kirchhoff's rules.

5.9.1 Polynomial and Electrical Network

Polynomial

A third-degree polynomial is specified by 4 coefficients:

$$y = f(x) = a + b \cdot x + c \cdot x^2 + d \cdot x^3 \tag{5.19}$$

They can be calculated if 4 points ($x^P{}_i$, $y^P{}_i$) are specified in a plane, leading to 4 linear equations:

$$y_i = a \cdot 1 + b \cdot x_i + c \cdot x_i^2 + d \cdot x_i^3 \quad i = 1, 2, 3, 4 \tag{5.20}$$

Two examples are shown in Fig. 5.25.

[32] G10:N10 as an X-series and G11:N11 as a Y-series.

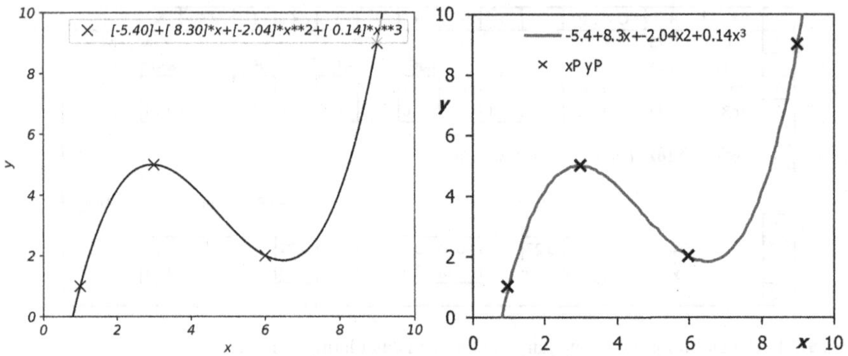

Fig. 5.25 3rd-order polynomials generated, **a** (left) with the Python program in Sect. 5.9.4 and **b** (right) with the EXCEL spreadsheet in Sect. 5.9.3

Matrix of powers of *x*

The system of linear equations is put into matrix form, with the matrix $\underline{P}$ composed of the powers (0, 1, 2, 3) of x_i ("matrix of powers") and the known y_i as a vector on the right side of the equation:

$$\begin{bmatrix} 1 \; x_1 \; x_1^2 \; x_1^3 \\ \cdot \\ \cdot \\ 1 \; x_4 \; x_4^2 \; x_4^3 \end{bmatrix} \cdot \begin{bmatrix} a \\ b \\ c \\ d \end{bmatrix} = \begin{bmatrix} y_1 \\ y_2 \\ y_3 \\ y_4 \end{bmatrix} \tag{5.21}$$

$$\underline{\underline{P}} \cdot \underline{coeff} = \underline{y}_P$$

The vector of unknowns contains the coefficients: $[a, b, c, d]$. The coefficients may be obtained by applying $\underline{P}^{-1}$, the inverse of $\underline{P}$, to $\underline{y}_P$:

$$\underline{coeff} = \underline{\underline{P}}^{-1} \cdot \underline{y}_P \tag{5.22}$$

The matrix of powers has to be non-singular; otherwise, its inverse cannot be built. A square matrix is nonsingular if and only if its determinant is nonzero. The determinant is obtained with MDET (EXCEL) or `npl.det (Python)`.

Question

What is the determinant (< 0, $= 0$, > 0) of the matrix of powers if two x values of the four defining points are identical?[33]

[33] The determinant is zero because it is impossible to draw a polynomial through two points when one is above the other one.

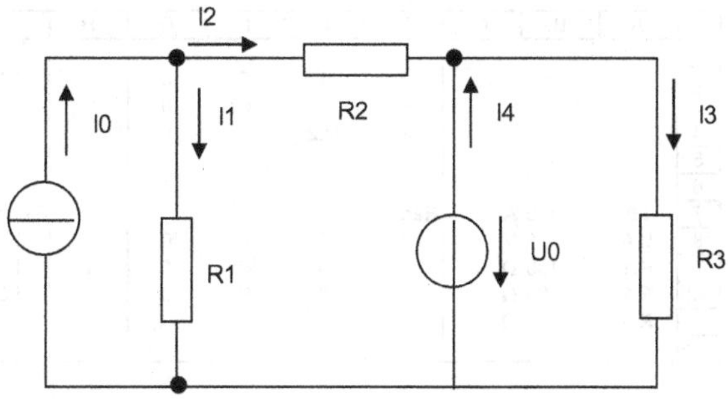

Fig. 5.26 Electrical circuit comprising a current source I0, a voltage source U0, and three ohmic resistors R1, R2, and R3

Electrical Networks

In Fig. 5.26, an electrical network with three ohmic resistors R_1, R_2, and R_3, a constant current source I_0, and a constant voltage source U_0 is shown. We are going to calculate the currents I_1, I_2, I_3, and I_4 in the different branches. The directions of these currents (i.e., the directions of the arrows in the figure) are arbitrary, but they must be chosen at the very beginning. The signs of the currents are with respect to these directions.

Kirchhoff's rules as matrix equation

The four linear equations for the four unknown currents I_1, I_2, I_3, and I_4 are obtained with the help of Kirchhoff's rules, namely, with two mesh rules (the voltage around a closed circuit must be zero):

$$R_3 \cdot I_3 - U_0 = 0 \text{ and } - R_1 \cdot I_1 + R_2 \cdot I_2 + U_0 = 0 \tag{5.23}$$

and two junction rules (the currents flowing into a junction must equal the currents flowing out of the junction):

$$-I_1 - I_2 + I_0 = 0 \text{ and } I_2 + I_4 - I_3 = 0 \tag{5.24}$$

These four equations are transformed so that the known source voltage U_0 (or 0) is on the right side of the mesh rules and the known source current I_0 (or 0) is on the right side of the junction rules so that they can be expressed as a matrix equation:

$$\underline{\underline{Res}} \cdot \underline{I} = \underline{Srcs} \tag{5.25}$$

$\underline{\underline{Res}}$ is a matrix with components 0 or 1 or the resistances of the circuit, $\underline{I}$ is the vector of the unknown currents, and $\underline{Srcs}$ is a vector whose coefficients are zero or the known source voltages and currents. An example for the network in Fig. 5.26 is given in the spreadsheet of Fig. 5.27 (S).

Questions

concerning Fig. 5.27 (S):

	A	B	C	D	E	F	G	H	I	J	K	L	M	N
1														
2				1	1	0	0		I.1		I.0			
3				0	0	R.3	0	•	I.2	=	U.0			
4				-R.1	R.2	0	0		I.3		-U.0			
5				0	1	-1	1		I.4		0			
6														
7	**I.0**	0.5 A		**Res**					Currents		**Srcs**			
8	**U.0**	5 V		1	1	0	0		I.1		0.5			
9	**R.1**	100 Ω		0	0	90	0	•	I.2	=	5			
10	**R.2**	68 Ω		-100	68	0	0		I.3		-5			
11	**R.3**	90 Ω		0	-1	1	-1		I.4		0			
12														

Fig. 5.27 (S) Eq. (5.25) applied to the network of Fig. 5.26. The matrix equation is presented in rows 1–5 in general form and in rows 8–11 with the concrete values from column B. The vertical lines, and the dots in column H, are intended to indicate matrix calculation. They have no computational function in the spreadsheet

Verify that the matrix calculation in rows 8–11 is identical to Eqs. 5.23 and 5.24!

What are the formulas in D10:G10?[34]

5.9.2 Data Structure and Nomenclature

Polynomial

x_P, y_P coordinates of 4 points in the plane, specified as two vectors

$\underline{\underline{A}}$ 4 × 4 matrix ("matrix of powers of x"), powers of x_P in rows

$\underline{\underline{A}}_{Inv}$ the inverse of matrix A

$\underline{coeff}$ solution a, b, c, d of the system of linear equations, coefficients of the polynomial

Electrical circuit

R_1, R_2, R_3, R_4 four ohmic resistances

I_0 current of a constant-current source

U_0 voltage of a constant-voltage source

$\underline{\underline{Res}}$ square matrix containing resistances, zeros, and ones

$\underline{\underline{ResInv}}$ inverse of Res

$\underline{Srcs}$ vector containing the voltage and current sources or zero

$\underline{Currents}$ vector containing the currents to be determined with the system of linear equations.

[34] D10 = [=-R.1], E10 = [=R.2], F10 = [0], G10 = [0].

We need functions to build the inverse of a square matrix and to multiply matrices.

EXCEL: MINVERSEMMULT
`Python:` `np.inv@` operator (or `np.matmul`)

5.9.3 Spreadsheet Solutions

Polynomial

The solution of Eq. 5.22 is implemented in the spreadsheet of Fig. 5.28 (S). The coordinates of the defining points are stored in column vectors named $\underline{x}_\text{P}$ and $\underline{y}_\text{P}$. The matrix $\underline{\underline{A}}$ is composed of column vectors obtained as powers of $\underline{x}_\text{P}$. To solve for the coefficients of the polynomial, we have to build the inverse $\underline{\underline{A}}_\text{Inv}$ of $\underline{\underline{A}}$ to be applied to $\underline{y}_\text{P}$ in a matrix multiplication.

Electrical network

For the network of Fig. 5.26, the spreadsheet layout in Fig. 5.27 (S) applies, in rows 2–4 in general form and in rows 8 to 11 with the values from A7:B11. The indices of the resistances and the constant current and constant voltage are separated from the letters by a dot, because a name like R1 would be interpreted as a cell address.

The unknown currents are obtained by forming the inverse matrix to $\underline{\underline{M}}$ to be applied to the vector $\underline{S}$ of sources:

$$\underline{\underline{M}}^{-1} \cdot \underline{\underline{M}} \cdot \underline{I} = \underline{\underline{M}}^{-1} \cdot \underline{S} \text{ or } \quad \underline{\underline{M}}^{-1} \cdot \underline{S} = \underline{I}$$

The solution for the matrix equation in Fig. 5.27 (S), applying the EXCEL matrix functions MINVERSE to get the inverse matrix and MMULT to perform matrix multiplication, is given in Fig. 5.29 (S) where, in column N, we check whether Kirchhoff's rules are satisfied.

	A	B	C	D	E	F	G	H	I	J	K	L	M	N	O	P
2				=xP^3	=xP^2	=xP	1		{=MINVERSE(A)}					=MMULT(Ainv;yP)		
3	xP	yP		A					Ainv					coeff		
4	1	1		1	1	1	1		-0.01	0.03	-0.02	0.01		0.14	dd	
5	3	5		27	9	3	1		0.23	-0.44	0.29	-0.07		-2.04	cc	
6	6	2		216	36	6	1		-1.24	1.92	-0.87	0.19		8.30	bb	
7	9	9		729	81	9	1		2.03	-1.50	0.60	-0.13		-5.40	aa	
8																

Fig. 5.28 (S) Spreadsheet implementation of Eq. 5.22; $\underline{\underline{A}}_\text{inv}$ is the inverse matrix of $\underline{\underline{A}}$; the resulting polynomial is presented in Fig. 5.25b

	C	D	E	F	G	H	I	J	K	L	M	N	O
12													
13		*{=MINVERSE(Res)}*							*{=MMULT(ResInv;Srcs)}*				
14		**ResInv**					**Scrs**		**Cur**				
15		0.40	0.00	-0.01	0.00		I.0		0.23	I.1		-5.00	*=-R.1*I.1+R.2*I.2*
16		0.60	0.00	0.01	0.00	·	U.0	=	0.27	I.2		-0.50	*=-I.1-I.2*
17		0.00	0.01	0.00	0.00		-U.0		0.06	I.3		5.00	*=R.3*I.3*
18		-0.60	0.01	-0.01	-1.00		0		-0.21	I.4		0.00	*=I.2+I.4-I.3*
19													

Fig. 5.29 (S) Solving the matrix equation in Fig. 5.27 (S) with the inverse of <u>*Res*</u>. In column N, Kirchhoff's rules are checked

Table 5.17 Calculating the coefficients *coeff* of a polynomial of the 3rd order when 4 points (in arrays (x_P, y_P)) are given

```
1  import numpy.linalg as npl
2  xP=np.array([1,3,6,9])         #Defining points
3  yP=np.array([[2,6,2,2]])
4  xP0_= np.array([1,1,1,1])
5  xP1= xP
6  xP2= xP**2
7  xP3= xP**3
8
9  A=np.array([xP0,xP1,xP2,xP3]).transpose(1,0)
10 yPt_= yP.transpose(1,0)
11 coeff_= npl.solve(A,yPt)
```

A	coeff	yPt
[[1 1 1 1]	[[-4.00]	[[2]
[1 3 9 27]	[7.67]	[6]
[1 6 36 216]	[-1.78]	[2]
[1 9 81 729]]	[0.11]]	[2]]

Question

The calculations in column N (reported in column O) of Fig. 5.29 (S) are intended to check whether Kirchhoff's rules are fulfilled. Which of the four equations Eq. 5.23 left or right and Eq. 5.24 left or right are checked in N15–N18?[35]

5.9.4 Python Programs

Polynomial from defining points

In Table 5.17, the coordinates of the defining points are specified in the arrays $\underline{x}_P$ and $\underline{y}_P$. The matrix $\underline{A}$ is put together in line 9, first as a list of the rows x_{P0}, x_{P1}, x_{P2}, x_{P3} that is then transposed to get a form as in Eq. (5.21); see also the printout in the lower cell. Likewise, the vector $\underline{y}_P$ is generated. In line 11, the system of linear equations

[35] N15: Eq. 5.23 right; N16: Eq. 5.24 left; N17: Eq. 5.23 left; N18: Eq. 5.24 right.

Table 5.18 *Python* program for plotting the polynomial specified by the coefficients in Table 5.17

```
x=np.linspace(0,10,100)
y=coeff[0]*x**3+coeff[1]*x**2+coeff[2]*x+coeff[3]
FigStd('x',0,10,2,'y',0,10,2)
plt.plot(x,y,'k-')
plt.plot(xP,yP,'kx')        #Defining points of the parabola
```

is solved with the function `solve` of the `numpy.linalg` library. Alternatively, the coefficients can be obtained by applying the inverse of the "powers of x" matrix to the vector of pre-specified y values:

$$\underline{coeff}_2 = \underline{\underline{P}}_{\text{Inv}} @ \underline{y}_{\text{P}}$$

The results of the two methods are identical. The coefficients are used to generate the coordinates of the polynomial shown in Fig. 5.25a. The entities in the lower cell of Table 5.17 are arranged so that they correspond to the equation:

$$\underline{\underline{P}} \cdot \underline{coeff} = \underline{y}_{\text{Pt}}$$

The entity $\underline{y}_{\text{Pt}}$ is the transposed row vector $\underline{y}_{\text{P}}$. At least the first coefficient of $\underline{y}_{\text{Pt}}$ can be checked by mental calculation.

Questions

The coefficients in Fig. 5.28 (S) are the same as in the `Python` solution in Table 5.17, however, in reverse order. What is the reason for this?[36]

What is the more versatile version: (a) `y = coeff[0]*x**3` + … or (b) `y = coeff[0] + coeff[1]*x`?[37]

The program for realizing the plot in Fig. 5.25a is shown in Table 5.18.

Electrical network

A `Python` program for solving Kirchhoff's equations corresponding to the spreadsheet layout in Fig. 5.27 (S) and Fig. 5.29 (S) is presented in Table 5.19. The matrix $\underline{\underline{Res}}$ is defined as in the spreadsheet in Fig. 5.27 (S), whereas the vector of the sources is first defined as a row vector $\underline{Src}$ and then transposed into a column vector $\underline{Srcs}$ equal to the one in Fig. 5.27 (S) and to $\underline{Srcs}$ in column K of Fig. 5.29 (S). Lines 9 and 10 can be combined into one instruction.

[36] The matrix of powers $\underline{\underline{P}}$ in Fig. 5.28 (S) contains the columns in reverse order to Table 5.17.

[37] Version (b) is more versatile, because it can easily be extended to higher order, … + `coeff[3]*x**4 +` ....

Table 5.19 Solving Kirchhoff's equation

```
1  import numpy.linalg as npl
2  R1,R2,R3=100,68,90
3  I0,U0=0.5,5
4  Res=np.array([[1,1,0,0],
5                [0,0,R3,0],
6                [-R1,R2,0,0],
7                [0,-1,1,-1]])
8  Src=np.array([[I0,U0,-U0,0]])
9  Srcs=Src.transpose(1,0)
10 ResInv = npl.inv(Res)
11 Cur=ResInv@Srcs
12 CurS=npl.solve(Res,Srcs)
```

```
ResInv:
 [[ 0.40 -0.00 -0.01 -0.00]
  [ 0.60  0.00  0.01  0.00]
  [ 0.00  0.01  0.00  0.00]
  [-0.60  0.01 -0.01 -1.00]]
```

Cur with ResInv (inverse matrix)	CurS with linalg (solve)
[[0.23]	[[0.23]
[0.27]	[0.27]
[0.06]	[0.06]
[-0.21]]	[-0.21]]

The system of linear equations is solved in two ways, in line 11 with the inverse matrix *ResInv* as in the spreadsheet calculation and in line 12 with the function `solve` of the `numpy.linalg` library. Both methods yield the same result (second cell in Table 5.19) as the spreadsheet procedure.

5.10 Some Mathematical Functions

Numpy, np.*	EXCEL	
Basic numeric information		
abs	ABS(x)	absolute value
sign	SIGN(x)	sign (+1, −1 or 0)
Basic mathematical operations		
sum	SUM(x; x_2)	$x + x_2$ (no matrix function)
prod	PRODUCT(x;x_2)	$x \cdot x_2$ (no matrix function)
	QUOTIENT(x;x_2)	integer portion of x/x_2, no matrix function
mod	(x; x_2)	Modulo, remainder from x/x_2
sqrt	SQRT(x)	Positive square root
Rounding functions		
round	ROUND(x; N)	Up or down, to N digits
	ROUNDUP(x; N)	Up to bigger absolute value, to N digits
	ROUNDDOWN(x; N)	Down to smaller absolute value, to N digits
	MROUND(x; x_2)	Up or down to a multiple of x_2, no matrix function

Numpy, np.*	EXCEL	
Ceil	CEILING(x; x_2)	Up versus ∞ to a multiple of x_2
Floor	FLOOR(x; x_2)	Down versus −∞ to a multiple of x_2
Rint	INT(x)	Down to the next integer
fix	TRUNC(x)	Towards zero to the next integer
Trigonometric functions		
pi	PI()	Constant value of pi
rad2deg	DEGREES	Converts radians to degrees
deg2rad	RADIANS	Converts degrees to radians
cos	COS(α)	Cosine of a given angle (rad)
arccos	ACOS	Inverse cosine [−1, 1]→ [0, π]
sin	SIN(α)	Sine of a given angle (rad)
arcsin	ASIN	Inverse sine [−1, 1]→ [−π/2, π/2]
tan	TAN(α)	Tangent of a given angle (rad)
arctan	ATAN	inverse tangent (*arcus tangens*) [−∞, ∞]→ [-π/2,π/2]
arctan2(y,x)	ATAN2(x;y)	Angle of a given pair of x and y coordinates; **attention**: Order of arguments is different in EXCEL and Python!
Exponents and logarithms		
Exp	EXP(x)	e raised to the power of x
Log	LN(x)	Natural logarithm, inverse of *exp*
	LOG(x;b)	Logarithm of x to base b LOG(X; B_1) = LOG(X; B_2) * LOG(B_2; B_1)
log10	LOG10	Base 10 logarithm
log2	LOG(x; 2)	Base 2 logarithm
Sums		
Footnote 38	SUMIF	Adds the cells in a supplied range that satisfy a given criterium
Footnote[a]	SUMIFS	Adds the cells in a supplied range that satisfy multiple criteria
sum(x * y)	SUMPRODUCT(x;y)	$x * y$ over arrays
sum(x * x)	SUMSQ(x)	x^2 over arrays
sum(x^2-y^2)	SUMX2MY2(x;y)	x^2 minus y^2 over arrays
sum(x^2 + y^2)	SUMX2PY2(x;y)	x^2 plus y^2 over arrays
sum((x-y)^2)	SUMXMY2(x;y)	$(x - y)^2$ r arrays
Cumsum	Footnote[b]	Cumulated sum
Functions of linear algebra		
npl.det	MDETERM	Determinant of a square matrix
npl.inv	MINVERSE	Inverse of a square matrix
np.matmul(A,B) or A@B	(A;B)	Matrix product of two matrices

Numpy, np.*	EXCEL	
np.dot(x,y)	SUMPRODUCT(x;y)	Dot product of two vectors

[a] Can be achieved with list comprehension, as in the following example: `Rng = np.linspace(1,20,20) ; x = [×1 for × 1 in Rng if ×1 > 3 if × 1 < 6]; sum(x)`

[b] In a spreadsheet operation, B1 = [=A1]; B2 = [=B1 + A2]; … cumulates in column B the values in column A

5.11 Questions and Tasks

1. How do you get the anchor points of the inverse function of $y = y(x)$?[38]

Differentiation and integration

2. You have calculated the difference quotient between the boundaries of an interval. Over which value do you plot the result?
3. What is the formula for the numerical second derivative of a function, specified at equidistant positions with distance dx?
4. Over which positions of an interval (beginning, middle, end) are values for the first derivative, the second derivative, and the integral to be plotted?
5. What does the trapezoidal rule of integration stand for?

Vectors

6. What are the Cartesian coordinates of vectors of length 1 (one) pointing in the x direction, the negative y direction, and 45° from the x-axis?
7. What are the components of a vector pointing from $(\underline{x}_1) = (4, 5)$ to $(\underline{x}_2) = (6, 3)$?
8. What are the lengths of the two vectors pointing from (0, 0) to (1, 1), and from (0, 0) to (3, 4)?
9. How do you form a scalar product of two vectors $\underline{A}$ and $\underline{B}$ in two different ways, (i) component by component, and (ii) using the angle ϕ between the two vectors?
10. What is the angle between two vectors if the scalar product is zero?
11. What are the coordinates of a unit vector in the direction of the segment from $[x_1, y_1]$ to $[x_2, y_2]$?
12. What are the coordinates of the perpendicular to the vector (x, y)?

[38] The inverse function is $x = x(y)$. To get its interpolation points, you only have to swap the two columns for x and y in the spreadsheet or the two lists in `Python`.

	A	B	C	D	E	F	G	H	I	J	K	L	M	N	O	P	Q	R	S	T	U	V
1	A				B				D													
2	1	0	1		1	0	0		1	2	3			=MMULT(E2:G4;I2:K4)								
3	0	1	0		0	1	0		4	5	6											
4					0	0	1		7	8	9								=MMULT(A2:C3;I2:K4)			
5																						

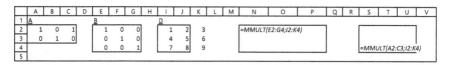

Fig. 5.30 (S) Five matrix ranges, named cell ranges: A = [A2:C3], B = [E2:G4], C = [I2:K4], D = [I2:J4], x = [I2:K2], y = [I3:K3]

System of equations

13. Which arguments does the EXCEL function MMULT(?) have, which multiplies two matrices? What is the equivalent in `Python`? What are the relations between the widths and heights of the two matrices?
14. You have set up a system of equations $\underline{\underline{M}} \cdot \underline{I} = \underline{I}$, with known matrix $\underline{\underline{M}}$ and known source vector $\underline{S}$, but unknown vector $\underline{I}$. By which instructions of type matrix operation do you get the coefficients of $\underline{I}$ in EXCEL and `Python`?[39]

Functional expressions

Calculate the values of the following expressions in EXCEL:

15. ARCTAN2(1;1), LOG10(0.001),
16. PRODUCT(2;3;4;5), POWER(10;3).
17. ROUND(3.74638,2), INT(17.453), REMAINDER(127; 2).

The functions in Questions 18 and 19 refer to Fig. 5.30 (S). What values result for:

18. SUMPRODUCT(x;y) , SUMXMY2(B;C)? What are the corresponding expressions in `Python`?
19. MMULT(B;C), MMULT(A;D)? What are the corresponding expressions in `Python`?
20. You are to determine the derivative dy/dx of a function defined by `x = np.linspace (0,10,11)` and y = x**2. How do you determine dy and dx with list slicing? What are the first two elements of dy/dx, and over which values do they have to be plotted?
21. Define the arrays $\underline{A}$, $\underline{B}$, and $\underline{C}$, in `Python` corresponding to the three ranges A2:C3, E2:G4, and I2:K4, respectively, in Fig. 5.30 (S)! Calculate $\underline{\underline{B}}$@$\underline{\underline{C}}$ and $\underline{\underline{A}}$@$\underline{\underline{C}}$!
22. What are the two arguments in the function to calculate the angle of the vector (1, 2) to the x axis in EXCEL and `Python`?

[39] EXCEL: I = MMULT (M);S). `Python`: I = `npl.inv`(M)@S; npl stands for `numpy.linalg`, M and S are of type `np.array`.

Superposition of Movements

6

We learn how to compose complicated movements from simple ones, namely translations and rotations in a plane. The exercises honor famous scientists: Bernoulli (cycloid), Foucault (pendulum) , and Steiner (moment of inertia). In spreadsheets, we systematically use sliders and macros with which we have familiarized ourselves in previous chapters.

6.1 Introduction: Translations and Rotations

Solutions of Exercises 6.2 (Python), 6.3 (Excel), 6.4 (Excel), and 6.5 (Python) can be found at the internet adress: go.sn.pub/or1CXF.

Simple movements
In this chapter, we put together movements in a plane from two simple movements:

Translations T, straight-line movements in one direction, generally defined by a two-dimensional velocity vector (v_x, v_y);

Rotations R, rotations in the *xy*-plane, described by an angular velocity ω_z and the radius r of the trajectory.

Polar coordinates
We use polar coordinates (r, φ) to describe rotations and convert them, e.g., for graphical representation in charts, to Cartesian coordinates (x, y):

$$x = r \cdot \cos(\phi);\ y = r \cdot \sin(\phi) \tag{6.1}$$

Projectile motion, T-T (Exercise 6.2)
A projectile trajectory is composed of two linear movements. If friction is not taken into account, these are a vertical one accelerated by gravity and a uniform horizontal

D. Mergel, *Physics with Excel and Python*,
https://doi.org/10.1007/978-3-030-82325-2_6

one. We attach a velocity vector and its vertical and horizontal component to the trajectory at a freely chosen point.

Cycloid, rolling curve, T-R (Exercise 6.3)
We calculate the trace of a point on a wheel rolling on a plane, resulting from a translation of the wheel axis on a straight line and a rotation about the wheel axis, with translation speed and angular speed being dependent on each other.

The resulting curve, called a *rolling curve* or *cycloid*, represents the brachistochrone (the fastest path to fall from one point to a lower point) , as shown by Johann Bernoulli. This will be treated in the follow-up book *Physics with Excel and Python, Using the Same Data Structure. Applications* in the chapter "Calculus of Variations".

Foucault's pendulum, T-R (Exercise 6.4)
What is the trace of a swinging pendulum on a rotating surface? It is obtained by superposing the movement of the linear oscillation in the laboratory system with a rotation of the base table on which the motion is recorded. Oscillation and rotation are independent of each other.

This experiment has historical significance. Michel Foucault demonstrated with a 67-m long pendulum suspended from the dome of the Pantheon in Paris that the earth rotates against the fixed stars.

Swinging anchor, R-R (Exercise 6.5)
We consider an anchor in the form of a hanging T with three mass points attached, one at each end of the T and one at the junction of the two lines. The mass points are supposed to be connected by massless struts. We hang the rigid anchor at the upper end of the stem or hold it at its center of gravity.

The anchor's motion results from a superposition of the rotation of a selected point of the anchor, e.g., the center of gravity, around the suspension point, and a rotation of the anchor around the selected point. A rotational matrix describes this motion.

We calculate the center of gravity of the anchor and, with Steiner's theorem,[1] its moment of inertia when rotating around the upper end of the stem and when rotating around its center of gravity.

Sound emitted from a moving source, T-T (Exercise 6.6)
In the last exercise of this chapter, we investigate the circular wavefronts emitted from a moving acoustic source.

[1] Also known as the *parallel axis* theorem or *Huygens-Steiner* theorem.

Animations

The movements treated in this chapter are well suited to be animated. The basic technique for this with `FuncAnimation` of the `matplotlib.animation` library is explained in Sect. 6.2.5 in connection with the projectile trajectory.

6.2 Projectile Trajectory with Velocity Vectors (T-T)

We calculate and plot the trajectory of a projectile composed of two linear motions (T), a uniform horizontal one and a vertical one accelerated by gravity. The parameters are launch height, angle, and speed. Vectors for the horizontal, vertical, and total velocities are attached at three points to the trajectory.

6.2.1 Projectile Trajectory and Velocity Vectors

Trajectory and attached arrows

In Fig. 6.1a, a projectile trajectory is shown with arrows representing velocity vectors attached for three different time points. The parameters are launch height and angle, and speed. In Fig. 6.1b, two trajectories for different launch angles are displayed. The trajectories are, in all cases, downwardly open parabolas. Figure 6.1b shows that maximum height and maximum distance depend on the launch angle.

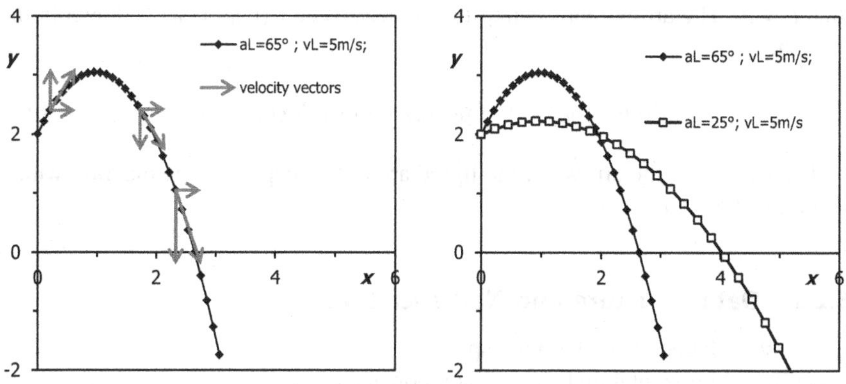

Fig. 6.1 **a** (left) Projectile trajectory, coordinates at different equidistant times; velocity vectors at three different times. **b** (right) Two trajectories for two different launch angles

Our task is to plot a projectile trajectory with given values for launch height y_L, angle α_L, and velocity v_L. In Fig. 6.1a, we have chosen $y_L = 2$ m, $\alpha_L = 65°$, and $v_L = 5$ m/s. The time distance dt between the calculation points is 0.05 s.

Coordinates of the projectile trajectory
The coordinates of the parabola are calculated from

$$v_x = v_{Lx} \quad \text{with} \quad v_{Lx} = v_L \cdot \cos(\alpha_L);$$
$$v_y = v_{Ly} - g \cdot t \quad \text{with} \quad v_{Ly} = v_L \cdot \sin(\alpha_L) \tag{6.2}$$

as

$$x(t) = v_{Lx} \cdot t \tag{6.3}$$

$$y(t) = y_L + v_{Ly} \cdot t - \frac{1}{2} g \cdot t^2 \tag{6.4}$$

In the x-direction, there is a uniform motion with the horizontal initial velocity v_{Lx}. The movement in the y-direction is the sum of a uniform motion with velocity v_{Ly}, determined by the initial velocity and a downward motion accelerated by gravity.

▶ **Tasks** For a given speed and launch height, change the angle such that (a) the height and (b) the width reached will be maximum!

Calculate the impact velocity for the two trajectories in Fig. 6.1b!

Determine the maximum height and width analytically and compare them with the value of the simulation!

Attach the tangential vector of the velocity and the decomposition of this vector into x and y components at the point of the trajectory corresponding to a specific time $t = tt$! The arrow representing the velocity vector (v_{Xtt}, v_{Ytt}) is drawn in the xy-plane

$$\text{from } (x_{tt}, y_{tt}) \text{ to } (x_{tt}, y_{tt}) + (V_{Xtt}, V_{Ytt}) \cdot scal \tag{6.5}$$

The lengths of the arrows in the figure are to be adapted to the diagram with a scaling factor *scal*.

6.2.2 Data Structure and Nomenclature

y_L, a_L, v_L	launch height, angle, speed
v_{Lx}, v_{Ly}	horizontal and vertical components of v_L
t	equidistant (dt) series of instants of time
$x(t)$, $y(t)$	trajectory as a function of t
tt	list of three specific times
x_{tt}, y_{tt}	position at tt

	A	B	C	D	E	F
1	Time interval	**dt**	0.05 s			
2	**Prespecifications**					
3	Launch height	**yL**	2 m			
4	Launch angle	**aL**	65 °			
5	Launch speed	**vL**	5.00 m/s			
6	Gravitational acceleration	**g**	9.81 m/s^2			
7	**Deduced therefrom**					
8		**vLx**	2.11 m/s			
9		**vLy**	4.53 m/s			

	B	C	D	E	F
11	="aL="&aL&"°; vL="&vL&"m/s;"				
12	**aL=65° ; vL=5m/s;**				
13	=B15+dt	=vLx*t	=yL+vLy*t-g/2*t^2		
14	**t**	**x**	**y**		
15	0	0.00	2.00		
16	**0.05**	0.11	2.21		
44	1.45	3.06	-1.74		

Fig. 6.2 **a** (left, S) The parameters for the task are defined and used to calculate the horizontal, v_{Lx}, and vertical, v_{Ly}, initial velocities. The launch angle α_L is set with a slider. **b** (right, S) Continuation of **a**. Coordinates of the parabola (*t, x, y*) in equidistant time steps *dt* in columns B, C and D; the label for the figure is generated in B12 with the formula reported in B11

v_{Xtt}, v_{Ytt} velocity vectors at *tt (3 instances)*

scal scaling factor [s] for the velocity vectors in the *xy*-plane

x_{Att}, y_{Att} coordinates of the tips of the arrows representing the vectors.

6.2.3 Spreadsheet

Trajectory

A possible spreadsheet calculation is shown in Fig. 6.2 (S).

The freely selectable parameters initial height, angle, and speed of the launch are in C3:C6. From this, the initial horizontal and vertical velocities, $v_{Lx,}$ and v_{Ly}, are calculated in C8:C9.

Questions

Which formulas are in C8 and C9 of Fig. 6.1a (S)?[2]

How is the legend “aL = ...” in Fig. 6.2b (S) generated?[3]

Velocity-vector coordinates

► **Task** Calculate the *x*- and *y*-components of the velocity for a given time *tt*! In Fig. 6.3 (S), this is done for *tt* = 0.818 s, a time set with a slider.

In Fig. 6.3a, the coordinates of the arrows representing the velocity vector at $t = tt$ and its horizontal and vertical components are calculated according to Eq. 6.2. This is repeated in Fig. 6.3b for two other time points. The resulting arrows are displayed in Fig. 6.1a. As of EXCEL 2007, line segments in charts can be provided with arrowheads (EXCEL 1019: FORMAT DATA SERIES/FILL & LINE/END ARROW TYPE), and consequently our vector arrows.

[2] vLx = C8 = [= vL*cos(aL/180*pi())]; vLy = C9 = [= vL*sin(aL/180*pi())].

[3] See cell B11! Concatenation of text and variables.

	H	I	J	K	L
14	**velocity vectors**				
15	scaling factor		**scal**	0.2	
16	**tt**	**xAtt**		**yAtt**	
17	818				
18	**0.818**	1.73	*=vLx*tt*	2.42	*=yL+vLy*tt-g/2*tt^2*
19	0.818	2.15	*=I18+scal*vLx*	1.73	*=K18+scal*(vLy-g*tt)*
20					
21		1.73	*=I18*	2.42	*=K18*
22		2.15	*=I19*	2.42	*=K18*
23					
24		1.73	*=I18*	2.42	*=K18*
25		1.73	*=I18*	1.73	*=K19*
26					

	H	I	J	K
27	0.1	0.21		2.40
28	0.1	0.63		3.11
29				
30		0.21		2.40
31		0.63		2.40
32				
33		0.21		2.40
34		0.21		3.11
35				
36	1.1	2.32		1.05
37	1.1	2.75		-0.20
38				
39		2.32		1.05

Fig. 6.3 **a** (left, S) Coordinates (in I18:K19) of an arrow representing a velocity vector and its vertical and horizontal components (I21:K25) at time *tt*, defined in H18; *tt*, x_{Att}, and y_{Att} are the names for the areas H18:H42, I18:I42, and K18: K42, containing the coordinates of the three vectors in Fig. 6.1a. **b** (right, S) Continuation of **a**. Another velocity vector at another time, this time set directly in H27 without a slider; the formulas are structurally the same as those reported in J18:J25 and L18:L25, but with references to different cells

Questions

concerning Fig. 6.3 (S):

The time at which velocity vectors are calculated and attached to the trajectory is set with a slider. What is the LINKED CELL of this slider, and what are probable MIN and MAX? What are the formulas in H18 and H19?[4]

Change the time *tt* with the slider so that the height $y = 0$ is reached for the discharge height and speed in Fig.6.1a and a launch angle 65°. At what time does this occur, and at what speed does the projectile hit the ground?[5]

What is the purpose of the quantity *scal* in the formulas in row 15? How big is it, and what physical unit does it have?[6]

An Excel trick

When you want to specify vectors for several instants of time, you can copy the range H18:K25 in Fig. 6.3a if the formulas are written with relative and absolute cell references so that they remain valid when copied. In Fig. 6.3 **b** (S), the formulas have been copied into the area H27:K34. Regarding cells H27:H28, corresponding

[4] In Fig. 6.3a (S), H17 is the cell linked to the slider (LINKED CELL). MIN = 0, MAX = 1500, as can be estimated from the position of the rider and the number in H17. H18 = H17/ 1000; H19 = H18.

[5] The projectile reaches the ground at $t_{\mathrm{i}} = 1.25$ s and hits with $v = 8.01$ m/s, calculated with.

$$v = \sqrt{v_{0x}^2 + \left(v_{0y} - g \cdot t_i\right)^2}$$

[6] *Scal* = 0.2 s. This parameter determines the length of the velocity vectors in their representation in the plane (x [m], y [m]); see also Exercise 5.5. It occurs in an equation of the kind x [m] = x_0 [m] + *scal* * v [m/s]; *scal* has the unit [s].

to H18:H19, the desired time is entered directly into H27 and H36. As the range H18:K25 has been copied twice, vectors are attached at a total of three points.

6.2.4 Python

Projectile trajectory
In the left cell of Table 6.1, the parameters of the exercise are specified. A label *lbl1* for the legend in a figure is generated. In the right cell, initial horizontal and vertical velocities are determined and the projectile trajectory ($x(t)$, $y(t)$) is calculated.

Drawing arrows that represent vectors
Table 6.2 draws a figure that is similar to Fig. 6.1a. The time instants for the three velocity vectors are specified in list t_2. The coordinates of the arrows representing these vectors are calculated in a for-loop.

In `Python`, in order to draw an arrow, we have to make use of the function `plt.arrow`, which requires, among others, the initial position (x_0, y_0) and the length of the vectors in the x- and y-directions as input. As we prefer to enter begin- and end-points, we have defined a new function *ArrowP*, reported in Table 6.3, with two positional arguments P_0 and P_1, and some keyword arguments with default values.

ArrowP has two positional arguments, tail point P_0 and head point P_1, and three keyword arguments, ls = line style, lw = line width, and hw = head width. In the current situation, the default head width is too large and we have specified $hw = 0.1$ in the function calls (lines 30–32 in Table 6.2).

Table 6.1 Projectile trajectory, with the same data structure as in the spreadsheet of Fig. 6.2

`dt=0.05 #Time interval`	`#Deduced`
`#Prespecified`	`aL*=np.pi/180 #Radian`
`yL=2.0  #Launch height`	`vLx=vL*np.cos(aL)`
`aL=65   #Launch angle`	`vLy=vL*np.sin(aL)`
`vL=5.0  #Launch speed`	`#Projectile trajectory:`
`g=9.81  #Gravit. accel.`	`t=np.arange(0,1.45+dt,dt)`
`lbl1="aL="+str(aL)+"; vL="+str(vL)`	`x=vLx*t`
	`y=yL+vLy*t-g/2*t**2`

Table 6.2 Python program for drawing arrows representing velocity vectors at the trajectory; the loop in the right cell corresponds to Fig. 6.3

```
FigStd('x',0,6,1,'y',-2,4,1)
plt.plot(x,y,'kD-',ms=2,label=lbl1)
plt.legend()

 #Velocity vectors at three time instants t2
t2=[0.1,0.8,1.1]
scal=0.2
for tt in t2:
    vXtt=vLx                        #Constant horizontal velocity
    vYtt=vLy-g*tt                   #Uniform vertical motion
    xtt=vXtt*tt
    ytt=yL+vLy*tt-g/2*tt**2
    P1=[xtt,ytt]                    #Foot position of arrow
    Ax=xtt+vXtt*scal                #Tip position of arrow
    Ay=ytt+vYtt*scal
    ArrowP(P1,[Ax,Ay], hw=0.1)
    ArrowP(P1,[Ax,ytt],hw=0.1)
    ArrowP(P1,[xtt,Ay],hw=0.1)
```

Table 6.3 User-defined function for drawing arrows

```
def ArrowP(P0, P1, c="k", ls='-' ,lw=1, hw=0.4):
    #c has to be given as c="k", not c='k' (2020)
    (x0,y0)=P0
    (x1,y1)=P1
    plt.arrow(x0,y0,x1-x0,y1-y0,
        length_includes_head=True,
        head_width=hw, fill=False,
        linestyle=ls, color=c, linewidth=lw)
```

6.2.5 Animation of Figures with FuncAnimation[7]

We are going to set up an animated version of Fig. 6.1a by extending the program presented in Table 6.1 that provides all data $x(t)$, $y(t)$, $v_x(t)$, and $v_y(t)$ that are accessed in the following program as global arrays.

Creating a figure and a subplot object

In Table 6.4, the sublibrary *animation* is imported from matplotlib. A figure object *fig* is set up and its default font size is set to 7 points (lines 4 and 5). In general, an array of *r* x *c* subplots can be introduced into the frame of a figure. The instruction is `add_subplot(rcn)` with *r* and *c* specifying the number of rows and columns. The index *n* indicates the individual subplot, starting at 1 in the upper left

[7] Matplotlib.pyplot.subplots—Matplotlib 3.4.1 documentation

Table 6.4 Setting up a figure and a subplot object

```
%matplotlib notebook
import matplotlib.animation as animation
cm = 1/2.54                    #Centimeter in inches
fig=plt.figure(figsize=(9*cm,9*cm))
plt.rcParams.update({'font.size': 7})
ax=fig.add_subplot(111)  #1 row, 1 column, 1=top left
ax.set_xlim(0,6)
ax.set_ylim(-2,4)
ax.set_xlabel('x [m]',size=9)
ax.set_ylabel('y [m]',size=9)
```

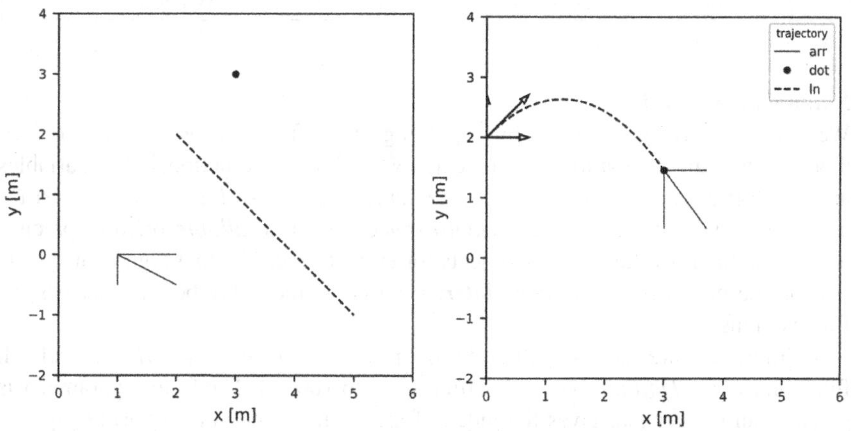

Fig. 6.4 Presenting the objects *arr*, *dot*, and *ln* **a** (left) with toy data, **b** (right) at the end of the animation

corner and increases to the right. We need only one subplot, so the instruction is `add_subplot(111)`. Resulting figures are shown in Fig. 6.4 with different plot objects.

`%matplotlib` is a magic function that renders the figure in a notebook instead of displaying a dump of the figure object.[8]

Creating plot objects

We first create some toy data. In Table 6.5, three two-dimensional print objects *arr*, *dot*, *ln* are created. In Table 6.5a, the lists provided for the x and y data are empty but the styles (line or dot) are already specified. Furthermore, a legend object *leg* is created with captions for the three plot objects. In Table 6.5b, the three plot objects are provided with some toy data. Running this cell, yields Fig. 6.4a.

[8] https://stackoverflow.com/questions/43027980/purpose-of-matplotlib-inline.

Table 6.5 Creating plot objects **a** (top) with empty lists **b** (bottom) with toy data, result shown in Fig. 6.4a

```
 #Creating plot objects
arr,=ax.plot([],[],'k-',lw=0.5)
dot,=ax.plot([],[],'ko',ms=3)
ln, =ax.plot([],[],'k--',lw=1)
leg = ax.legend(['arr','dot','ln'])

leg.set_title('trajectory',prop={'size':6})
```

```
arr,=ax.plot([1,2,1,1,1,2],[0,0,0,-0.5,0,-0.5],
                                        'k-',lw=0.5)
dot,=ax.plot([3],[3],'ko',ms=3)
ln, =ax.plot([2,3,4,5],[2,1,0,-1],'k--',lw=1)
plt.savefig("PhEx 6-2 Trajectory initial",dpi=1200)
```

Making an animation

We get a "living figure" by applying the function `FuncAnimation` that creates an animation by repeatedly calling a function.[9] Its variables are `FuncAnimation(`*`fig`*, *`func`*, *`frames = None`*, *`interval = 200`*, *`init_func = None`*, ...). *Fig* is a *figure* object. *func* is a *callable*, or, more precise, a function that refreshes the plot objects for every frame. The first argument in *func* will be the next value in *frames*. *Interval* specifies the delay between successive frames in ms.

In Table 6.4, we have specified the figure as *fig* and in Table 6.7a, the refresh function as *animDotLine*. The delay time is set to 100 ms. The frames parameter is given as an integer that gives the index of the position and velocity arrays up to a certain time. The final result is shown in Fig. 6.4b (from Table 6.6).

The refresh function *animDotLine* is given in Table 6.7a. The variable in the function header is taken as index for the global arrays for position and velocity.

The plot objects are refreshed with the instruction `set_data`. For the projectile (*dot*) it is just the current position (one point) . The trajectory itself (*ln*) is represented by the curve traversed so far; the data are obtained by appending the current position to the lists x_{Lin} and y_{Lin}. The data x_{Arr} and y_{Arr} for the velocity arrows are calculated in another function *coArr*, reported in Table 6.7b.

We take the plot objects in this exercise as prototypes for other exercises:

dot single point,
arr new picture,
ln continued curve.

[9] https://matplotlib.org/stable/api/_as_gen/matplotlib.animation.FuncAnimation.html

Table 6.6 Creating an animation object recurring to the figure object *fig* and the function *animDotLine*; plotting arrows for indices in a list (here only one element)

```
xLin, yLin =[],[]
xArr, yArr =[],[]
scal=0.2
dotAnim=animation.FuncAnimation(fig,animDotLine,
     frames=int(len(t)*0.6),interval=100, repeat=False)

for k in [2,11,20]:
    xArr,yArr=coArr(k)
    for i in np.arange(0,5,2):
        ArrowP([xArr[i],yArr[i]],[xArr[i+1],yArr[i+1]])
plt.show()
plt.savefig("PhEx 6-2 Trajectory Anim partly",dpi=1200)
```

Table 6.7 **a** (left) Refresh function in the animation **b** (right) Function called in *animDotLine* for each frame

```
def animDotLine(i):
    xLin.append(x[i])
    yLin.append(y[i])
    dot.set_data(x[i],y[i])
    ln.set_data(xLin,yLin)
    xArr,yArr=coArr(i)
    arr.set_data(xArr,yArr)
 #x,y from Table 6.1
 #xLin,yLin from Table 6.6
```

```
def coArr(i):#Coeff. arrow
    xi,yi=x[i],y[i]
    vxi,vyi=vx[i],vy[i]
    xArr=[xi,xi+vxi*scal]
    #2 elements
    yArr=[yi,yi+vyi*scal]
    xArr+=(xi,xi+vxi*scal)
    #4 elements
    yArr+=(yi,yi)
    xArr+=(xi,xi)
    #6 elements
    yArr+=(yi,yi+vyi*scal)
    return xArr, yArr
```

Questions

The font size for the axis titles in Fig. 6.4b seems to be too small. What is the reason and how can you change it?[10]

How to provide the velocity vectors with arrow heads?[11]

[10] The default font size was set with `plt.rcParams.update({font.size':7})`. With such an instruction you can change the whole font, not only its size.

[11] Use unit-line and perpendicular vectors, see Chap. 5.

Does the time development of the animation reflect the true development?[12]

In this exercise and most easily in all other exercises, all coordinates of the plot objects are calculated in the main program and stored in arrays that are then accessed for the animation. The animation is thus an addition to the main program. The advanced programmer can calculate the coordinates on the run, individually and temporarily for every frame.

6.3 Cycloid, Rolling Curve (R-T)

We consider the movement of a point fixed on a wheel rolling along a straight horizontal line. Viewed from the laboratory system, it is composed of a rotation about the wheel axis and a uniformly progressing translation of the axis parallel to the line. The speed of the point depends on its current altitude above the line.

6.3.1 Trace of a Writing Point Fixed at a Rolling Wheel

We are going to examine the movement of a point ("writing point") on a wheel that rolls along a straight line in a plane as, e.g., shown in Fig. 6.5. The movement of the rolling wheel is composed of a rotation (R) about its axis and a translation (T) of the axis parallel to the plane ("the road"). Rotation period T_W of the wheel and velocity v_A of its axis are related as follows:

$$v_A = \frac{2\pi r_W}{T_W} \text{ with } r_W = \text{radius of the wheel} \tag{6.6}$$

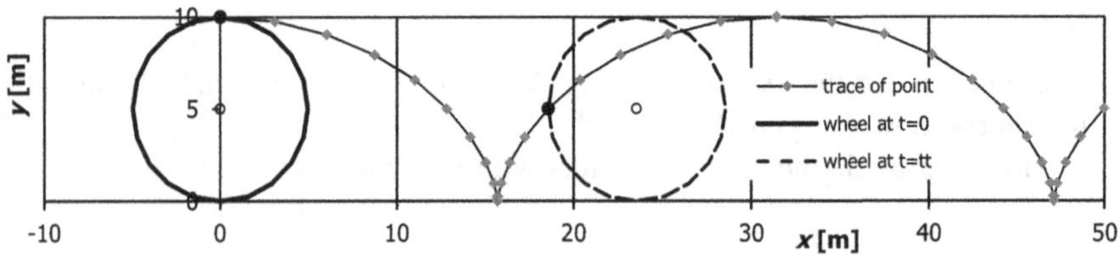

Fig. 6.5 Trace of a writing point on a wheel rolling along a straight line, always with the same time interval between two adjacent marks

[12] The time development of the animation reflects the true time development because the frames variable i is proportional to the time.

During a circulation period T_W, the wheel unrolls once on the road so that the axis covers a distance $2\pi r_W$, the length of the circumference.

For the coordinates (x_P, y_P) of the writing point and its speed v_T along its trajectory, the following equations hold:

$$\begin{aligned} x_P &= x_{ArA} + x_C = r_p \cdot \cos(\omega_W t) + x_A \\ y_P &= y_{ArA} = r_P \cdot \sin(\omega_W t) \end{aligned} \tag{6.7}$$

$$v_t = \frac{\Delta s}{\Delta t} = \frac{\sqrt{\Delta x^2 + \Delta y^2}}{\Delta t} \tag{6.8}$$

The index “ArA” designates rotation around the center (the axis), and x_A the translation of the axis.

► **Tasks** Represent the wheel and the writing point graphically at any time, to be set with a slider in the spreadsheet! Set-up a corresponding `Python` program with animation!

Show the trajectory $y = y(x)$ of the writing point in the same diagram!

Determine the writing point's speed along its trajectory by numerical differentiation!

Speed of the writing point

The speed of a writing point on the wheel's rim along its trajectory is shown in Fig. 6.6a, together with its average speed.

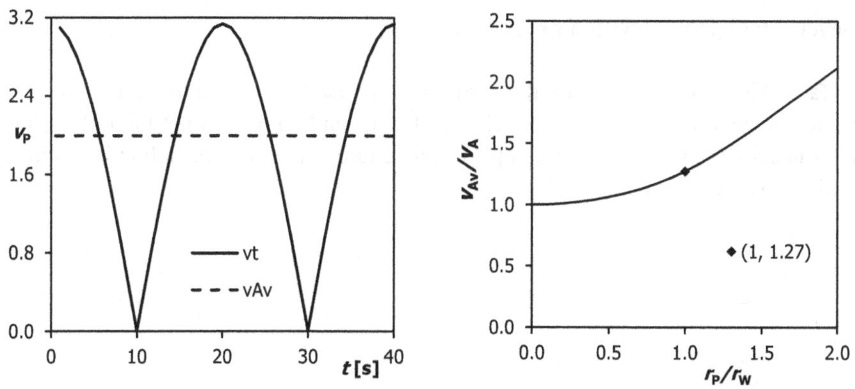

Fig. 6.6 **a** (left) Speed v_P of a point on the rim of the wheel in [m/s] as a function of time ($r_P = r_W = 5$ m). **b** (right) Mean trajectory velocity v_{Av} of the point relative to the velocity v_A of the axis, as a function of the distance r_P of the point P to the wheel axis relative to the radius r_W of the wheel

Questions

At which points of the wheel is the writing point's speed minimum and at which is it maximum?[13]

How do you calculate the speed $v_T(t)$ of the writing point along its trajectory numerically from $x_P(t)$ and $y_P(t)$ at time t, when the distance between the interpolation points is Δt?[14]

Over which t value is the numerically calculated path velocity to be plotted?[15]

What is the orbital period of the wheel in Fig. 6.6a?[16]

What is the speed v_A of the axis in Fig. 6.6a?[17]

When the writing point touches the ground, its speed disappears, here, at $T =$ 10 s and 30 s. Its speed is maximum ($v = 2v_{Axis} = \pi$ m/s) when it is at the rotating wheel's highest point.

The average speed of the writing point is indicated in Fig. 6.6a by the horizontal dashed line. To get the average speed numerically, it is essential to average over whole cycle times, e.g., over two cycle times, as in Fig. 6.6a.

The average speed depends on the distance of the writing point from the axis of the wheel. In Fig. 6.6b, the average speed of the point relative to the speed of the axis is plotted as a function of the distance of the writing point from the axis. When the writing point is on the axis, its average speed is equal to the speed of the axis. When it is on the rim, the speed ratio is $1.273 = 4/\pi$.

▶ **Task** Determine the point's mean path velocity as a fraction of the axis velocity for different distances of the point to the axis as in Fig. 6.6b!

Several trajectories with a procedure

▶ **Task** Vary the distance of the writing point r_P from the axis at time $t = 0$ systematically, e.g., as in Fig. 6.7. The distance can be greater than the radius, and also negative. In EXCEL, use a rep-log procedure; in `Python`, define a function with r_P as the argument!

[13] Minimum speed: point on the road ($v = 0$); maximum speed ($v = 2v_{Axis}$): at the highest point of the wheel.

[14] $\vec{v}_T = (v_x; v_y) = ((x(t) - x(t - \Delta t))/\Delta t; (y(t) - y(t - \Delta t))/\Delta t)$ then $|v_T| = \sqrt{v_x^2 + v_y^2}$.

[15] The speed is plotted over the center of the interval for which the velocity is numerically calculated.

[16] $T = 20$ s, period of the speed profile, speed of the axis $= 2\cdot\pi \cdot 5/20 = 1{,}57$ [m/s].

[17] Speed of the axis $v_A = \frac{2\pi \cdot r_A}{T_w} = \frac{2\pi \cdot 5}{20} = \frac{\pi}{2}\left(\frac{m}{s}\right)$

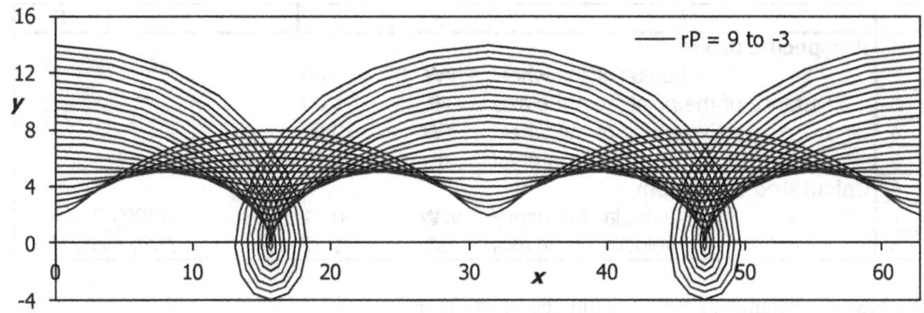

Fig. 6.7 Trajectories of writing points on a wheel with various distances r_P to the axis

Brachistochrone

Johann Bernoulli discovered, in 1696, that cycloids solve the brachistochrone problem. A brachistochrone curve is a trajectory on which a body in the homogeneous gravitational field of the earth glides fastest from a starting point to a lower endpoint.

To describe such trajectories with our data structure, we would have to set ω_W positive and $y_R = -r_R$, so that the wheel, hanging from a ceiling, unrolls to the right. The resulting curve is essentially part of one of the curves in Fig. 6.7 reflected at the x-axis. It indeed has the property of a brachistochrone, as we shall see in the chapter "Variational Calculus" of the follow-up textbook "Physics with Excel and Python. Applications".

6.3.2 Data Structure and Nomenclature

r_W	radius of the wheel
r_P	distance of the writing point to the wheel axis
T_W, w_W	cycle time, and angular velocity of the wheel
$v_A=(2\pi r_W)/T_W$	velocity of the axis
t	series of equidistant (dt) time instants
t_C	center of the time intervals of t
x_{ArA}, y_{ArA}	rotation of the point around the axis
x_A	horizontal position of the axis
x_P, y_P	trace of the writing point in the xy-plane
v_P	velocity of the writing point along its trajectory
tt	one specific time point.

	A	B	C	D	E	F
1	**Prespecifications**					
2	Radius of the wheel	**rW**	5.00 m			
3	Distance of the point from the axis	**rP**	5.00 m			
4	Cycle time of the wheel	**TW**	20.00 s			
5	Time interval	**dt**	1.00 s			
6	**Calculated therefrom**					
7	Angular frequency	**wW**	-0.31 1/s		*=-2*PI()/TW*	
8	Velocity of the axis	**vA**	1.57 m/s		*=2*PI()*rW/TW*	

Fig. 6.8 (S) Parameters for the rolling curve in Fig. 6.5

6.3.3 Excel

Parameters of the motion

In Fig. 6.8 (S), the movement parameters are specified in named cells, in particular, the radius r_W and the cycle time T_R of the wheel. From the given parameters, we derive angular frequency ω_W, height $y_{P,}$ and velocity v_A of the axis. The distance of the writing point to the axis of the wheel r_P need not be equal to the radius r_R of the wheel. It may be bigger or smaller. The wheel will move on the plane to the right, rotating clockwise. The angular frequency ω_R, with which the polar angle is calculated, is therefore negative.

Questions

From which axis is the angle φ of the plane polar coordinates measured?[18]

What is the angle φ when the writing point is at the highest point of the wheel?[19]

What is the angle φ when the writing point is at the height of the axis?[20]

What sign does ω_W have in a right-handed coordinate system when the wheel rotates clockwise?[21] Compare with Fig. 6.8 (S)!

Trace of the writing point

In Fig. 6.9 (S) , the coordinates (x_P y_P) of the writing point are calculated for 41 instants of time, with intermediate calculations for the rotation (x_{ArA}, y_{AA}) of the point about the axis of rotation (C:D) and for the translational motion (x_A) of the axis (column E).

[18] The angle φ is measured from the positive x axis.

[19] $\varphi = \pi/2 = 90°, 90° + 360°, 90° + 360° + 360°, \ldots$

[20] $\varphi = 0 = 0°$.

[21] The angular frequency is negative when the point runs clockwise.

	B	C	D	E	F	G	H	I	J
9						vAv	1.99 m/s		=AVERAGE(vP)
10		wheel at t=0.5		trace of point			1.268		=vAv/vA
11	=B15+dt	=rP*COS(wW*t+PI()/2)	=rP*SIN(wW*t+PI()/2)+rW	=vA*t	=xArA+xA	=yArA	=SQRT((xP-F13)^2+(yP-G13)^2)/dt	=(t+B13)/2	
12	t	xArA	yArA	xA	xP	yP	vP	tC	
13	0.5	0.78	9.94	0.79	1.57	9.94			
14	**1.5**	2.27	9.46	2.36	4.63	9.46	**3.10**	**1**	
53	40.5	0.78	9.94	63.62	64.40	9.94	3.14	40	

Fig. 6.9 (S) Coordinates of a writing point when unrolling the wheel; rotation of the point around the wheel axis (x_{ArA}; y_{ArA}); translation of the axis of the wheel (x_A); addition of the two movements yields ($x(t)$; $y(t)$); speed along the path v_t

The corresponding equations are

$$x_{arA}(t) = r_P \cos(\omega t) + x_R;\ y_{arA}(t) = r_P \sin(\omega t) + r_W \tag{6.9}$$

Here, it is assumed that, at $t = 0$, the writing point is vertically above the axis. An example can be found in Fig. 6.5. You can extend the solution by allowing the selected point to assume any position at $t = 0$ by choosing the phase shift within the circular functions!

In column E, there is the displacement of the x-coordinate of the center of the wheel over time (translation in the x-direction). The coordinates of the rotating point in the laboratory system are in columns F and G, calculated from the superposition (component-wise addition of the Cartesian coordinates) of the rotation about the wheel axis (x_{arA}, y_{arA}) and the translation of the axis (x_M, 0):

$$x(t) = x_A + x_{arA} \quad y(t) = y_{arA} \tag{6.10}$$

The resulting trajectory is called a cycloid (rolling curve).

Question

In which cell of Fig. 6.9 (S) is the mean path velocity calculated?[22]

▶ **Task** Determine the mean path speed of the point relative to the axis' speed for different distances of the point to the axis! A typical evaluation can be seen in Fig. 6.6a, b. It is best to use a VBA rep-log procedure that varies the distance and logs the average speed!

[22] The mean speed v_{Av} is calculated in H9 of Fig. 6.9 (S).

Fig. 6.10 (S) Position of the point $(x(t_2); y(t_2))$ at time tt; wheel at $t = 0$ (x_{rad}) and $tt = t_2$ (x_{t2}); tt is set with the slider. The formulas are the same as in Table 6.10

	K	L	M	N	O	P
10		point at t=tt		axis at tt	wheel at t=tt	
11		=rP*COS(wW*tt+PI()/2)+xAtt	=rP*SIN(wW*tt+PI()/2)+rW	=vA*tt	=xArA+xAtt	
12	tt	xPtt	yPtt	xAtt	xWtt	
13	15	18.56	5.00	23.56	24.34	
14					25.83	
15					27.10	
53					24.34	

Wheel at time tt

In Fig. 6.5, the wheel is shown at $t = 0$ and at a second time tt. In Fig. 6.10 (S), the coordinates of the wheel are determined for tt specified with a slider. The writing point has the coordinates (x_{Ptt}, y_{Ptt}). The coordinates of the wheel are (x_{Wtt}, y_W), with y_W being the y-coordinates calculated for $t = 0$. If you put the pointer into the right part of the slider and keep it pressed down, the wheel rolls to the right.

6.3.4 Python

The `Python` program exhibits the same structure as the formula network in the spreadsheet. In Table 6.8, the parameters of the task are set.

In Table 6.9, the wheel's coordinates at the start and the trace of the writing point are calculated. The velocity $v_P = ds/dt$ along the trace of the point is calculated in line 15. The length ds of the trace sections is obtained by slicing x_P and y_P.

The coordinates of the wheel and the writing point at a specific time $t = tt$ are obtained with the function *WheelAt* in Table 6.10, which has only tt as an argument and resorts otherwise to global variables.

The plot program for yielding a figure like Fig. 6.5, with the arrays calculated in Table 6.9, is given in Table 6.11. The coordinates for the wheel and the writing point at a specific time $t = tt$ are obtained by calling the function *WheelAt(...)* in Table 6.10.

Table 6.8 Setting the parameters for the rolling wheel, the same as in Fig. 6.8 (S)

```
 #Prespecified:
rW=5.0              #Radius of the wheel
rP=5.0              #Distance of the point from the axis
TW=20               #Period of wheel rotation
dt=1.0              #Time increment
 #Calculated therefrom:
wW=-2*np.pi/TW      #Angular frequency
vA=2*np.pi*rW/TW    #Velocity of the axis
```

Table 6.9 Calculation of the coordinates of the wheel at its start and of the trace of the writing point

```
 #Coordinates of point vs. time:
t=np.arange(0,40+dt,dt)
 #Coordinates of the wheel at start:
xArA=rW*np.cos(wW*t+np.pi/2)
yArA=rW*np.sin(wW*t+np.pi/2)+rW
xA=vA*t                  #From velocity of axis
xP=xArA/rW*rP+xA         #Writing point
yP=rP*np.sin(wW*t+np.pi/2)+rW
 #Veloc. along the trajectory of the point:
vP=np.sqrt((xP[1:]-xP[:-1])**2+(yP[1:]-yP[:-1])**2)/dt
tC=(t[1:]+t[:-1])/2  #Valid for vP
```

Table 6.10 Function for specifying the coordinates of the wheel and the writing point at $t = tt$

```
def WheelAt(tt):
    #Wheel and point at t = tt:
    xAtt=vA*tt              #Position of axis
    xPtt=rP*np.cos(wW*tt+np.pi/2)+xAtt
    yPtt=rP*np.sin(wW*tt+np.pi/2)+rW
    xWtt=xArA+xAtt          #x Coordinates of the rim
    return xAtt,xPtt,yPtt,xWtt
```

Table 6.11 Plot program yielding a figure like Fig. 6.5

```
FigStd('x',-10,50,10,'y',0,10.0,2.5,xlength=12,ylength=4)
plt.plot(xArA,yArA,'k-')    #Wheel at t = 0
plt.plot(xP[0],yP[0],'ko') #Point at t = 0
plt.plot(xP,yP,'k-')        #Trace of point

xAtt,xPtt,yPtt,xWtt=WheelAt(tt=15)
plt.plot([xP[0],xAtt],[rW,rW],'ko', fillstyle='none')
                            #Pos. of axis
plt.plot(xWtt,yArA,'k--')   #Wheel
plt.plot(xPtt,yPtt,'ko')    #Selected point
plt.axis('scaled')
```

A meaningful animation could comprise the functions *dot* for the writing point, *ln* for the cycloid, *arr* for the wheel from Sect. 6.2.5.

6.4 Foucault's Pendulum (T-R)

We calculate the trace of a pendulum swinging in the laboratory system (T) on a rotating table (R).

6.4.1 A Lecture Experiment

In a lecture experiment about Foucault's pendulum, a thread pendulum swings over a rotating plate, writing a trace thereon. In the laboratory system, the pendulum swings in a plane, with its suspension point being located in the axis of rotation of the plate.

Figure 6.11a shows the trace of the pendulum for a period of oscillation of $T_P = 1.2$ s and a rotation time of the table of $T_r = 9$ s. The partial circle "Stylo" represents the trace of a pen, resting in the laboratory system, along the rotating plate to indicate the sense of rotation.

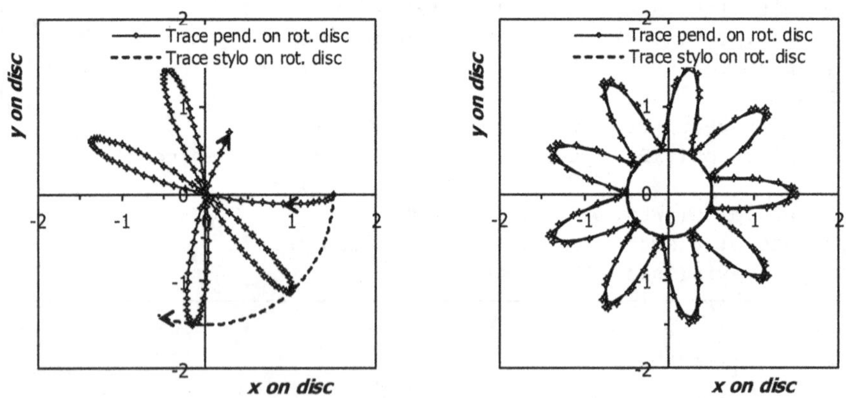

Fig. 6.11 **a** (left) Traces of a pendulum oscillating in the laboratory system ($T_P = 1.2$ s) and of a stylus at rest in the laboratory system on a rotating plate ($T_T = 9$ s), the suspension point of the pendulum being in the axis of rotation of the plate; the unit length is 1 cm, as explained in the main text. **b** (right) Closed track of a pendulum whose suspension point is not in the axis of rotation ($T_P = T_T/9$)

Questions

A thread pendulum is swinging with a period of 12.7 s. How long is the pendulum?[23]

In what period of time does the earth rotate by 1°?[24]

What are the amplitude of the pendulum and the horizontal displacement of the suspension point against the rotation axis in Fig. 6.11b?[25]

Under what condition do closed tracks occur in the lecture experiment on Foucault's pendulum? What is the concrete condition in Fig. 6.11b?[26]

For simplification, we assume that the trace of the pendulum on the plate at rest or, more generally, in the *laboratory system* is described by

$$x_P = A_P \cdot \cos(\omega_P t) \tag{6.11}$$

Then, we let the plate rotate around its vertical axis. The trace of the pendulum on the rotating table is composed of the oscillation in the x-direction (T, in the laboratory system) and an angular displacement on the table according to its rotation (R). The equations for the conversion of the coordinates (x_L, y_L) in the laboratory system into the coordinates (x_T, y_T) of the rotating table are

$$x_T = x_P \cdot \cos(\omega_T t) \quad \text{and} \quad y_T = x_P \cdot \sin(\omega_T t) \tag{6.12}$$

where ω_T is the angular frequency of the rotating table. These equations are a special case of the general form for a counter-clockwise rotation by ϕ applying a rotational matrix:

$$\begin{bmatrix} x_T \\ y_T \end{bmatrix} = \begin{bmatrix} \cos\phi & -\sin\phi \\ \sin\phi & \cos\phi \end{bmatrix} \cdot \begin{bmatrix} x_P \\ y_P \end{bmatrix} \tag{6.13}$$

Where have all the units gone?

► **Tim** In Fig. 6.11, we have not specified any physical units for the lengths.

► **Alac** That's no problem: Times in seconds, lengths in meters. That's standard.

---

[23] $T = 12{,}7$ s, $\omega = 2\pi/T = \sqrt{g/l}$, → length l of the pendulum = 40 m.

[24] The earth rotates by 360° in one day, → in 4 min, by 1°; $\Delta t = 1°/360° * 24 * 60 * 60$ s = 240 s.

[25] The deflection of the pendulum is from 0.5 cm to 1.5 cm. The suspension point of the pendulum is shifted by a distance 1 cm against the center. The amplitude is 0.5 cm.

[26] The ratio of the period of the pendulum and the circulation time of the plate must be a natural number. In Fig. 6.11b, the ratio is 9 to 1. The pendulum makes nine oscillations during one turn of the plate.

▶ **Mag** Concerning calculation, everything is clear. But does that make sense physically? How long is the pendulum?

▶ **Tim** From the oscillation period, $T_P = 1.2$ s, as stated in the caption, it follows that $l = 36$ cm.

▶ **Mag** How does the maximum swing fit in with that?

▶ **Alac** I admit: A deflection of 1.8 m does not fit with the pendulum length. So, let's decide that the pendulum should be deflected by just 1.8 cm.

▶ **Mag** So, the unit of length in Fig. 6.11 is 1 cm. Indeed, this is experimentally difficult to record, but at least our calculation is consistent.

▶ **Tasks** Create a spreadsheet calculation/a `Python` program for the experiment described above and vary the pendulum's oscillation duration and the table's cycle time!

Check if the pendulum track is as expected (a) when the rotation period of the table is large compared to the period of oscillation and (b) when the oscillation period and rotation period are identical!

Change the calculation for the case in which the table's rotational axis is still in the plane of the swinging pendulum but no longer passes through the suspension point of the pendulum! An example can be seen in Fig. 6.11b.

6.4.2 Data Structure and Nomenclature

A_P	amplitude of the pendulum
T_P, w_P	oscillation period of the pendulum and corresponding angular velocity
T_r, w_r	rotation period of the table and corresponding angular velocity
x_{Sh}	shift of the suspension point with respect to the rotational axis
t	series of equidistant (dt) time instants
x_P	position of the pendulum at t in the lab system
x_T, y_T	trace of the pendulum on the table
x_{St}, y_{St}	trace of a stylus at rest in the lab system on the table, to check the direction of rotation of the table.

6.4.3 Excel

Setting the parameters
The parameters for the movement in Fig. 6.11 are specified in Fig. 6.12 (S). The quantity x_{Sh} (in C6) determines the displacement of the suspension point against the plate's axis of rotation.

	A	B	C	D	E
1	**Prespecifications**				
2	Amplitude of pendulum	**Ap**	1.50		0.50
3	Period of pendulum	**Tp**	1.20		1.00
4	Period of rotation	**Tr**	9.00		9.00
5	Time interval	**dt**	0.0173		0.09
6	Suspension point vs rot. axis	**xSh**	0.00		1.00
7	**Calculated therefrom**				
8	Angular frequency pendulum	**wP**	5.24	*=2*PI()/Tp*	
9	Angular frequency rotating disc	**wR**	-0.70	*=-2*PI()/Td*	

Fig. 6.12 (S) Specifications for the movement presented in Fig. 6.11; values in column C for partial picture **a**, those in column E for partial picture **b**

	B	C	D	E	F	G	H
12	=B14+dt	=Ap*COS(wP*t)+xSh	=xP*COS(wR*t)	=xP*SIN(wR*t)	=Ap*COS(wR*t)	=Ap*SIN(wR*t)	
13	**t**	**xP**	**xT**	**yT**	**xSt**	**ySt**	
14	0.000	1.50	1.50	0.00	1.50	0.00	
15	**0.017**	1.49	1.49	-0.02	1.50	-0.02	
174	2.768	-0.52	0.18	0.49	-0.53	-1.40	

Fig. 6.13 (S) In column C, the pendulum motion $x_P(t)$ is calculated in the laboratory system. In columns D and E, this movement is transformed (to x_T, y_T) into the coordinate system of the rotating table. In columns F and G, the coordinates on the rotating plate of a point ("Stylo") fixed in the laboratory system are calculated

Question

How long is the pendulum in Fig. 6.12 (S) when the oscillation period T_P is given in seconds?[27]

Trace of pendulum

The movement itself is calculated in Fig. 6.13 (S) for the pendulum swinging in the x-direction.

▶ **Task** Complete the diagram with two points representing the pendulum's positions and the pen at a selectable time! In Fig. 6.11a, this was done for $t = 0.2249$ (arrow close to $x = 1$). A suggestion: Use a slider to select a row from 14 to 174 and copy the coordinates from that line to an area added to the diagram as a point. You can use the reference type INDIRECT for this purpose.

[27] $T = 1.2$ s, $\omega = 2\pi/T = \sqrt{g/l}$, length l of the pendulum = 0.36 m.

Table 6.12 Foucault's pendulum, specification of the parameters, the same as in Fig. 6.12 (S)

```
 #Prespecified:
Ap=1.5                  #Amplitude of the pendulum
Tp=1.2                  #Period of the pendulum
Tr=9.0                  #Period of the rot. disc
dt=0.0173               #Time interval
xSh=0.0                 #Shift of suspension point
 #Calculated therefrom:
wP=2*np.pi/Tp           #Circ. freq. of the pendulum
wR=-2*np.pi/Tr          #Circ. freq. of the disc
```

Table 6.13 Setting up the arrays describing the motion of the pendulum in the laboratory system and its trace on the rotating table

```
t=np.arange(0,2.768+dt,dt)
 #Traces:
xP=Ap*np.cos(wP*t)+xSh #Pendulum in lab
xT=xP*np.cos(wR*t)     #Trace of pendulum on table
yT=xP*np.sin(wR*t)
xSt=Ap*np.cos(wR*t)    #Trace of stylus on table
ySt=Ap*np.sin(wR*t)
```

6.4.4 Python

In Table 6.12, the parameters of the swinging pendulum are specified, with the same values as in Fig. 6.12 (S).

The arrays describing the motion of the pendulum in the laboratory system and its trace on the rotating table are set up in Table 6.13, together with the trace of a stylus fixed in the laboratory system

To get figures such as those in Fig. 6.11, we apply the program in Table 6.14. The parameters in Table 6.12 are for Fig. 6.11a. To get Fig. 6.11b, the parameters in column E of Fig. 6.12 (S) have to be inserted into Table 6.12.

Remember: Within brackets or parentheses, line breaks are allowed after punctuation marks, as is applied in Table 6.14, line 5. Explicit line breaks are possible after a backslash (\) as in line 2.

Questions

concerning Table 6.14:
How many positional arguments are in the header of *ArrowP*?[28]
What is the first argument in the header *of ArrowP* in line 9?[29]

[28] *ArrowP(P0,P1,...)* has two positional arguments, foot point P_0 and head point P_1.
[29] *P0* = [xT[−2],yT[−2]].

Table 6.14 Plotting the arrays obtained in Table 6.13 to get a picture like that in Fig. 6.11a, *ArrowP* from Table 6.3

```
FigStd('x',-2.0,2.0,0.5,'y',-2.0,2.0,0.5)
plt.plot(xT,yT,'k-x',ms=3,label='pend\
                              ulum on disc')
plt.plot(xSt,ySt,'k--',label='stylo on disc')
ArrowP([xSt[-2],ySt[-2]],[xSt[-1],
                              ySt[-1]],hw=0.1)
i=10
ArrowP([xT[i],yT[i]],[xT[i+1],yT[i+1]],hw=0.1)
ArrowP([xT[-2],yT[-2]],[xT[-1],yT[-1]],hw=0.1)
plt.legend()
```

To which arrows in Fig. 6.11a do the three calls of the function *ArrowP* correspond?[30]

▶ **Task** Set up an animation in the laboratory system, with the pendulum swinging horizontally and the table rotating! The frames should be the equidistant time instants t to mimic the oscillation.

6.5 Anchor, Deflected Out of Its Rest Position (R-R)

The rotation of an anchor about its suspension point is described as a rotation (R) of the center of gravity about the suspension point and a rotation (R) of the anchor about the center of gravity (if the anchor is not suspended there). The moment of inertia is calculated using Steiner's theorem.

6.5.1 Deflected Anchor

Coordinates of the deflected anchor

We consider the rotation of an anchor about a point located in the origin of the coordinate system. In Fig. 6.14a, the anchor is held at the end S of the stem, in Fig. 6.14b, at its center of gravity C_g. The construction of the anchor is simplified with four mass points attached to the ends of a hanging T (see the inset in Fig. 6.14b).

[30] First: end of trace "stylo"; second: initial phase ($i = 10$) of trace "pend"; third: end of trace "pend".

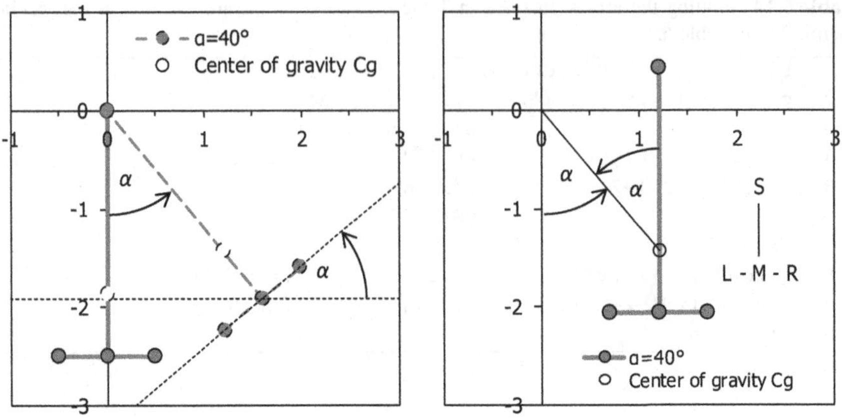

Fig. 6.14 An anchor is deflected by an angle $\alpha = 40°$, held **a** (left) at the upper end of the stem, **b** (right) at its center of gravity. The four characteristic point masses are designated by the letters S, L-M-R. The center of gravity is marked by an open circle

In the following programs, the anchor's characteristic points are listed as [L, M, S, M, R] (Left, middle, top, middle, right; see Fig. 6.14b). We have doubled M because the anchor can then be drawn in one uninterrupted line. The distances L-M = M-R and S-M are specified, respectively, as r_A and l_A, so that the coordinates of the anchor at rest are defined as

$$x_A = [-r_A, 0, 0, 0, r_A]$$
$$y_A = [-l_A, -I_A, 0, -l_A, l_A] \tag{6.14}$$

Rotation about the suspension point S

If the anchor is rotated about its top point S (Fig. 6.14a), the displacement of any point R of the anchor, with Cartesian coordinates $(-r_A, -l_A)$ at rest, may be considered the sum of two rotations:

$$\begin{array}{lll} \text{Rotation of point } R \text{ around } M: (-r_A, 0) & \rightarrow & (-r_A \cdot \cos(\alpha), r_A \cdot \sin(\alpha)) \\ \text{Rotation of point } M \text{ around } S: (0, -l_A) & \rightarrow & (l_A \cdot \sin(\alpha], -l_A \cdot \cos(\alpha) \\ \text{With the sum}: \quad (-r_A, -l_A) & \rightarrow & (-r_A \cdot \cos(\alpha) + l_A \cdot \sin(\alpha), \\ & & r_A \cdot \sin(\alpha) - l_A \cdot \cos(\alpha)) \end{array}$$

The total rotation can be presented as a matrix multiplication:

$$= (-r_A, -l_A) \cdot [RotMat] \tag{6.15}$$

with the rotational matrix defined as

$$RotMat = \begin{bmatrix} \cos(\alpha) & \sin(\alpha) \\ -\sin(\alpha) & \cos(\alpha) \end{bmatrix} \tag{6.16}$$

The rotational matrix can be applied to any point in the plane:

$$(x_R, y_R) = (x, y) \cdot [RotMat] \tag{6.17}$$

Here, we have chosen to present the coordinates as row vectors, because this is more convenient in spreadsheet calculations where coordinate vectors (a pair of two numbers) can more clearly be stored in successive rows.

▶ **Task** Calculate the coordinates of the anchor for the freely selectable parameters: length l_A of the stem (A-M), half the length r_A of the crossbar (M-R or M-L), and angle α!

Center of gravity as weighted sum

The center of gravity is defined as the sum over the coordinates of the characteristic points weighted with their mass:

$$\begin{aligned} x_G &= \frac{\left(\sum_i m_i \cdot x_i\right)}{\sum_i m_i} \\ y_G &= \frac{\left(\sum_i m_i \cdot y_i\right)}{\sum_i m_i} \end{aligned} \tag{6.18}$$

This is achieved in EXCEL with

y_G = SUMPRODUCT(y, m)/SUM(m)

and in `Python` with

```
yG = np.dot(y, m)/np.sum(m)
```

The function `np.dot` returns the dot product of two arrays. For 1-D arrays, it is the inner product of the vectors, i.e., the sum of the products of the components, the equivalent of SUMPRODUCT. For 2-D arrays, it is equivalent to matrix multiplication.

The anchor is held at its center of gravity

In Fig. 6.14b, the anchor is held at its center of gravity by a rod (considered to be massless) that can rotate around S and is currently rotated by an angle α. The coordinates of the deflected anchor are the coordinates of the anchor at rest shifted by the coordinates of the center of gravity after rotation by α according to Eq. 6.17 applied to (x_G, y_G). Now, only one rotation is effective.

Moment of inertia

The moment of inertia I is defined as

$$I = \sum_i r_i^2 \cdot m_i \tag{6.19}$$

where r_i is the distance to the axis of rotation, here, $r_i^2 = x_i^2 + y_i^2$.

It is calculated

in Python as: np.sum((x**2+y**2)*m)
in EXCEL as: SUMPRODUCT($x; x; m$) + SUMPRODUCT($y; y; m$)

We calculate the moment of inertia by applying the formulas:

- (a) to a rotation of the anchor about the point S (corresponding to Fig. 6.14a),
- (b) to a rotation of the anchor about its center of gravity,
- (c) to a rotation of the total mass, concentrated in the center of gravity, about S (corresponding to Fig. 6.14b).

The sum of (b) and (c) should be equal to (a), according to Steiner's law.

Animation
We set up an animation with the anchor swinging correctly with sinusoidal time dependence about its suspension point.

6.5.2 Data Structure and Nomenclature

r_A	distance L-M = M-R
l_A	distance M-S
x_A, y_A	coordinates of the anchor at rest (5 elements, combinations of r_A and l_A)
xy_A	= [x_A, y_A] (2D range)
α	angle of rotation
x_R, y_R	coordinates of the deflected anchor, rotated around the suspension point
x_G, y_G	coordinates of the center of gravity for the anchor at rest
x_Gr, y_Gr	coordinates of the center of gravity rotated around S, at the origin of the coordinate system
RotMat	rotational matrix

Equivalence:

EXCEL: [x_R,y_R] = MMULT(xy_A,*RotMat*), xy_A = [x_A, y_A]
Python: [x_R,y_R] = *RotMat* @ [x_A,y_A] or np.matmul (RotMat,[xA,yA])

6.5.3 Excel

Specs

The parameters of the exercise are specified in columns A:E of Fig. 6.15 (S), with the rotation angle being adjusted with a slider. The coordinates of the anchor at rest are specified in columns G:H by inserting—$\pm l_A$ or $\pm r_A$, or 0 when appropriate. The masses attached to these points are given in column I. The coordinates of point M appear twice in x and y, but the second time with zero mass so that the center of gravity and moment of inertia may be calculated in formulas in Fig. 6.16 (S), taking the whole arrays x_A, y_A, and m as input.

The anchor can be made to oscillate by running the time t with a slider and calculating the deflection angle α in C4 with $\alpha = A \cdot \cos(\omega \cdot t)$ with suitable A and ω.

Rotational matrix

The coordinates (x_R, y_R) of the deflected anchor are calculated by applying the rotational matrix *RotMat* in Fig. 6.16 (S) to the coordinates (x_A, y_A) of the anchor at rest.

Center of gravity

The center of gravity C_g and the moments of inertia are calculated in Fig. 6.17 (S). B5 = ["Deflection angle ="&C3&"°". The moment of inertia for a rotation about the suspension point S is calculated twice:

	A	B	C	D	E	F	G	H	I
1	Length of rod	**lA**	2.50	m			**xA**	**yA**	**m**
2	Half-length of cross bar	**rA**	0.50	m		L	-0.50	-2.50	1.0
3		220	40.00	°	*=B3-180*	M	0.00	-2.50	1.0
4		**a**	0.70	rad	*=C3/180*PI()*	S	0.00	0.00	1.0
5	Deflection angle α=40°					M	0.00	-2.50	0.0
6						R	0.50	-2.50	1.0

Fig. 6.15 (S) Anchor parameters in B1:C4; the deflection angle α is determined with the slider in A3 (linked cell = B3) . he coordinates x, y of the line connecting the characteristic points are in columns G and H with the associated masses in column I, set to 1 except for the second reference to point M. B5 = "Deflection angle ="&C3&"°"

	K	L	M	N	O	P
1	**RotMat**			**xR**	**yR**	
2	0.77	0.64		1.22	-2.24	*=MMULT(xyA;RotMat)*
3	-0.64	0.77		1.61	-1.92	
4				0.00	0.00	
5	*=COS(a)*	*=SIN(a)*		1.61	-1.92	
6	*=-SIN(a)*	*=COS(a)*		1.99	-1.59	

Fig. 6.16 (S) Continuation of Fig. 6.15 (S). The rotational matrix *RotMat* is applied to the coordinates x_A and y_A, bound together in Fig. 6.15 (S) in one matrix xy_A = [G2:H6] in Fig. 6.15 (S)

	AA	AB	AC	AD	AE
1	**Center of gravity Cg**			**Moment of inertia for rotation about Cg**	
2	**xG**	**yG**		5.19	*=SUMPRODUCT(xA-xG;xA-xG;m)*
3	0.00	**-1.88**	*=SUMPRODUCT(yA;m)/SUM(m)*		*+SUMPRODUCT(yA-yG;yA-yG;m)*
4	**xGr**	**yGr**		**for rotation of Cg**	
5	**1.21**	**-1.44**	*=MMULT(xGyG;RotMat)*	14.06	*=(xG^2+yG^2)*SUM(m)*
6					
7	**Moment of inertia for rotation about S**				
8		**19.25**	*=SUMPRODUCT(x;x;m)*	**19.25**	*=AD2+AD5*
9			*+SUMPRODUCT(y;y;m)*		

Fig. 6.17 (S) Continuation of Figs. 6.15 (S) and 6.16 (S). AA:AC: center of gravity, moment of inertia for rotation about the point $S = (0, 0)$; AD:AE: moment of inertia with Steiner's theorem as the sum of two rotations calculated with the spreadsheet functions SUMPRODUCT and SUM

- in AB8, with the coordinates (x, y) of the anchor at rest,
- in AD2:AD8, as the sum of the moments of inertia for rotation about the center of gravity C_g and a rotation of the total mass in the center of gravity about (0,0).

Both calculations should yield, according to Steiner's theorem, the same result, and they do.

6.5.4 Python

Specs and rotational matrix

A Python solution corresponding to Figs. 6.15 (S) and 6.16 (S) is given in Table 6.15.

Table 6.15 Swinging anchor, specs as in Fig. 6.15 (S)

```
 #Prespecified:
lA=2.5               #Length of stem
rA=0.5               #Half-length of crossbar
a=40                 #Angle of deflection in °
a*=np.pi/180         #    in rad

 #Coordinates of the anchor at rest:
 #                    L-M-S-M-R
xA=np.array([-rA,0,0,0,rA])
yA=np.array([-lA,-lA,0,-lA,-lA])

 #Rotational matrix:
RotMat=np.array([[np.cos(a),-np.sin(a)],
                 [np.sin(a), np.cos(a)]])
 #Coordinates of the deflected anchor:
[xR,yR]=RotMat@[xA,yA]
```

Animation

In Table 6.16, a function *ancRot* is defined that is applied in the animation. It calculates the coordinates of the anchor and its center of gravity for an angle α. The function *anim* called within *FuncAnimation* converts the frame number i into the angle α through a sine function that mimics an oscillation.

Center of gravity

In Table 6.17, the coordinates of the center of gravity C_g are calculated, at rest and rotated. The dot product is used to calculate the nominator in the fraction for the center of gravity (Eq. 6.18).

A program for drawing the rotated anchor as shown in Fig. 6.14a is presented in Table 6.18.

In Table 6.19, the moments of inertia are calculated when the anchor is rotated:

- (a) about S, the upper end of the stem,
- (b) about Cg, the center of gravity; and
- (c) when the center of gravity is rotated about (0, 0).

Table 6.16 The anchor swings with sinusoidal time dependence about S

```
def ancRot(a):                  #Angle a=alpha
     #Rotational matrix:
    RotMat=np.array([[np.cos(a),-np.sin(a)],
                     [np.sin(a), np.cos(a)]])
     #Coordinates of the deflected anchor:
    [xR,yR]=RotMat@[x,y]
    [xGr,yGr]=RotMat@[xG,yG]
    return xR, yR, yGr, yGr
anc,=plt.plot(xR,yR,'k-o')

def anim(i):
    a=np.pi/4*np.sin(i*0.1)
    xR, yR, xGr, yGr = ancRot(a)
    anc.set_data(xR,yR)

AnchorAnim=animation.FuncAnimation(figA,anim,
    frames=range(180),interval=100,repeat=False)
```

Table 6.17 Coordinates of the center of gravity, at rest and rotated

```
 #Center of gravity Cg
m=np.array([1,1,1,0,1])         #Masses of L-M-S-M-R
xG=np.dot(xA,m)/np.sum(m)       #Coordinates of c of g
yG=np.dot(yA,m)/np.sum(m)
[xGr,yGr]=RotMat@[xG,yG]        #Coordinates of rotated anchor
```

Table 6.18 Program for drawing a figure like Fig. 6.14a

```
FigStd('x',-1.0,3.0,1,'y',-3.0,1.0,1.0)
plt.plot(xA,yA,'k-o',label="α=0")          #At rest
plt.plot(xR,yR,'k:o',fillstyle='none',
    lw=2,label="α="+str(round(a,2)))  #Rotated by α
plt.plot(xG,yG,'ko',ms=4,
              label='c of grav.')          #Center of gravity
plt.plot(xGr,yGr,'ko',ms=4)
plt.legend()
plt.axis('scaled')
```

Table 6.19 Moment of inertia

`#Momentum of inertia, around S`	
`IS=np.sum((x**2+y**2)*m)`	IS 19.25
`#Rotated coord. of center of gravity:`	xG 0.0
`[xGr,yGr]=RotMat@[xG,yG]`	yG -1.875
`#Moment. of inertia, around c of grav`	
`IarCg=np.dot((x-xG)**2,m) \`	IarCg 5.19
`     +np.dot((y-yG)**2,m)`	
`#Rotation of center of gravity`	
`ICg=(xG**2+yG**2)*np.sum(m)`	ICg 14.06
`Itot=IarCg+ICg`	Itot 19.25

We confirm again that (b) is smaller than (a) and that (b) + (c) equals (a).

Table 6.20 displays a program for plotting the anchor held at its center of gravity, as in Fig. 6.14b.

6.6 Wavefronts, Sound Barriers, and Mach Cone (T-T)

We draw the wave crests of sound waves emitted by a source moving at a certain speed and direction in the *xy*-plane and demonstrate the breaking of the sound barrier for supersonic speed. Polar coordinates are used for the calculation, and Cartesian coordinates for the scatter diagrams.

Table 6.20 Program for plotting the anchor gripped at its center of gravity (Fig. 6.14b)

```
x_Cg=xA+(xGr-xG)
y_Cg=yA+(yGr-yG)

FigStd('x',-1.0,3.0,1,'y',-3.0,1.0,1.0)
plt.plot(xA,yA,'k-o')                          #Anchor at rest
plt.plot(xG,yG,'kx',ms=8,label='c of g')       #Cent. of gravity
plt.plot(x_Cg,y_Cg,'k:o',fillstyle="none")     #Rotated anchor
plt.plot([0,xGr],[0,yGr],'k--',label="α="+str(a))
                                               #Rotated rod
plt.plot(xGr,yGr,'kx',ms=8)
plt.plot()
plt.legend()
plt.axis('scaled')
```

6.6.1 Emitting Sound Waves

In Fig. 6.18, crests of sound waves in a plane emitted from a moving source are shown, in **b**, for supersonic speed. One wave crest is emitted in every period of the sound signal.

Sound barrier and Mach cone

▶ **Mag** Are the motions linear or rotational?

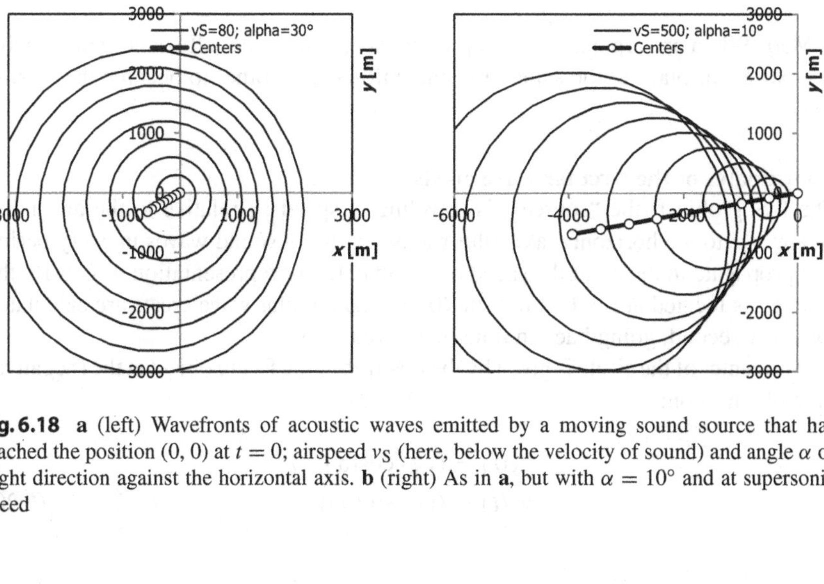

Fig. 6.18 **a** (left) Wavefronts of acoustic waves emitted by a moving sound source that has reached the position (0, 0) at $t = 0$; airspeed v_S (here, below the velocity of sound) and angle α of flight direction against the horizontal axis. **b** (right) As in **a**, but with $\alpha = 10°$ and at supersonic speed

▶ **Alac** The motion of the airplane is along a straight line.

▶ **Mag** Ok, so it's translational. But what about the sound?

▶ **Tim** Sound is a longitudinal wave, so it's a linear motion.

▶ **Alac** Ok. But how do the circles arise?

▶ **Mag** As wave propagation is isotropic, the wave crests are circles.

▶ **Tim** And the circles are described with polar coordinates.

▶ **Alac** I understand, polar coordinates but no circular motion.

▶ **Tim** Ok, that's clear. The airplane, modeled as a point in the plane, moves linearly in two dimensions. Sound is a wave and propagates isotropically in air.

▶ **Mag** Another point: What happens if the aircraft flies at exactly the speed of sound?

▶ **Alac** Then a sound barrier builds up, and there is a loud bang.

▶ **Mag** Simulate this situation! The best way to do so in a spreadsheet is to install a slider and increase the speed v_S, of the sound source slowly, starting from zero up to the speed of sound!

▶ **Tim** What does "breaking the sound barrier " mean?

▶ **Mag** When an airplane speeds up to the speed of sound, all sound waves arrive at a particular place at the same time and enforce each other to become the "sound wall".

Coordinates of the circular wave crests

The flying object (the "source") is traveling at speed v_S relative to the air and at an angle α to the horizontal axis (the x-axis), emitting sound waves in every period that propagate in the air at the speed of sound c. In our representation, Fig. 6.18, the source t is located at $t = 0$ at the site (0, 0), and circular wave crests are calculated for every second, going back in time (negative time).

The center of the circle is given by the position of the flying object at the (negative) time of emission:

$$\begin{aligned} x_S(t) &= (v_S \cdot \cos(\alpha)) \cdot t \\ y_S(t) &= (v_S \cdot \sin(\alpha)) \cdot t \end{aligned} \tag{6.20}$$

The expressions in parentheses decompose the distance $v_Q \cdot t$ traveled into horizontal *(x)* and vertical *(y)* components. $x_S(t)$ and $y_S(t)$ are smaller than 0 because $t <$ 0 for our settings.

Starting from the trajectory of the object, waves are spreading with velocity *c* and have covered a distance *r* up to time *t*:

$$r = -c \cdot t \text{ for } t \leq 0 \tag{6.21}$$

so that (in our drawing plane) a circular wavefront arises. In three-dimensional reality, the wavefronts are, of course, spherical surfaces. For our two-dimensional representation, the quantities *r* become the radii of the circles in Fig. 6.18.

6.6.2 Data Structure and Nomenclature

c	speed of sound
v_S	speed of sound source
α	angular deviation of the linear track of the source from the *x*-axis
t	array of instants of time, negative; the current time is 0
r	array of radii of wave crests emitted at times *t*
x_S, y_S	arrays of coordinates of the source at times *t*; the position at $t = 0$ is (0, 0)
phi	list of the polar angles for drawing the wave crests
x, *y*	2D matrices containing the coordinates of a set of wave crests, shape *size(phi)* · *size(t)*.

6.6.3 Spreadsheet Solution

Calculating circles

In the spreadsheet of Fig. 6.19 (S), we produce eight circles to represent the crests of waves that have been sent out at instants $t = -1, -2, \ldots, -8$ s. Two examples are shown in Fig. 6.18.

The definition of the coordinates of a circle is best done in polar coordinates. In Fig. 6.19 (S), the angle is defined in [A13:A43] from $\varphi = 0$ to 2π in 31 steps of $d\varphi = 0.209 = 2\pi/30$, set in A10. The three parameters of the task (sound speed *c*, speed v_S of source, flight angle α) are defined in B1:B4 and get named in A1, A2, A4.

The worksheet in Fig. 6.19 (S) has the typical Γ structure indicated by the bold ┌ - shaped line. The calculation range below Γ spans B13:R43. The column-specific parameter set, coded as row vectors r, x_S, y_S, for the eight functions is in B7:I9 above Γ, controlled by the time *t* (see the formulas reported in J7:J9). The independent variable φ, the polar angle of the circles, is in A13:A43, to the left of Γ.

	A	B	C	D	E	F	G	H	I	J	K	L	R	S
1	c_	340	m/s	*Sound speed*										
2	vS	500	m/s	*Speed of source*				*="vS="&vS&"; alpha="&B3&"°"*						
3		10	°	*Direction angle*				vS=500; alpha=10°						
4	alpha	0.175	rad	*=$B$3/180*PI()*										
5														
6	t	-1	-2	-3	-4	-5	-6	-7	-8					
7	r_	340	680	1020	1360	1700	2040	2380	2720	*=-c_*t*				
8	xS	-492.4	-985	-1477	-1970	-2462	-2954	-3447	-3939	*=vS*COS(alpha)*t*				
9	yS	-86.82	-174	-260.5	-347.3	-434.1	-520.9	-607.8	-694.6	*=vS*SIN(alpha)*t*				
10	0.209	*=2*PI()/30*												
11	*=A13+$A$10*	*=r_*COS(phi)+xS*									*{=r_*SIN(phi)+yS}*			
12	phi	x									y			
13	0.000	-152	-305	-457	-610	-762	-914	-1067	-1219		-87	-174	-695	
14	0.209	-160	-320	-480	-639	-799	-959	-1119	-1279		-16	-32	-129	
43	6.283	-152	-305	-457	-610	-762	-914	-1067	-1219		-87	-174	-695	

Fig. 6.19 (S) Coordinates (x, y) of circles with their centers shifted in the xy-plane (below Γ); Γ starts at B13 and is extended to the right and downwards; the x-coordinates in B13:I43 are generated with the formula reproduced in B11. The corresponding y values are in columns K:R. The angular coordinates φ for all curves are in A13:A43 (independent variable left of Γ). The time in row 6 controls the radius (row 7) and the coordinates of the center (in rows 8 and 9)

This table can be enlarged row by row beyond row 43, for example, to allow for a finer angular scale, because it is downwards open. It cannot be broadened column by column, because, starting with column K, another calculation range follows, in which the y-coordinates of the circles are calculated.

▶ Remember: If you have written the formula in a cell correctly with relative and absolute references or with variable names, you can drag it into a larger cell range without changing it.

Questions

concerning Fig. 6.19 (S):

What is the meaning of the formula in cell B4, reported in D4?[31]

What is the formula in cell D7?[32]

What is the formula in cell D9?[33]

Wouldn't it be nicer to have the wave crests cover the whole cone in Fig. 6.18b? How would you do that?[34]

---

[31] Transformation of the angle from degrees to radians: $360° = 2\pi$, $180° = \pi$.

[32] [D7] = [= −c_*t], see J7!

[33] [D9] = [=vS*sin(alpha)*t], see J9!

[34] Specify the number of time instants to go from −1 to −12 in 12 steps.

Below Γ, eight circles around the origin with the radii r from line 7 are calculated from the column vector polar angle ϕ and the row vectors r, x_S and y_S.

The x values in B13:I43 are calculated with the simple formula [=r_*cos(phi) + xS]; the current values are taken from the same row or the same column as the current cell.

Matrix formula

The formulas in range B13:I43 of Fig. 6.19 (S) refer, for each cell, to the same column (when named row vectors are addressed) and the same row (when the named column vectors are addressed). The same formula network can be generated in any range with a matrix formula. An example is given in K13:I43 (= [{=r_*cos(phi) + yS}]). To do so, activate the range, enter the formula, and complete with Ψ*Ctrl + Shift + Enter!*

▶ **Task sliders** Install two sliders to adjust the source's direction and speed and observe how the diagram reacts to changes in both parameters and a change in sound velocity.

Discussion EXCEL

▶ **Tim** The wave crests in my diagram do not change when I change the speed of the aircraft.

▶ **Mag** You have copied the *numbers* from Fig. 6.19 (S) into range B7:I9. However, these cells have to contain *formulas*, not just numbers. Then, the values in these cells change as the parameters of the problem are changed. The formulas for column I can be found in column J. If you drag cells I7:I9 (with the formulas in J7:J9) to the left all the way to column B, the formula network for the parameters is complete.

▶ **Tim** What exactly is our task?

▶ **Mag** The polar angle is in column A, independent of the time. Your task is to enter a formula into cell B13 that creates the x coordinates of all eight circles at time t by dragging to the right and down to cell I43. To get the y coordinates, you have to apply a matrix formula, such as in K11 in Fig. 6.19.

At time $t = 0$, the aircraft shall be at the origin (0, 0) of the coordinate system. The time in row 6 of Fig. 6.19 (S) is counted backward. So, the coordinates (x_S; y_S) in rows 8 and 9 indicate where the flying object was 1, 2, etc., seconds ago. They are calculated, as discussed above, from the speed v_S of the source, the angle α, and the time t. The respective location of the flying object at this past instant is also the center of the circles representing the crests of the emitted sound waves.

Table 6.21 Coordinates of wave crests

```
c=340                     #[m/s] speed of sound
vS=500                    #[m/s] speed of source
alpha=10                  #[Degree]
alpha*=np.pi/180          #[rad]

t=np.linspace(-1,-8,8,endpoint=True)
r=-c*t                    #Radius
xS=vS*np.cos(alpha)*t #Center
yS=vS*np.sin(alpha)*t

ph=np.array([np.linspace(0,2*np.pi,31)])
phi=ph.transpose(1,0)

x=r*np.cos(phi)+xS        #Coord. of circle
y=r*np.sin(phi)+yS
```

6.6.4 Python

The `Python` program for calculating the coordinates of the wave crests is shown in Table 6.21. After specifying the parameters speed of sound c, speed of source v_S, and deviation of the track of the source from the horizontal by an angle α, we specify the instants t of time at which the source is supposed to emit a signal. From t, we get the radii r of the circular sound crests and the Cartesian coordinates of their centers x_S and y_S, i.e., the current positions of the sender, all as arrays broadcast from t.

In *ph*, we define, as an array, the 31 angles with which circles are to be drawn as regular polygons. This array is transposed into a column vector *phi*. With these constructs and the two instructions

```
x = r*np.cos(phi) + xS
y = r*np.sin(phi) + yS
```

we reproduce the spreadsheet data structure of Fig. 6.19 (S) with the four row vectors t, r, x_S, y_S, and one column vector *phi*. The values of the different arrays are reported in Table 6.22. They coincide with those of the spreadsheet calculation in Fig. 6.19 (S).

Tangent to a circle

With Fig. 6.20, we complete Fig. 6.18b with tangents to the circles that represent the Mach cone. The construction of the tangent to a circle is illustrated in Fig. 6.20a and realized in Table 6.23.

▶ **Task** Verify the instructions in Table 6.23 with the help of Fig. 6.20a!

Table 6.22 Structure of the arrays in Table 6.21

```
alpha 0.175
t       [-1 -2 -3 -4 -5 -6 -7 -8]
r       [ 340   680   1020 ...  2040   2380   2720]
xS      [-492 -985 -1477 ... -2954 -3447 -3939]
yS      [ -87 -174   -260 ...  -521   -608   -695]
x[0]    [-152 -305   -457 ...  -914 -1067 -1219]
```

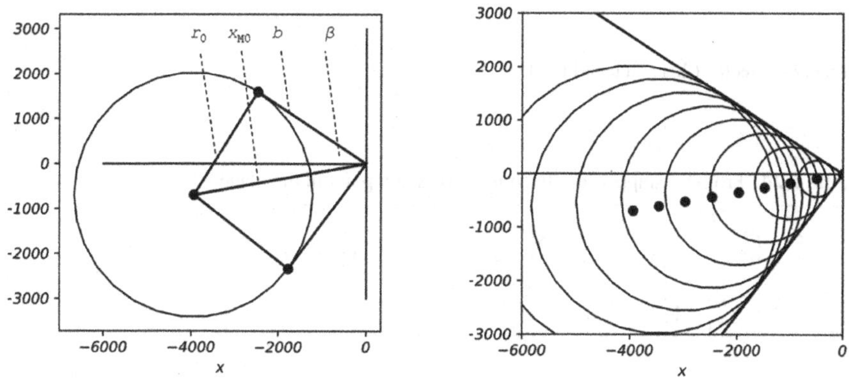

Fig. 6.20 **a** (left) Construction of a tangent to a circle. **b** (right) Mach cone with the sound barrier drawn as a tangent to the circles

Table 6.23 Tangent to a circle, drawn as straight lines from (0, 0) to (x_T, y_T) and from (0, 0) to (x_{Tu}, y_{Tu})

```
#Tangent, angle beta
xM0=vS*t[-1]
r0=r[-1]
b=np.sqrt(xM0**2-r0**2)
beta=np.arcsin(r0/-xM0)
xT=-np.cos(-alpha+beta)*b
yT=np.sin(-alpha+beta)*b
xTu=-np.cos(-alpha-beta)*b
yTu=np.sin(-alpha-beta)*b
```

6.7 Questions and Tasks

1. What are the polar coordinates for the Cartesian coordinates (0, 5) and (1, 1)?
2. What are the Cartesian coordinates for the polar coordinates $r = 2$, and $\phi = 45°$ or $\phi = 135°$?
3. The spreadsheet formula = COS(90) returns −0.44807362. Is EXCEL thus wrong? However, `np.cos(90)` similarly returns −0.4480736161291701. Why?
4. Given the vector (3, 4), you are to attach this vector and its x- and y-components as arrows to point (1, 1) . What does the data series in a spreadsheet look like? What are the three lists in `Python` for the coordinates?

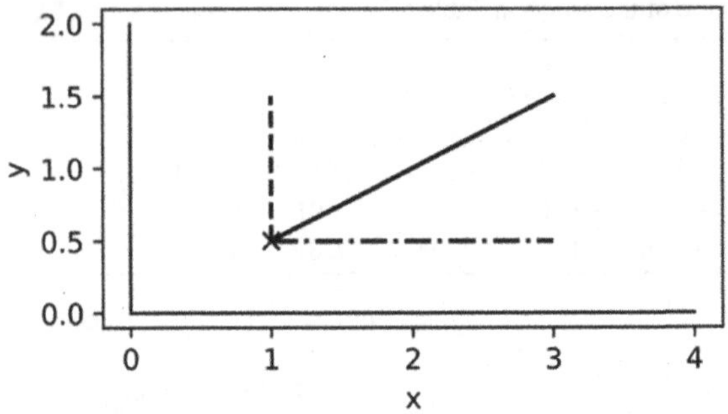

Fig. 6.21 Vector (2, 1) attached to (1, 0.5)

Table 6.24 Program snippet for attaching arrows at a point in the plane

```
FigStd('x',0,4,1,'y',0,2,0.5)
plt.plot(xO[0],xO[1],'kx')
plt.plot([xO[0],Arrow[0]],[xO[1],Arrow[1]],'k-')
plt.plot( … )
plt.plot([…,'k-.')
plt.axis('…')
```

5. In Fig. 6.21, the vector $v_A = (2, 1)$ and its components are attached to the point $x_O = (1, 0.5)$. Specify the variables x_O, v_A, and *Arrow* and complete the program snippet in Table 6.24!
6. You are to draw the path (x, y) of a point in the plane, which moves with speed v in the xy-plane at an angle $\alpha = 30°$ to the x axis. What are the formulas in the parameter representation $(x, y) = f(v,\alpha)$?
7. Describe the rotation of a point on a circle around the origin with rotation time T, both in polar and in Cartesian coordinates!
8. Which formulas apply to the Cartesian coordinates of a circle with diameter d, moving with velocity v along the x-axis?
9. A point moves with constant velocity v along the y axis of a laboratory system. What are its polar and Cartesian coordinates, in the laboratory system, and in a system moving relative to the laboratory system, with constant angular velocity ω_D, around an axis through the origin of the laboratory system?
10. Calculate the moments of inertia of the two dumbbells in Fig. 6.22! The two points represent masses of equal size. The connections between the masses and to the suspension points are supposed to be massless.

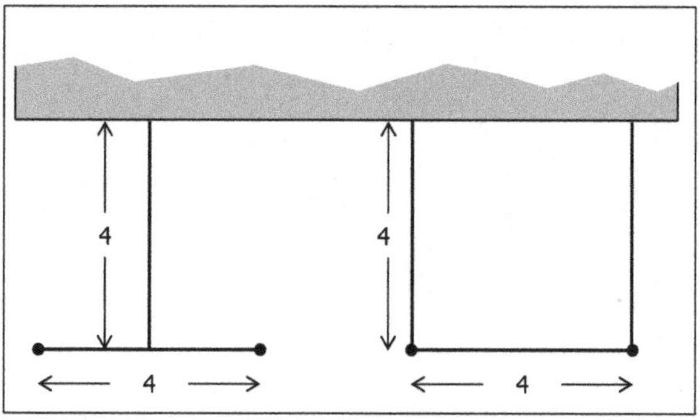

Fig. 6.22 A dumbbell, (left) suspended on the (massless) middle stem fixed to it, (right) suspended on two threads at its ends

11. Explain Steiner's theorem on the basis of Fig. 6.22a, b, together with Fig. 6.14!
12. How do you calculate the center of gravity of an arrangement of point masses m_i at (x_i, y_i)?

Integration of Newton's Equation of Motion

7

The Newtonian equation for one-dimensional motions of point masses is numerically integrated by estimating the mean acceleration $<a>$ in the time intervals between the supporting points of the time trajectory. Four methods are used: *Euler*, *half step*, "*Progress with look-ahead*" and *Runge-Kutta of the 4th order*. Three examples from an adventurous life are simulated realistically: Stratosphere jumping, bungee jumping, and car in a racing start. Different friction models (dry, viscous, Newtonian) are applied.

7.1 Introduction: Approximated Mean Value Instead of Exact Integration

Solutions of Exercises 7.2 (Python), 7.3 (Excel), 7.4 (Python), and 7.6 (Excel) can be found at the internet adress: go.sn.pub/THdoLU.

7.1.1 Newton's Equation of Motion

We investigate the motion of a point mass on a straight line. According to Newton's laws, the acceleration at a certain point in time is determined by the forces acting on the body at that point in time. These forces F are generally dependent on the location x, e.g., due to a spring, or on the velocity $\dot{x}$, e.g., due to friction, or on both, so that, generally, we have $F = F(x, \dot{x})$.

Newton's equation of motion for one dimension is

$$\ddot{x}(t) = \frac{F(x, \dot{x})}{m} \tag{7.1}$$

D. Mergel, *Physics with Excel and Python*,
https://doi.org/10.1007/978-3-030-82325-2_7

where $\ddot{x} = a$ is the acceleration, $\dot{x} = v$ the velocity, x the location, and m the mass of the body. If all forces within a period of time are known, as well as the location and velocity of the point mass at the beginning of the period, the differential equation of motion, Eq. 7.1, can be solved. Then, the values of x and $\dot{x}$ at the end of the time period can be calculated.

The forces in some of the exercises in this chapter depend only on velocity or only on location, making the calculations simpler and easier to follow.

We learn how Newton's equation of motion can be solved numerically in a sufficiently precise way over a period of time. To introduce the different methods, we work on tasks that can be analytically solved so that we can check the results of our calculations: Vibrations of a mass-spring system without friction (Exercise 7.2) and falling through air with friction (v^2-proportional, Exercise 7.3).

Mean value instead of integral

The numerical solution we are striving for is based on difference equations in which time progresses by finite amounts Δt. If the location and the velocity of the body at the beginning t_0 of a time interval are known, the values at the end $t_0 + \Delta t$ of the interval follow as

$$v(t_0 + \Delta t) = v(t_0) + \int_{t_0}^{t_0+\Delta t} a(t)dt = v(t_0) + \langle a \rangle \cdot \Delta t \qquad (7.2)$$

$$x(t_0 + \Delta t) = x(t_0) + \int_{t_0}^{t_0+\Delta t} v(t)dt = x(t_0) + \langle v \rangle \cdot \Delta t \qquad (7.3)$$

Equations 7.2 and 7.3 are exact. The expressions with the mean values <*a*> and <*v*> correspond exactly to the integral, because the mean value is simply defined like that.

Our task is to estimate the *mean acceleration* <*a*> and *the mean velocity* <*v*> in the time interval under consideration. The numerical estimation is an approximation for which we present four different methods in the next section.

Ψ *Approximated mean instead of exact integral.*

Friction

To consider friction, we use the general formula:

$$a_{Fric} = -sgn(v) \cdot a_f \cdot |v|^n$$

The acceleration due to friction is proportional to a power n of the speed (the absolute value of the velocity). The first term −sgn(v) guarantees that friction always tries to decrease the speed. The exponent n depends on the type of friction:

- $n = 0$ for internal friction, applied for the losses in a bungee rope (Exercise 7.6)
- $n = 1$ for viscous friction, applied for the damping of a harmonic oscillator (Exercise 7.2)

- $n = 2$ for Newtonian friction, active when dragging a body rapidly through a viscous medium, in our exercises, air (Exercises 7.3, 7.4, and 7.5).

Observe the influence of the different types of friction on the time trajectory!

7.1.2 Four Methods for Estimating the Average Acceleration in a Time Segment

In the simple *Euler method*, the mean acceleration in the time segment $[t_{\mathrm{n}}, t_{\mathrm{n+1}}]$ is approximated by the acceleration at the beginning of the time segment:

$$v(t_{n+1}) = v(t_n) + a(t_n) \cdot \Delta t \tag{7.4}$$

$$x(t_{n+1}) = x(t_n) + v(t_n) \cdot \Delta t \tag{7.5}$$

This procedure can be improved if the values from Eq. 7.4 and Eq. 7.5 are taken only as a "preview" for the velocity $v_p(t_n + \Delta t)$ and the location $x_p(t_n + \Delta t)$, from which the acceleration $a_p(t_n + \Delta t)$ at the end of the considered time period is calculated. For this "*Progress with look-ahead*", the quantities at the beginning of the next interval are estimated as:

$$v(t_{n+1}) = v(t_n) + \frac{a(t_n) + a_p(t_n + \Delta t)}{2} \cdot \Delta t \tag{7.6}$$

$$x(t_{n+1}) = x(t_n) + \frac{v(t_n) + v_p(t_{\mathrm{n}} + \Delta t)}{2} \cdot \Delta t \tag{7.7}$$

The *half-step procedure* has a similar structure; the values at the beginning of an interval are used to estimate the values in the middle of the interval, which then represent the entire interval.

A further improvement can be achieved with the fourth-order *Runge-Kutta method*, in which three projections are made in an interval, the first two into the middle and the third to the end of the interval. Then, a weighted average of four values is taken as representative of the whole interval.

All methods are trained

We shall apply all four methods in Exercise 7.3 (Falling from a great height; the force depends only on the speed) and compare the results with each other. There are analytical solutions for this task, so that we can check the precision of our numerical methods.

All methods provide sufficiently accurate solutions if the distance between the supporting points, which mark the time segments' boundaries, is made sufficiently small. The bigger the effort with which the mean acceleration in a time segment is calculated, the longer the time segments can be. Consequently, it must be checked, for each method, whether the selected length of the time segment is short enough.

As standard procedure in EXCEL, we shall use *Progress with look-ahead*, because it proves to be sufficiently efficient and can be implemented clearly in a spreadsheet. However, experienced readers may also use the *half-step procedure*, which provides the same accuracy and is also explained in detail in Exercise 7.3.

In `Python`, we shall use a function that implements the 4th-order Runge–Kutta method.

Adventurous life

In the course of this chapter, we will deal with three cases from an adventurous life: stratospheric jumping, bungee jumping, and a car in a racing start. In the follow-up volume *Physics with Excel and Python, Using the Same Data Structure. Applications*, the procedures from this chapter come into full fruition, with the treatment of motions in the plane, all kinds of oscillation, field lines, and wave functions of the Schrödinger equation.

7.1.3 Tactical Approaches in Python and Excel

In `Python`, the steps from t to $t + dt$ are done in a progress loop that calls a progress function, invariably called *progr* in all exercises. It does not have a proper code, but is assigned to one of the existing functions for *Euler*, *look-ahead*, or *Runge-Kutta*. Within the progress functions, the acceleration is invariably called *acc*, again, without proper code, but with an assignment to a function that is specific for the physical problem under consideration (*accSpring*, *accFall*, *accJump*, *accPwr*, *accBungee*). This approach is typical of Software solutions in which functions are embedded within a larger body.

The EXCEL approach is less general and must be adapted individually to each problem. The preview calculations for the current time interval are done in a row with specific formulas for acceleration. The values at the beginning of a new time interval in a new row have to be calculated from values in the preceding row.

Animations

The motions calculated in this chapter can easily be animated with the methods presented in Sect. 6.2.5.

7.2 Harmonic Oscillation with "Progress with Look-Ahead" and "Runge–Kutta"

We integrate Newton's equation of motion for a mass-spring system by calculating the acceleration in a time interval, (a) as an average of two values, one at the beginning of the interval and the other calculated for the end of the interval with the values at the beginning, and (b) as an average over four values with the Runge-Kutta method.

7.2.1 Equation of Motion

The equation of motion for a mass fixed to a linear-elastic spring is as follows:

$$a = \ddot{x} = \frac{D}{m} - d \cdot v = -f \cdot x - d \cdot v \tag{7.8}$$

Here, x is the deflection out of the rest position, and D and m are the spring constant and the mass. Damping is set proportional to the velocity v and opposite to it: $-d \cdot v$.

Questions

What is the physical unit of the "spring constant" f in Eq. 7.8 in SI units?[1]
What is the physical unit of the damping constant d in Eq. 7.8 in SI units?[2]
What are the initial conditions $x(0)$ and $v(0)$ in Fig. 7.1a?[3]

The solution for vanishing friction ($d = 0$) is a stationary oscillation with constant amplitude. We use this knowledge to check whether our numerical solutions are good enough. An example is shown in Fig. 7.1 for an oscillation with $f = 0.10$ calculated with *LookAhead*.

The zoom in Fig. 7.1b shows that the amplitude for $n = 100$ has increased over the initial value, indicating that 100 points for a time span of 80 are not good enough, but that $n = 400$ does the job. The position of the fourth maximum is at $t = 79.5$, corresponding reasonably well to the theoretical value of $4\pi/\sqrt{(0.1)} = 79.48$. We may therefore rely on *LookAhead* with this segmentation of time and can play around with the parameters.

In Fig. 7.2, we have doubled the spring constant and introduced damping, causing a strong decay of the amplitude. In the zoom, we see furthermore that the period of the damped oscillation is bigger than for the undamped oscillation.

▶ **Task** Vary the constant $f = D/m$ and observe whether the period duration behaves as predicted by the formula $T = (2\pi)/\sqrt{f}$ (for vanishing friction)!

[1] $[f] = [a/x] = (\mathrm{m/s^2})/\mathrm{m} = 1/\mathrm{s}^2$.
[2] $[d] = [a/v] = (\mathrm{m/s^2})/(\mathrm{m/s}) = 1/\mathrm{s}$.
[3] $x(0) = 1$ (maximum deflection), $v(0) = 0$ (slope of $x(t)$).

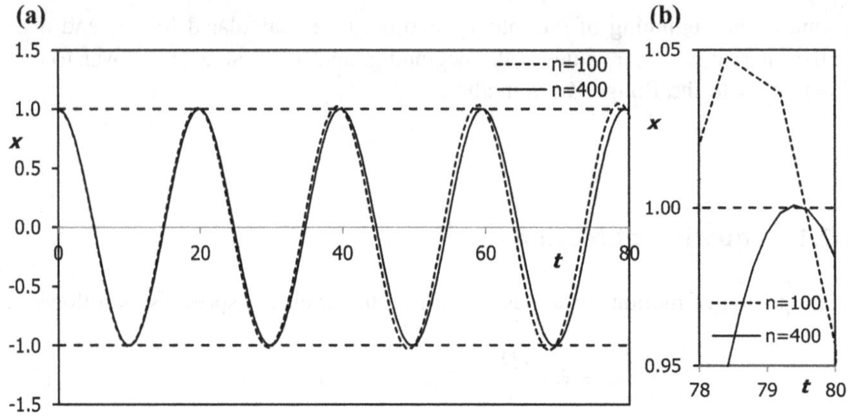

Fig. 7.1 **a** (left) Oscillation calculated with *LookAhead*. **b** (right) Zoom of **a**

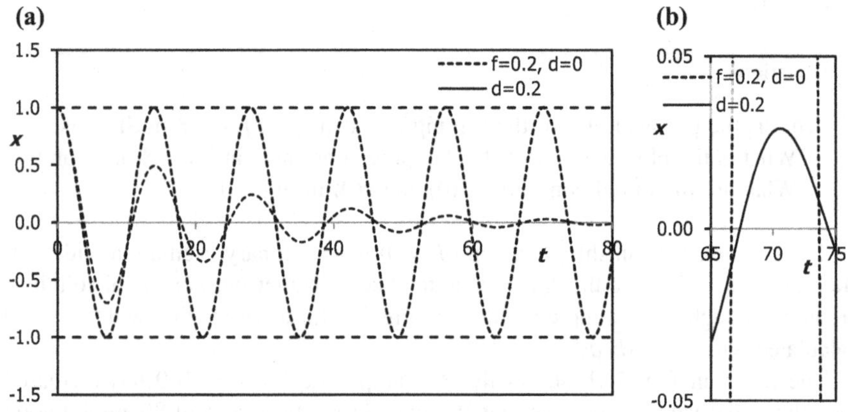

Fig. 7.2 Oscillation of a harmonic oscillator. **a** (left) Undamped and damped. **b** (right) Zoom of damped oscillation

In the following sections, we present *LookAhead* in a spreadsheet and *LookAhead*, *Euler*, and *Runge-Kutta* in `Python`. Varying the number *n* of points as in Fig. 7.1, reliable results are obtained with:

`n = 20,000` for *Euler*,
`n = 400` for *Progress with Look-ahead*,
`n = 60` for *Runge-Kutta.*

There is a reduction of about 50 going from *Euler* to *LookAhead* and about 8 going further *to Runge-Kutta*. Therefore, we choose *LookAhead* as standard in spreadsheet calculations with only four additional columns (see Fig. 7.3), thus keeping the programming effort to be repeated in every new spreadsheet small. We choose *Runge-Kutta* as standard in `Python` because the 12 necessary statements (Table 7.2) have to be implemented only once in a function that can be used in all further exercises.

	A	B	C	D	E	F	G	H
1	n	400			f	0.10	*spring constant*	
2	dt	**0.20**	*=80/n*		d	0.00	*damping constant*	
3		**n=400**	*="n="&n*					
4	*=A6+dt*	*=B6+(C6+F6)/2*dt*	*=C6+(D6+G6)/2*dt*	*=-f*x-d*v*	*=x+v*dt*	*=v+aA*dt*	*=-f*xD-d*vD*	
5	**t**	**x**	**v**	**aA**	**xD**	**vD**	**aD**	
6	0.00	1.00	0.00	-0.10	1.00	-0.02	-0.10	
7	**0.20**	**1.00**	**-0.02**	-0.10	0.99	-0.04	-0.10	
406	80.00	0.98	-0.06	-0.10	0.97	-0.08	-0.10	

Fig. 7.3 (S) An oscillation of a mass-spring system is calculated with LookAhead. The time runs *vertically* down, the look-ahead to the end of the current time segment is performed *horizontally* to the right. The mass is assumed to be $m = 1$

7.2.2 Data Structure and Nomenclature

n	number of points for a time span of 80 s
f, d	spring and damping constants
dt	$= 80/n$, length of the time interval
t	array of time points, $(n + 1)$ elements, dt apart
x	deflection of the mass at t (array)
v	velocity of the mass at t (array)
a	acceleration at t (array)
x_A, v_A	deflection and velocity at a specific time t_A
x_D, v_D	deflection and velocity predicted for time $t + \mathrm{d}t$
x_B, x_C	deflection predicted at time $t + \mathrm{d}t/2$ (Runge-Kutta)
v_B, v_C	velocity predicted for time $t + \mathrm{d}t/2$ (Runge-Kutta)

All calculations are performed for $t = 0$ to $t = 80$ s.

7.2.3 Spreadsheet Calculation

The basic spreadsheet layout for integrating Newton's equation of motion with *LookAhead* is shown in Fig. 7.3 (S), with the resulting deflection displayed in Fig. 7.1.

The time t in column A runs from top to bottom, and thus so does the deflection $x(t)$ of the mass in column B and its velocity $v(t)$ in column C. For a step from t_n to t_{n+1}:

- the acceleration a at the beginning of the time segment beginning with t_n is calculated in each row from the values of x and v at that time (column D),
- with these values, the deflection x_D and the velocity v_D at the end $t_n + \mathrm{d}t$ of the time segment are predicted (columns E and F),
- with these new values, the acceleration a_D at the end of the time interval is estimated in column G.

All calculations reported for columns D to G use formulas with named variables referring to the same row. The new values for x and v at the next instant t_{n+1} of time are calculated in the following row with information from the previous row; see the formulas in B4 and C4 valid for B7 and C7, e.g., B7 = [=B6 + (C6 + F6)/2*dt].

The estimation of the mean value at the time $t_{n+1} = t_n + dt$ can be improved by predicting further values for speed and acceleration in the current time segment $[t_n, t_n + dt)$ in further columns, especially with the more efficient fourth-order *Runge-Kutta* method. However, due to the simpler table layout, we use *LookAhead* as our standard in spreadsheet calculations.

7.2.4 Python

Progress functions

In the second cell of Table 7.1, we have implemented the *LookAhead* method as a function, similar to the spreadsheet calculation of Fig. 7.2 (S). In the first cell of Table 7.1, the *Euler* method is implemented. In both functions, the acceleration is not calculated explicitly, but is rather outsourced to another function *acc* that must be specified in the main program.

The Runge-Kutta method of the 4th order is implemented in a function reported in Table 7.2, with two jumps into the center of the interval (x_B, v_B and x_C, v_C) and one jump to the end of the interval (x_D, v_D). The values (x_R, v_R) for the next time instants to be returned are calculated as a weighted average over the velocities or the accelerations.

Ψ Half, half, whole; the halves count twice.

The main program is shown in Table 7.3. In the first 4 lines, the parameters of the exercise are defined, followed by the definition of the function *accSpring* for the acceleration which makes use of the global parameters f and d. The arrays t, x, v are defined with their length corresponding to the specified number n of time instants, with the first entries containing the initial conditions set in lines 12–14.

Table 7.1 Functions performing the *Euler* and the *lookAhead* methods; the current values of $x(t)$ and $v(t)$ are passed as x_A and v_A

```
def Euler(xA,vA):
    aA=acc(xA,vA)
    xR=xA+vA*dt
    vR=vA+aA*dt
    return xR,vR
```

```
def lookAhead(xA,vA):
    aA=acc(xA,vA)
    xD=xA+vA*dt
    vD=vA+aA*dt
    aD=acc(xD,vD)
    xR=xA+(vA+vD)/2*dt
    vR=vA+(aA+aD)/2*dt
    return xR,vR
```

Table 7.2 Function implementing the Runge-Kutta method

```
def RungeKutta(xA,vA):
    aA=acc(xA,vA)
    xB=xA+vA*dt/2 #Half step into center of interval
    vB=vA+aA*dt/2
    aB=acc(xB,vB) #Accel. at center of interval
    xC=xA+vB*dt/2 #Second half step
    vC=vA+aB*dt/2 #
    aC=acc(xC,vC) #Accel. at center of interval
    xD=xA+vC*dt   #Full step
    vD=vA+aC*dt   #
    aD=acc(xD,vD) #Accel. at end of interval
    xR=xA+(vA+2*vB+2*vC+vD)*dt/6
    vR=vA+(aA+2*aB+2*aC+aD)*dt/6
    return xR,vR  #Position, velocity at end of interval
```

Table 7.3 Calculating the motion of an oscillator

```
1   n=30                    #Number of time instants
2   dt=80/n
3   f=0.10                  #Spring constant
4   d=0.0                   #Damping constant
5
6   def accSpring(x,v):
7       return -f*x-d*v
8
9   t=np.zeros(n)
10  x=np.zeros(n)
11  v=np.zeros(n)
12  t[0]=0                  #Initial conditions
13  x[0]=1
14  v[0]=0
15  acc=accSpring           #acc = Name in progress function
16
17  for i in range(n-1):    #Progress loop
18      t[i+1]=t[i]+dt
19      #progr=Euler; lbl1="Euler, n="+str(n)
20      #progr=lookAhead; lbl1="lookAhead, n="+str(n)
21      progr=RungeKutta; lbl1="RungeKutta, n="+str(n)
22      x[i+1],v[i+1]=progr(x[i],v[i])
```

Progress loop

The integration is performed in a progress loop (line 17–22) progressing by time steps d*t*. It recurs to the progress function *progr* that has no code of its own but is linked to one of the progress functions *Euler*, l*ookAhead*, or *RungeKutta*. To keep

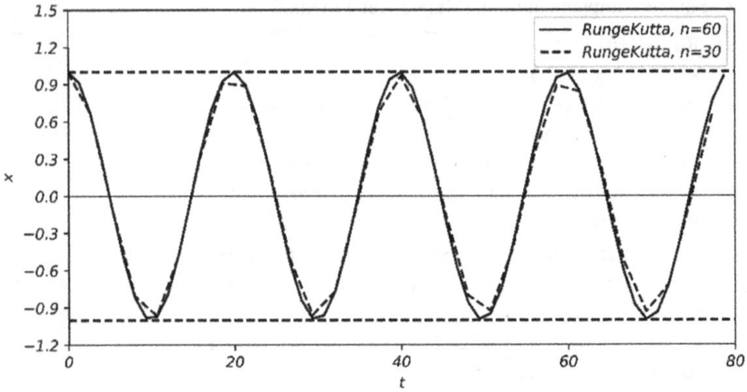

Fig. 7.4 Oscillation of a mass-spring system calculated with 4th order Runge–Kutta (with Table 7.3)

track of the program flow it is necessary to create the labels for the corresponding figures in the same program cell.

Within the progress functions, in Table 7.1, the acceleration is invariably taken from a function *acc* that, again, has no code of its own but must be linked to a function that is specific to the physical problem under consideration, here, in line 15, to *accSpring*. This increases the versatility of programming in that different approaches can be tried out.

To get stable results, we have to choose $n = 20{,}000$ for *Euler*, 400 for *LookAhead*, and 60 for 4th order *Runge-Kutta* ($n = 30$ is not good enough). The results of *Runge-Kutta* calculations are shown in Fig. 7.4.

7.3 Falling from a (Not Too) Great Height

Two forces act on a body falling from a great height: constant gravity and friction proportional to the square of the velocity. For these settings, the acceleration depends only on the velocity of the body, so that the location does not have to be calculated synchronously but can be determined subsequently by integration over the velocity. Initial acceleration and stationary velocity are compared with analytical solutions.

7.3.1 Limiting Cases, Analytically Solved

We drop a body from a great height and let two forces act on it: the gravity $-m \cdot g$, assumed to be independent of the height, and a friction force $k \cdot v^2$, proportional to the square of the speed (Newtonian friction) and opposite to its direction. With these assumptions, the equation of motion results as

$$a = -g - \frac{k}{m} v^2 \cdot sgn(v) = -g - \frac{k}{m} v \cdot |v| \tag{7.9}$$

where *sgn* (signum) is a function that calculates the sign of the argument. The force $m \cdot a$ depends only on the speed, because we assume, simplifying, that gravity and friction forces do not depend on the height. So, we don't have to consider the location in the calculation and may calculate it afterward by integrating the velocity trajectory. For a fall through air, $k = \rho A/2$ is valid with the density ρ of the air and the cross-sectional area A of the body.

In Fig. 7.5a, velocities $v(t)$ obtained with *Runge-Kutta of the 4th-order* ("RK4") for two interval lengths dt = 2 and dt = 4 and with *LookAhead* for dt = 0.5 coincide at their points of calculation. Halving the interval length does not change the RK4 solution, so dt = 4 is good enough. For *LookAhead*, an interval length eight times smaller is necessary to obtain the same result.

In Fig. 7.5b, v(t) trajectories are shown for three different initial velocities: zero, upwards, and downwards. The final speed is independent of the initial velocity.

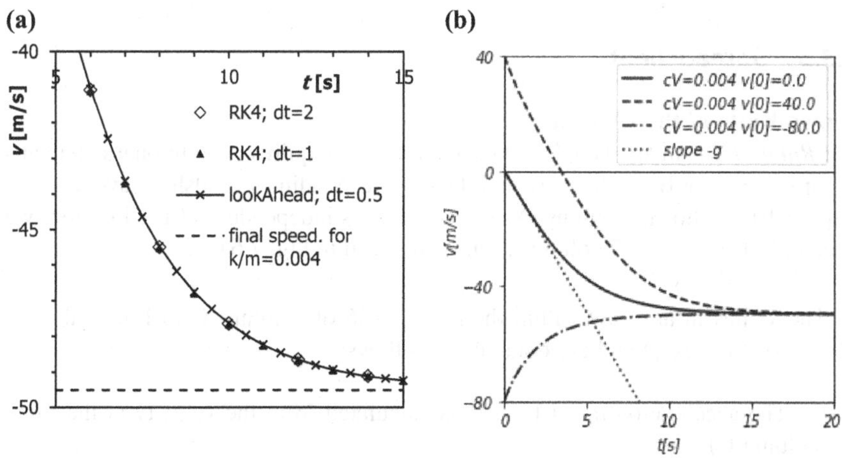

Fig. 7.5 **a** (left) Velocity of the falling object calculated in a spreadsheet with Runge-Kutta of the 4th order (*RK4*) and l*ookAhead*. **b** (right) Velocity of the falling object as a function of time for different initial velocities, calculated with *RK4* in `Python`

If, at the beginning of the fall, the velocity is 0, $v(0) = 0$, the initial gradient of the velocity curve is $-g$; the initial acceleration is the acceleration due to gravity close to the surface of the earth.

After a certain time, the body falls at a constant speed, because the frictional force equals gravity. This final velocity results from the condition $a = 0$ (no further acceleration):

$$v_{Fin} = -\sqrt{\frac{m \cdot g}{k}} \tag{7.10}$$

7.3.2 Data Structure and Nomenclature

g	acceleration due to gravity
cV	$= k/m$, (see Eq. 7.9) pre-factor of friction
t	array of time instants dt apart
v_A	velocity at t

Accelerations

a_A	at the beginning of the calculation interval
a_B, a_C	in the center
a_D	at the end.

7.3.3 Spreadsheet

Runge-Kutta of the 4th order

The *Runge-Kutta* method implemented in a function requires 10 additional statements (1 for A, 3 × 3 for B, C, D, see Table 7.1). A corresponding spreadsheet layout would require 10 additional columns. As our problem is independent of the location, we need only $1 + 3 \times 2 = 7$ columns. This is realized in Fig. 7.6 (S).

The values in any row within the range 6–26 of columns C to I are valid for the interval $[t_n, t_{n+1})$ and are obtained as follows:

- (1) The acceleration a_A at time t_n is calculated from the velocity at that time (column C).
- (2) The velocity v_B is calculated with the values v_A and a_A for the middle of the interval, i.e., for t_n + dt/2, and from that, the corresponding acceleration a_B (first half-step, column E) is obtained.

	A	B	C	D	E	F	G	H	I	J
1	**g**	9.81	m/s²	cV=k/m:=0.004						
2	**cV**	0.004	1/m							
3	**dt**	2.00		**RK4; dt=2**	*="RK4; dt="&dt*					
4	=A6+dt	=B6+(C6+2*E6+2*G6+I6)*dt/6	=-g-cV*vA^2*SIGN(vA)	=vA+aA*dt/2	=-g-cV*vB^2*SIGN(vB)	=vA+aB*dt/2	=-g-cV*vC^2*SIGN(vC)	=vA+aC*dt	=-g-cV*vD*ABS(vD)	
5	***t***	***vA***	***aA***	**vB**	**aB**	**vC**	**aC**	**vD**	**aD**	
6	0.0	0.00	-9.81	-9.8	-9.43	-9.4	-9.45	-18.9	-8.38	
7	**2.0**	**-18.65**	-8.42	-27.1	-6.88	-25.5	-7.20	-33.1	-5.44	
26	40.0	-49.52	0.00	-49.5	0.00	-49.5	0.00	-49.5	0.00	

Fig. 7.6 (S) Calculation with 4th order *Runge-Kutta*; v_B, a_B, v_C and a_C are predicted values in the middle of the time interval, v_D and a_D are predicted values at the end of the time interval

- (3) The velocity v_C and, from that, the acceleration a_C for $t_n + dt/2$ are calculated a second time (second half step, columns F and G), now with the values a_B and v_B.
- (4) The velocity v_D at the end of the interval, i.e., for $t_n + dt$, is calculated with the values a_C and v_C and, from that, the acceleration a_D (whole step, columns H and I)

The new value of the velocity at t_{n+1} is calculated in the next row (e.g., in B7 with the formula in B5) from the velocity at t_n and a weighted average of the four accelerations calculated in the previous row (here row 6) for the interval $[t_n, t_{n+1}]$:

$$v_{n+1} = v_n + \langle a \rangle \cdot dt \text{ with } \langle a \rangle = (1a_A + 2a_B + 2a_C + 1a_D)/6 \tag{7.11}$$

The accelerations from the half steps count twice. We remember the procedure with a broom rule:

► Ψ *Half, half, whole, the halves count twice.* (4th order Runge-Kutta). "Halves" means half-steps.

Questions

concerning Fig. 7.6 (S):

Which accelerations from steps (1)–(4) are used in the formulas for velocities reported in F4, H4, and B4?[4]

Why is the sum in B7 divided by 6 to get the average, although only four accelerations enter the formula?[5]

---

[4] a_B in the center of the time interval after the first jump; a_c in the center after the second; a_A, $2a_B$, $2a_C$, a_D for $t = 2$.

[5] Ψ *Half, half, whole; the halves count twice.* The sum of the weights is 6.

Fig. 7.5**a** shows two solutions obtained with Fig. 7.6 (S) for $\mathrm{d}t = 4$ and $\mathrm{d}t = 2$.

▶ **Task** Determine the height fallen until you reach 95% of the final speed! You have to integrate the speed over time!

To get the three functions $v(t)$ displayed in Fig. 10.5b, we have to run the calculation three times with different initial values written into B6 and store the results $v_A(t)$ in extra columns (COPY, PASTE SPECIAL/CONTENTS).

7.3.4 Python

Table 7.4 reports the preparatory instructions for the progress loop. A special feature of the code is the stacking of x and v into a 2D matrix to store the fall trajectories for three initial conditions.

In lines 1–4 of Table 7.4, the fall parameters and the function for the acceleration are defined. In lines 8–10, the elementary arrays t, x, v are specified. In order to calculate the motion for three different initial velocities, position x and velocity v are replicated in `np.stack` constructions so that they can be addressed by the same index as the curve itself. Labels and line styles for the three curves in a diagram are also organized as lists with three entries in lines 16 to 19 in Table 7.5.

The progress of motion for the three initial velocities 0, 40, and—80 m/s is calculated in Table 7.5 in a nested loop. In the first loop over k, line 17, labels are created for the three initial velocities and in line 19, line styles before the progress is calculated.

The program for plotting the three time-curves is given in Table 7.6. The line styles and the labels have been specified in Table 7.5 so that curves, styles, and labels can all be addressed with the same index.

Table 7.4 Parameters for the fall and function for calculating the acceleration during the fall

```
1   g=9.81                     #[m/s²] Gravitational acceleration
2   cV=0.004                   #[1/m]  Friction coefficient
3   def accFall(x,v):
4       return -g-cV*v**2*np.sign(v)
5   acc=accFall                #Link to acc in progress function
6
7   dt=0.05                    #[s] Time increment
8   t=np.arange(0,20+dt,dt)
     #t sets the structure also for x and v, and determines the
    length of the progress loop.
9   x=np.zeros(len(t))
10  v=np.zeros(len(t))
11  xM=np.stack((x,x,x))   #For 3 initial conditions
12  vM=np.stack((v,v,v))
```

Table 7.5 Progress loop for the fall; "Runge–Kutta" is the function defined in Table 7.2

```
13  vM[0,0]=0
14  vM[1,0]=40
15  vM[2,0]=-80
16  lbl1=[0,1,2]
17  for k in range(3):               #Labels for three curves
18      lbl1[k]='cV='+str(cV)+' v[0]='+str(vM[k,0])
19  linStyle=['k-','k--','k-.']
20  for k in range(3):               #Over all initial conditions
21      for i in range(len(t)-1): #Progress loop over t
22          xM[k,i+1],vM[k,i+1]= RungeKutta(xM[k,i],vM[k,i])
```

Table 7.6 Plotting three time-curves

```
23  FigStd('t[s]',0,20,5,'v[m/s]',-80,40,40)
24  for k in range(3):  #Over initial conditions
25      plt.plot(t,vM[k],linStyle[k],label=lbl1[k])
26  plt.plot([0,20],[0,-20*g],'k:', label = "slope -g")
     #Straight line with slope -g
27  plt.legend()
```

Questions

What is the shape of x_M and v_M in Table 7.4 and what are the first and last two elements of t?[6]

What does "linStyle[k]" in line 25 of Table 7.6 do?[7]

How can the two k-loops in Table 7.5 be merged into one?[8]

7.4 Stratospheric Jump

When considering jumps from great altitudes, it must be taken into account that the coefficient of friction changes with height according to changes in air density. The friction force is, therefore, a function of both location and velocity. We model the air density in a simplified way with the barometric formula for a constant temperature (15 °C). In this model, the maximum speed for a jump from 39 km altitude is reached after a 50 s fall.

[6] The shape of x_M and v_M is (401, 3), `len`(t) = 401, t = [0, 0.05, ..., 19.95, 20.00].

[7] `linStyle[k]` selects the k-th element of the list *linStyle* defined in line 19.

[8] Insert line 18 between lines 20 and 21. The indentation with respect to line 20 is already correct.

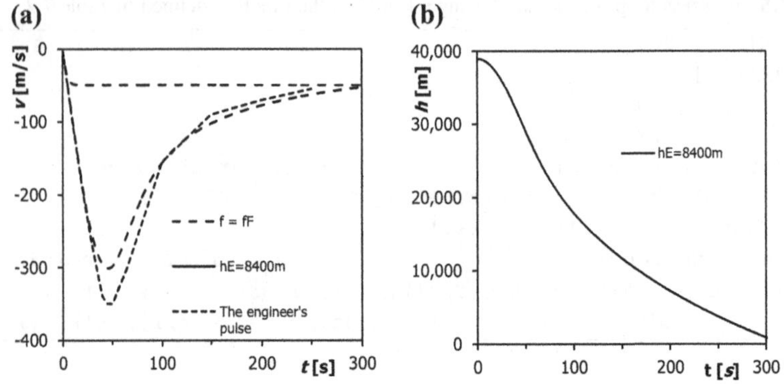

Fig. 7.7 **a** (left) Velocity for a fall from 38,969 m, compared with the data from Baumgartner's jump ("The engineer's pulse", Fig. 7.8a); the dashed line "f=fF" is valid for constant friction. **b** (right) Altitude as a function of time for the velocity calculated in **a**

t [s]	0	10	20	30	40	45	50	65	100	150	200	250
v [m/s]	0	-100	-190	-270	-340	-350	-350	-300	-155	-90	-70	-55

	A	B	C	D	E	F	G	H	I	J
3	**dt**	0.4 s			**fF**	0.004 1/m				
4	**g**	9.81 m/s²			**hE**	8400 m		hE=8400m		
5	=A7+dt	=B7+(D7+H7)/2*dt	=fF*EXP(-h/hE)	=D7+(E7+I7)/2*dt	=-g+f*v^2	=h+v*dt	=fF*EXP(-hD/hE)	=v+a*dt	=-g+fD*vD^2	
6	*t*	*h*	*f*	*v*	*a*	*hD*	*fD*	*vD*	*aD*	
7	0.0	38969	0.0000	0.00	-9.81	38969	0.0000	-3.92	-9.81	
8	**0.4**	**38968**	0.0000	**-3.92**	-9.81	38967	0.0000	-7.85	-9.81	
807	320.0	-150	0.0041	-49.45	0.15	-170	0.0041	-49.39	0.15	

Fig. 7.8 **a** (top) Measured data of the jump of October 14, 2012 (The engineer's pulse, Oct.15, 2012, Mechanical analysis of Baumgartner's dive (Part B)). **b** (bottom, S) Spreadsheet layout for a jump from a high altitude, when the friction coefficient (f and f_D in columns C and G) is height-dependent

On October 14, 2012, the Austrian adventurer Felix Baumgartner ascended into the stratosphere in a helium balloon. He jumped out at a height of about 39 km and fell about 34 km in free flight, reaching supersonic speed before opening his parachute. Some measurement data are reported in Fig. 7.8a and plotted in Fig. 7.7a as a dashed line. We are going to simulate this jump with a simple model for height-dependent friction.

Height-dependent friction force

The coefficient of friction depends on the altitude, because the density of air, and, with it, the coefficient of friction, decreases with increasing altitude. We apply a height-dependent friction coefficient that directly corresponds to the barometric formula:

$$f(h) = f_F \cdot \exp\left(-\frac{h}{h_e}\right) \tag{7.12}$$

with $f_F = 0.004$ 1/m valid for $h = 0$ (the same as in Exercise 7.3). With $h_e = 8400$ m (valid for an air temperature of 15 °C), we get the best coincidence with the empirical data for the velocity as a function of altitude (see Fig. 7.7a).

Our simple model can well reproduce the essential characteristics of the time trajectory of the velocity. In the beginning, the fall is free, i.e., without friction, showing up as a linear increase in speed. After about 50 s, the maximum fall speed is reached; in the real experience, it exceeded the speed of sound. After that, the speed decreases sharply because of increasing friction due to increasing air density and approaches the stationary speed we calculated for $f = f_H = c_0$ in Exercise 7.3.

For a more realistic simulation, we have to consider that the atmosphere consists of different air layers with different temperatures. But we are content to find the main characteristics of the jump when applying just the barometric formula with constant temperature.

7.4.1 Data Structure and Nomenclature

t	array of time points, dt apart, independent variable
h_E	characteristic height of the barometric formula
f_F	friction at $h = 0$
h	height for t
f	friction coefficient at h
v	velocity at t
a	acceleration at t
h_D, f_D, v_D, a_D	values predicted for the end of the interval.

7.4.2 Spreadsheet Calculation

The spreadsheet layout for the integration of the equation of motion is presented in Fig. 7.8b. The integration method is *LookAhead*. The main change from Exercise 7.3.2 is that the height is now calculated simultaneously with the velocity. This is necessary because the coefficient of friction is height-dependent.

The height is calculated in column B from the height and the speed in the preceding rows; from that, in column C, the coefficient of friction f, which is then used to calculate the acceleration a in column E. Analogously, the same applies to the "looked-ahead" values in columns F to I.

Table 7.7 Parameters for the stratospheric jump, together with a function for the acceleration

```
n=801            #Number of calculation points
dt=320/(n-1)  #Time span 320 s
g=9.81           #[m/s²], Gravitational acceleration
hE=8400          #Characteristic barometric height
fF=0.004         #[1/m] Damping constant

def accJump(xi,vi):          #x represents height
    f=fF*np.exp(-xi/hE)   #Barometric formula
    ac=-g+f*vi**2
    return ac
```

Questions

Which thermodynamic quantities determine the characteristic height h_E?[9]

The formulas for a and a_D in Fig. 7.8b (S) are valid only when the velocity is always negative, so that the body is always falling. Sloppy programming! How do the formulas have to be changed when the body can also rise?[10]

Interpret the formula in B8:B807 in the form $h(t+dt)=$![11]

The calculated velocity is shown in Fig. 7.7a as a function of time and compared with the empirical data of the stratospheric jump by Baumgartner.

7.4.3 Python

The parameters for the Python program are specified in Table 7.7, together with a function for the acceleration, *accJump*.

The main program is reported in Table 7.8 with the specification of the arrays t, h, and v and their initial values for $t = 0$. The for-loop is identical to all our programs for integrating the equation of motion. We have to assign the name *acc*, called in our standard function *RungeKutta*, to the acceleration function specific to the current problem.

7.5 A Car Drives with Variable Power

A driver wants to accelerate his vehicle at full throttle, but making sure that the wheels are not spinning (a "racing start"). We consider two approximations for this behavior, constant power and speed-proportional power. We

[9] The height h_E arises from the Boltzmann distribution of air molecules in the atmosphere: air density $\rho(h) = \rho(0) \cdot \exp(-mgh/k_BT)$; $h_E = k_BT/mg$, where m is the average mass of the air molecules.

[10] Instead of [=... + v²] one sets [=... + abs(v)^2*SIGN(v)] or [=...−v*abs(v)].

[11] Formula in B8:B807: $h(t + dt) = h(t) + (v(t) + v_D(t))/2 \cdot dt$.

Table 7.8 Progress loop for the stratospheric jump

```
t=np.empty(n)        #n determines the length of t, h, and v
h=np.empty(n)
v=np.empty(n)
t[0]=0
h[0]=38969           #[m] Jump height
v[0]=0

acc = accJump        #acc() accessed within RungeKutta
for i in range(n-1):
    t[i+1]=t[i]+dt
    h[i+1],v[i+1]=RungeKutta(h[i],v[i])
```

answer the following two questions: How does the final speed depend on the frictional resistance? How do longer-term fluctuations in power become noticeable? The programming challenge is to realize case distinctions.

7.5.1 Various Types of Power

Analytical solution for vanishing sliding friction

In this section, we characterize the racing start of a car ("full throttle") by a constant power P during the whole process (not by constant acceleration, as is usually assumed in textbook exercises). Without friction, the work W is converted into kinetic energy:

$$W = P \cdot t = \frac{m}{2} v^2 \rightarrow v = \sqrt{\frac{2P}{m}} \cdot \sqrt{t} \tag{7.13}$$

The resulting velocity v and the acceleration $a = dv/dt$ are shown in Fig. 7.9a for $P = 100$ kW and a car with mass $m = 1500$ kg. The velocity increases with the square root of time, and the acceleration becomes infinitely large at $t = 0$.

Non-zero motion friction and limited acceleration

In the real case, two effects have to be considered: (1) motion-inhibiting friction (rolling friction, driving resistance) and (2) the condition that the driving force must never be greater than the static friction force between the tire and the road. The second effect is taken care of by limiting the acceleration to a maximum value a_{Max}. We generally set the frictional force that inhibits motion proportional to v^{nF} so that the exponent n_F can later be selected as a parameter, e.g., $n_F = 2$ for friction caused by an air stream.

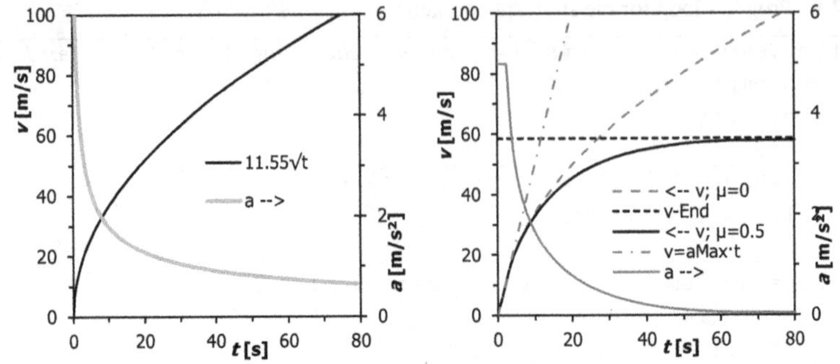

Fig. 7.9 **a** (left) Speed and acceleration for a constant power of $P = 100$ kW, without friction losses, calculated according to 7.13. **b** (right) Speed and power if driving friction is taken into account and acceleration is limited to a_{max}; numerically calculated

The power needed to increase the kinetic energy and to overcome friction is

$$P = \frac{d}{dt}\left(\frac{m}{2}v^2\right) + v \cdot F_R \tag{7.14}$$

$$F_R = \mu \cdot v^{n_F} \;\rightarrow\; P = m \cdot v \cdot \frac{dv}{dt} + \mu \cdot v^{n_F+1} \tag{7.15}$$

The acceleration

$$a(t) = \frac{dv}{dt} = \frac{P}{m \cdot v(t)} - \frac{\mu}{m} \cdot v(t)^{n_F} \tag{7.16}$$

does not depend on the location. It can be calculated from the power and the current velocity alone.

Question

The velocity in this task is always greater than or equal to zero. How do Eqs. 7.15 and 7.16 have to be changed if negative velocities are allowed?[12]

For stationary motion ($v = $ const.), we get

$$a(t) = \frac{dv}{dt} = 0 \;\rightarrow\; v_s(t) = \sqrt[n_F+1]{P/\mu} \tag{7.17}$$

The stationary velocity v_s can easily be calculated analytically with Eq. 7.17, which we use to check the numerical calculation. This check also works the other

[12] $v(t) \rightarrow Abs(v(t); v(t)^{n_F} \rightarrow Abs(v(t))^{n_F} \cdot sgn(v(t))$

way around, because, if numerically and analytically calculated values do not match, this may also indicate that our mathematical derivations are erroneous. It is, therefore, doubly good to compare the results of the two methods; we check whether our logical reasoning and our numerical calculation are consistent.

The calculated curves for the acceleration and the speed under friction can be seen in Fig. 7.9b. The acceleration at the beginning of the journey is at its pre-set maximum value 5 m/s^2 (for the acceleration, the right ordinate is valid) and goes practically to zero within 80 s. The speed is limited; the numerically calculated curve converges towards the stationary value obtained from Eq. 7.17. We can, therefore, assume that we did not make any gross programming mistakes. Without friction ($\mu = 0$), the velocity would continue to increase as in Fig. 7.9a.

Fluctuating power

What is the impact of power fluctuations on the speed? The answer is found in Fig. 7.10a, similar to Fig. 7.9b, yet with the power always fluctuating by 10% after a time span $DelT = 3.5$ s. Although the acceleration fluctuates on this scale, the speed hardly fluctuates. Speed is the integral of acceleration and, as such, averages over fluctuations. The fluctuations of the acceleration influence the speed more strongly when they remain constant over a longer duration.

▶ **Task** Change the time period *DelT* during which the power remains constant and observe the time course of *a* and *v*!

Speed-proportional power

For internal combustion engines, the power depends on the number of revolutions of the engine, and thus on the speed. If a function is known for this dependence, it can be included in the formula for the power. We investigate the simple case in which

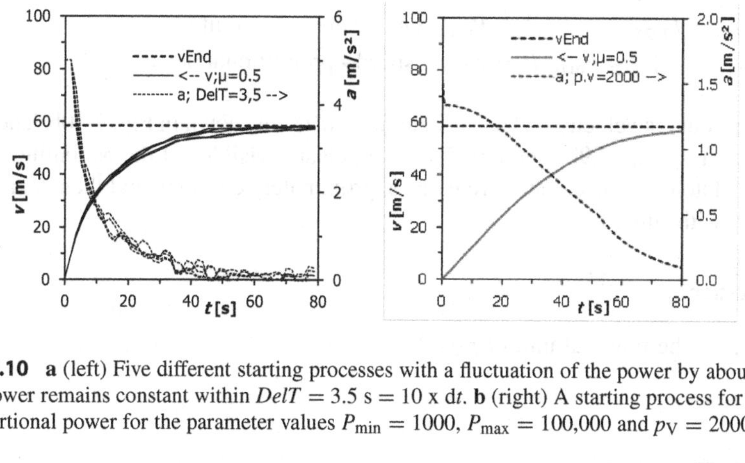

Fig. 7.10 **a** (left) Five different starting processes with a fluctuation of the power by about 10%; the power remains constant within $DelT = 3.5$ s $= 10$ x d*t*. **b** (right) A starting process for speed-proportional power for the parameter values $P_{min} = 1000$, $P_{max} = 100{,}000$ and $p_V = 2000$

the power P is proportional to the speed v:

$$P = p_V \cdot v \tag{7.18}$$

▶ **Tim** If the car is at a standstill, the power is zero, i.e., the car does not start at all.

▶ **Alac** That's why you put your foot down when you're idling and then let the clutch come.

▶ **Tim** How are we supposed to include that in our calculation model?

▶ **Mag** With a case distinction: If the power, according to Eq. 7.18, is too small, we use a constant value $P_{\min}$; in real life, this would be achieved by letting the clutch slide.

▶ **Alac** That's for low speed. On the other end, at high revs, the engine runs out of breath.

▶ **Mag** Then we apply a constant power when exceeding a certain speed.

▶ **Tim** Is that really the way to do it?

▶ **Mag** Anyway, this is our model assumption. If this were a research project, we would have to compare the curves calculated according to the model with measured curves, and probably also consider that the gear would be changed at a certain speed.

In our model, we distinguish three cases:

- $v \cdot p_V \leq P_{\min}$ $P = P_{\min}$, clutch slides
- $v \cdot p_V \geq P_{\max}$ $P = P_{\max}$, engine at power limit
- $P_{\min} < v \cdot p_V \leq P_{\max}$ $P = v \cdot p_V$, speed - proportional power

The time curves velocity $v(t)$ and acceleration $a(t)$ are shown in Fig. 7.10b. The differences from Fig. 7.9b (and Fig. 7.10a) are clearly visible. At the beginning, with sliding clutch, the acceleration remains approximately constant, and then decreases linearly with time.

Question

What is the physical unit of p_V?[13]

[13] $[p_V] = [P/v] =$ Ws/m = J/m.

7.5.2 Data Structure and Nomenclature

m	mass of the car
μ	friction coefficient for air resistance
n_F	power in friction law
P	default constant power
P_{min}	power when the clutch slides
P_{max}	engine at the power limit
p_V	coefficient for speed-proportional power
a_{Max}	maximum acceleration to avoid sliding
t	sequence of times, dt apart, independent variable
P_{var}	varying power as a function of t
a	acceleration at t
v	velocity at t
v_S	stationary velocity
x	distance travelled since $t = 0$

7.5.3 Excel

Constant power

For the numerical calculation, we use our standard method, "Progress with look-ahead" (see Fig. 7.11 (S)). The condition "driving force < static friction force" is taken into account by limiting the acceleration using a MIN function, $a = Min(a(t); a_{Max})$, where, for *a(t)*, Eq. 7.16 is to be used. The corresponding spreadsheet formula, valid for C6:C235, is reported in C4.

We do not set the velocity at time $t = 0$ in B6 to zero, but rather to 0.00001, so that, in C6, there is no division by 0. The value calculated in column C is

	A	B	C	D	E	F	G	H
1	dt	m	μ	n	P	aMax		
2	0.35	1500	0.5	2	100000	5.00		
3	<-- v; μ=0.5		a -->					
4	=A6+dt	=B6+(C6+E6)/2*dt	=MIN(P/m/v-μ/m*v^n;aMax)	=v+a*dt	=MIN(P/m/vD-μ/m*vD^n;aMax)			
5	t	v	a	vD	aD		<-- v; μ=0	
6	0.00	0.00001	5.00	1.75	5.00		0.00001	
7	**0.35**	**1.75**	5.00	3.50	5.00		1.75	
235	80.15	58.25	0.01	58.26	0.02		102.09	

Fig. 7.11 (S) Acceleration and velocity are calculated using the method "Progress with look-ahead". The parameters of the problem are defined in row 2 with their names in row 1. The values in column G were copied from column B when μ was set to 0

thereby not affected, because, from the MIN condition, the maximum permissible acceleration a_{Max} results anyway.

Remember: In the calculation of the velocity v in a row (time t), only values from the previous row (time $t-dt$) are used. Therefore, the formula in B7 (reported in B4) gets input only from B6, C6, and E6. Formulas should only be entered into the cells in bold, here in A7 and B7. The rest of the respective columns is obtained by copying ("dragging down"). The initial values in A6 and B6 are entered as numbers. The formulas reported in C4:E4 are valid in their entire columns, e.g., C6:C235, because they do not refer to cell addresses, but rather to names for variables (cells and column ranges).

Questions

concerning Fig. 7.11 (S):

When is the acceleration in column C of constant?[14]

Examine and describe the expressions for a and a_D![15]

Another technical note for the spreadsheet calculation: 10/5/2 = 1. So, the formula is divided consecutively: 10/5 = 2, then 2/2 = 1. According to the rules of fractional calculation, the formula could be interpreted as 10/(5/2) = 4, which is not so in EXCEL.

▶ **Task** Determine time and distance as a function of the power needed to accelerate to 95% of the final speed! To calculate the distance, you must integrate the velocity.

Temporally fluctuating power with the Mod function

How do the curves $a(t)$ and $v(t)$ change when the power does not remain constant, but rather fluctuates over time?

When the power fluctuates over time, we cannot treat it as a global constant P, but must insert an extra column in the spreadsheet, P_{var}, into which the power can be entered at any time. To calculate the acceleration, we must access this variable, and not the global parameter P.

In Fig. 7.12 (S), a calculation is presented in which the power fluctuates over a longer period of time, i.e., remains constant over a time interval Δt (Del$T$). The two variables r_T and P_{var} have been inserted into columns B and C for this purpose. The formulas for v, a, v_p and a_p are exactly the same as in Fig. 7.11 (S), but with the time-dependent power P_{Var} instead of the constant power P.

[14] As long as the acceleration calculated in column C according to Eq. 7.16 is greater than the value a_{Max} given in F2.

[15] Think about it and read on!

	A	B	C	D	E	F	G	H	I	J	K
1	dt	DelT	P	DelP	m	μ	n	aMax			
2	0.35	3.4	100000	10000	1500	0.5	2	5.00		<-- v;μ=0.5	
3										a; DelT=3,5 -->	
4	=A6+dt	=ABS(MOD(t;DelT))	=IF(rT <=0.2;NORM.INV(RAND();P;DelP);C6)	=D6+(E6+G6)/2*dt	=MIN(Pvar/m/v-μ/m*v^n;aMax)	=v+a*dt	=MIN(Pvar/m/vD-μ/m*vD^n;aMax)		a; DelT=3,5		
5	t	rT	Pvar	v	a	vD	aD		t.R	v.R	a.R
6	0.00	0.00	115530	0.00001	5.00	1.75	5.00		0.00	0.00	5.00
7	**0.35**	0.35	**115530**	**1.75**	5.00	3.50	5.00		1.75	8.75	5.00
15	3.15	3.15	115530	15.72	4.82	17.40	4.32		15.75	40.17	1.23
16	3.50	0.10	125449	17.32	4.73	18.97	4.29		17.50	42.20	0.98
235	80.15	1.95	104604	58.13	0.07	58.16	0.07		71.75	56.94	-0.11

Fig. 7.12 (S) The calculation of Fig. 7.11 (S) is extended with columns B and C, in which a time-dependent power is calculated, remaining constant over the time period *DelT*. Columns I, J and K are written by SUB *Vari* in Fig. 7.13 (P) with a selection of the data from columns A, B and C

In our model, we require that the power remain constant over a period of time Del$T$. This condition can be fulfilled with the MOD function (performing a *modulo* operation) , whose action is visible in column B of Fig. 7.12 (S). The time t is divided by *DelT* and the remainder is returned; e.g., in row 16 *t/DelT* = 3.5/3.4 = 0.1, remainder (*modulus*) is 0.1, and 0.1 is returned.

The expression in C4 of Fig. 7.12 (S), valid for C7, outputs a new value for P_{var} only at certain times, otherwise, the value from the previous time is carried over. It is structured as follows:

IF ($rT \leq 0.2$) THEN (P_{var} new) ELSE (old value from the previous cell C6)
with P_{var} new = NORM.INV(RAND (); P; *DelP*), setting a new power, fluctuating around the mean value P with a standard deviation Del$P$.

In columns I, J, and K, time t_{R}, velocity v_{R}, and acceleration a_{R} are entered successively for five start processes using a macro (protocol procedure SUB *Vari* in Fig. 7.13 (P)): The curves are graphically represented in Fig. 7.10a. In every run,

```
1  Sub Vari()
2  'The course of velocities is
3  'calculated several times and
4  'stored in columns 9-11.
5  r2 = 6
6  Cells(3, 10) = "a; DelT=" & Round(Cells(2, 2), 2)
7  For rep = 1 To 5
8    Application.Calculation = xlCalculationManual
9
10
11   For r = 6 To 235 Step 5
12     Cells(r2, 9) = Cells(r, 1) 't.R
13     Cells(r2, 10) = Cells(r, 4) 'v.R
14     Cells(r2, 11) = Cells(r, 5) 'a.R
15     r2 = r2 + 1
16   Next r
17   r2 = r2 + 1
18   Application.Calculation = xlCalculationAutomatic
19 Next rep
20 End Sub
```

Fig. 7.13 (P) Rep-log procedure, copies five start operations (FOR REP = 1 TO 5) continuously into columns 9 to 11 (= I, J, K) of Fig. 7.12 (S). Only the data for every fifth time are transferred (line 11, … STEP 5)

	A	B	C	D	E	F	G	H	I
1	**dt**	**m**	**μ**	**nF**	**P**	**aMax**		**Pmin**	1000
2	0.35	1500	0.5	2	100000	1.50		**Pmax**	100000
3								**p.v**	2000
4	=A6+dt	=IF(v*p.v>Pmin;IF(v*p.v<Pmax;v*p.v;Pmax);Pmin)	=C6+(D6+F6)/2*dt	=MIN(Pv/m/v-μ/m*v^nF;aMax)	=v+a*dt	=MIN(Pv/m/vD-μ/m*vD^nF;aMax)			
5	**t**	**Pv**	**v**	**a**	**vD**	**aD**			
6	0.00	1000	0.00001	1.50	0.53	1.27			
7	**0.35**	1000	**0.48**	1.38	0.97	0.69			
235	80.15	100000	56.87	0.09	56.91	0.09			

Fig. 7.14 (S) Speed-proportional power $P_v = v\cdot p_V$ if it is bigger than P_{min} and smaller than P_{max}

only every fifth data point ("Step 5") is transmitted.

Questions

What would the curves of v in Fig. 7.10a look like if, in Fig. 7.13 (P), line 17 were missing?[16]

How does *delP* appear as a standard deviation in the fluctuating power in Fig. 7.12 (S)?[17]

Speed-proportional power

To get a power that is proportional to the speed of the car, Fig. 7.14 (S), we replace the formula in C4 of Fig. 7.11 (S) , valid for C6:C235, with a nested if-loop with the outputs $v\cdot p_V$, P_{Max}, P_{Min}:

[= IF(v* p_v > Pmin;
  (IF(v* p_v < Pmax; **v*pV**;
    Pmax));
    Pmin)]

The power is $v\cdot p_V$, i.e., proportional to the speed, if it is bigger than P_{min} and smaller than P_{max}.

[16] There would be no empty line separating the data sets, so that the last point of a curve would be connected to the first point of the following curve.

[17] See formula in C4 where a formula for normal noise is applied.

Table 7.9 Parameter specification for driving with constant power

```
dt=0.35
m=1500                  #[kg] Mass of the car
mu=0.5                  #Friction coefficient
nF=2                    #Power coefficient in friction law
P=100000                #[W] Power
aMax=5.0                #[m/s²] max. acceleration

def accPwr(x,v):   #v is always positive
    a=min(P/m/v-mu/m*v**nF,aMax)
    return a

t=np.arange(0,80.15+dt,dt)
x=np.zeros(len(t))
v=np.zeros(len(t))
a=np.zeros(len(t))
```

Table 7.10 Progress loop with *RungeKutta* for constant power

```
16  v[0]=0.00001
17  acc=accPwr    #acc is name for acceler. within RungeKutta
18  for i in range(len(t)-1):
19      a[i]=accPwr(x[i],v[i])
20      x[i+1],v[i+1]=RungeKutta(x[i],v[i])
```

7.5.4 Python

Constant power

The setting of the parameters for driving with constant power is shown in Table 7.9. The size of the arrays is specified with `np.arange(0,80.15 + dt,dt)` to yield exactly the same range as in the EXCEL spreadsheet of Fig. 7.11 (S). The acceleration is executed in a function *accPwr* that is assigned the additional name *acc* (line 17 in Table 7.10) expected in *RungeKutta.*

Progress of motion is achieved in Table 7.10 with the usual progress loop.

Question

Why do we calculate a[i] separately in line 19 of Table 7.10, although it is already calculated within Runge-Kutta?[18]

---

[18] We want to plot $a(t)$, e.g., in Figs. 7.9 and 7.10.

Fluctuating power

To calculate the car's velocity for fluctuating power, the additional parameters *DelT*, *DelP*, and *Pc* are specified in Table 7.11. To store the results of five different runs, matrices v_{M} and a_{M} are built by stacking five times the arrays v and a, respectively (lines 6 and 7). Their shape is 5 rows × 231 columns. The usual progress-loop contains an inner loop ($k =$) running over the 5 rows and calling *RungeKutta* successively with the individual `vM[k,i]` so that, in the end, five time series of velocity and acceleration will have been calculated.

The contents of a_{M} and v_{M} are completely calculated in Table 7.11 and plotted with the program in Table 7.12.

Table 7.11 Calculation of five time series of v and a for motion with fluctuating power

```
1  import numpy.random as npr
2  DelT=3.5                                   #Fluctuation period
3  DelP=10000                                 #Fluctuation amplitude
4  Pc=100000                                  #Mean power
5  P=Pc
6  vM=np.stack((v,v,v,v,v))
7  aM=np.stack((a,a,a,a,a))
8  for i in range(len(t)-1):
9      for k in range(5):
10         aM[k,i]=accPwr(x[i],vM[k,i])*10
11         PBef=P        #Setting the (new?) power
12         rem=np.mod(t[i],DelT)    #0 ≤ rem ≤ DelT
13         P=Pc+DelP*npr.randn() if rem<=DelT else PBef
14         x[i+1],vM[k,i+1]=RungeKutta(x[i],vM[k,i])
```

```
np.shape(vM)       (5, 231)
np.shape(vM[0])    (231,)
```

Table 7.12 Plotting five time series of velocity and acceleration for motion with fluctuating power

```
FigStd('t',0,80,20,'v',0,60,20)
print('np.shape(t)        ',np.shape(t))
print('np.shape(vM[0]) ',np.shape(vM[0]))
for k in range(5):
    plt.plot(t,vM[k],'k-')       #Full black
    plt.plot(t,aM[k],'k:')       #Dotted black
```

```
np.shape(t)        (231,)
np.shape(vM[0])    (231,)
```

Table 7.13 Definitions for speed-proportional power

```
Pv=np.zeros(len(t))
Pmin=1000              #Min. power
Pmax=100000            #Max. power
pv=2000                #Pn=v*pv, proportionality factor
aMax=1.5               #Max. acceleration

def accPwrV(x,v):   #Pn is specified in progress loop
    a=min(Pn/m/v-mu/m*v**nF,aMax)
    return a
```

Table 7.14 Loop for progress with *RungeKutta* for speed-proportional power

```
v[0]=0.00001
acc=accPwrV              #acc is name in RungeKutta
for i in range(len(t)-1):
    if (Pmin<v[i]*pv<Pmax): Pn=v[i]*pv
    elif (v[i]*pv<Pmin): Pn=Pmin
    else: Pn=Pmax
    Pv[i]=Pn
    a[i]=accPwrV(x[i],v[i])
    x[i+1],v[i+1]=RungeKutta(x[i],v[i])
```

Questions

How do you change the arguments in line 1 of Table 7.12 to get the same axis titles as in Fig. 7.10?[19]

How does the changing power P enter *accPwr(...)*?[20]

Speed-proportional power

The additional parameters for driving with speed-proportional power are given in Table 7.13, together with a function *accPwrV(x, v)* for calculating the acceleration with the input variables x and v, as is expected for the function *acc* applied in *RungeKutta*. The function *accPwrV* makes use of global parameters. To keep control of the situation, the global parameters and the function should be defined in the same program cell. This is actually the case in Tab. 7.13, except for P_v, the speed-dependent power that depends on the speed calculated in the progress loop.

The progress loop for speed-proportional power in Table 7.14 contains a nested `if ... elif ... else` query to assign the correct value of the power to the variable P_n accessed as a global parameter in *accPwrV*.

[19] `FigStd('t [s]',0,80,20,'v [m/s]',0,60,20).`

[20] The power P is a global variable changed within the progress loop.

7.6 Bungee Jump

A bungee jumper falls in free fall until the rope becomes tight; only gravity acts. When the rope is stretched, two additional forces come into play: the back-driving rope force and the friction force due to the inner friction of the rope. We neglect air friction so that the force depends only on the location, and not on the speed. The form of motion is, at times, a free fall or a damped vibration.

7.6.1 Simulation of the Motion

In a bungee jump, a person hangs on an elastic rope and lets her/himself fall from a great height into the depths. As an example, we consider a bungee jumper (mass m = 60 kg, size 1.65 m) jumping from a height of 50 m. The bungee rope is l = 25 m long in relaxed condition and has an elastic constant (corresponding to a spring constant) of k = 100 N/m. The zero of the z-axis is at the point where the rope is fixed. The jumpers move in free fall until the elastic rope is fully expanded but not yet stretched. For l = 25 m, this is at $z = -25$ m.

No friction
We first consider the situation without friction in which the jumper starts at height $z = -25$ m (fully elongated rope, not yet stretched; see Fig. 7.15a). The jumper oscillates like a mass-spring system with a period of $T = 2\pi/\sqrt{k/m} = 4.87s$ or $4 \cdot T$ = 19.47 s.

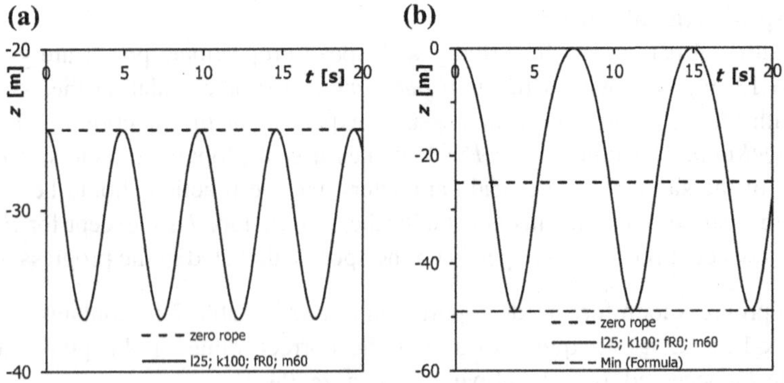

Fig. 7.15 **a** (left) Oscillation when the jumper starts with a fully elongated rope. **b** (right) Sequence of free fall and oscillation when the jumper starts at the fixing point of the rope

When the jumpers start at the fixing point of the rope at the height of $z = 0$ m, Fig. 7.15b, they:

- fall freely until the rope is fully elongated ($z = -25$ m),
- complete a half-period oscillation driven back by the elasticity of the rope to the zero-rope position ($z = -25$ m),
- and overshoot up to the original height $z = 0$ in free fall (starting upwards because of the initial upward velocity), where the height is a parabolic function of time.

This sequence of motions repeats periodically.

Friction force of the rope

We assume a *friction work* W_F proportional to the elongation or relaxation path of the rope, because, during elongation and relaxation, rope components are shifted against each other, thus dissipating energy. Consequently, the frictional force $f_R = dW_F/dh$ is constant. For the example of Fig. 7.16, we have chosen $f_R = 220$ N.

The motion with unexpanded rope (above $z = -l$) is still a free fall, whereas, with expanded rope, a dampened oscillation is observed.

The horizontal lines in Fig. 7.16 are obtained from analytical calculations of:

- the maximum height fallen,
- the top speed of the jumper,
- the rest position at the stretched rope.

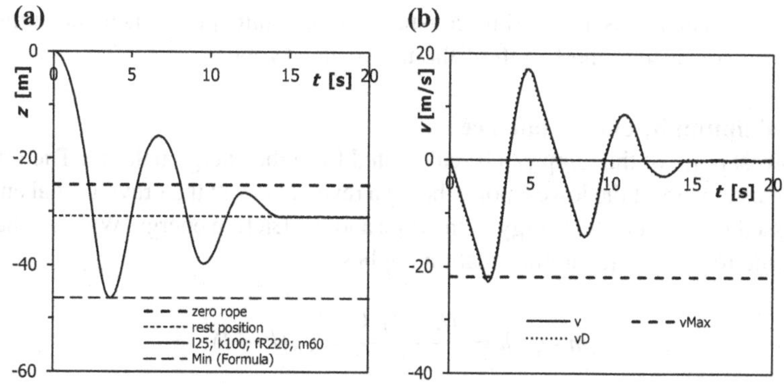

Fig. 7.16 **a** (left) Height as a function of time during a bungee jump, "zero rope" = length of unexpanded rope. **b** (right) Velocity as a function of time; v_D (dotted line) = look-ahead to end of the time segment, calculated with the acceleration at the beginning. Parameters specified in Fig. 7.18 and Table 7.15.

Question

What do you think: is rope friction effective only for elongation or also for relaxation of the rope?[21]

Reverse problem: The properties of the rope are deduced from the observed motion

In this exercise, we have specified values for the spring constant and the friction force of the rope and therewith calculated the time curves for the position and velocity of the jumper.

More interesting from a physical point of view is the reversal of the problem, in which the properties of the rope are deduced from an actually possible observation. A possible approach is to observe the oscillation period and measure the time it takes for the jumper to come to rest, and then adjust the parameters of the simulation model so that the simulation fits the measured results. We could then possibly see whether our assumptions about the rope's elastic and frictional properties and our neglect of air friction are justified.

7.6.2 Analytical Calculations

We calculate the minimum and the rest position, as well as the maximum speed by setting up the equation of motion with the starting point at $z = 0$:

$$m \cdot \ddot{z} = -m \cdot g \text{ for } z > -l \tag{7.19}$$

$$m \cdot \ddot{z} = -m \cdot g - (z - l) \cdot k - f_R \text{ for } z \leq -l \tag{7.20}$$

The parameter k is the spring constant coming into play when the jumper's position is lower than the length of the unexpanded rope.

First minimum by energy balance

The lowest point of the jump can be calculated from the energy balance. The kinetic energy disappears at the lowest point, being a reversal point; the gravitational energy has passed into the elastic energy of the rope and the friction energy. With $h =$ height fallen, the following quadratic equation applies:

$$m \cdot g \cdot h = \frac{(h - l)^2 k}{2} + f_R(h - l) \tag{7.21}$$

[21] Elongation: gravitational energy is converted into kinetic, elastic, and friction energy.
Relaxation (rope becomes shorter): elastic and kinetic energy is converted into gravitational and frictional energy.

$$h^2 + h \cdot \left(-2l + \frac{2f_R}{k} - \frac{2mg}{k}\right) + \left(l^2 - f_R\frac{2l}{k}\right) = 0 \tag{7.22}$$

It is solved in Fig. 7.17 (S), Eq. A, for the parameters of Fig. 7.18.

Maximum speed of the Bungee jumper

The maximum speed of the jump is obtained by considering the dependence of the kinetic energy on the height of the fall:

$$E_{kin} = \frac{m}{2}v^2 = mgh - \frac{k}{2}(h-l)^2 - f_R(h-l) \tag{7.23}$$

The gravitational energy goes into the kinetic energy of the jumper, the elastic energy and the friction work of the rope.

The derivative of the kinetic energy with respect to the height gives the position of the maximum velocity:

$$\frac{dE_{kin}}{dh} = mg - k(h-l) + f_R = 0 \tag{7.24}$$

$$h = \frac{mg + kl + f_R}{k} \tag{7.25}$$

The maximum speed is calculated from the kinetic energy $T_{\max}$ at this altitude:

$$v_{max} = \sqrt{\frac{2T_{max}}{m}} \tag{7.26}$$

The value of $v_{\max}$ is calculated in Fig. 7.17 (S), Eq. (B), for the parameters of Fig. 7.16, specified in Fig. 7.18 (S).

Eq. (A)	Terms of the quadratic equation	**p**	-57.37	=-2*l+2*fR/k-2*m*g/k
		q	515.0	=l^2-fR*2*l/k
		h+	-11.14	=p/2+SQRT((p/2)^2-q)
	Deepest point	**h-**	-46.23	=p/2-SQRT((p/2)^2-q)
Eq. (B)	Hight for max. speed	**hVm**	33.09	=(m*g+k*l+fR)/k
	Max. kinetic energy	**Tmax**	14426	=m*g*hVm-k/2*(hVm-l)^2-fR*(hVm-l)
	Max. speed	**vMax**	21.93	=SQRT(2*Tmax/m)
Eq. (C)	Rest position	**hRest**	30.89	=m*g/k+l

Fig. 7.17 (S) Analytical calculation of the lowest point, the maximum speed, and the rest position of the bungee jump; p is the first parenthesis of Eq. 7.22, q the second. Eq. (A) = Eq. 7.22, Eq. (B) = Eq. 7.25 and Eq. 7.26, Eq. (C) = Eq. 7.28

Final position of the bungee jumper

When the jumper has come to rest, there is a balance between the weight force and the elastic rope force:

$$m \cdot g = k \cdot (h - l) \tag{7.27}$$

$$h = \frac{m \cdot g}{k} + l \tag{7.28}$$

calculated in Eq. (C) of Fig. 7.17 for the parameters of Fig. 7.18 (S).

▶ **Task** Determine the time until the jumper is at rest as a function of jump height!

▶ **Task** Vary k and F_R and observe how the number of oscillations and the time until rest change!

7.6.3 Data Structure and Nomenclature

g	gravitational constant 9.81 m/s^2
l	length of the rope
k	elastic constant of the rope
f_R	friction force of the rope, constant, energy $= f_R \cdot (h - l)$
m	mass of the jumper
t	sequence of times, dt apart
z	position of the jumper at t, rope fixed at $z = 0$
z_{Min}	minimum position of the jumper
z_{End}	final rest position of the jumper
v	velocity of the jumper as a function of t
v_{Max}	maximum speed
T_{Max}	maximum kinetic energy.

7.6.4 Excel

The parameters m, l, k, f_R, and g (= 9.81 m/s^2) are specified in named cells in A1:F2 of Fig. 7.18 (S). H3 displays the formula (concatenation of text and variables) for obtaining the legend in H2.

The equation of motion of the bungee jump is integrated with our standard method, "Progress with look-ahead". The result of the calculation is shown in Fig. 7.16.

	A	B	C	D	E	F	G	H	I	J	K	L	M
1	dt	g	l	k	m	fR		;					
2	0.05	9.81	25	100	60	220		l25; k100; fR220; m60					
3	s	m/s²	m	N/m	kg	N		=C1&l&H1&D1&k&H1&"fR"&fR&H1&"m"&m					
4													
5		a=IF(z<-l;-g-SIGN(v)*fR/m-k/m*(z+l);-g)											
6			aD=IF(zD<-l;-g-k/m*(zD+l)-SIGN(vD)*fR/m;-g)										
7	=A9+dt	=B9+(D9+G9)/2*dt	=C9+(B9+E9)/2*dt		=v+a*dt	=z+(v+vD)/2*dt				=MIN(v)	=MIN(z)	=INDEX(z;401)	
8	t	v	z	a	vD	zD	aD			vMax	zMin	zEnd	
9	0.00	0.00	0.00	-9.81	-0.49	-0.01	-9.81		Simu	**-22.70**	**-46.28**	**-30.95**	
10	**0.05**	**-0.49**	**-0.01**	-9.81	-0.98	-0.05	-9.81		Formula	-21.93	-46.23	-30.89	
409	20.0	0.15	-30.9	-3.56	-0.03	-30.9	3.77						

Fig. 7.18 (S) Calculation model for a bungee jump; l = length of the rope; k = spring constant; m = mass of the jumper; f_R = friction constant of the rope; acceleration is inhibited by friction only when the rope is stretched; the formulas a = IF(...) in row 5 and a_D = IF(...) in row 6 are valid for the column vectors a and a_D, respectively. Bewegung wird gehemmt

Questions

Analyze the formula "$a = \ldots$ " in B5 of Fig. 7.18 (S) applied in column D. What is the logical structure of the formula? When does only gravity act?[22]

Which forms of the time trajectory do you observe for an unstretched and a stretched rope?[23]

The spreadsheet set-up presented here applies the same friction force for relaxing and stretching rope. Friction losses are likely to occur only when the rope is longer than in the unstretched state (discuss!); this is taken into account through an IF query.

The calculated height is shown in Fig. 7.16a as a function of time. For checking purposes, the prominent points of the jump calculated analytically, namely, maximum depth (= minimum height) and rest position, are entered into the diagram as horizontal lines. The agreement with the simulated curve is quite good, which indicates that neither our analytical calculations nor our simulation within the framework of our model contain gross errors;-) or, less likely, both errors cancel each other out.

In Fig. 7.16b, the velocity is plotted as a function of time, and also, as a horizontal line, the maximum velocity analytically calculated according to Eq. 7.26. The dashed curve represents the velocity foreseen at the end of the interval calculated with the acceleration at the beginning of the interval. The full curve shows the velocity at the beginning of the next interval, calculated from the mean of the two accelerations, the one at the beginning of the interval and the estimated one

---

[22] IF(LOGICAL_TEST, [VALUE_IF_TRUE], [VALUE_IF_FALSE]). If the amount of the deflection is smaller than the rope length, then only the acceleration due to gravity acts; neither the elastic rope force nor the friction force are active. In the IF function, this is the [VALUE_IF_FALSE] case, i.e., when the condition ($z < -l$) is not fulfilled.

[23] Unstretched: free fall, parabola; stretched: damped oscillation around the rest position.

Table 7.15 Parameters and function for calculating the acceleration during a bungee jump

```
g=9.81                  #[m/s²]
l=25                    #[m]    Length of rope
k=100                   #[N/m]  Spring constant
m=60                    #[kg]   Mass of jumper
Fr=220                  #[N]    Frictional force

def accBungee(z,v):     #v positive, negative, or 0
    if z<-l:            #Tensioned rope
        a=-g-k/m*(z+l)-np.sign(v)*Fr/m    #np.sign(0)==0
    else: a=-g
    return a

lbl_1=("l"+str(l)+"; k"+str(k)+"; fR"+str(Fr)+"; m"+str(m))
```

lbl_1	l25; k100; fR220; m60

Table 7.16 Progress of the bungee jump

```
dt=0.251                    #[s] Time increment
t=np.arange(0,100+dt,dt)
v=np.zeros(len(t))          #Velocity
z=np.zeros(len(t))          #Height
acc=accBungee               #acc(...) is accessed in RungeKutta
v[0]=0                      #Initial conditions
z[0]=0
for i in range(len(t)-1):
    z[i+1],v[i+1]=RungeKutta(z[i],v[i])
```

at the end of the interval. The difference between the two curves illustrates the difference between the *Euler* and the *lookAhead* methods.

7.6.5 Python

The parameters of the bungee jump and the function *accBungee* for calculating the acceleration of the jumper are given in Table 7.15. The progress of the motion of the jumper is calculated in Table 7.16.

7.7 Questions and Tasks

From integration to numerical averaging

1. How is the velocity $v(t)$ related to the acceleration $a(t)$?

2. How is the mean value of a continuous function $f(t)$ within the range t_1 to t_2 defined?
3. Interpret the formula $v(t_n + dt) = v(t_n) + [a(t_n) + a(t_n + dt)]/2 \cdot dt$ with respect to the previous two questions!
4. What does the broom rule (Ψ *half, half, whole, the halves count twice*) tell us about the 4th order Runge-Kutta method?

Half-step procedure in a spreadsheet
In Fig. 7.19 (S), you can see a spreadsheet structure for the numerical integration of Newton's equation of motion for a fall with friction using the half-step method. The spreadsheet calculation is structured in the same way as for our "Progress with Forecast" procedure. However, the predicted acceleration a_P is calculated for the middle of the interval and is considered representative of the entire time interval.

5. What are the formulas for x_P, v_P, and a_P?
6. What are the formulas for x and v at the beginning of the next time interval (in C7 and D7)?

Function for "Progress with look-ahead" or "half step"
In Table 7.17, you can see the code of a `Python` function for calculating the progress from t to $t + dt$. However, in lines 4 to 8, the formulas have been replaced by 1.

7. Insert the appropriate formulas if "Progress with look-ahead" is to be used!

	B	C	D	E	F	G	H	I	J
1	**g**	9.81 m/s^2		*gravitational acceleration*					
2	**cV**	1.5 1/m		*friction coefficient*					
3	**dt**	0.005 s		*time increment*					
4	*=B6+dt*	*=C6+G6*dt*	*=D6+H6*dt*	*=-g-cV*v^2*SIGN(v)*	*=x+v*dt/2+a/8*dt^2*	*=v+a*dt/2*	*=-g-cV*vP^2*SIGN(vP)*		
5	**t**	**x**	**v**	**a**	**xP**	**vP**	**aP**		
6	0.000	4.000	1.000	-11.31	4.00	0.97	-11.23		
7	**0.005**	4.005	**0.944**	-11.15	4.01	0.89	-10.99		
406	2.000	-0.361	-2.557	0.00	-0.37	-2.56	0.00		

Fig. 7.19 (S) Falling with friction, half-step procedure: The velocity v_P and the acceleration a_P in the middle of the interval are calculated. Formula in C4 is valid for C7

Table 7.17 Function for numerical integration of the Newtonian equation of motion

```
1  dt=0.1
2  def ForcA(x,v):
3      a=acc(x,v)
4      xA=1
5      vA=1
6      aA=1
7      xN=1
8      vN=1
9      return(xN,vN)
```

8. Insert the appropriate formulas if the half-step method is to be used!

Power and work

9. How are work W and power P related to the force F, the body's displacement x, and the velocity v?

Bungee jumping

10. What are the formulas for: a) the elastic energy of a spring with constant k, b) the gravitational energy of a mass m near the earth's surface, and c) the kinetic energy of a mass m?

Consider the statement: "The frictional energy when stretching or relaxing a rope is proportional to the change in length (absolute value) of the rope."

11. How is this approach justified?
12. Which function results for the frictional force?

Mathematical pendulum

The oscillation equation for a mass-spring system is

$$\frac{\partial^2 x(t)}{\partial t^2} = -\frac{k}{m} \cdot x(t) \tag{7.29}$$

The oscillation equation for a mathematical pendulum is

$$\frac{\partial^2 \phi(t)}{\partial t^2} = -\frac{g}{l} \cdot \sin(\phi) \tag{7.30}$$

13. Why does no mass appear in Eq. 7.30 as does in Eq. 7.29?
14. Why does *sin*(ϕ) turn up on the right side of Eq. 7.30, and not simply ϕ, as one might expect in analogy to Eq. 7.29?

Random Numbers and Statistical Reasoning

8

We perform statistical experiments by applying functions that generate uniformly distributed, normally distributed, and $\cos^2$ distributed random numbers. We set up frequency distributions of a set of random numbers and compare them quantitatively with model distributions by means of the Chi^2 test, the interpretation of which will be explored. (Random-number generators are used in later chapters to simulate measurement inaccuracies and noise, e.g., in Chap. 9 (Evaluation of measurements) and Chap. 10 (Trend lines).) Required spreadsheet functions are: RAND(), FREQUENCY() as a matrix function, and CHISQ.TEST(), as well as the logical functions AND and OR, together with their counterparts in the Python libraries numpy.random and scipy.stats.

8.1 Introduction: Statistical Experiments Instead of Theoretical Derivations

Solutions of Exercises 8.2 (Excel), 8.3 (Python), 8.5 (Excel), and 8.8 (Python) can be found at the internet address: go.sn.pub/zHV7Ko.

Russian proverb: *Доверяй, но проверяй! Trust but verify!*

German variant: Vertrauen ist gut, Kontrolle ist besser. *Trust is good, control is even better.*

Ψ If in doubt, count!

Ψ Mostly, not always. "Fundamental rule of statistical reasoning".

Random-number generators

In Python, two libraries have to be imported to make statistical functions available:

D. Mergel, *Physics with Excel and Python*,
https://doi.org/10.1007/978-3-030-82325-2_8

`numpy.random (npr)` for the functions `rand` (random numbers between 0 and 1) and `randn` (normally distributed random numbers).

`scipy.stats (sct)` for the functions `norm(0,1)` (normal distribution) and `chisquare` (Chi^2 test).

The spreadsheet function RAND and the *Python* function `npr.rand` return random numbers x that are equally distributed between 0 and 1 ($0 \leq x < 1$). Other distributions are obtained when these random numbers are redistributed with suitable functions, namely, with the inverse of the distribution function of the desired probability density. For the Gaussian, exponential and Cauchy-Lorentz distributions, these inverse functions are available as NORM.INV, LN and TAN in EXCEL, and `sct.norm(0,1).ppf`, `np.log`, and `np.tan` in `Python`.

The distribution function can be approximated with a polyline in a *user-defined function*. This is important for cases in which no predefined function for the inverse of a cumulated distribution function exists. As an example, we calculate the diffraction image of photons that have passed through a double-slit where they are distributed along a line perpendicular to the slits in $\cos^2$-shaped maxima.

What is to be learned?

After having worked through this chapter, you should be able to handle the following safely:

- generating a set of random numbers that obey a given distribution model,
- determining the frequencies of occurrence of a data set, and displaying them graphically,
- quantitatively comparing observed frequency distributions with model distributions. To this end, we practice the cautious use of the Chi^2 test with the correct degree of freedom *dof*.

Statistical experiments instead of theoretical derivations

▶ **Alac** Numbers, numbers, numbers. They're quite tiring. Is it worth the effort?

▶ **Mag** Yes. This is the basis of the evaluation of experiments.

▶ **Alac** Evaluation of Lab-course experiments? I'm happy when I have the annoying protocols behind me. I just want to memorize a few answers from the introduction to the exercises to get through the exam talks scot-free.

▶ **Mag** We only practice what is absolutely necessary—but understanding that well is quite important, not only for studying physics, but for all empirical sciences and political statements.

▶ **Alac** Understand it well? Do we have to reproduce mathematical proofs?

▶ **Mag** No. We perform experiments to grasp the essence of certain theorems of statistics.

▶ **Tim** “Statistical experiments”, that sounds interesting. Nevertheless, probability theory is not my thing. Nobody in my study group understands the Chi^2 test yet.

▶ **Mag** I can understand that. Appearances are often deceptive, and no statement is a hundred percent certain. Nevertheless, you have to be able to move on this unsteady ground. In this chapter, we will start with gait exercises.
We use two types of statistical test in which a random experiment is repeated many times:

- multiple tests for equal distribution and
- multiple tests for error probability.

In Exercise 8.2, multiple tests for equal distribution will be used to check whether the results of Chi^2 tests are equally distributed, as it should be when the model distribution corresponds to the population from which the samples are taken.

Multiple tests for error probability are used in Chap. 9, “Evaluation of measurements”, and Chap. 10, “Fitting of trend curves to measurement points”, to check whether the experimental confidence intervals of a measurement result match the assumed confidence levels.

Statistical functions in Python and Excel

```
import numpy.random as npr
  import scipy.stats as sct.
```

Functions concerning the normal distribution:

`sct.norm(xm,xd).pdf(x)` NORM.DIST (x;xm;xd; FALSE)
$-\infty < x < \infty$, probability density function
`sct.norm(xm,xd).cdf(x)` NORM.DIST(x;xm;xd; TRUE)
$-\infty < x < \infty$, cumulative density function (distribution function)
`sct.norm(xm, xd).ppf(p)` NORM.INV(p; xm; xd)
$0 \le p \le 1$, percent point function, inverse of *cdf*.

Functions generating random numbers:

`npr.rand(n)` RAND() (matrix function)
random numbers equi-distributed between [0 and 1)
`npr.randn(n)` NORM.INV(RAND();0;1) (matrix function)
normally distributed
RANDBETWEEN (no matrix function)
between two given integers,
`npr.choice(2, 100, p = [0.2, 0.8])`

chooses numbers from a list with specified probabilities, here, 100 numbers 0 or 1
`npr.choice (10, 10, replace = False)`
without replacement; every number occurs only once.

Nomenclature

pdf probability density function
cdf cumulative density function (distribution function)
ppf percent point function, inverse of *cdf*
dof degrees of freedom, by default number of measurements minus 1
ddof delta degrees of freedom, 0 by default, greater than 0 if parameters of the model distribution other than the sample mean are estimated from the empirical distribution of the frequencies of occurrence.

8.2 Equi-Distributed Random Numbers, Frequencies of Occurrence, Chi^2 Test

We generate numbers randomly between 0 and 1, determine their frequency of occurrence in intervals of width 0.1, and check, with the *Chi^2 test,* whether they are equally distributed.

8.2.1 A Spreadsheet Experiment with Random Numbers

In this exercise, we shall perform experiments with random numbers. They will be explained by means of the spreadsheet in Fig. 8.1, where the data structure is clearly laid out, and repeated in Sect. 8.2.3 in `Python`.

(1) We generate 1000 random numbers named *Rnd* in column A, equally distributed between 0 and 1, with the spreadsheet function RAND().
(2) We determine the empirical frequency distribution *FrqObs* of these numbers in intervals with 11 specified boundaries $I_b = 0.0, 0.1, \ldots., 0.9, 1.0$ in column C. The 10 intervals between 0 and 1 all have the same width of 0.1. Principally, however, the intervals may be of different widths.
(3) We perform a Chi^2 test to check how close the observed frequencies *FrqObs* are to *FrqXpt*, the frequencies expected for an equi-distribution. This is done in Range H6:I9. In H9, the Chi^2 test for the equi-distribution. It yields the same value as performed with the function ChiSq.Test on *FrqObs* and *FrqXpt*. In H8, this is performed via the value of *ChiSqr* in I6 and the function CHISQ.DIST.RT.

	A	B	C	D	E	F	G	H	I	J	K
2	=RAND()			{=FREQUENCY(Rnd;Ib)}		=(FrqObs-FrqXpt)^2/FrqXpt					
3	Rnd		Ib	FrqObs	FrqXpt	CSq					FrqStep
4	0.38		0	0							
5	0.57		0.1	111	100	1.21		dof	ChiSqr		90
6	0.80		0.2	92	100	0.64		9	10.10	=SUM(CSq)	90
7	0.34		0.3	90	100	1					90
8	0.72		0.4	85	100	2.25		0.34	=CHISQ.DIST.RT(ChiSqr;dof)		90
9	0.35		0.5	98	100	0.04		0.34	=CHISQ.TEST(FrqObs;FrqXpt)		90
10	0.95		0.6	117	100	2.89					110
11	0.44		0.7	102	100	0.04		0.30	=CHISQ.TEST(FrqObs;Frqstep)		110
12	0.04		0.8	97	100	0.09					110
13	0.54		0.9	95	100	0.25					110
14	0.05		1	113	100	1.69					110
15	0.04			0							
1003	0.69										

Fig. 8.1 (S) In A4:A1003, named *Rnd*, 1000 random numbers are generated. In D4:D15, the spreadsheet function FREQUENCY is used to count how many of these random numbers fall into intervals whose limits I_b are specified in C4:C14. Column E shows the frequencies expected for equal distribution. In H8, the Chi² test with *FrqXpt* is performed. In H11, the Chi² test is performed against a step function *FrqStep* in column K, *dof* (in H6) = degrees of freedom

Empirical frequency distribution

In EXCEL, empirical frequencies of occurrence are determined with the matrix function FREQUENCY(DATA_ARRAY; BINS_ARRAY). It outputs 12 frequencies (in D4:D15 in Fig. 8.1 (S)) for the 11 boundaries I_b (BINS_ARRAY), because it also determines the frequencies of the data below the lowest and above the highest boundaries.

The first value in BINS_ARRAY gives the number of occurrences in the examined data set that lie below the first interval boundary, that is, from $-\infty$ on. The last value specifies the number of occurrences that lie above the last interval limit, i.e., up to ∞.

▶ *Ψ Always one more in Excel! O.k., but of what and than* what?[1]

The spreadsheet function FREQUENCY is a *matrix function*. Before it is entered, a spreadsheet range must be activated that can capture all output data, and the operation must be concluded with a *magic chord:-)* , *Ψ (Ctl + Shift) + Enter.*

Read the box to get an answer to the important question: "What is a matrix function?"

What is a matrix function in Excel?

[1] For the output of the matrix function *Frequency*, a range has to be activated comprising one cell more than the number of interval boundaries.

FREQUENCY is a matrix function. In terms of spreadsheet calculation, this means that a cell range must be selected that is large enough to contain the amount of the returned data. In the spreadsheet of Fig. 8.1 (S), the area D4:D15 was marked, then the function was entered (see D2), and the process was completed with the "magic chord" for matrix functions: (CTL + SHIFT) + ENTER.

Questions

What is the formula in A4:A1003?[2]

What set of numbers is sorted into the intervals in Fig. 8.1 (S).[3]

The Chi^2 test as a judge

If the random numbers are evenly distributed between 0 and 1, we may expect that, below 0 and above 1, there will be no values, and that in each of the ten intervals of width 0.1, 100 values will occur. This is specified in the variables *FreqXpt*. A Chi^2 test can be used to check how well the observed distribution, here, *FrqObs*, and the expected one, here, *FrqXpt*, match.

To this end, we calculate the value of χ^2 (Chi^2, "Chi square") defined as

$$\chi^2 = \sum\nolimits_{i=1}^{N} \frac{(O_i - E_i)^2}{E_i} \tag{8.1}$$

with O_i and E_i being the observed and expected frequencies, respectively, and N the number of intervals considered for the comparison. In our implementations, O_i = *FreqObs* = [D5:D14], excluding the intervals with zero occurrence, and E_i = *FreqExp* = [E5:E14] (only non-zero numbers), with the same size as *FreqObs*. χ^2 is termed *ChiSqr* and calculated in I6 as the sum over the individual terms *CSq* in column F.

The *Chi^2 test* is listed in the literature under various names, e.g., χ^2-Test, Chitest. We call it the *Chi^2 test*, because it is based on a distribution of the quantity χ^2 = chi^2 ("chi squared"), defined in Eq. 8.1.

We perform such a Chi^2 test in two ways:

- by introducing the calculated value of *ChiSqr* into the distribution function CHISQ.DIST.RT (in H8), and
- by applying a special function for that test, CHISQ.TEST (in H9).

[2] = RAND().

[3] The 1000 numbers in A4:A1003.

The values of such tests are between 0 and 1, giving the probability that χ^2 of the *current sample* fits better with the model distribution of the *whole population* than *any other sample*. If the Chi^2 test results in values significantly smaller than 0.01, it is generally concluded that observation and expectation do not match, i.e., the sample is not from the assumed model population. The values in H8 and H9 are 0.34, so that there is no reason to doubt that the random numbers are equally distributed.

The result 0.34 means that, for 34% of other samples of the same population, χ^2 will be greater, i.e., the frequency distribution of these samples fits worse with the theoretical one. A more precise explanation is given in Exercise 8.8.

In H8, the Chi^2 test is performed by explicitly entering the value of *ChiSqr* and the degrees of freedom *dof* into the inverse function of the integral distribution function of Chi^2. In EXCEL, we use CHISQ.DIST.RT(CHISQR,DOF). The suffix.RT means that a right-tailed distribution function is to be used. The left-tailed distribution function is addressed with CHISQ.DIST. The meaning of these terms will become clearer in Exercise 8.8.

DOF is the degree of freedom. For our test, $dof = 9$, corresponding to the 10 intervals in which the frequencies are compared minus 1; minus 1 because the frequency in the tenth interval is no longer independent of the other frequencies, but is rather determined by the total number of events.

Questions

FrqObs in Fig. 8.1 (S) is defined for 12 intervals. Why is the degree of freedom for the Chi^2 test not 11, but rather 9?[4]

The spreadsheet function ChiSq.Test

EXCEL provides the spreadsheet function CHISQ.TEST(ACTUAL_RANGE, EXPECTED_RANGE) for the Chi^2 test. The test is employed in H9 of Fig. 8.1 (S) and results in the same value $p = 0.34$ as the first test. This function calculates the value of *ChiSqr* internally and assumes a degree of freedom one less than the number of intervals. It must not be applied when parameters of the expected distribution are calculated from the data set.

To repeat: The spreadsheet function CHISQ.TEST may only be used if the parameters of the model distributions are defined in advance, as is the case in our example, and not estimated from the sample.

Does a step distribution also fit?

In addition to comparing our observed frequencies with an equi-distribution, we also compare them with the step distribution *FrqStep* in column K of Fig. 8.1 (S).

[4] The comparison is only between numbers in the 10 intervals with nonzero value, D5:D14 and E5:E14; $dof = 10 - 1 = 9$.

▶ **Tim** Looking at Fig. 8.1 (S), we see that the Chi^2 test provides a larger value ($p = 0.34$ in H9) for the correct distribution than for the incorrect step distribution ($p = 0.30$ in H11).

▶ **Alac** Well, that's to be expected. A good reliability check must deliver a lower probability value for a wrong model distribution than for the right one!

▶ **Mag** Let's wait and see! The correct distribution does not always fit best.

Repetition of the statistical experiment

We are going to generate the random numbers anew. To this end, we place the cursor into an empty cell and "delete" its content. EXCEL reacts as if the calculation had been changed, generates all 1000 random numbers anew, and recalculates the distribution *FreqObs*. The Chi^2 test sometimes gives higher values for the comparison with *FrqStep* than for the one with *FrqXpt*.

Within the VBA rep-log procedure in Fig. 8.2b, we copy 1000 results of Chi^2 tests into columns 13 and 14. Furthermore, we count how often the step function fits better with the observed data than the equi-distribution. This holds true in about 10% of the tests.

The frequency distribution of the results of the two times 1000 Chi^2 tests is shown in Fig. 8.2a. The values for the comparison with the equi-distribution scatter around 100, i.e., they seem to be equi-distributed, whereas the test against the step distribution delivers mostly values below 0.1. These results are also reflected in the average values of the Chi^2 tests: 0.5 for the equi-distribution and 0.13 for the step distribution.

Repetitions of statistical experiments are generally not possible in real life, in which there is mostly only one sample and one result of a Chi^2 test. This is, however, possible in our exercises, because we invent our population ourselves with the

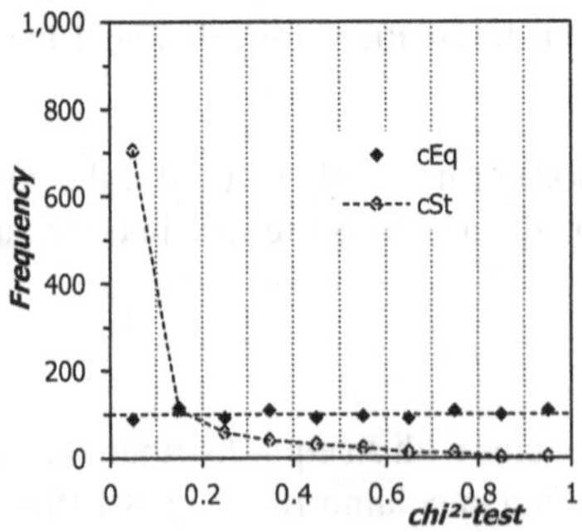


```
Sub protoc()
StepBetter = 0
For r = 4 To 1003
  ChiSqr_Eq = Range("H9")
  Cells(r, 13) = ChiSqr_Eq
  ChiSqr_Step = Range("H11")
  Cells(r, 14) = ChiSqr_Step
  If ChiSqr_Step > ChiSqr_Eq _
    Then StepBetter = StepBetter + 1
  Next r
Range("N1") = StepBetter
End Sub
```

Fig. 8.2 **a** (left) Frequency distribution of the results of Chi2 tests against an equi-distribution (*cEq*) and a step distribution (*cSt*). **b** (P, right) The VBA procedure for 1000 repetitions of the two Chi2 tests in Fig. 8.1 (S)

Table 8.1 Interval boundaries and theoretical frequencies of occurrence

```
db=0.1
Ib=np.arange(0,1+db,db)
FrqXpt=np.ones(10)*100       #Equidistribution, expected freqs.
FrqStep=np.ones(10)          #Step distribution
FrqStep[5:]*=110             #Entries 5 to 9
FrqStep[:5]*=90              #Entries 0 to 4
```

function RAND(), meaning we have a didactical method at hand for demonstrating the peculiarities of statistics.

8.2.2 Data Structure and Nomenclature

Rnd	1000 numbers supposedly randomly distributed between 0 and 1
I_b	11 boundaries for 10 intervals ("bins") between 0 and 1
FrqObs	frequencies of occurrence for *Rnd*
FrqXpc	frequencies expected for an equi-distribution
FrqStep	frequencies expected for a step distribution
dof	Degrees Of Freedom, number of intervals minus 1
ddof	Delta Degrees Of Freedom, to be deduced from *dof* if parameters of the distribution are estimated from the sample.

8.2.3 Python

Standard histogram in Python

The interval boundaries I_b and the expected frequencies of occurrence for an equi-distribution (*FrqXpt*) and a step function (*FrqStep*) are specified in Table 8.1. We know that there are 10 intervals with numbers greater than zero, so that we specify *FrqXpt* and *FreqStep* with size 10.

Questions

What is the value of *FrqStep[5]* in Table 8.1?[5]

In Table 8.2, 1000 Chi^2 tests are performed in a for loop on *Rnd*, an array with 1000 random numbers created with the Python function `npr.rand (1000)`. The frequency distribution is obtained with `np.histogram(Data;boundaries)`. This function returns an array of shape (2, 10) with the frequencies reported in row number 0 and the boundaries repeated

[5] *FrqStep[5]* is 110. *FrqStep[:5]* relates to the first 5 elements, *FrqStep[0]* to *FrqStep[4]*.

Table 8.2 Performing 1000 Chi² tests of 1000 random numbers *Rnd* versus equi-distribution *FrqXpc* and versus step function *FrqStep*

```
7  import scipy.stats as sct
8  import numpy.random as npr
9  N=1000
10 cEq=np.zeros(N)                          #Chi² values for equi
11 cSt=np.zeros(N)                          #Chi² values for step
12 StepBetter=0
13 for i in range(N):                       #N Chi² tests
14     Rnd=npr.rand(1000)                   #Sample
15     FrqObs=np.histogram(Rnd,Ib)[0] #Freq. of occurrence
16     ChiSqr_Eq=sct.chisquare(FrqObs[0],FrqXpt,ddof=0)[1]
17     ChiSqr_Step=sct.chisquare(
   FrqObs[0],FrqStep,ddof=0)[1]
18     cEq[i]=ChiSqr_Eq
19     cSt[i]=ChiSqr_Step
20     if ChiSqr_Step > ChiSqr_Eq: StepBetter+=1
21 print(„StepBetter  ", StepBetter)
```
```
StepBetter 57      #Better Chi² test for step distribution
```

in row number 1. We therefore specifiy the returned variable with index 0: `FrqObs = np.histogram (Rnd,Ib)[0]` to get the observed number of frequencies.

Contrary to the EXCEL function FREQUENCY, only the 10 frequencies in the intervals between the lowest and highest boundaries are returned, and not the frequencies of occurrence below the lowest and above the highest interval boundaries. The size of *`FreqObs`*`[0]` is, therefore, one less than the number of the specified boundaries (`Ib`, reproduced in *`FreqObs`*`[1]`).In a diagram, we would plot them over the 10 centers of the intervals with, e.g., `plt.plot(Ic, FrqObs)`.

The Chi² test is performed with the function `sct.chisquare (FreqObs,FreqExp,ddof)`.The parameter `ddof` specifies a reduction in *dof*, the degrees of freedom. If no parameter of the theoretical distribution is calculated from the observed sample, `ddof` = 0. This is the case for our model distributions because we estimate only the mean from the sample. So, `dof` is set internally to the number of sorting intervals ("bins") minus 1.

The tests are performed within a for-loop with $N = 1000$ iterations and their results are stored in arrays c_{Eq} and c_{St} (lines 20, 21). Plots of c_{Eq} and c_{St} over the center of the intervals look like Fig. 8.2a. We get qualitatively the same results: The chi² values of the "Test of the 1000" versus the equi-distribution (*FrqEq*) are equally distributed between 0 and 1, while those vs. the step distribution (*FrqStep*) increase strongly for Chi² test approaching zero.

► *Ψ Always one more in Excel! O.k., but from what and than* what?[6]

[6] For the output of the matrix function FREQUENCY, a range has to be activated comprising one cell more than the number of interval boundaries. Frequencies of occurrence are returned for values below the lowest interval boundary and for values above the highest boundary.

▶ *Ψ Always one less in Python! O.k., but from what and than* what?[7]

All-including intervals

The `Python` function `np.histogram` can cover, similarly to the EXCEL function FREQUENCY, the whole range of real numbers from minus infinite to infinite by extending the definition of the interval boundaries with [`-np.inf,` ….,]. This is recommended, because it allows for checking whether the sum of all frequencies is equal to the total number of data points considered.

Questions

Regarding Table 8.2: In what percentage of cases does the Chi^2 test in our `Python` program give a higher value for *Step* than for *Eq*?[8]

How does the code have to be changed if 10,000 random numbers are to be considered?[9]

Degree of freedom

What is the degree of freedom for the Chi^2 tests in lines 16 and 17 of Table 8.2? From Fig. 8.1 (S) and Fig. 8.2a (S), we deduce that the degree of freedom is 9. Is it correct to set `ddof` = 0? We may not be sure what the description means and check this question with the program in Table 8.3, a modified version of parts of Tables 8.1 and 8.2. In lines 8 and 9, we sum up the results of Chi^2 tests of 10,000 repetitions of the statistical experiment, in ChiS0 with `ddof` = 0 and ChiS1 with `ddof` = 1. The average values in the lower cell show that, with `ddof` = 0, we are closer to the theoretical average 0.5 of Chi^2 tests than with `ddof` = 1. So, `ddof` = 0 seems to be true, so that we may conclude that the default degree of freedom in `sct.chisquare` is 9, number of intervals minus 1, $dof = n - 1$, as for the EXCEL function CHISQ.TEST.

Consider also the improvement of the code in Table 8.3 over that in Tables 8.1 and 8.2: The number of repetitions can be changed by changing *N* alone.

---

[7] Numpy's np.histogram outputs a number of frequencies that is one less than the number of boundaries; frequencies of occurrence are returned only for internal intervals.

[8] Bottom cell of Table 8.2: *StepBetter* = 57, of 1000 runs, makes about 6%. *StepBetter* is updated in line 20 within the for-loop.

[9] Line 9: N = 10,000; line 14: npr.rand(10,000), line 16 FrqXpc*10; line 17: FrqStep*10. It would be better to use the variable *N* instead of specific numbers.

Table 8.3 Checking the degree of freedom with $N = 10{,}000$

```
N=10000
FrqXpt=np.ones(10)/10 #Freq. in each of ten bins = FrqXpt*N
ChiS0=0
ChiS1=0
for i in range(N):
    Rnd=npr.rand(N)
    FrqObs=np.histogram(Rnd,Ib)[0]
    ChiS0+=sct.chisquare(FrqObs,FrqXpt*N,ddof=0)[1]
    ChiS1+=sct.chisquare(FrqObs,FrqXpt*N,ddof=1)[1]

print("ChiS0   ", np.round(ChiS0/N,3)) #Average value of
print("ChiS1   ", np.round(ChiS1/N,3)) #       Chi² tests
```

```
ChiS0   0.501           #0.50 expected for correct dof
ChiS1   0.429
```

8.3 Points Randomly Distributed in a Unit Square

We create coordinates of points randomly distributed in a unit square, use the logical functions AND, OR, NOT to separate points in sub-regions of the unit square, and illustrate the broom rule: Ψ *Chance is blind and checkered.*

8.3.1 Creation and Distribution of the Points

Chance is blind and checkered

Figure 8.3a displays a sample of 2000 points (x, y), randomly distributed in the unit square. The cross + indicates the center of the distribution obtained with the mean (x_m, y_m) of the coordinates. The vertical bars ("devi bars") are at $x = x_m \pm x_{Sd}$, with x_{Sd} being the standard deviation of the sample. The region within the vertical bars is called *V*. Figure 8.3b displays the same points, with the addition of horizontal bars at $y = y_m \pm y_{Sd}$ (y devi bars), and marks points in the inner rectangle, called *VH*, with open diamonds. The region within the horizontal bars is called *H*.

Questions

concerning Fig. 8.3:

Do the points give the impression of being equally distributed?

Do the white spots and the point clusters disturb the impression of equal distribution?

What fraction of points in Fig. 8.3a lies within the vertical lines? More or less than 50%?

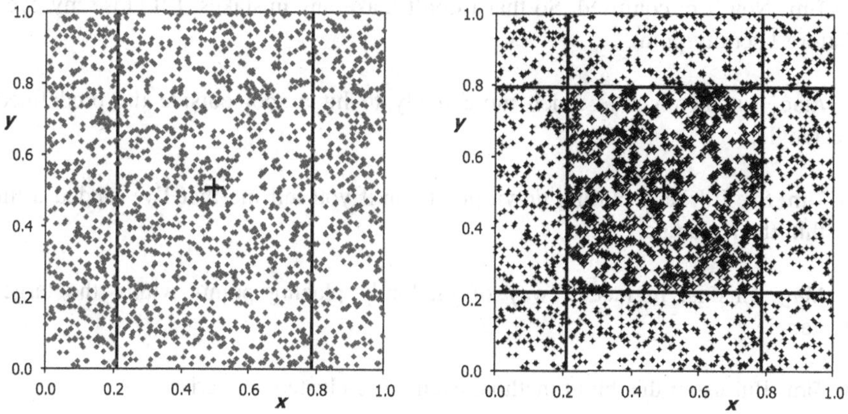

Fig. 8.3 **a** (left) 2000 points (x, y) are randomly distributed in the unit square; the empirically determined center is marked by "+". The vertical lines mark the standard deviation from the center in the x-direction. **b** (right) As in **a**, but with the addition of horizontal lines marking the standard deviation from the center in the y-direction and the points in the inner rectangle marked with open diamonds

How many points lie within the inner rectangle vH in Fig. 8.3b? Is it more or less than 25%?[10]

Calculate a theoretical estimate for the number of points in the rectangle with $x_{Sd} = 1/\sqrt{12}$![11]

How do you generate the sets of numbers x and y anew, in `Python` and in EXCEL?[12]

▶ **Mag** Take a look at Fig. 8.3, where the distribution of points is intended to be random. You need to place another point, but in such a way that the distribution looks genuinely random. Where would you put it?

▶ **Alac** In no case would I place it where there are already many points heaped up, but rather into one of the white areas, where there aren't any points yet.

▶ **Mag** You must not do that under any circumstances. Chance is blind and does not see where it has already struck before.

---

[10] Solution in Fig. 8.5 (S), cell P3, 672/2000 = 0.336.

[11] $\frac{2}{\sqrt{12}} \cdot \frac{2}{\sqrt{12}} = \frac{1}{3}$(see also Eq. 8.6).

[12] EXCEL: Place the cursor in an empty cell and press "Del", i.e., "delete" the contents of an already empty cell. Then, all formulas in the spreadsheet are recalculated, including the spreadsheet function RAND(). -`Python`:The `numpy` function `np.rand(..)` has to be executed in a loop or in a user-defined function.

▶ **Tim** Now I'm confused. So that I don't make any mistakes, I'll close my eyes before I set my point.

▶ **Alac** But then you're much more likely to hit a spot outside the predefined rectangle.

▶ **Tim** So, I'll keep putting down points until you tell me that I've landed a hit within the rectangle.

▶ **Mag** This is, in fact, a way of randomly placing points within the target rectangle.

▶ **Tim** But using this blind method, even more clusters can arise.

▶ **Mag** That's all right. Ψ: *Chance is blind and checkered.*

Statistical characteristics of a sample

The average value or *mean* x_m of a sample is defined, with the number n of elements in the sample, as

$$x_m = \frac{\sum_i x_i}{n} \tag{8.2}$$

The *variance* σ^2 (or var_X in the notation that we will later use) of a sample is calculated with the mean as

$$\sigma^2 = var_x = \frac{\sum_i (x_i - x_m)^2}{(n-1)} \tag{8.3}$$

and therefrom the *standard deviation* as the square root:

$$x_{Sd} = \sigma = \sqrt{\frac{\sum_i (x_i - x_m)^2}{(n-1)}} \tag{8.4}$$

The number of elements in the denominator of the variance is reduced by 1, ($n - 1$), because one parameter, the mean, has been estimated from the sample and not been fixed a priori.

In mathematical literature, the standard deviation is usually called σ. We use the term x_{Sd} (or y_{Sd}) instead, because it has the same physical unit as the sample members x_i (y_i).

In EXCEL, the corresponding functions are AVERAGE for the mean, VAR.S for the variance of a *sample* with (n-1) in the denominator, and, correspondingly, STDEV.S for the standard deviation. In `Python`, the functions are `np.mean` and `np.std(x, ddof = 1)`, `ddof` = 1 being necessary to get the correct denominator ($n-$ 1). When the standard deviation of an *entire population* is to be determined, we have to use STDEV in EXCEL and `ddof` = 0 in the `Python` function.

According to statistical theory, for a uniform distribution between 0 and 1, we get

$$Mean\ x_m = \frac{1}{2} \tag{8.5}$$

$$Standard\ deviation\ x_{Sd} = \frac{1}{\sqrt{12}} \approx 0.289 \tag{8.6}$$

8.3.2 Data Structure and Nomenclature

x, y	random coordinates of points in the unit square
x_m, y_m	mean of x and y
x_{Sd}, y_{Sd}	standard deviation of x and y
in_X	true if $x_m - x_{Sd} \le x < x_m + x_{Sd}$
in_Y	true if $y_m - y_{Sd} \le y < y_m + y_{Sd}$
in_{XY}	true if in_X == True and in_Y == True
V	inner vertical stripe of points with in_X == True
H	inner horizontal stripe of points with in_Y == True
VH	inner rectangle, points with in_{XY} == True
in_V, in_H, in_{VH}	number of points in *V, H, VH*
x_I, y_I	coordinates of points in *VH*
L, R, B, T	left, right, bottom, top stripe defined by $x \le x_m - x_{Sd}$, $x \ge x_m + x_{Sd}$, …

8.3.3 Excel

With the spreadsheet function RAND(), 2000 random numbers are generated each in columns A and B of Fig. 8.4 (s), designated x and y, and displayed as 2000 points (x, y) in the plane (Fig. 8.3a).

The mean values x_m and y_m of 2000 between 0 and 1 equally distributed random numbers are (not surprisingly) about 0.5. The standard deviations are found to be $x_{Sd} = 0.29$ and $y_{Sd} = 0.29$, corresponding within 2% to the theoretical value of 0.289 (Eq. 8.6). So, there is a fraction $2 \times 0.29 = 0.58$ to be expected within the vertical bars or within the horizontal bars, and a fraction $0.58^2 \approx 0.33$ $\left(\frac{2}{\sqrt{12}} \cdot \frac{2}{\sqrt{12}} = \frac{1}{3}\right)$ within the inner rectangle of Fig. 8.3b.

	A	B	C	D	E	F	G	H	I	J	K	L
1	=RAND()	=RAND()					=AND(xm-xSd<=x;x<xm+xSd)	=AND(ym-ySd<=y;y<ym+ySd)	=AND(G4;H4)	=IF(I4;x;NA())	=IF(I4;y;NA())	=COUNT(yI)
2	**x**	**y**	**xm**	0.51	=*AVERAGE(x)*		**inX**	**inY**	**inXY**	**xI**	**yI**	
3	0.24	0.57	**xSd**	0.29	=*STDEV.S(x)*		TRUE	TRUE	TRUE	0.24	0.57	**672**
4	0.38	0.55	**ym**	0.49	=*AVERAGE(y)*		TRUE	TRUE	**TRUE**	**0.38**	**0.55**	**0**
5	0.89	0.22	**ySd**	0.29	=*STDEV.S(y)*		FALSE	TRUE	FALSE	#N/A	#N/A	=*SUM(inY)*
2002	0.12	0.02					FALSE	FALSE	FALSE	#N/A	#N/A	

Fig. 8.4 (S) In columns A and B, each 2000 random numbers between 0 and 1 are generated, and in C3:F3, their mean value (x_m, x_m) and standard deviations x_{Sd} and y_{Sd} are determined. The columns G to I check whether the points are within certain boundaries. L4 contains the formula SUM(INY). "#N/A" means "Not available". The name of this worksheet is "calc" and is addressed as such in VBA procedures

	M	N	O	P	Q	R	S	T	U
1	=COUNTIF(inX;TRUE)	=COUNTIF(inY;TRUE)	=COUNTIF(inXY;TRUE)	=inVH/2000	=COUNTIF(x;"<="&xm-xSd)	=COUNTIF(x;">"&xm+xSd)	=COUNTIF(y;"<="&(ym-ySd))	=COUNTIF(y;">"&ym+ySd)	
2	**inV**	**inH**	**inVH**	**fVH**	**nL**	**nR**	**nB**	**nT**	
3	1157	1179	672	0.336	409	434	408	413	

Fig. 8.5 (S) COUNTIF is applied, counting how often the data x and y in Fig. 8.4 (S) are less away from the center than the standard deviation, in M:O with the logical data in x_{in}, y_{in}, and xy_{in}. In Q:T, the data x and y themselves are taken for the logical query if the data are further away. We get $inV + nL + nR = 2000$

Check with And

In column G, for each individual x-coordinate, we check whether it deviates less than the standard deviation downwards or upwards from the mean value x_m. In column H, the same is done for the y-component. For these checks, the logical function AND(LOGICAL1, [LOGICAL2], ...) is used. The corresponding formulas are.

$$in_X = \text{AND}(x_m\text{-}x_{Sd} <= x;\ x < x_m + x_{Sd})$$
$$in_y = \text{AND}\ (y_m\text{-}y_{Sd} <= y;\ y < y_m + y_{Sd})$$

In column I, we mark the points within the *devi* limits of both x_m and y_m. We cannot use the statement AND(x_{in}; y_{in}), because it returns TRUE if *all* elements of the column vectors x_{in} and y_{in} are true. So, we have to write, e.g., I4 = AND(G4;H4), and so on, to check *individually* whether the contents of both cells are true.

Logical functions

In columns J and K, we extract the coordinates of the points within the inner rectangle with a statement like K4 = [=If(I4; x; NA())], returning the value of x if I4 = = True and #N/A else. The column ranges in K and J can be entered as data series into a figure with the cells with #N/A being ignored.

A check COUNT(yI) (in L3) yields 672, although the queried range y_I comprises 2000 entries, because only the cells with numbers, and not those with #N/A, are counted. The function SUM(inY) returns 0 because only numbers are summed up and logical values are ignored.

Logical Functions

We have used two logical functions so far:

IF(LOGICAL_TEST, [VALUE_IF_TRUE], [VALUE_IF_FALSE]);

in [VALUE_IF_TRUE] and [VALUE_IF_FALSE] may appear numbers, or functions whose result is then processed.

AND(LOGICAL1, [LOGICAL2], ...)

expects logical values as input and outputs TRUE or FALSE.

Please refer to EXCEL help to inform yourself about the other functions in the category LOGIC: OR; FALSE; TRUE; NOT; IFERROR.

Count the Trues

In Fig. 8.5 (S), the logical calculations of Fig. 8.4 (S) are continued. The spreadsheet function COUNTIF(RANGE, CRITERIA) counts the non-empty cells of a range whose contents match the search criteria. In M3:O3, the ranges in_X, in_Y, *and* in_{XY} are queried as to how often their cells contain TRUE. In P3, the relative frequency f_{VH} of points within the inner rectangle in Fig. 8.3b is reported ($f_{VH} = 0.336$).

In the subsequent columns Q to T, the number of points in the lower n_L, right n_R, bottom n_B, and top n_T stripe is calculated. Here, logical expressions formulated as a character string are used as criteria. In the formula S3 = [=COUNTIF(y&" < = "&$y_m - y_d$)], the criterion is [" < = "&$y_m - y_d$]. The expression is of the form Ψ *"Text"&Variable*, composed of the comparison symbol in quotation marks "<" and the arithmetic expression $y_m - y_d$.

In Fig. 8.6 (S), the results of Fig. 8.5 (S) are evaluated to estimate the relative amount of points within the inner rectangle with a probabilistic calculation. In columns V and W, the number of points outside the x and y devi boundaries is obtained through addition; in X and Y, it is converted into the numbers inside the devi bars (the same as in Fig. 8.5 (S)) from which the relative amount is obtained in Z and AA. These relative amounts are interpreted as probabilities.

Fig. 8.6 (S) Continuation of Fig. 8.5 (S)

	V	W	X	Y	Z	AA	AB
1	=nL+nR	=nB+nT	=2000-U3	=2000-V3	=W3/2000	=X3/2000	=pV*pH
2			==inV	==inH	**pV**	**pH**	**pVH**
3	843	821	1157	1179	0.58	0.59	0.341

Product of probabilities
As the x and y coordinates are independent of each other, the probability of finding a point in the inner rectangle may be calculated as the product of the two probabilities mentioned above:

$$p_{VH} = p_V \cdot p_H$$

We get $p_{VH} = 0.341$ (AB3). This is to be compared with the relative frequency of occurrence $f_{VH} = 0.336$ in Fig. 8.5 (S) and with the theoretical value of $1/3 = 0.333$ (from Eq. 8.6).

Keep in mind the differences:

- The theoretical probability is *logically derived* from the theoretical values of the mean and the standard deviation of an equi-distribution independent of the current experiment with 2000 points.
- The quantity p_{VH} *is estimated* from the relative occurrence of points in V and H in our current experiment.
- The relative frequency of occurrence in VH *is empirically determined* in our current experiment.

Due to these methodological differences, the three values are not necessarily the same; but, as they evaluate the same set of points, their values should be close together.

8.3.4 Python

In `Python`, we have to import the `numpy.random` library as `npr` (see Table 8.4) for our exercise. The generation of the 2000 points is straightforward with `npr.random (2000)`.

The logical queries as to whether the points are in the areas V or H are performed in three different ways, all resulting in a Boolean array with length 2000:

- in line 6, "`inX = np.logical_and ((xm-xSd) < x, x < (xm + xSd))`", identical to the spreadsheet formulas in Fig. 8.4 (S), with the result being shown in the bottom cell of Table 8.4,
- in line 7, "`inX2 =` " as list comprehension with identical results to those in line 6, as checked with the instruction "`inX == inX2`", resulting in a Boolean array of `Trues`,
- in line 8, "`inY =` " as a different type of list comprehension.

The query as to whether a point is in the inner rectangle is done by means of the list comprehension in line 9.

Table 8.4 Python: 2000 points (x, y) randomly distributed in a unit square, the points (x, y) are depicted in Fig. 8.3a, as are the points (x_I, y_I) in Fig. 8.3b

```
1  import numpy.random as npr
2  x=npr.random(2000)
3  xm,xSd = np.mean(x),np.std(x)
4  y=npr.random(2000)
5  ym,ySd = np.mean(y),np.std(y)
6  inX=np.logical_and(xm-xSd<x,x<xm+xSd)     #Within std error?
7  inX2=[(xm-xSd<x) & (x<xm+xSd)]
8  inY=[(ym-ySd < yi < ym+ySd) for yi in y]
9  inXY=[inX & inY]
    #Arrays of coordinates of points within error range:
10 xI=np.extract(inXY,x)
11 yI=np.extract(inXY,y)
```

```
inX    [ True  True  True ... False False  True]
len(inX)     2000
inX==inX2
      [[ True  True  True ...  True  True  True]]
len(yI)       649
```

The coordinates of the points in the inner rectangle are obtained, in lines10 and 11, with the function `np.extract(condition, array)`. The condition here is that *inXY* == `True`. The length of x_I, i.e., the number of points in the inner rectangle, is 649, to be compared to 672 = COUNT(YI) in the EXCEL sheet of Fig. 8.4 (S). The theoretical expectation for a binomial question (point in VH or not in VH) is $x_m = n \cdot p = 667$ and $x_{Sd} = \sqrt{npq} = 21$, so that the above values are within the standard error range. The meaning of these statements will be discussed in Chap. 9.

Logical operations on lists and arrays in Python

The operators & (and), | (or), and ~ (not) operate item-wise on lists of Booleans and create new lists of Booleans. In Table 8.4, however, the `numpy functions` `np.logical_and`, `np.logical_or`, and `np.logical_not` are applied; they do the same, except that they create arrays. The difference becomes apparent when we try to combine two of them, e.g., with &. This is not possible for lists; in trying to do so, we get the error message `"Unsupported operand type(s) for &: 'list' and 'list'"`. But it works well with arrays, e.g.,

```
inXY=[inX & inY]
```

yielding a Boolean array with `"True"` if the point lies within the inner rectangle in Fig. 8.3b. With the instruction.

```
xI = np.extract(inXY,x)
```

Table 8.5 Size of the Boolean arrays denoting the various partial areas in Fig. 8.3b

`inV=sum(inX)`	inV 1124
`inH=sum(inY)`	inH 1153
`inVH=sum(inXY)`	inVH 649
	fVH 0.32
`inL=sum(xi<xm-xSd for xi in x)`	inL 437
`inR=sum(xm+xSd<xi for xi in x)`	inR 439
`inB=sum(yi<ym-ySd for yi in y)`	inB 433
`inT=sum(ym+ySd<yi for yi in y)`	inT 414

a new array is created containing the x coordinates of the points in the inner rectangle.

The points in the inner square are plotted with an open diamond (see Fig. 8.3b) with the instruction

```
plt.plot(xI,yI,'kd', fillstyle='none').
```

In Table 8.5, we have calculated the numbers of points within the various partial areas *V*, *H*, *VH*, *L*, *R*, *B*, *T* of the unit square, defined in Sect. 8.3.2.

The values TRUE and FALSE in a Boolean array are coded as 1 and 0 in a binary list. So, the number of TRUEs is equal to the sum over the list. This is not possible in Excel; = SUM(INY) in L4 of Fig. 8.4 (S) yields 0.

Questions

How are in_V and in_H related to in_L, in_R, in_B, in_T?[13]

How do you estimate the probability of finding points in VH from in_V and in_H? Compare the constructions in lines 6 to 9 in Table 8.4![14]

8.3.5 Why Calculate Twice?

► **Alac** In Sect. 8.3.3, we calculate the probability of finding a point in the inner rectangle one time too many, once as f_{VH} in O3 of Fig. 8.5 (S) and a second time as $p_V \cdot p_H$ in Fig. 8.6 (S). We could have saved work.

► **Tim** There is a difference: f_{VH} estimates the probability from the relative frequency in *VH*, whereas p_{VH} relies on the product rule of probability.

► **Mag** Tim is right. Anyhow, it is always good to follow two calculation paths that are expected to deliver the same result. In this way, we can check whether

[13] $in_V = 2000 - in_L - in_R$; $in_H = 2000 - in_T - in_B$.

[14] $p_{VH} = p_V \cdot p_H = in_V/2000 \cdot in_H/2000 = 0.324$.

our thoughts have been logical and whether we have correctly transferred our thoughts into the calculation. This is especially advisable when using functions and formulas for the first time.

▶ **Practical advice** Before you seriously use functions for the first time, check their operation with examples for which you can do the arithmetic mentally!

Product rule of probability
As can be seen in Fig. 8.5 (S) (and Table 8.5), only about $p_{\mathrm{H}} = 59\%$ ($inH/2000 = 0.59$) of the points lie within the two horizontal lines in Fig. 8.3b. Within the vertical lines, the proportion is also 0.59 (0.56). Within the rectangle (p_{VH}), it is less. According to the rules of probability, it should be the product of the probabilities within the vertical and horizontal lines, $p_{\mathrm{VH}} = p_{\mathrm{V}} \cdot p_{\mathrm{H}}$, provided the two events are independent of each other. This is indeed the case, because the x- and y-coordinates have been created independent of each other. We estimate the probabilities with the point frequencies $(0.59 \times 0.59) = 0.348$. If there is a difference between this probability and that obtained with the number of points in the inner rectangle, it is due to the fact that the probabilities are not exact, but simply estimated from the number of points in the considered areas.

8.4 Set Operations in Numpy

Python provides a data format for sets, and the operations that will be performed on them. With such sets, we illustrate basic rules for the probabilities of unions and intersections, as well as Bayes' rule on conditional probabilities and apply logical queries.

8.4.1 Sets

In the program underlying Fig. 8.7, 400 points are created in the unit square, with their x- and y-coordinates being two arrays of independent, equally distributed random numbers. In Fig. 8.7a, only the points in three subsets are displayed: the top stripe T with $y > 0.8$ (marked with $+$), the right stripe R for $x > 0.6$ (marked with x, and the points in the right upper rectangle RT (additionally marked with open squares □).

The points in the ranges R and T are defined in Python as sets with their coordinates obtained through logical queries $x > 0.6$ and $y > 0.8$ (see Table 8.7). The number of points in the sets is obtained with the function `len`:

```
len(R) → 158, len(T) → 85.
```

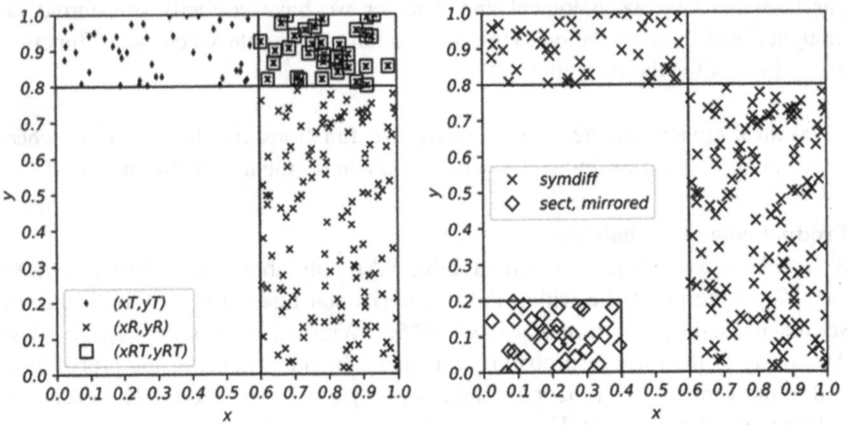

Fig. 8.7 Points in *T*, *R*, and *RT*. **a** (left) Plotted with their coordinates, e.g., `plt.plot(xT, yT, ...);` **b** (right) Plotted as two sets, *symdiff* and *sect* = *RT* mirrored at (0.5, 0.5) (see Table 8.8)

The probability that a new point will be in one of the sets may be estimated as:

```
p(R) = 158/400 = 0.395, p(T) = 85/400 = 0.213
```

Now, we perform set operations on *R* and *T* (Table 8.8):

- *Union*, defined in `Python` as `R|T`; comprising the points in *R* **or** *T*
- *Section*, defined in `Python` as `R&T`, comprising the points in *R* **and** *T*, marked above in Fig. 8.7a with open squares.

For the situation in Fig. 8.7, we get:

```
len(R|T) = 211, p(R|T) = 211/400 =0.528
len(R&T) = 32, p(R&T) = 32/400 = 0.080
```

Applying the addition rule for the union yields:

$$p(R|T) = p(R) + p(T) - p(R\&T) \tag{8.7}$$

which holds exactly true for our special case:

```
211/400 == 158/400 + 85/400 - 32/400
```

The multiplication rule for independent sets reads:

$$p(R\&T) = p(R) \cdot p(T) \tag{8.8}$$

The precondition for the multiplication rule is that the two events are independent of each other. *R* and *T* are indeed independent sets, because *x* and *y* are generated independently, and the membership in *R* and *T* depends exclusively on *x* or *y*. As an illustrative counter-example, consider *RT* where *x* and *y* are not independent of each other.

Checking the multiplication rule for our special case gives us

$$\mathrm{p(R\&T)} = 32/400 = 0.080$$
$$\mathrm{p(R)} \cdot \mathrm{p(T)} = 158/400 \times 85/400 = 0.084$$

There is no exact coincidence between these probabilities, because they are based on an estimate using a sample size of 400. Increasing the number of points lets the two probabilities come closer together.

Other set operations are:

- *Difference*, defined as *R-T*, points in *R* but not in *T*:
- `len(R-T)= len(R) - len(R&T)`
- *Symmetric difference*, defined in `Python` as `R^T`, points either in *R* or in *T*, indicated in Fig. 8.7b with x:
- `len(R^T)=len(R|T)-len(R&T).`

For our specific case, we get: `len(R-T) = 126, len(R^T) = 179.`

Conditional probabilities, Bayes' rule

What is the probability *p(T\R)* of finding a point in set *T* when we know that it is in set *R*? Graphically, this is the area of *RT* divided by the area of *T*, formulated with probabilities as

$$p(T \backslash R) = \frac{p(R\&T)}{p(R)} \tag{8.9}$$

For our specific case, we get

$$\mathrm{p(T\backslash R)} = (32/400)/(158/400) = 0.202$$
$$\mathrm{p(R\backslash T)} = (32/400)/(85/400) = 0.376$$

The two conditional probabilities are interconnected by Bayes' rule:

$$p(R \backslash T) = p(T \backslash R) * \frac{p(R)}{p(T)} \tag{8.10}$$

Questions

What is the theoretical probability of finding a point in the upper-right rectangle RT of Fig. 8.7?[15]

Check the rules for difference and symmetric difference for the numbers of our specific case![16]

Check Bayes' rule for the numbers of our specific case![17]

8.4.2 Data Structure and Nomenclature

x, y	coordinates, random between 0 and 1
R	set of points in the right stripe
x_{Rt}	left boundary of R
in_{R}	Boolean array, TRUE if the point is in the right stripe
x_{R}, y_{R}	coordinates of points in R
R_{tup}	array of tuples (x_{R}, y_{R})
T	set of points in the top stripe
y_{Tp}	bottom boundary of T
RT	section of R and T
x_{T}, y_{T}	coordinates of points in T.

8.4.3 Python

In Table 8.6, random coordinates (x, y) of $N = 400$ points are generated from which those points are selected that lie in a vertical stripe R right of x_{Rt}. To achieve this, first, a logical array in_{R} is determined containing `True` when the condition $x_{\mathrm{Rt}} < x$ is fulfilled. Then, the coordinates x_{R} and y_{R} are extracted from the `True` positions.

In Table 8.7, two sets R and T are built based on the coordinate lists x_{R}, y_{R} and x_{T}, y_{T}. We first zip the coordinates into tuples (result in Table 8.10) and then define R and T as sets with these tuples as elements. In our specific case, an array with 158 (out of 400) entries results. The same is done for a horizontal stripe T above y_{Tp}. In the following, the set operations *union*, *section*, *difference*, and *symmetric difference* are performed. The size of the resulting sets is reported in the right cell of Table 8.7.

[15] The probability of finding a point in RT is its area $0.2 \times 0.4 = 0.08$ in the unit square.

[16] `len(R-T) = len(R) - len(R&T); 126 = 158 -32.`
`len(R^T) = len(R|T)-len(R&T); 179 = 211 - 32.`

[17] `32/85 = 32/158 * 158 / 85.`

Table 8.6 Coordinates of the three sections *R*, *T*, and *RT* in Fig. 8.7

```
N=400
x= npr.random(N)
y= npr.random(N)

xRt = 0.6                                #Right border
inR=[xRt<x]                              #Right stripe
xR=np.extract(inR,x)
yR=np.extract(inR,y)

yTp = 0.8                                #Top border
inT=[yTp<y]                              #Top stripe
xT=np.extract(inT,x)
yT=np.extract(inT,y)

inRT=np.logical_and(inR,inT)             #Top-right rectangle
xRT=np.extract(inRT,x)
yRT=np.extract(inRT,y)
```

Table 8.7 Defining the sets R and T and combined sets

`Rtup=zip(xR,yR) #Right str.`		
`R=set(Rtup)`	`len(R)`	158
`Ttup=zip(xT,yT) #Top stripe`		
`T=set(Ttup)`	`len(T)`	85
`union   = R|T`	`len(R|T)`	211
`sect    = R&T`	`len(R&T)`	32
`differ  = R-T`	`len(R-T)`	126
`symdiff = R^T`	`len(R^T)`	179
`un_m_sc = union-sect`	`len(un_m_sc)`	179

Questions

In Table 8.6, two variables with nearly identical names occur, *x*Rt (a scalar) and *x*RT (an array). Why is there no name conflict?[18]

Table 8.8 reports the instructions for plotting Fig. 8.7b.
Bayes' rule is numerically checked in Table 8.9.

[18] Python is case-sensitive; xRt and xRT are two different names. Nevertheless, this is bad naming.

Table 8.8 Plotting the sets in Fig. 8.7b

```
FigStd('x',0,1,0.1,'y',0,1,0.1)
x4,y4=zip(*symdiff)
 #For effect of zip(* see:
https://stackoverflow.com/questions/29139350/difference-
between-ziplist-and-ziplist
plt.plot(x4,y4,'kx',label="symdiff")
x5,y5=zip(*sect)
x6,y6=np.array(x5),np.array(y5)
plt.plot(1-x6,1-y6,'kD',fillstyle="none",
                    label="sect, mirrored")
```

Table 8.9 Bayes' rule

`T_given_R=len(sect)/len(R)`	0.203
`R_given_T=len(sect)/len(T)`	0.376
`Bayes=T_given_R*len(R)/len(T)`	0.376

Table 8.10 Data structure of R_{tup} and R defined in Table 8.7

```
xR     [ 0.86  0.78  0.86 ...  0.99  0.73  0.97]
yR     [ 0.79  0.24  0.14 ...  0.09  0.43  0.48]
*zip(xR,yR)   (0.86, 0.79) (0.78, 0.24) …
Rtup   <zip object at 0x000001E3C57F7B48>
R      {(0.85, 0.17), (0.9, 0.09), … }
```

Some Python constructs

The function `np.extract (Boolean_array, Value_array)`, applied in Table 8.6, extracts those values of the `Value_array` for which the corresponding position in a `Boolean_array` contains `True`.

Table 8.10 elucidates the data structure of a zip object and a set. The `zip()` function pairs together the items of iterators passed as arguments. An iterator is an object that contains a countable number of values; in our case, we pair the two arrays x_R and y_R. The set R contains the same tuples as R_{tup}, but in a different order because the order does not play any role in sets.

8.5 Normally Distributed Random Numbers

The functions NORM.INV(RAND();0;1) (EXCEL), as well as `sct.norm.ppf(npr.random(…))` and `npr.randomn(…)`

(Python), generate numbers that are standard-normally distributed. We check this statement by comparing a frequency distribution of a set of such numbers with the theoretical distribution using the Chi^2 test. The expected frequency of occurrence in an interval is calculated in two ways: *Exactly*, with the values of the cumulative distribution function (*cdf*) at the interval boundaries, and *approximately*, with the probability density (*pdf*) in the middle of the interval. The inaccuracy of the approximation will only become visible with more than 10,000 random numbers.

8.5.1 Normal Distribution, Probability Density and Distribution Function

This section presents the mathematical background of the formulas applied later in this exercise.

Normal distribution
The following nominations and specifications hold:

x_m, x_Sd	mean value and standard deviation
$-\infty < x < \infty$	argument range
$0 < p < 1$	probability, value range

EXCEL provides two spreadsheet functions, one with a Boolean parameter CUMULATIVE? for calculations with normal distributions, corresponding to three Python functions available from the scipy library:

- probability density function *pdf*, $x \rightarrow p$

  NORM.DIST (X, $X_\text{M,}$ X_SD, CUM), *cum* = *False* or *0*
  `sct.norm(xm, xsd).pdf(x)`

- cumulated density function *cdf* (distribution function), $x \rightarrow p$

  NORM.DIST (X, X_M, X_SD,CUM),), *cum* = *True* or *1*,
  `sct.norm(xm, xsd).cdf(x)`

- the inverse of the distribution function, $p \rightarrow x$

  NORM.INV(P, X_M, X_SD) ,
  `sct.norm(xm, xsd). ppf(p)`

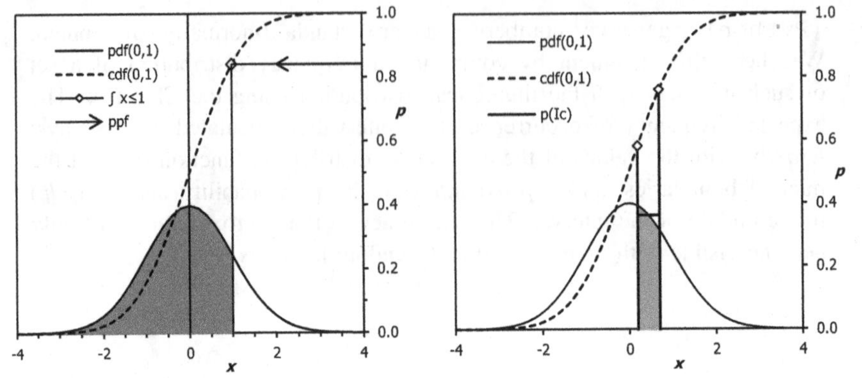

Fig. 8.8 **a** (left) Solid line (bell curve): *pdf*(0, 1)) , probability density of the standard normal distribution (i.e., $x_m = 0$ and $x_d = 1$); dashed line: *cdf*(0, 1), distribution function of the normal distribution; the content of the grey area under the probability density is indicated by the diamond at the end of the vertical bar ending in a diamond on the distribution function *cdf*. **b** (right) The grey area below the probability density *pdf* is equal to the difference of the two *y*-values of the diamonds on the distribution function *cdf*. The horizontal bar indicates the value of the probability density in the middle of the considered interval

The extension *ppf* stands for "percent point function". However, most people in statistics use "quantile function".

In mathematical textbooks, the normal distribution is often referred to as $N(\mu, \sigma^2)$, where μ is the mean value and σ^2 is the variance. However, since it is the standard deviation that has to be entered as a parameter into the spreadsheet and Python functions, we shall use the abbreviations $pdf(x_m, x_{Sd})$ in the text and in legends for the probability density of the standard normal distribution that is defined by the equation

$$pdf(x_m, x_{Sd}) = \frac{1}{\sqrt{2\pi} \cdot x_{Sd}} \cdot \exp\left(-\frac{1}{2}\left(\frac{x - x_m}{x_{Sd}}\right)^2\right) \tag{8.11}$$

The bell-shaped *pdf*(0, 1) is shown in Fig. 8.8, together with the corresponding *S*-shaped *cdf*(0, 1) (dashed line).

Questions

What is the physical unit of the argument of *exp* in Eq. 8.11?[19]

What is the physical unit of *pdf* and *pdf*·dx?[20]

[19] The argument in *exp* is dimensionless, because the nominator and denominator have the same unit.

[20] [*pdf*] = 1/[x_d], a density; [*pdf*·dx] = unit-less, a probability.

Do the following two spellings yield different results in EXCEL: Y = EXP(-2^2) and Y = EXP(-(2^2))?

Same question for `np.exp (-2**2)` and `np.exp (-(2**2))` in `Python`.[21]

Distribution function or cdf

In mathematical literature, the integral of a probability density is called the distribution function; it is monotonously increasing and spans the value range (the range of the output values) from 0 to 1. In `Python`, it is called *cdf*, "cumulated (probability) density function". We shall adopt this notation together with *pdf* and *ppf* in this text.

The distribution function of the normal distribution cannot be represented in a closed form, but this is not a disadvantage for us, since EXCEL offers a spreadsheet function NORM.DIST($x; x_m; x_d$; TRUE) and `Python` offers `sct.norm(`x_M`, `x_{sd}`).cdf(x)`, which are good approximations.

In Fig. 8.8a, the probability density of the normal distribution is shown up as a bell curve and the distribution function as a monotonously increasing S-shaped curve. The open diamond marks the distribution function at $x = 1$, corresponding to the value of the integral over the probability density from $-\infty$ to 1, which is represented as a grey-filled area under the bell curve, representing the probability of finding a value of the random number below $x = 1$.

The function *ppf(p)* is the inverse of the distribution function *cdf(x)*, giving the value of x for a specified p, as demonstrated by the kinked line in Fig. 8.8a: A value p_0 on the vertical axis between 0 and 1 is assigned a value x_0 on the horizontal axis (see arrow). It is the probability of finding a value $x \leq x_0$, corresponding to the grey area under the probability density.

Probability that a random number falls into an interval

What is the probability $p(x_1, x_2)$ that a normally distributed random number lies in the interval $[x_1, x_2)$? To calculate this, we have to integrate over the probability density *pdf(x)* :

$$p(x_1, x_2) = \int_{x1}^{x2} pdf(x)dx = cdf(x_2) - cdf(x_1) \tag{8.12}$$

The integral in Eq. 8.12 is the difference of the distribution function at the two interval limits. To give an example: in Fig. 8.8b, an interval from 0.2 to 0.7 has been selected. The probability of finding a normally distributed random number in this interval corresponds to the grey area under the probability density *pdf*, which, in turn, corresponds to the difference of the values of the distribution function *cdf* at the two positions marked with open diamonds.

[21] Yes, there is a difference, but only in EXCEL. In a spreadsheet, the argument of the Gaussian function must be spelled as (-(x^2)), because negation has operator precedence over potentiation, quite surprisingly for mathematically-educated readers and an annoying error source, so: –2^2 = 4; –(2^2) = –4.

The integral over an interval can be *approximated* by the product of the probability density in the center x_C of the interval and the width Δx of the interval:

$$p(x_1, x_2) \approx pdf\left(\frac{x_2 - x_1}{2}\right) \cdot (x_2 - x_1) = pdf(x_C) \cdot \Delta x \qquad (8.13)$$

In Fig. 8.8b, this value is indicated by the horizontal line between the interval limits. The more linear the probability density between the selected interval limits and the narrower the interval, the better this approximation. The frequency of occurrence in that interval for a sample of size N is $pdf(x_C) \cdot \Delta x \cdot N$.

8.5.2 Random-Number Generator and Frequencies of Occurrence

The function `npr.randn(N)` of the library `numpy.random` returns an array of N numbers randomly distributed according to the standard normal distribution with $x_m = 0$ and $x_{Sd} = 1$. In EXCEL, there is no simple function for doing this, and thus a nested function, NORM.INV(RAND();0;1), has to be used. The literal equivalent in `Python` is `sct.norm.ppf (npr.random(N))`.

Figure 8.9 displays the two frequency distributions of 10,000 random numbers created with the two `Python` functions `randn` and `ppf (random)`, together with the theoretically expected distribution. We see that, indeed, the two functions act alike, and the numbers provided seem to be normally distributed.

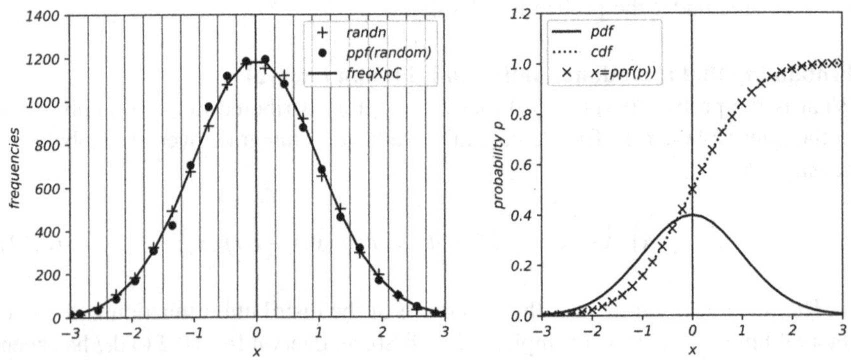

Fig. 8.9 **a** (left) Frequency distribution of the numbers generated with the two generators of normally distributed random numbers in Table 8.12 (*randn* and *ppf (random)*); the vertical bars represent the interval boundaries; the polyline connects the frequencies *freq*x_{pC} expected for a normal distribution (obtained with *cdf*(0, 1)). **b** (right) Data for *pdf*(*x*) , *cdf*(*x*), and *ppf*(*p*) of the standard normal distribution obtained from Table 8.11.

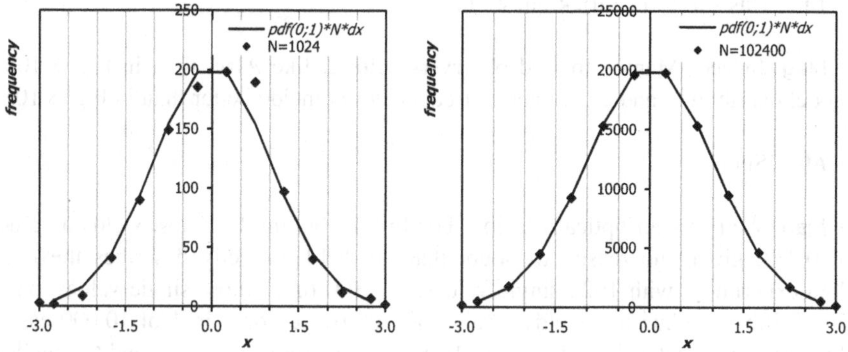

Fig. 8.10 **a** (left) Frequency distribution for 1024 random numbers. **b** (right) Frequency distribution for 100 times more random numbers than in **a**

8.5.3 Where Do Observed and Theoretical Frequencies Fit Better Together?

Figure 8.10a displays experimentally determined frequencies of occurrence for 1024 random numbers generated in EXCEL with NORM.INV(RAND();0;1) as data points over the interval centers x_c. The numbers of events outside of the minimum and maximum interval boundaries are shown in the picture on these interval boundaries (here, −3 and 3). Figure 8.10b shows a similar picture for a hundred times more random numbers.

The polylines in Figs. 8.10a and Fig. 8.10b represent the frequencies of occurrence in the interval $[x_1, x_2)$:

$$freq(x_1\ to\ x_2) = ppf(x_c) \cdot N \cdot \Delta x \tag{8.14}$$

where N is the total number of data and x_c and Δx are the center and width, respectively, of the sorting intervals, also called the "bins".

Questions

What additional information would significantly increase the information content of the legends Fig. 8.10a, b? Compare Fig. 8.16b![22]

▶ **Mag** Where do observed and theoretical frequencies fit better together, in the left or right picture of Fig. 8.10?

▶ **Alac** Quite clearly, in the right picture. The experimental and theoretical points are much closer together than in the left picture.

[22] Indicating the result of a Chi^2 test would considerably increase the information in the figure.

▶ **Tim** This surely is a trick question.

▶ **Mag** Indeed. Many unbiased observers believe, like Alac, that, in Fig. 8.10b, model distribution and experimental frequencies coincide better than in Fig. 8.10a.

▶ **Alac** Sure.

▶ **Mag** Well, it's an optical illusion. For Fig. 8.10a, the Chi^2 test yields a value of 0.333, giving no reason to doubt that Eq. 8.14 correctly describes the random experiment with 1024 single values. For 100 times more single values, e.g., Fig. 8.10b, the Chi^2 test yields a value of 0.0001, i.e., only in 1 of 10,000 cases does an even greater deviation occur between the data of a sample and the model for the population.

▶ **Alac** Then, we can presumably no longer claim that the theoretical distribution describes the experimental distribution well. Why not?

▶ **Mag** For 102,400 individual values, the approximation of Eqs. 8.13 and 8.14 is not good enough. For so many data, statistical science expects a deviation that is even smaller than we can note with the eye.

Exactly calculated frequencies of occurrence

The probability of finding a random number in an interval is determined exactly by the difference of the cumulative distribution function *cdf* at the interval boundaries, as prescribed in Eq. 8.12. To determine the frequency in the interval, the probability must be multiplied by the total number *N* of the analyzed data:

$$freq(x_1\ to\ x_2) = (cdf(x_2) - cdf(x_1)) \cdot N \tag{8.15}$$

In this formula, the interval width Δx does not show up, in contrast to Eq. 8.14.

A comparison of the frequencies expected according to Eq. 8.15 with the experimental frequencies in the intervals in Fig. 8.10b results in CHISQ.TEST = 0.37. So, everything is fine again.

▶ The more individual data are available for a statistical test, i.e., the larger *N* is, the easier it is to detect discrepancies between experimental frequencies and those predicted by a model distribution. For only a few measurements, deviations from a wrong model distribution may not be noticeable, because deviations from the correct model distribution can also be large.

What is exact, and what is practical?

▶ **Mag** We have often calculated the theoretical frequency in an interval with Eq. 8.14 as (*pdf* in the center of the interval) times (interval width) times (total number of data). What can you say about this approach?

▶ **Tim** It's wrong. You have to take the difference of the distribution function at the interval limits times the total number, according to Eq. 8.15.

▶ **Alac** However, this is a little cumbersome to program. The approximate method gives faster results, and you can graphically display the probability density with a polygon line.

▶ **Mag** That's true. The path via probability density is an approximation. But in many cases, especially in experiments with smaller amounts of data, it does not lead to noticeable differences from the exact method with the distribution function *cdf*.

▶ **Alac** That's what I'm saying. So, it's more practical.

▶ **Tim** Agreed! Nevertheless, I only use this as a first approximation and will take the trouble to apply the method with the *cdf* at the end of my work.

What can we learn philosophically ;-)?

The philosopher Karl Popper stated that scientific theories cannot be confirmed, but only falsified. We find an example of this in our exercise. Let us forget the arguments in the above dialogue and again accept Eq. 8.14 as a valid theory.

In Chi^2 tests, this model provides sufficiently large values for small amounts of data. However, this finding *does not confirm* the theory; it only gives us *no reason to reject* it. It is only for large amounts of data that the Chi^2 test yields such small values that we may reject the theory with a very small probability of error.

In the history of physics, there have always been cases in which only a refined measuring technique was able to falsify a theory that then turned out to be merely an approximation. An example: In Newtonian mechanics, the mass of a body is independent of the velocity of the body, but this is not so in Einstein's theory of special relativity.

Chi^2 test with ever bigger sample sizes

We are going to perform an experiment in which, four times for each integer *Smp100N* (= 1 to 100), we:

- create an array with 1024**Smp100N* normally distributed random numbers
- determine their observed frequencies of occurrence f_{Obs}
- compare f_{Obs} with frequencies f_{Xpt} expected from the probability density $pdf(x_c)$ in the center x_c of the intervals by means of Chi^2 tests.

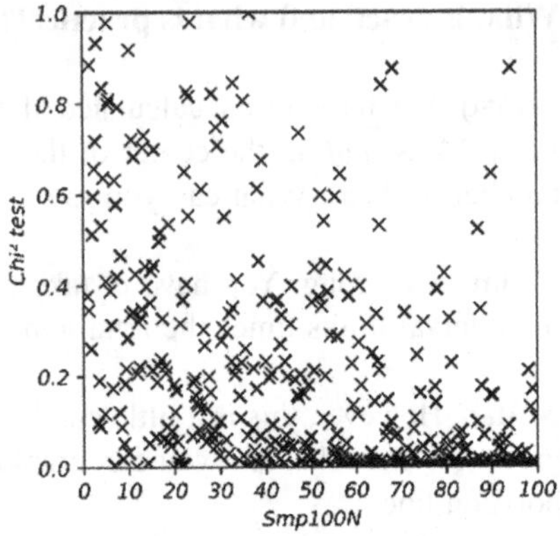

Fig. 8.11 Results of Chi^2 tests for sets with ever more samples (1024*N)

The results of the Chi^2 test are shown in Fig. 8.11 as a function of N. The number of random numbers is 1024* *Smp100N*. For a small size of the set ($\leq 10*1024$ for $Smp100N \leq 10$), the values of the Chi^2 test seem to be equally distributed, so that there is no reason to doubt that our approximation describes the empirical results correctly. For increasing size, the values of the Chi^2 test tend to be low, indicating that the theoretical model is not correct. The bigger the size of the set, the less probable the high values of the Chi^2 test are. Nevertheless, values above 0.5 may occur.

Take some time to think about this exercise! It will tell you a lot about the reliability of a Chi^2 test, a model for other statistical tests, about statistical reasoning and the Philosophy of Science.

8.5.4 Data Structure and Nomenclature

The normal distribution

x_m, x_{Sd}	mean and standard deviation of the normal distribution
$pdf(x_m, x_{Sd})$	$x \rightarrow p$, probability density function
$cdf(x_m, x_{Sd})$	$x \rightarrow p$, distribution function
$ppf(x_m, x_{Sd})$	$p \rightarrow x$, inverse *of cdf*, "percent point function"
x	array of arguments, equally spaced
p_D	*pdf* of x
p_C	*cdf* of x
p	array of probabilities, equally spaced between 0 and 1.

Normally distributed random numbers

N	size of the set of random numbers
x_{N}	array of numbers created with `randn(N)`
$freq_{\mathrm{N}}$	frequency of occurrence of x
x_{P}	array of numbers created with `ppf(npr.random(N))`
x_{b}	boundaries of sorting intervals (bins)
$freq_{\mathrm{P}}$	frequencies of occurrence of x_{P}
$freq_{\mathrm{XpC}}$	frequencies of occurrence expected from *cdf*
$freq_{\mathrm{XpP}}$	frequencies of occurrence expected from *ppf*.

8.5.5 Python

Normal distribution

Table 8.11 shows the program for generating arrays with values of these functions to be displayed as three data series in Fig. 8.9. The functions concerning the normal distribution have to be made available from the library `scipy.stats`, usually imported under the abbreviated name `sct` with `import scipy.stats as sct`.

Random Gaussian generator

The `Python` program in Table 8.12 creates a random sample of 10,000 normally distributed numbers in two ways, first, with the function `npr.randn`(10,000), and second, with the function `npr.random` (generating random numbers between 0 and 1) inserted into `sct.norm.ppf`, the inverse of the distribution function. The second version corresponds to the nested EXCEL function and can generally be applied to other distributions, e.g., to get the $\cos^2$-distribution needed in Exercise 8.7.

Next, the empirical frequency distribution of the two samples is determined within 23 intervals with the boundaries `Ib` $= -\infty, -3, -2.7, \ldots, 2.7, 3, \infty$. This is done with the function `np.histogram` that expects the sample and the interval boundaries as input and returns the empirical frequencies on row 0 (*freq[0]* in Table 8.12) and the unchanged interval boundaries on row 1. The data obtained on x and x_{P} are displayed in Fig. 8.9a, together with the theoretically expected frequencies $freq_{\mathrm{XpC}}$.

Table 8.11 `Python` functions related to the normal distribution

```
import scipy.stats as sct
xN=np.linspace(-4,4,41)
pD=sct.norm(0,1).pdf(xN)        #Probability density function
pC=sct.norm(0,1).cdf(xN)        #Cumulated density function
xP=sct.norm(0,1).ppf(pC)        #Percent point function
```

Table 8.12 Normally distributed random numbers x and x_P

```
N=10000
xN=npr.randn(N)
xP= sct.norm.ppf(npr.random(N))  #Corresp. to Excel formula

b=np.linspace(-3,3,21)
dx=b[1]-b[0]                        #Interval width
xb=np.empty(len(b)+2)               #Interval boundaries
(xb[0], xb[1:-1], xb[-1])=(-np.inf,b, np.inf)
freqXpC=(sct.norm(0,1).cdf(xb[1:])  #Expected freqs.
        -sct.norm(0,1).cdf(xb[:-1]))*N
xc=np.ones(len(xb)-1)         #Centers of intervals ("bins")
xc[1:-1]=(xb[1:-2]+xb[2:-1])/2
xc[0],xc[-1]=xb[1],xb[-2]

freqN =np.histogram(xN,bins=xb)
freqP =np.histogram(xP,bins=xb)
```

```
x          [ 1.20  0.88  1.67 ...   -0.78 -1.82]
xb         [ -inf -3.00 -2.70 ...    3.00   inf]
freqN[0]   [     17     22    50 ... 16    18]
freqXpC     [   13.5  21.2  47.3 .. 21.2 13.5]
```

8.5.6 Excel

Normal distribution

We create maps of the probability density of the normal distribution and its distribution function. In the following spreadsheet, Fig. 8.12 (S), 41 values of the probability density *pdf* and the distribution function *cdf* are generated and shown in the diagrams of Fig. 8.8.

	A	B	C	D	E	F	G
2	**dx**	0.2					
3	*=A5+dx*	*=NORM.DIST(x;0;1;FALSE)*	*=NORM.DIST(x;0;1;TRUE)*			*=NORM.INV(p;0;1)*	
4	***x***	***pdf(0,1)***	***cdf(0,1)***		***p***		
5	-4.0	0.000	0.000		0.0	#NUM!	
6	**-3.8**	0.000	0.000		0.1	-1.28	
14	-2.2	0.035	0.014		0.9	1.28	
15	-2.0	0.054	0.023		1.0	#NUM!	
45	4.0	0.000	1.000				

Fig. 8.12 (S) Function table for the normal distribution (probability density *pdf(0,1)*) with the mean value 0 and the standard deviation 1, column B), the associated cumulative (integral) normal distribution or distribution function (*cdf(0,1)*, column C), and the inverse of the distribution function in column F

	A	B	C	D	E	F	G	H
1	=COUNT(xN)		dx	0.5				
2	1024			N	1024	=SUM(freq)		
3	=NORM.INV(RAND();0;1)		=AVERAGE(D5:D6)	=D5+dx	{=FREQUENCY(xN;xb)}	=NORM.DIST(xc;0;1;0)*N*dx	=CHISQ.TEST(E6:E17;freqXpC)	
4	xN		xc	xb	freq	freqXpP		
5	-0.38		-3	-3	0			
6	-1.72		**-2.75**	**-2.5**	4	4.7	**0.51**	
17	1.41		2.75	3	6	4.7		
18	-0.64		3		0			
1028	-1.14							

Fig. 8.13 (S) Table layout used to generate normally distributed random numbers and to determine their experimental (column E) and theoretical (column F) frequencies; the Chi^2 test is performed in G6

A spreadsheet function for a Gaussian random generator

To create a set of normally distributed numbers in EXCEL, we use

Norm.Inv(Rand(); 0; 1)

corresponding to `sct.norm(0,1).ppf(npr.rand())` in `Python`. It claims to return random numbers distributed according to the probability density of a Gaussian bell curve with the mean value 0 and the standard deviation 1 (standard normal distribution).

We check this claim in Fig. 8.13 (S) by entering the above function into 1024 cells named x_N, thus generating 1024 random numbers. A frequency distribution *freq* is calculated from them for the interval boundaries x_b and displayed in a diagram over the interval centers x_c (see Fig. 8.10a). The expected theoretical frequency distribution $freq_{XpP}$ is calculated *with the probability density* in the center of the intervals.

For a normal distribution with mean value 0 and standard deviation 1 (standard normal distribution) , frequencies as those for $freq_{XpP}$ in column F are expected. The corresponding formula (F3) is composed of three terms:

- NORM.DIST(x_C; 0; 1; 0) , the probability density *pdf* in the center x_c of the interval, where the second and third positions indicate the mean value (here, 0) and standard deviation (here, 1) of the Gaussian curve. The fourth position (FALSE or 0, TRUE or 1) determines whether the probability density or, otherwise, the distribution function is to be returned.
- The width dx of an interval (here, 0.5).
- The total number N of random numbers (here, 1024).

These expected frequencies are shown in Fig. 8.10a as a polyline. The Chi^2 test in G6 gives a value of 0.51. So, we have no reason to doubt that our function NORM.INV(RAND();0;1) produces standard normally distributed (=

	A	B	C	D	E	F
1				50	15000	=SUM(fSum)
2	=RAND()			{=FREQUENCY(x;xb)}	from Macro	
3	x		Ib	Freq	fSum	
4	0.30		0.1	3	1511	
5	0.31		0.2	6	1530	
11	0.73		0.8	9	1523	
12	0.17		0.9	7	1468	
13	0.71			6	1538	
53	0.76					

```
1 Sub SumHist()
2 For rep = 1 To 100
3   Application.Calculation = xlCalculationManual
4   For r = 4 To 13
5     Cells(r, 5) = Cells(r, 5) + Cells(r, 4)
6   Next r
7   Application.Calculation = xlCalculationAutomatic
8 Next rep
9 End Sub
10
```

Fig. 8.14 **a** (left, S) The frequency of 50 numbers x in column A is determined in column D. The sum of the frequencies in column E has been obtained with SUB *SumHist* in **b** (P). **b** (right, P) Procedure adding up the frequencies in column D into f_{Sum}. Two nested loops apply: (FOR r = ...) adds up the frequencies once, the superordinate loop (FOR rep = ...) repeats this process. CELLS(4,5) IN THE LOOP (FOR R = 4 ...) in the procedure is cell E4 in the spreadsheet

Gaussian-distributed) random numbers, and that their frequency of occurrence is theoretically described using the probability density in the center of the intervals.

We shall use this function in later chapters with various standard deviations to simulate noise during measurements.

Questions

The distribution of 102,400 random numbers is displayed in Fig. 8.10b. How can you supplement the EXCEL solution in Fig. 8.13 (S) designed for 1024 random numbers so as to also get such a large number?[23]

Summing Up Frequencies with a VBA Routine

It is often practical to set up a calculation model in which frequencies are first determined on a small sample. If the spreadsheet calculation runs without errors, then the statistics can be made more extensive by repeating the random experiment several times and adding up the frequencies found. An example is given in Fig. 8.14.

The procedure *SumHist* adds, in the inner loop (r = ...), the values in *Freq* to the values in f_{Sum} and repeats this 100 times. Before the (r = ...) loop, the automatic calculation is switched off (line 3), because, otherwise, all random numbers would be generated anew and every entry in the spreadsheet and all frequencies recalculated. This would make the check number in E1 unequal to the total number of sample points as it should not be. Such a discrepancy could therefore be used to identify incorrect programming. After finishing the (r = ...) loop, the automatic calculation is switched on again to generate the random numbers x anew.

Questions

Questions concerning Fig. 8.14:

[23] With a rep-log procedure adding up the frequencies in *freq* 100 times. Continue reading!

How often was *SumHist* of (P) called to yield the results in Fig. 8.14a (S)?[24]

Why is the automatic calculation in SUB *SumHist* Fig. 8.14b (P) switched off before adding the frequencies?[25]

What would happen if the automatic calculation were not switched on again after a summation of the frequencies?[26]

▶ **Task** Write a log procedure for repeating the random experiment of the previous task in Fig. 8.13 one hundred times and add up the frequencies! Do not forget to switch off the automatic calculation while the current frequency distribution is added to the sum! Next, create a frequency distribution and do a Chi^2 test!

8.6 Random-Number Generator, General Principle

A random-number generator for a desired probability distribution *pdf* can be created if the inverse function *ppf* of the associated cumulative distribution function *cdf* exists and is "fed" with random numbers equally distributed between 0 and 1.

Why do Norm.Inv(Rand();0;1) and sct.norm.ppf(npr.random) generate normally distributed random numbers?

The answer to this question is made plausible with Fig. 8.15. There, the probability density *pdf* of the standard normal distribution is represented with a bell curve, and the kink points of the right angles lie on the distribution function *cdf*. The interval limits of the uniformly subdivided x-axis are transferred to the y-axis by the *cdf* of the normal distribution.

The width of the intervals on the y-axis is $\Delta y = \mathrm{d}cdf(x)/\mathrm{d}x \cdot \Delta \mathrm{x}$. So, if the y-axis is "fed" with random numbers between 0 and 1 and transferred by *ppf* to the x-axis, a distribution of the x-values corresponding to $\mathrm{d}cdf(x)/\mathrm{d}x$ results, i.e., according to the derivative of the *cdf*, and thus a distribution proportional to the associated probability density $pdf(x) = \mathrm{d}cdf(x)/\mathrm{d}x$.

[24] SUB *SumHist* ran three times. Each time, the random experiment was repeated one hundred times, so that the summed frequency distribution captures $3 \times 100 \times 50 = 15{,}000$ numbers (see cell E1 in Fig. 8.14a).

[25] Every single summation would otherwise lead to a recalculation of all random numbers, and thus to changed frequencies of occurrence. This would be noticed thanks to the fact that, in cell E1 of Fig. 8.14a (S), there would be no number corresponding to 3 x 100 x 50. The sum in cell E1 is therefore a check as to whether an error has been made in the spreadsheet or in the procedure.

[26] The random numbers would not be determined anew, and the same frequencies would always be added up.

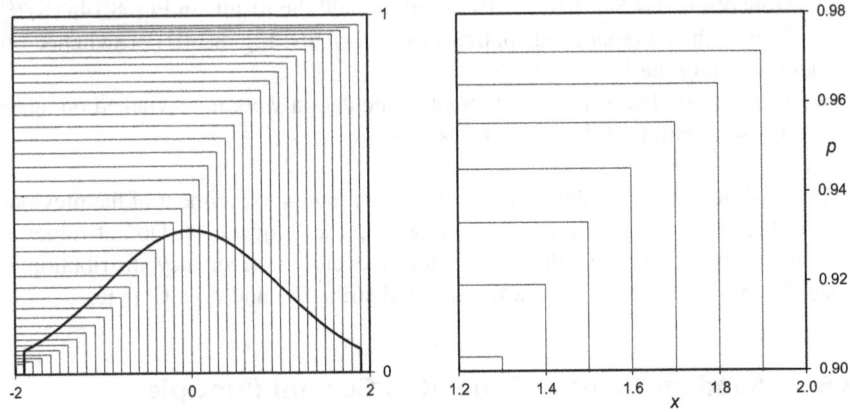

Fig. 8.15 **a** (left) Bell curve: Normal distribution. **b** (right) Blow-up of **a**, $\Delta y = (dy/dx)\cdot\Delta x$

	A	B	C	D	E	F	G	H
1			dx	0.3	N	10000	=SUM(fSum)	
2	=-LN(RAND())		=(D4+xb)/2	=D4+dx	{=FREQUENCY(x;xb)}	by log-procedure	=EXP(-xc/1)*N*dx	
3	x		xc	xb	freq	fSum	fTheo	Chi² test
4	4.60			0	0	0		
5	1.30		**0.15**	**0.3**	301	2579	2582	0.67
6	0.24		0.45	0.6	196	1932	1913	
13	0.58		2.55	2.7	16	254	234	
14	0.15		2.85	3	14	171	174	
15	0.04				47	465		
1003	0.34							

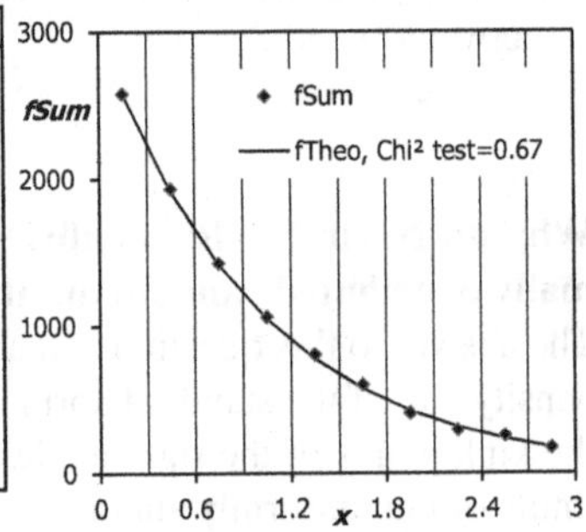

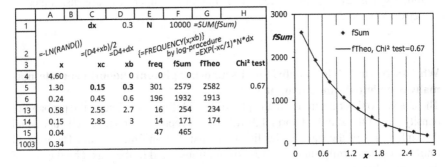

Fig. 8.16 **a** (S) Generating exponentially distributed random numbers. **b** (right) Display of data from **a**

Distributions generated with simple mathematical functions

We generalize our findings: The inverse function $ppf(p)$ of a distribution function $cdf(x)$ can be used as a random generator for the associated probability density $ppf(x)$:

$$y = P_I(x) \quad \text{(distribution function, } cdf\text{)}$$

$$\frac{dy}{dx} = p(x) \quad \text{(probability density, } pdf\text{)}$$

$$x = P_I^{-1}(y) \text{ (ppf, inverse of the distribution function)}$$

All distributions for which the inverse function of its distribution function can be built can be generated using standard functions. We still have to make sure that the argument range of the inverse function is correctly covered by the range from 0 to 1. Examples can be found in Table 8.13.

Table 8.13 Some functions, their integrals, and the inverse functions of the integral, suited to generating random number distributions (EXCEL)

pdf(x)	cdf(x)	ppf(p)
Equi-distribution 0 to 1		RAND()
Normal distribution	NORM.DIST(X;XM;XD;TRUE)	NORM.INV(RAND())
Exponential	Exponential	-LN(RAND()), natural logarithm
Cosinus	Sinus	Arcus sinus(2*(Rand()-0.5)
$2x$ for $x = 0$ to 1	x^2	$(Rand())^{\frac{1}{2}}$
$3x^2$ for $x = 0$ to 1	x^3	$(Rand())^{\frac{1}{3}}$
$1/(\pi \cdot (1+x^2))$ Cauchy-Lorentz	Arcus tangens(x)	Tan(π(Rand()-0.5)

Questions

How do you construct a distribution $p(x) = c{\cdot}x^3$ for $0 \leq x < 1$? How big is c?[27]

Exponential distribution

As an example, we construct a random-number generator that produces a decreasing exponential distribution from $x = 0$ to ∞. We know that

$$\mathrm{pdf(x)} = \exp(-\mathrm{x}) \rightarrow \mathrm{cdf(x)} = 1 - \exp(-\mathrm{x}) \rightarrow \mathrm{ppf(p)} = -\ln(\mathrm{x})$$

So, -IN(RAND()) should do the job in EXCEL; it is implemented in Fig. 8.16a (S) with the results shown in **b**.

Questions

How many random numbers are generated in Fig. 8.16a under the name x?[28]
What is the total size of f_{Sum} displayed in Fig. 8.16b?[29]
How was f_{Sum} most likely calculated from *freq*?[30]

▶ **Task** Generate random numbers that are distributed according to a cosine arc and check the results with a Chi^2 test!

[27] The antiderivative of $p(x)$ is $P(x) = c/4 \cdot x^4$; $P(1)$ must be 1 → c = ¼. The inverse function of $P(x)$ is $\sqrt[4]{Rand()}$. The random function is thus [=RAND()^0.25] in EXCEL or `np.random(N)**0.25` in Python.

[28] The array x comprises 1000 numbers (rows 4 to 1003).

[29] The array f_{Sum} comprises $N = 10{,}000$ numbers (cell F1).

[30] By a rep-log procedure summing up *freq* 10 times (similar to Fig. 8.14b).

▶ **Task** Generate random numbers distributed between 0 and 1 according to $p(x) = 3x^2$. Determine the theoretically expected frequencies, approximated with the probability density, Eq. 8.14, as well as exactly with the distribution function, Eq. 8.15. Perform Chi^2 tests to see if the formulas given in the table are correct and for what number of random numbers the approximation of Eq. 8.14 is good enough! In EXCEL, use a *rep-log procedure* that repeats the random experiment!

8.7 Diffraction of Photons at a Double-Slit

We simulate the diffraction of photons at a double-slit, intending to demonstrate the wave-particle duality of light. For this, we need a random-number generator distributing the impact points of the phonons on a screen according to a cos^2 probability density. The goal concerning programming is to learn how to implement such a random-number generator using a finite polyline.

8.7.1 Physical Background: Wave-Particle Dualism

In Fig. 8.17a, b, the diffraction of electrons or photons at a double slit is simulated to illustrate the wave-particle dualism. The diffraction image is created by many particles that hit a screen behind the double-slit randomly. The image does not consist of stripes behind each slit, but is rather an interference pattern with a maximum behind the middle position between the two slits.

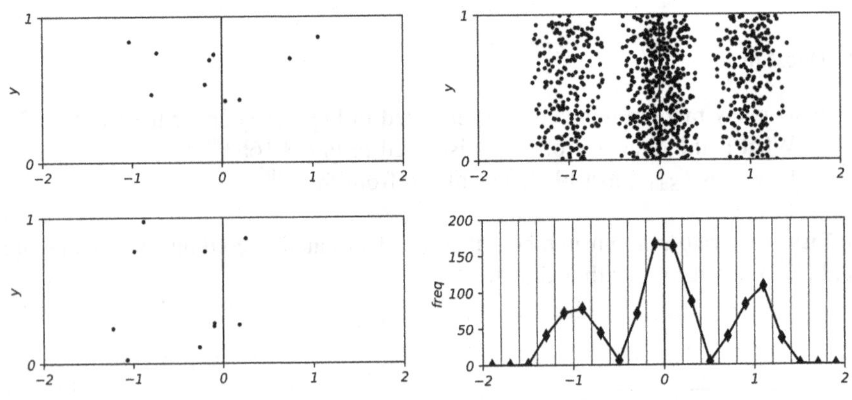

Fig. 8.17 **a** (left) Two diffraction images of ten photons each. **b** (right) Top: Diffraction image of 1000 photons, bottom: Distribution of the x-coordinate of the 1000 photons (the y-coordinate is evenly distributed)

In physical reality, the distance of the screen to the slits is large against the distance between the slits, and the independent variable of the interference pattern is the angle with respect to the mid-perpendicular of the double-slit.

We may imagine the screen upon which the diffraction image is captured as a gelatine film with silver grains or as a modern CCD detector with pixels. The impact of a photon is shown by the fact that a single silver grain or a single pixel is activated. Figure 8.17a shows two typical images after the impact of 10 photons. Such experiments in the real world show that light is "granular", i.e., consists of energy packets that are locally confined.

The events in Fig. 8.17a seem to be evenly but randomly distributed over the entirety of the detector surface. However, after 1000 photons have been detected, a pattern like that in Fig. 8.17b emerges; the positions $x = -1/2$ and $x = 1/2$ have never been hit by any photon. This is a consequence of the wave character of light resulting in a diffraction image consisting of a central maximum and secondary maxima of the same width. The distance between the maxima is determined by the reciprocal of the distance between the two slits. The ratio of the intensities depends on the width of a single slit. Here, we have arbitrarily chosen a ratio of 2 to 1 between the central maximum and the side maxima.

If … then … else …

The x-coordinates x_{Cos} generated by a cos^2-generator are to be distributed according to the diffraction figure:

- 50% in the central maximum of width 1 (position at $x = 0$),
- 25% in the right side-maximum of width 1 (position at $x = 1$), and
- 25% in the left side-maximum of width 1 (position at $x = -1$).

This can be achieved with a random number *rnd* between 0 and 1 and a logical query:

If 0.5 < *rnd*, then $x = x_{Cos}$ (in 50% of cases).
    else If *rnd* < 0,75, Then x = $x_{Cos} + x_0$ (in 75% – 50% = 25% of cases).
        else, x = x_{Cos} - x_0 (in the remaining 25% of cases).

Questions

How do the critical numbers in the IF query have to be changed when the maxima are to occur at x_0, $x_0 + 1$, and $x_0 + 2$ and intensity ratios of 6:3:1 are to be obtained?[31]

In a spreadsheet, we can make a "living picture"; with each change of the spreadsheet (e.g., when an already empty cell is "deleted"), the coordinates are

[31] If *rnd* < 0.6, then $x = x_{Cos} + x_0$; Else, if *rnd* < 0.9, then $x = x_{Cos} + x_0 + 1$; Else, $x = x_{Cos} + x_0 + 2$.

recalculated and the points in the diffraction image with ten photons, Fig. 8.17a, jump around erratically. The attentive observer may suspect that the points stay away from the straight lines $x = -1$ and $x = 1$, but this observation is not convincing.

In Python, we can create such a "living picture" through animation as shown in Sect. 6.2.5.

8.7.2 Cos² Distribution

Our task is to build a random generator that distributes the particles according to the diffraction pattern. We only want to simulate the central maximum of the diffraction figure at $x = 0$ and the first two secondary maxima around $-x_0$ and $+\ x_0$ (here $x_0 = 1$, arbitrary units on the screen). The intensity distribution in each maximum shall be approximated by the same cos^2 probability density (see Fig. 8.18a),

$$CosSq\_pdf(x) = 2 \cdot cos^2(\pi x) \tag{8.16}$$

however, with different amplitudes.

The distribution function *cdf*(*x*) is the integral of this probability density function *pdf*(*x*) with an argument range from -0.5 to 0.5 and a value range from 0 to 1:

$$CosSq\_cdf(x) = \frac{\sin(2\pi x)}{2\pi} + x + \frac{1}{2} \tag{8.17}$$

It is shown in Fig. 8.18a ("CosSq_cdf").

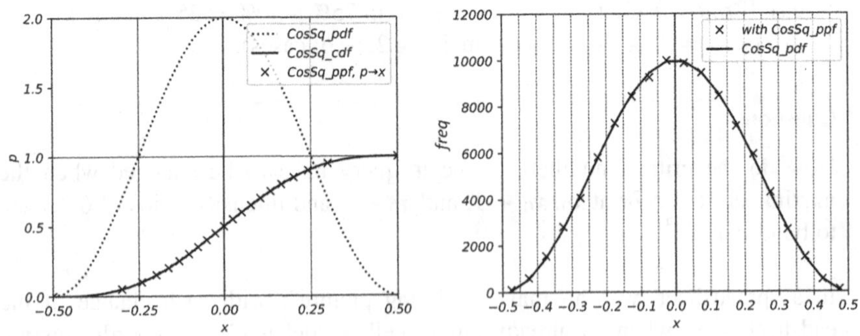

Fig. 8.18 **a** (left) Functions related to the cos² distribution, data from Table 8.15. **b** (right) The frequencies of occurrence of 100,000 outputs of the function *CosSq_ppf(rand))* are actually cos² distributed

Questions

How big are the values $CosSq\_cdf(-0.5)$, $CosSq\_cdf(0)$, $CosSq\_cdf(0.5)$ in Eq. 8.17[32]

How big is the area under the curve *CosSq_pdf* in Fig. 8.18a?[33]

For our random-number generator, we need the inverse function of *CosSq_cdf(x)* in Eq. 8.17.

Solution in three steps

- We create a function that distributes $\cos^2$-distributed values in the range −0.5 to 0.5.
- We generate ten points according to the desired diffraction image (as in Fig. 8.17a).
- We repeat the random experiment with the ten points one hundred times and determine the frequency distribution of the x-values (as in Fig. 8.17b).

The gradual emergence of the diffraction pattern can be demonstrated with animation.

8.7.3 Data Structure and Nomenclature

$CosSq\_pdf(x)$	$p(x) = 2{*}cos^2(\pi \cdot x)$
$CosSq\_cdf(x)$	integral of $CosSq\_pdf(x)$
$CosSq\_ppf(p)$	inverse of $CosSq\_cdf(x)$
x_I	array of x, 33 values from −0.5 to 0.5
p_I	$CosSq\_cdf(x_I)$
$m[0], \ldots$	slope of p_I in interval $x_I[0]$ to $x_I[1]$
x_{Cos}	x-values, random according to a cos^2 distribution.

8.7.4 Python

We define the three functions related to the $\cos^2$ distribution:

- *CosSq_pdf(x)* , probability density function, argument range from $-0.5 \le x \le 0.5$, normalized so that the area under the curve is 1.

[32] $p(-0.5) = 0$; $p(0) = ½$; $p(0.5) = 1$.

[33] (Area under $\cos^2$) = (triangle $(0.5 - (-0.5)) \cdot 2/2$) = 1. This is the condition for a probability density function.

Table 8.14 Python functions related to the $\cos^2$ distribution

```
def CosSq_pdf(x):
    return 2*(np.cos(x*np.pi)**2)

def CosSq_cdf(x):
    return np.sin(2*np.pi*x)/(2*np.pi)+x+1/2

xI=np.linspace(-0.5,0.5,33)
pI=CosSq_cdf(xI)
m=((pI[1:]-pI[:-1])/
    (xI[1:]-xI[:-1]))        #Slope
#m[0] is slope of interval pI[0] to pI[1]

def CosSq_ppf(p):
    for i in range(1,33):
        x0=0
        if pI[i-1]<=p<=pI[i]:
            x0=xI[i-1]+1/m[i-1]*(p-pI[i-1])
            return x0        #Why return within if?
```

- *CosSq_cdf(x)*, cumulative density function (distribution function), argument range from $-0.5 \leq x \leq 0.5$, monotonously increasing from 0 to 1.
- *CosSq_ppf(p)*, percent point function, the inverse of *cdf*, argument range $0 \leq p \leq 1$.

These are all displayed in Fig. 8.18a.

The three functions are defined in Table 8.14. The probability density function *CosSq_pdf* essentially returns a value of $\cos^2$; *CosSq_cdf* is its antiderivative. For the inverse of the distribution function (*cdf*), a closed expression does not exist. We therefore approximate it with a polyline with 33 vertices. *CosSq_cdf* for the *x*-values of the vertices is calculated in array p_I and the slope within the 32 intervals in *m*. The three arrays thus obtained are used to get the linear approximation for *Cos²_ppf*, with *p* as the independent variable and *x* as the dependent variable, and the slope $1/m$ between the vertices.

Questions

Is there a variant of *CosSq_ppf* that can suffice with 5 if queries for the 32 intervals?[34]

How do you plot the results of Table 8.15?[35]

[34] See the program code in Table 8.16.

[35] Plot with: `plt.plot(xC,hist[0],'kx'); plt.plot(xC,theo,'k- ')`.

Table 8.15 $N = 100{,}000$ outputs of the $\cos^2$-generator

```
N=100000
rn=npr.random(N)
xCos=list(CosSq_ppf(r)for r in rn)
dx=0.05                               #Interval width
xb=np.arange(-0.5,0.5+dx,dx) #Endpoint not included
xc=(xb[1:]+xb[:-1])/2
hist=np.histogram(xCos,xb)       #Empirical frequencies
theo=N*dx*CosSq_pdf(xc)          #Theoretical frequencies
```

Table 8.16 Interval search with five queries

```
def CosSq_ppf2(p):
    iLow=0
    iHigh=32
    for i in range(5):
        iMid=np.int((iLow+iHigh)/2)
        if p <= pI[iMid]:iHigh=iMid
        else:iLow=iMid
    x0=xI[iLow]+1/m[iLow]*(p-pI[iLow])
    return x0
```

CosSq_ppf is used in Table 8.15 to create 100,000 random numbers, the frequency distribution of which is displayed in Fig. 8.18b, together with the theoretical curve obtained with *CosSq_ppf*.

Interval search with five queries

Table 8.16 realizes the search in 32 intervals with 5 queries.

8.7.5 Excel

User-defined spreadsheet function

We develop a user-defined spreadsheet function that approximates the random-number generator for a $\cos^2$-distribution piecewise linearly (see Fig. 8.19 (P)), as has been done in the `Python` program. Remember: a user-defined spreadsheet function must be in a module. It must not be assigned to a specific worksheet (see Exercise 4.9).

Global data arrays for x_I, p_I, and m are defined at the top of the VBA sheet, in lines 1 to 3 of Fig. 8.19 (P), and can be called by all procedures and functions in the same VBA sheet or module. They must be initialized with the correct numbers before the first call of the function.

```
1 Public xI(32) As Single
2 Public pI(32) As Single
3 Public m(32) As Single
4
5 Function CosSq_ppf(p0)
6 For i = 1 To 32
7   If (pI(i - 1) <= p0 And p0 <= pI(i)) Then _
8   co = xI(i - 1) + m(i) * (p0 - pI(i - 1))
9 Next i
10 CosSq_ppf = co
11 End Function
12
13
14 Sub init()
15 xI(0) = -0.5
16 xI(1) = -0.46875
17 ........
18 xI(32) = 0.5
19
20 pI(0) = 0
21 pI(32) = 1
22
23 For i = 1 To 32
24  m(i) = (xI(i) - xI(i - 1)) / (pI(i) - pI(i - 1))
25  Next i
26 End Sub
```

Fig. 8.19 (P) Function COSSQ_PPF approaches the inverse function of $P_I(X)$, *CosSq_cdf*(x) in Eq. 8.17, piecewise linearly as a polyline. The global fields x_I, p_I, and m must be initialized with the correct values, here, with SUB *init*, before the first call of the function. The text of the statements in *init* is generated in a spreadsheet (see Fig. 8.20 (S), columns E and F)

The function COSSQ_PPF corresponds to the `Python` function of the same name. The (FOR i = ..) loop queries in lines 7 and 8 for each of the 32 intervals whether p_0 lies in this interval. If this is the case, the value of the function is calculated as a linear interpolation between the vertices of the segment, the coordinates of which are stored globally in the data arrays $x_I(32)$ and $p_I(32)$. The slopes in the intervals are in the global array $m(32)$.

Questions

Is there a variant of *Cos2_ppf* that can suffice with five If queries for the 32 intervals?[36]

VBA code generated in a spreadsheet

Initialization of the global arrays x_I and p_I is done in a total of 66 lines in the procedure *INIT()*, also shown in Fig. 8.19 (P). The array *m(I)* contains the slope of the curve between points $i-1$ and i and is calculated from x and p_I within *init* (lines 23 to 25).

We could enter all 66 lines by hand, which is, of course, tedious and unpleasant, although it would not be much more time-consuming than tracking down errors in spreadsheets and other programs. But there is a more elegant way: we let EXCEL work for us. In various tasks, we have had a VBA procedure write formulas into a spreadsheet. Text in cells preceded by an = sign is interpreted as a formula. Now, we will simply do it the other way around: A text is assembled in a spreadsheet and copied into a VBA procedure to be interpreted there as a formula. This is done in Fig. 8.20 (S).

[36] Compare the `Python` code in Table 8.16.

	A	B	C	D	E	F	G	H
1	**0.031**	*=2/32*						
2	*=A4+$A$1*	*=SIN(2*PI()*xI)/2/PI()+xI+0.5*			*="xI("&n&") = "&xI*	*="pI("&n&") = "&pI*	*=cosSq_ppf(B5)*	
3	**xI**	**pI**		**n**	**Array x**	**Aray y**		
4	-0.500	0.000		0	xI(0) = -0.5	pI(0) = 0	0.000	
5	**-0.469**	0.000		1	xI(1) = -0.46875	pI(1) = 0.00020041090	**0.000**	
6	-0.438	0.002		2	xI(2) = -0.4375	pI(2) = 0.00159404009	0.000	
35	0.469	1.000		31	xI(31) = 0.46875	pI(31) = 0.9997995890	0.000	
36	0.500	1.000		32	xI(32) = 0.5	pI(32) = 1	0.000	

Fig. 8.20 (S) In columns A and B, there are 33 points on the curve $p_{\mathrm{I}}(x)$, Eq. 8.17; in columns D, E, and F, the VBA code for initialization of the arrays x_{I} and p_{I} is generated from columns A and B. This text is to be copied into the VBA editor, SUB *Init* in Fig. 8.19 (P)

A program is just a text, composed of code words, that is translated into computer instructions.

The column vectors x_{I} and p_{I} correspond to the lists with the same names as in the `Python` program (Table 8.14). The code for the VBA -function of the inverse *ppf*(p) of the distribution function *cdf*(x) ($x_{\mathrm{I}}(0) = \ldots, p_{\mathrm{I}}0) = \ldots$) in columns E and F is generated in the spreadsheet with text processing and copied into SUB *INIT()* IN Fig. 8.19 (P), which initializes the global data arrays x_{I}, p_{I}, and m.

In columns E and F of Fig. 8.20 (S), text corresponding to VBA code is generated. The spreadsheet formulas in the individual cells consist of text elements and numbers, e.g., in E5: [="xI("&n&") = "&xI], and yields [xI(1) = −0.4687]. The ranges E4:E36 and F4:F36 written in this way are then transferred to the Visual Basic editor by text copying.

▶ **Alac** The values can be calculated in the routine itself using Eq. 8.17.

▶ **Mag** Yes, that is possible. But by detouring via the spreadsheet, we have practiced the way in which code is generated as text in a spreadsheet and transformed into formulas. Code is nothing more than structured text interpreted by a programming language interpreter.

8.7.6 Simulation in a Spreadsheet

We will simulate the evolution of an interference pattern in a spreadsheet with two figures like the upper two in Fig. 8.17. Ten new photon impacts are shown every second in a snapshot (left); they are accumulated for the right figure.

Random experiment with 10 to 1000 photons

In Fig. 8.21 (S), we calculate the coordinates of ten photon impacts to be presented in the snapshot. The preliminary to the x-component is calculated as x_{Cos} with the user-defined spreadsheet function [= *COSSQ_PPF(RAND())*]. The random numbers x_0

	A	B	C	D	E	F	G	H	I	J	K
1			=IF(rn<0.5;xCos;IF(rn<0.75;xCos+1				1004	4	0;bnd)	0.16	1000
2	=cosSq_ppf(RAND())	=RAND()	=IF(rn<0.5;xCos;IF(rn<0.7...	=RAND()					=AVERAGE(J5;J4)	=J4+$J$1	=FREQ
3	**xCos**	**rn**	**x**	**y**		**x.1000**	**y.1000**		**bndC**	**bnd**	**freq**
4	-0.22	0.09	-0.22	0.08		-0.28	0.51		-2.00	-2.00	0
5	-0.09	0.84	-1.09	0.75		-0.22	0.29		**-1.92**	**-1.84**	0
13	-0.32	0.90	-1.32	0.91		-0.04	0.56		-0.64	-0.56	20
29						0.98	0.23		1.92	2.00	0
30						1.14	0.81		2.00		0
31						-0.07	0.86				

Fig. 8.21 (S) Ten random points within the diffraction image of a double-slit are generated in C:D. The total 1000 points in F, G have been accumulated (a 100 times) by SUB *More10* in Fig. 8.22 (P). The formula in K2 is = FREQUENCY (x.1000;bnd)

are distributed over the three maxima at 0, 1, and −1 by means of a second random number *rn* with (see formula in C2 and C1).

$$\begin{aligned} x = [= \text{IF}(0,5 < \text{rn}; \text{xCos}; \\ \text{IF}(\text{m} < 0,75; \text{xCos} + 1; \\ \text{xCos} - 1))] \end{aligned}$$

according to the specified ratio of the intensity of the maxima 0.5: 0.25: 0.25. Remember: the structure of the logical query in EXCEL is [=IF(LOGICAL_TEST; VALUE_IF_TRUE; VALUE_IF_FALSE)].

The *y*-component is distributed uniformly between 0 and 1. The points (*x, y*) are shown in Fig. 8.17a; they change their position with every change in the spreadsheet.

Sub More 10

The upper picture in Fig. 8.17b shows the accumulated photon impacts with the coordinates (x.1000, y.1000) from Fig. 8.21 (S). This range (columns F and G) is successively filled by SUB *More10* in Fig. 8.22 (P). It transfers the ten random coordinate pairs (*x, y*) from columns C and D consecutively to columns F and G using an index r_2 updated in G1 of the spreadsheet.

Sub Run, random experiment with 1000 photons

With a rep-log procedure, SUB *Run* in Fig. 8.23 (P), we repeat the random experiment *More 10* with 10 photons 100 times and display the increasing number of points in the cumulating diagram, Fig. 8.17b (top), getting at recognizing the pattern better and better.

Over the course of time, we recognize the diffraction pattern more and more clearly. The frequency distribution of the 1000× values in Fig. 8.17b reveals the $\cos^2$ distribution in the main maximum and in the two side maxima.

```
Sub More10()
r2 = Sheets("Coord").Cells(1, 7) 'G1
Application.Calculation = xlCalculationManual
For i = 0 To 9
    Sheets("Coord").Cells(i + r2, 6) = Sheets("Coord").Cells(i + 4, 3) 'C4 -> F4, etc.
    Sheets("Coord").Cells(i + r2, 7) = Sheets("Coord").Cells(i + 4, 4) 'D4 -> G4, etc.
    Next i
Application.Calculation = xlCalculationAutomatic
Wait
Sheets("Coord").Cells(1, 7) = r2 + 10 'index for columns F und G
End Sub
```

Fig. 8.22 (P) SUB *More10* writes the ten random coordinate pairs in columns C and D of Fig. 8.21 (S) *successively* into columns F, and G. SUB *Wait* is reported in Fig. 8.23(P). The pointer r_2 for the next free row is updated in CELLS(1,7) = G1

```
1 Sub Run()
2 init
3 Sheets("Coord").Range("F4:G6000").ClearContents
4 Sheets("Coord").Cells(1, 7) = 4
5 For n = 1 To 100
6     More10
7 Next n
8 End Sub

9  Sub Wait()
10 Dim m As Integer
11     h = Hour(Now)
12     m = Minute(Now)
13     s = Second(Now) + 1
14     waittime = TimeSerial(h, m, s)
15     Application.Wait waittime
16 End Sub
```

Fig. 8.23 (P) In the master procedure SUB *Run*, the coefficients for *CosSq_ppf* are generated by calling SUB *Init* of Fig. 8.19 (P); the old coordinates are deleted in line 3. SUB *More10* (Fig. 8.22 (P)) is called in a loop a hundred times

SUB *Run* is a log procedure that first deletes the old data in columns F and G in Fig. 8.21 (S). These are the coordinates of the points in Fig. 8.17a (top), so that the chart is now empty, and then calls the procedures *More10* and *Wait* a 100 times so that the chart fills up again (within 100 s) with 1000 points. We may call it a "master procedure", because it controls the program flow.

SUB *Wait* in Fig. 8.23 (P) stops the calculation for 1 s, so that the viewer can better follow how Fig. 8.21b top is filled up with new points and the frequency distribution in Fig. 8.21b bottom takes shape.

Questions

concerning Fig. 8.22 (P):

The VBA procedure reads the index r_2 of the next free line in the range into which the new coordinates of the photon impacts are to be written from the spreadsheet (G1). Alternatively, the index r_2 could also be updated in the main program SUB RUN() in Fig. 8.23(P). What is the difference between the two alternatives?[37]

How would you change Sub More10 if only 5 photons are to pass the double-slit?[38]

8.8 Chi² Distribution and Degrees of Freedom

Distributions of χ^2 values, obtained from statistical experiments on normally-distributed random numbers, are compared with theoretical distributions with appropriate degrees of freedom. The degree of freedom is the number of intervals reduced by 3 if the mean and standard deviation are estimated from the sample and reduced by 1 if they are fixed a-priori.

In this experiment, we generate $N = 1000$ standard-normally distributed random numbers, determine their frequencies of occurrence in 10 intervals, and compare them in Chi² tests with theoretically expected frequencies obtained from two models.

In the first model, we use the values 0 and 1 for mean x_m and standard deviation x_{Sd} so that no parameters of the distribution are estimated from the sample, and the degree of freedom is $dof = 9 =$ number of intervals – 1. The reduction by 1 is due to the fact that the frequency in the last interval is not free but determined by the total number and the frequencies in the other intervals.

In the second model, we estimate x_m and x_{Sd} from the sample so that $dof = 9 - 2 = 7$. We repeat the statistical experiment with 1000 data $N_c = 10{,}000$ times and determine the frequencies of occurrence of the values of Chi².

The result of the statistical experiments is shown in Fig. 8.24a. The lines represent frequencies obtained with the theoretical probability density function of Chi², available from the library `scipy.stats`, usually imported as `sct`, here, `sct.chi2.pdf(xc,dof = 9)` and `sct.chi2.pdf(xc,dof = 7)`, where x_c are the centers of the sorting intervals.[39] We see that the results of our simulation fit very well with the theoretical curves. So, we have got the *dof* right.

Fig. 8.24b presents the *pdf* of the Chi² distribution for $dof = 5$ and $dof = 15$. The shaded areas represent the results of Chi² tests. They are the accumulated

---

[37] Update of r_2 *in Sub Run()*: initial index always starts at a fixed value of r_2; old data are overwritten.

Read r_2 from the spreadsheet: old data remain; new data are appended below the already existing data.

[38] Change: line 4 → `i = 0 to 4`; line 10 … `r2 + 5`. The program becomes more flexible if an additional variable N is introduced: `i = 0 to N-1`; ... `r2 + N`.

[39] The sorting intervals are often called *bins* into which the data set is to be sorted.

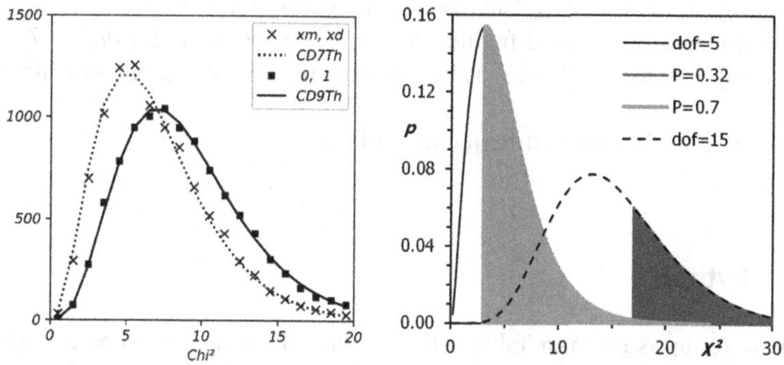

Fig. 8.24 **a** (left) Empirical (x, ·) and theoretical (…, —) distribution of values of Chi^2 for 7 and 9 degrees of freedom. **b** (right) Pdf of Chi^2; the *P*-value returned in a Chi^2 test corresponds to the shaded area ($P = 0.32$ for $dof = 5$ and $P = 0.7$ for $dof = 15$)

probabilities that a sample drawn from the population yields a higher value of Chi^2 when compared to the theoretical distribution of the population.

Keep in mind that, for 10 sorting intervals for normally distributed random numbers,

- $dof = 9$ when $x_m = 0$ and $x_{Sd} = 1$ are a-priori fixed,
- $dof = 7$ when x_m and x_{Sd} are estimated from the sample.

What is the Chi^2 test good for?
We check the hypothesis that empirical frequencies of occurrence arise from a theoretical distribution. The error probability to reject that hypothesis is given by the result of a Chi^2 test.

8.8.1 Data Structure, Nomenclature

N	number of random numbers
X_N	sample of N random numbers, standard-normally distributed
x_m	estimated mean of the sample
x_{Sd}	estimated standard deviation of the sample
x_b	interval boundaries, including $-\infty$ and ∞, defining 10 sorting intervals ("bins")
fo	observed frequencies of x_N in the sorting intervals
fx_9	frequencies expected for a normal distribution $norm(0, 1)$ in 10 intervals
fx_7	frequencies expected for $norm(x_m, x_{Sd})$ with x_m and x_{Sd} estimated from the sample.
N_{rep}	number of repetitions of the statistical experiment
dof	degree of freedom

cS_7 array for storing N_{rep} Chi^2 values comparing f_O and $fx7$
cD_7 theoretically expected frequencies of cS_7, *pdf* of Chi^2 for $dof = 7$
cS_9 array for storing N_c Chi^2 values comparing f_O and fx_9, *pdf* of Chi^2 for $dof = 9$
cD_9 theoretically expected frequencies of cS_9.

8.8.2 Python

The basic set-up is given in Table 8.17. 10 intervals from $-\infty$ to ∞ are specified and the frequencies of occurrence for a standard-normal distribution therein for $N = 1000$.

The interval boundaries x_b specified in lines 6 and 7 range from $-\infty$ to ∞, to capture the whole value range of normally distributed random numbers. They are reported in the lower cell of Table 8.17. The frequencies expected in the intervals are calculated with the cumulated density function at the interval boundaries. For the first and last interval, we have to consider that $cdf(-\infty) = 0$ (line 9) and $cdf(\infty) = 1$ (line 12), which is true, by the way, for any *cdf*.

The program in Table 8.18 organizes the repetition ($N_{rep} = 10{,}000$ times) of the statistical experiment in a for-loop. For every iteration *i*, a sample x_n of N random numbers is generated, and their frequency distribution f_O, mean x_m, and standard deviation x_d are determined. The frequencies f_O are compared with a Chi^2 test to:

- *fx9*, expected when 0 and 1 are taken as the mean and standard deviation, calculated before the loop (Table 8.17), lines 8 to 11;

Table 8.17 Expected frequencies fx_9 of a normal distribution in 10 intervals with boundaries x_b, $x_m = 0$, $x_{Sd} = 1$ set a-priori

```
1  import numpy.random as npr
2  import scipy.stats as sct
3  N=1000                     #Size of the set of random numbers
4  db=0.5                     #Width of the sorting intervals
5  xbb=np.arange(-2.0,2.0+db,db)
6  xb=np.zeros(len(xbb)+2) #Interval boundaries
7  (xb[0],xb[1:-1],xb[-1])=(-np.inf,xbb,np.inf)
8  fx9=np.zeros(len(xb)-1)
9  fx9[0]=sct.norm(0,1).cdf(xb[1])*N            #Below first bound.
10 fx9[1:-1]=(sct.norm(0,1).cdf(xb[2:-1])
11              -sct.norm(0,1).cdf(xb[1:-2]))*N
12 fx9[-1]=(1-sct.norm(0,1).cdf(xb[-2]))*N #Above last bound.
```

```
xb    [-inf -2.00 -1.50 -1.00 -0.50  0.00
                   0.50  1.00  1.50  2.00  inf]
```

Table 8.18 Histograms of cS9 and cS7, together with the theoretical probability densities CD9Th and CH7Th of Chi^2 for $dof = 9$ and 7 degrees of freedom

```
Nrep=10000                         #Number of repetitions
cS9=np.zeros(Nrep)
cT9=np.zeros(Nrep)
cS7=np.zeros(Nrep)
for i in range(Nrep):
    xn=npr.randn(N)                #Std-normal distribution
    fO=np.histogram(xn,xb)         #Empirical frequencies

    ChiSq=sct.chisquare(fO[0],fx9,ddof=0)
    cS9[i]=ChiSq[0]

    fx7=np.zeros(len(fO[0])) #For expected frequencies
    xm=np.average(xn)
    xd=np.std(xn,ddof=1)
    fx7[0]=sct.norm(xm,xd).cdf(Ib[1])*N
    fx7[1:-1]=(sct.norm(xm,xd).cdf(Ib[2:-1])
             -sct.norm(xm,xd).cdf(Ib[1:-2]))*N
    fx7[-1]=(1-sct.norm(xm,xd).cdf(Ib[-2]))*N
    ChiSq7=sct.chisquare(fO[0],fx7,ddof=2)
    cS7[i]=ChiSq7[0]
xbC=np.linspace(0,20,21)              #For distrib. of Chi²
xc=(xbC[1:]+xbC[:-1])/2               #Centers of intervals
CD9=np.histogram(cS9,xbC)             #Empirical freqs.
CD9Th=sct.chi2.pdf(xc,df=9)*Nrep      #Theoretical freqs.
CD7=np.histogram(cS7,xbC)
CD7Th=sct.chi2.pdf(xc,df=7)*Nrep
```

– *fx7*, expected when x_m and x_{Sd} are estimated from the sample, calculated individually for every iteration within the loop.

The results of the Chi^2 tests are stored in cS9 and cS7. The frequency distributions CD9 and CD7 thereof are determined in Table 8.18. The results are shown in Fig. 8.24a.

8.9 Questions and Tasks

Explain the following broom rules:

1. Ψ *Chance is blind and checkered*
2. Ψ *Always one more! But of what and than what? (Concerning frequency distribution.)*
3. Ψ *Come to a decision! Sometimes, it will be wrong.*

Frequencies of occurrence
Initial situation: In the range A1:A1000 of a spreadsheet, named *data*, there are 1000 numbers. In range D2:D10, three interval boundaries $I_b = 0.1$; 0.2 and 0.3 are specified.

4. How many intervals are defined by the interval boundaries?
5. Which numbers are captured in the second and last intervals?
6. The frequencies are to be calculated in a column range starting with E2. Which range must be activated for the spreadsheet function FREQUENCY if all numbers are to be sorted into intervals? With which "chord" do you complete the formula input?
7. Over which x values should the frequencies be displayed in a diagram?
8. We want to determine the same frequency distribution with the numpy function `np.histogram`(*data; Ib_{np}*). How are the interval boundaries Ib_{np} to be defined?

The 1000 numbers are now supposed to be random numbers equally distributed between 0 and 0.5.

9. Which equations do you use in EXCEL and numpy to generate such random numbers?
10. What mean value do you expect for the 1000 numbers?
11. What frequencies do you expect in the first and last intervals for interval limits of 0.1, 0.2, and 0.3 in the EXCEL function?

Normal distribution
Figure 8.25a shows the distribution function *norm.cdf(0;1)* and the probability density *norm.pdf(0;1)* of the standard normal distribution. Assume that you have generated 1 million normally distributed random numbers and answer the following questions within the reading accuracy of the figure:

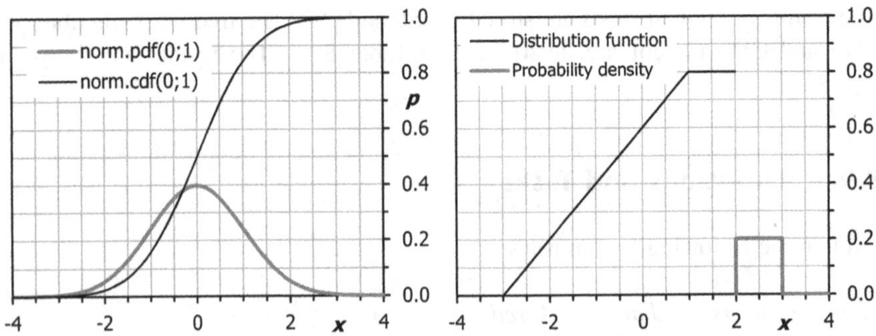

Fig. 8.25 **a** (left) Distribution function *norm.cdf(0;1)* and probability density *norm.pdf(0;1)* of the standard normal distribution. **b** (right) A distribution function and its probability density; the distribution function is not displayed above, and the probability density not below $x = 2$

12. How many numbers may be expected to have precisely the value 0?
13. How many numbers are expected to have a value between −0.1 and 0.1?
14. How many numbers are expected to have a value between −0.5 and 0.5?

Distribution function and random-number generators

15. Within what range do you expect 95% of the output of a function generating normally distributed random numbers?
16. Consider the EXCEL function NORM.INV(X;5;3)! What is its argument range? Where is the maximum of the distribution that results if $x =$ RAND()? How do you generate 1000 such numbers in a Range B1:B1000? What is the equivalent function for 1000 numbers in the `numpy.random` library?
17. Figure 8.25b shows an incomplete distribution function and an incomplete probability density. Complete the two functions!
18. What are the inverse functions of $y = cos(x)$ for $x = 0$ to π and of $y = exp(-x)$ for $x \geq 0$? Which are their argument ranges and which are their value ranges?
19. Which distribution is generated by the spreadsheet function (*arcus cosinus*) ACOS((RAND()-1)*2)?

User-defined spreadsheet function

In Fig. 8.26a, you find a spreadsheet layout for calculating a house-shaped polyline (see Fig. 8.26b). Implement a function *House(x)* for calculating the *y*-values from the *x*-values!

The entry in cell A25 of Fig. 8.26a is 6.38E–16, instead of the expected zero. This deviation is due to the sum of the rounding errors in the binary addition of $dx = 0.10$. In the user-defined spreadsheet function *House(x)*, the value of *x* must therefore be rounded, with ROUND(X,14).

	A	B
1	dx	0.10
2		
3	=A5+dx	=House(A6)
4	x	y
5	-2.00E+00	0
6	**-1.90E+00**	**0**
24	-1.00E-01	1.9
25	**6.38E-16**	2
45	2.00E+00	0

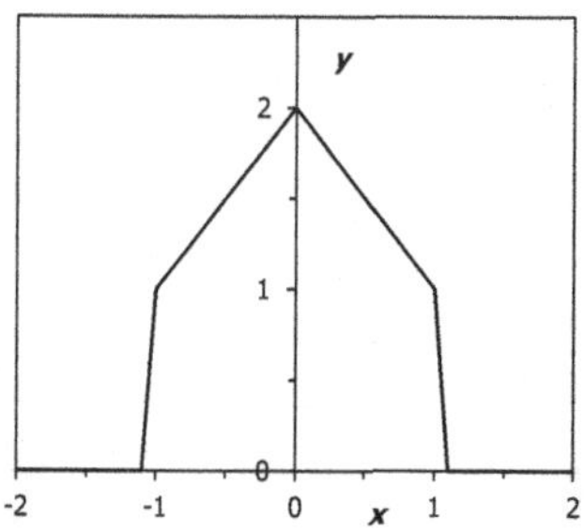

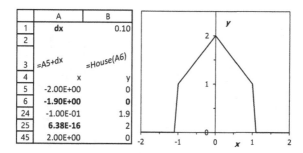

Fig. 8.26 **a** (left) The values of *y* are calculated with a user-defined spreadsheet function *House(x)*. **b** (right) Graph of the *House(x)* function with the data from Fig. 8.26a

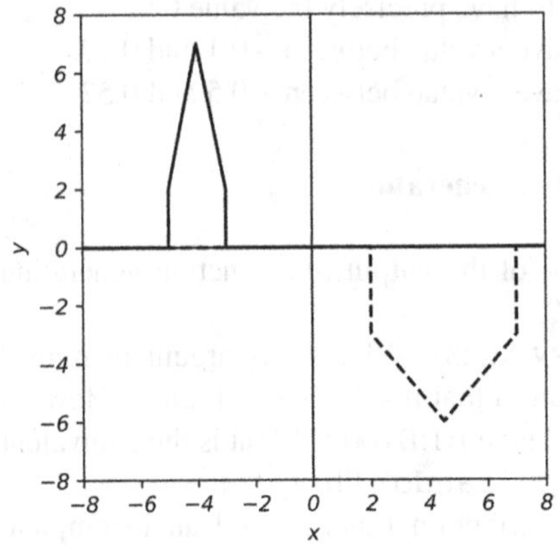


```
def House(x0,x1,y0,y1):
    x=[]
    y=[]
    ...
    return x, y


xMin=-8
xMax=8
x1,y1=House(-5,-3,2,7)
```

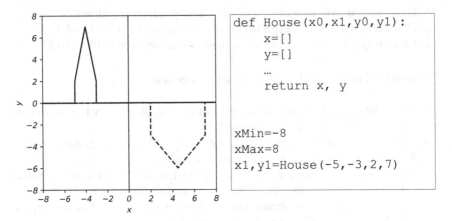

Fig. 8.27 a (left) Two shapes produced by a function *House* partly shown in **b** (right)

20. Why does the 20-fold addition of 0.10 to −2.00 not yield the smooth value 0.00, but rather 6×10^{-16}?[40]
21. Which statement can be used to make the function *House(x)* zero for $x \leq -1$ and $x \geq 1$?
22. Write a user-defined spreadsheet function that produces a shape like the "House" in Fig. 8.26b!
23. Figure 8.27a shows two shapes produced with a `Python` function *House* partly displayed in Fig. 8.27b. Complete the code!

[40] The numbers are binary coded so that decimal "smooth" numbers are not binary "smooth".

Evaluation of Measurements 9

We simulate measurement experiments by assuming "true" (known) values of a quantity and masking them with different noise levels that enter as standard deviation into normally distributed random numbers. Then, we "forget" the true values and estimate the measurand's now unknown value from the noisy signal. ψ *We know everything and play stupid*. This way, the students should get an impression of how unreliable measurements can be, learn to indicate the reliability, and understand how measurement uncertainties propagate. The importance of t statistics, connecting confidence interval and confidence level (or error range and error probability), is illustrated in several exercises.

9.1 Introduction: We Know Everything and Play Stupid

Solutions of Exercises 9.2 (Excel), 9.3 (Python), 9.4 (Excel), and 9.5 (Python) can be found at the internet adress: go.sn.pub/E3k8Ps.

► **Alac** Statistics? That's all Greek to me.

► **Mag** *Ψ If in doubt, count!*
Ψ Calculate with variances, report the standard error!

A simple measurement?
We shall determine the mass of a thin film on a glass substrate by weighing the substrate before and after coating. Estimating the mass is straightforward. However, stating a confidence interval requires this whole chapter, with t statistics and error propagation.

D. Mergel, *Physics with Excel and Python*,
https://doi.org/10.1007/978-3-030-82325-2_9

How should the poor student learn statistics?
Many students find it challenging to understand statistical statements. They are confronted with mathematically accurate but awkward-looking formulations and have to work according to strict rules.

In this chapter, the rules for statistical evaluation are not derived mathematically, but illustrated through simulations, particularly by *multiple tests for hit rates*. By repeating a noisy measurement series several times, we determine how often the standard error of the series's mean value captures the true value to specify the confidence level.

We know everything and play stupid
A writer of crime novels thinks up a criminal case, blurs the traces, and lets the detective reconstruct the case. So do we, by specifying the "true" values ourselves, turning them into measured values by adding noise, and then evaluating the noisy data. Finally, we report the measurement result with specified uncertainty. The true value may lie outside the confidence interval (the "error range" in laboratory jargon) of the measurement series' mean value. However, since we know the "true" value, we notice this and can observe regularities for our error.

Sherlock Holmes or Inspector G. Lestrade?
Some crime novel authors who want to be well received by the reader let the sharp-witted gentleman detective (Sherlock Holmes) or the curious lady detective (Miss Marple) appear alongside police officers, who stubbornly follow the rules and come to false results.

► **Tim** To which type does our evaluation belong?

► **Alac** To the smart detective, I hope.

► **Mag** No, unfortunately not! We must proceed strictly according to the rules. After all, intuition is often misleading when it comes to statistical issues. To get the best of both worlds, we follow the broom rule: Ψ *Trust (your intuition), but verify!*

Standard formulas
We apply the following formulas for a set of n measurements $x = \{x_i\}$.

The *mean* x_m of the set:

$$x_m = \frac{\sum_{i=1}^{n} x_i}{n} \tag{9.1}$$

The *variance* var_x of a set $x = \{x_i\}$:

$$var_x = \left(\sum_{i=1}^{n} (x_i - x_m)^2 \right) \cdot \frac{1}{n - n_P}, \quad n_P = 1 \text{ in this chapter} \tag{9.2}$$

where n_P is the number of parameters estimated from the data set. In this chapter, $n_P = 1$, because only one parameter, the mean, is estimated. In Chap. 10, trendlines are calculated and n_P is equal to the number of their parameters.

The *standard deviation* of the set:

$$x_{Sd} = \sqrt{var_x} \tag{9.3}$$

The *standard error* of the mean x_m of the set:

$$x_{Se} = \frac{x_{Sd}}{\sqrt{n}} \tag{9.4}$$

Standard error range

The standard error has the same physical unit as the elements of the sample and their mean value. Therefore, the mean x_m plus-minus the standard error x_{Se} is often reported as the result of a series of measurements:

$$Standard\,error\,range = x_m \pm x_{Se} \tag{9.5}$$

The standard error x_{Se} is calculated straightforward from Eqs. 9.2 (variance) and 9.3 (standard error).

Not to be forgotten: in your report, you also have to state the number of measurements. If at least eight measurements have been made, the following broom rule applies:

Ψ *Two within* and one out *of (the standard error range).*

The standard error range captures the true value of the measured quantity with a probability of about 2/3.

C-spec error

For statistically more precise reasoning, we have to connect confidence level and confidence interval by Student's[1] t statistics. We specify:

$$C - spec\,error\,range = x_m \pm t \cdot x_{Se} \tag{9.6}$$

For example, with eight measurements, the error range $x_m \pm 2.4x_{Se}$ captures the true value of the measurand with a probability of 95%. The interval $[x_m - t\,x_{Se}, x_m + t\,x_{Se}]$ is called the *confidence interval* (or *error range*) of the mean value x_m and the associated probability (95% in the example) *confidence level* C. The complementary probability (5% in the example) is called the *error probability* E:

$$C + E = 100\%$$

[1] Student is the pseudonym of W.S. Gosset.

We call the error for a well-specified confidence level *C-spec error* ("confidence-specified error").

The parameter determining the t distribution is the degree of freedom dof defined as

$$dof = n - n_P \tag{9.7}$$

where n is the number of measurements and n_P the number of parameters estimated from the data series. In the exercises in this chapter, n_P is 1 because only 1 parameter, the mean, is estimated from the data series. In the next chapter, coefficients of trend lines are estimated from the data and $n_P > 1$.

Multiple tests for obtaining hit rates

Statistical laws specify error probabilities for hypotheses. In this chapter's exercises, the hypothesis is that the estimated error range does not capture the true value. We do not derive such laws from axioms, but apply *multiple tests for hit rates* instead. They repeat statistical experiments multiple times and check whether the empirical hit or miss rates are compatible with the corresponding theoretical probabilities.

▶ **Tim** If this is not the case, then we have falsified the theoretical assumption.

▶ **Mag** Or have made a programming error. More about that later.

Simulation-based t adaptation

Many statistical laws are based on assumptions that are often not justified in our exercises or in real-life experiments, e.g., because the number of measurements is too small or the noise is too large. In such cases, the formally calculated error ranges may not exhibit the expected confidence level, e.g., in Exercise 9.9 where error propagation plays a role. In Exercise 9.8, we present a method to get C-spec errors, by adapting the t-factor so that a hit rate corresponding to a pre-specified confidence level is achieved. This method is suited also for real-world experiments.

Ψ Calculate with variances, report the C-spec error!

Mathematical formulas concerning confidence are formulated for variances. Statistical rules are based on properties of variances. We calculate with them to grasp the basic mathematical dependencies. For the final result, we strive to report the C-spec error because it is related to a confidence level and has the same physical unit as the measurand. In this way, we simulate the error propagation in sums, products and powers (Exercises 9.7 and 9.8).

Broom rules for measurements

We start with an exercise on a simple experiment, weighing a glass substrate, and end with a more challenging one, determining the mass of a thin film on such a glass substrate. In between, we have to improve our understanding of statistics with the following broom rules:

Ψ *Mostly, but not always.* No statistical statement is 100% certain.

▶ **Mag** This is the fundamental rule of statistical reasoning. More rules:

Ψ Always round to *relevant* digits; to convey the result concisely! (Exercises 9.2 and 9.3).

Ψ *Twice as good with four times the effort.* The measurement inaccuracy is halved if measurements are made four times as often (Exercise 9.4).

Ψ *Two within and one out of.* In one-third of all tests, the measurand's true value is outside of the standard error range if at least 8 measurements have been made (Exercise 9.5).

Ψ *Bad makes good even better, usually, but not always.* Even measurement results with a relatively large statistical uncertainty may improve the overall result by entering a weighted mean (Exercise 9.6).

9.2 Weighing a Glass Substrate

The weighing process is simulated by adding noise to the glass substrate's true weight and estimating the weight from the noisy data. We determine the standard error of the result and round the result to the relevant number of digits.

9.2.1 Discussion on the Accuracy of a Balance

▶ **Mag** We are going to determine the weight of a glass substrate with a typical weight of 1 g. How precise is this procedure if the measuring precision of the balance is 1 mg?

▶ **Alac** A simple calculation: the substrate's mass is about 1 g $\pm$ 1 mg, in other words, it can be determined with a relative accuracy of 1 per mill.

▶ **Mag** Be careful! The balance's specification only states that the mass is displayed in grams with 3 digits after the decimal point. However, the accuracy of the measurement may be lower than the precision of the balance's display, for example, due to ventilation or building vibrations.

▶ **Tim** But we cannot take such influences into account, because they are random and out of our control.

▶ **Mag** That's a good point, because when they are simply random, their influence on the weighing accuracy can be reduced if we repeat the measurement several times. With these multiple values, we can also specify the uncertainty of the result, in addition to the value of the mass.

▶ **Tim** I've experienced that before. In the physics lab course for beginners, we always have to report all results with measurement errors.

▶ **Alac** It's quite simple. *We already know the true values from our fellow students* in previous semesters, thus we only need to adjust the error range for our measured values so that the true value lies within the error bars. In this way, we avoid annoying questions from the supervisors.

▶ **Mag** That is similar to, but not exactly like, the way that we will do it in this exercise. As we generate the measurement data ourselves, we know the true values in advance, but then estimate them again from noisy data. However (!), we rigorously (!) calculate the standard error according to the rules.

▶ **Tim** Textbooks on measuring theory emphasize "confidence interval" and "confidence level".

▶ **Mag** We will learn the meaning of these terms later in Exercises 9.4 and 9.5, in which we will show that if 9 groups present their lab results with standard error ranges capturing the true value, 3 will probably have cheated.

9.2.2 Data Structure and Nomenclature

m_S true mass of the substrate
m_{Ns} measurement noise (standard deviation of a normal distribution)
dsp display precision of the scales
m_X measurement series
n number of measurements in m_X
m_M mass of the substrate estimated from the measurements
m_{Se} standard error of m_M.

9.2.3 Excel

Fig. 9.1 (S) presents an EXCEL solution, with the parameter specification in **a** and the process simulation in **b**. The raw data are generated in E6:E12 with normally distributed noise obtained with NORMINV(RAND();0;1) (see Exercise 8.5). They are transformed into the balance display, our measurement data, in column F. In G6:J6, the data are evaluated with the preliminary result for the estimated mass:

$$m_S\,(esti.) = (m_M \pm m_{Se})$$
$$m_s\,(esti.) = (0.995 \pm 0.022)\mathrm{g}$$

	A	B	C	D
1	**mS**	1	g	"true" mass of the substrate
2	**dsp**	3		display accuracy of the balance, typical display: 1.001 g
3	**mNs**	0.05	g	noise of the weighing process

	E	F	G	H	I	J
4	=mS+mNs*NORM.INV(RAND();0;1)	=ROUND(E7;dsp)	=AVERAGE(mX)	=STDEV.S(mX)	=COUNT(mX)	=mSd/SQRT(n)
5		**mX**	**mM**	**mSd**	**n**	**mSe**
6	1.0062099	1.006	0.995	0.057	7	0.022
7	1.0303005	**1.030**				
12	1.0878611	1.088				

Fig. 9.1 (S) Weighing a glass substrate. **a** (top) Specifications. **b** (bottom) Simulation of the process

Table 9.1 Weighing the mass of a substrate

```
mS=1          #[g]
dsp=3         #Digits displayed
mNs=0.05      #[g] Noise
n=7           #Number of measurements
mX=np.round(mS+mNs*npr.randn(n),3)
```

```
mX    [1.076 1.050 1.049 0.990 1.010 0.993 1.018]
```

```
mM=np.mean(mX)                           mM    1.027
mSd=np.std(mX)                           mSd   0.030
mSe=mSd/np.sqrt(n) #Standard error       mSe   0.011
```

still to be rounded to the relevant digits:

$$m_s(\text{ rounded }) = (1.00 \pm 0.02)\text{g}.$$

First, the standard error is rounded to one digit, and then, the mean value is rounded to the same number of decimal places as the standard error.

9.2.4 Python

The `Python` solution in Table 9.1 is straightforward.

Question

How do you report the results of Table 9.1 sensibly rounded?[2]

[2] 1.027 ± 0.011 becomes 1.03 ± 0.01.

9.3 A Procedure for Rounding to Relevant Digits

A formula network and a `Python` function are presented that round a measurement result to the relevant number of digits determined by the standard error. The equations rely on logarithms and powers.

9.3.1 Numerical Evaluations

The numerical evaluation of measurement series generally results in numbers with many digits, e.g., $x_{\mathrm{m}} = 0.008702$ g and $x_{\mathrm{Se}} = 0.000602$ g. As we have learned in Exercise 9.2, the final result should be $(0.87 \pm 0.06) \times 10^{-2}$ g or (8.7 ± 0.0) mg. We will obtain such results with formulas. To achieve this, we have to calculate with logarithms and powers of ten and link text and numbers.

9.3.2 Spreadsheet Calculation

The method is shown as a formula network in Fig. 9.2 (S), based on the example of Exercise 9.2, in which the mass of a glass substrate is determined. The estimated mean value x_{M} and its standard error x_{Se} are specified with 3 digits, the display accuracy of the balance.

The value of x_{Se} $(0.022 = 2.22 \times 10^{-2})$ is broken down into power of ten $(n_{\mathrm{Se}} = -2)$ and first digit $(\mathrm{x}_{\mathrm{SeR}} = 2)$ and reproduced as x_{SeRR} in decimal form with only one non-zero digit. The value for x_{M} is transformed into an integer x_{Mr}. Thus, the number of relevant digits corresponds to the precision of x_{SeR}, and is reproduced in decimal form with the reduced number of digits. The final result displayed in D7 is obtained with the formula in D8, concatenating text and variables.

Another variant, reporting the final result in exponential form, i.e., (9.8 ± 0.2)E-1 g, is shown in Fig. 9.3 (S). The value of x_{M} is transformed into a number x_{MrP} greater or equal to 1 and smaller than 10 and then rounded to the first non-zero digit of x_{Se}.

	A	B	C	D	E	F	G
1	**Name**	Weight		Power of xSe	**nSe**	-2	*=INT(LOG10(xSe))*
2	**xM**	0.977		First digit of xSe	**xSeR**	2	*=ROUND(xSe*10^(-nSe);0)*
3	**xSe**	0.022		Reduced xSe	**xSeRR**	0.02	*=ROUND(xSeR*10^nSe;ABS(nSe))*
4	**Unit**	g			**xMr**	98	*=ROUND(xM*10^(-nSe);0)*
5				Reduced xM	**xMrr**	0.98	*=ROUND(xMr*10^nSe;ABS(nSe))*
6							
7				**Weight = (0.98 ± 0.02) g**			
8				*=Name&" = ("&xMrr&" ± "&xSeRR&") "&Unit*			

Fig. 9.2 (S) The final result is obtained by rounding with formulas using logarithms

	A	B	C	D	E	F	G
10	**Name**	Weight		Power of xM	**nM**	-1	*=INT(LOG10(xM))*
11	**xM**	0.977		Reduced xM	**xMrP**	9.77	*=xM/(10^nM)*
12	**xSe**	0.022		Power of xSe	**nSe**	-2	*=INT(LOG10(xSe))*
13	**Unit**	g		Reduced xSe	**xSeRP**	2.2	*=xSe/(10^nSe)*
14				Rounded to relevant	**xMred**	9.8	*=ROUND(xMrP;nM-nSe)*
15				Rounded to relevant	**xSeRed**	0.2	*=ROUND(xSeRP;0)*10^(nSe-nM)*
16							
17				**Weight = (9.8 ± 0.2) E-1 g**			
18				=Name&" = ("&xMred&" ± "&xSeRed&") E"&nR&" "&Unit			

Fig. 9.3 (S) Same as Fig. 9.2 (S); however, separating the power in the result

Table 9.2 Function for returning the final result rounded to the relevant digits

```
def FinRes(Name,xM,xSe,Unit):
    #Power of the standard error:
     n=int(np.floor(np.log10(xSe)))
    #First digit of xSe:
     xSeR=np.round(xSe*10**-n,0)
    #Rounding to the certain digit:
     xSeRR=np.round(xSeR*10**n,np.abs(n))
    #Rounding xM to the certain digits:
     xMr=np.round(xM*10**-n,0)
     xMrr=np.round(xMr*10**n,np.abs(n))
     return str(Name) + '=' + str(xMrr) + "±" \
                            + str(xSeRR)+ Unit
```

9.3.3 Python Function

A user-defined function, performing the calculation of Fig. 9.2 (S), is introduced in Table 9.2 and applied in Table 9.3 in which two series of (x_M, x_{Se}) pairs are evaluated to check the validity of the formulas implemented in *FinRes*.

9.3.4 VBA Function

The rounding can also be performed with the user-defined VBA function *FinRes* in Fig. 9.4a (P), which reads-in the name, the value, and the uncertainty of the measurement result, and outputs the rounded result as shown in Fig. 9.4b. If this function is inserted into a VBA project module for user-defined functions (in our case, VBA PROJECT (*Dieters functions*.XLAM, Sect. 4.9.1), then it can be called in any EXCEL file.

Table 9.3 a (top) Continuation; applying the program in Table 9.2; b (bottom) Results

```
xM=[100.234, 1.234, -2.334,-0.004, 10000.1233]
xSe= [0.5,    0.0002, 0.02,    0.001,     0.0004]
for i in range(len(xM)):
    outpl=FinRes("a"+str(i),xM[i],xSe[i]," mg")
    print(outpl)
print("\n")          #Second series
xM=[1234.345,-12345.678,2.383,-0.0991297, 0.000930]
xSe=[ 35.023,     34.56, 0.01,  0.00223046,0.000017]
```

a0=100.2±0.5 mg	a6=1230.0±40.0 mg
a1=1.234±0.0002 mg	a7=-12350.0±30.0 mg
a2=-2.33±0.02 mg	a8=2.38±0.01 mg
a3=-0.004±0.001 mg	a9=-0.099±0.002 mg
a4=10000.1233±0.0004 m	a10=0.00093±2e-05 mg

```
Function FinRes(Name, xM, xSe, Unit)
'Input: Name of the variable, mean, uncertainty, physical unit
n = Int(Log(xSe) / Log(10)) 'power of the uncertainty
xSeR = Round(xSe * 10 ^ -n, 0) 'first digit of the uncertainty
xSeRR = xSeR * 10 ^ n 'rounding to the certain digit
xMr = Round(xM * 10 ^ -n, 0) 'rounding the measured value
xMrr = Round(xMr * 10 ^ n, Abs(n))
FinRes = Name & "=" & xMrr & "±" & xSeRR & Unit
End Function
```

	A	B	C	D	E
2					=FinRes(A4;B4;C4;D4)
3	a0	100.234	0.5	mg	a0=100,2±0,5 mg
4	a1	1.234	0.0002	mg	**a1=1,234±0,0002 mg**
5	a2	-2.334	0.02	mg	a2=-2,33±0,02 mg

1	=FinRes(A4;B4;C4;D4)
2	a0=100,2±0,5 mg
3	**a1=1,234±0,0002 mg**
4	a2=-2,33±0,02 mg
5	a3=-0,004±0,001 mg
6	a4=10000,1233±0,0004 mg
7	a5=-234,6±0,3 mg
8	a6=12350±40 mg
9	a7=-12350±30 mg
	a8=2,38±0,01 mg
	a9=-0,099±0,002 mg
	a10=0,0093±0,00002 mg

Fig. 9.4 a (left, P) The user-defined function *FinRes* (top) reads in the name and the value of a measured quantity (columns A and B in the spreadsheet), as well as its measurement error (column C), and (bottom) outputs the rounded measurement result as text (column E). The spreadsheet has to be treated with a main program that calls *FinRes*. b (right) Ten results of the FUNCTION *FinRes*

9.4 Increasing the Measuring Accuracy Through Repetition

We illustrate the meaning of the standard error with repetition procedures to get the hit rate, i.e., how often the error range captures the true value. The more often a quantity is measured, the more accurately its value can be determined. If the same measurement setup is always used, the standard error of the mean value of the measurement series is inversely proportional to the root of the number of measurements. Ψ *Twice as good with four times the effort.*

9.4.1 Standard Deviation and Standard Error of the Mean Value of a Measurement Series

How can we halve the standard error?

► **Alac** If we want to be twice as good, we just have to measure twice as often.

► **Mag** No, keep the following broom rule in mind and study the next section:
Ψ *Twice as good with four times the effort.*

► **Tim** I remember the reason for this. It is the variance of the mean value of a measurement series that is inversely proportional to the number of measurements. The standard error is the square root of the variance.

► **Mag** One more hint: All mathematically justified theorems make statements about variances. To estimate the measurement error, however, the standard error with the same physical unit as the mean must be quoted:
Ψ *Calculate* with variances, report the *C-spec error*!

In this section, we learn about the quantities: variance var_X of a set of values $x = \{x_i\}$, standard deviation x_{Sd} of the data in the set, and standard error x_{Se} of the mean x_m of the set.

The standard error x_{Se} of the mean value x_m of a set of measured values, used to indicate the measurement result's uncertainty, is smaller than the standard deviation x_{Sd} of the set, measuring the scatter in the set. The standard deviation x_{Sd} is, in principle, independent of the number n of measurements, while the standard error x_{Se} decreases with increasing square root of n.

Visual estimate

Question

How big do you estimate the uncertainty of the mean value in Fig. 9.5a to be?[3]

By what factor is the standard error of the mean smaller than the standard deviation of the set of measurements?[4]

Within which range do you expect a 10th measured point in Fig. 9.5?[5]

Calculated error range

Figure 9.6 displays measured data with calculated means (vertical bar), standard deviations of the set of measured values (horizontal square brackets above), and standard errors of the mean (horizontal square brackets below).

[3] To check your judgement, have a look at Fig. 9.6!

[4] By the factor $1/\sqrt{n}$, where n is the number of measurements, as you will learn in the course of this exercise.

[5] The 10th measured point should be within the standard deviation of about 1 with a probability of about 2/3.

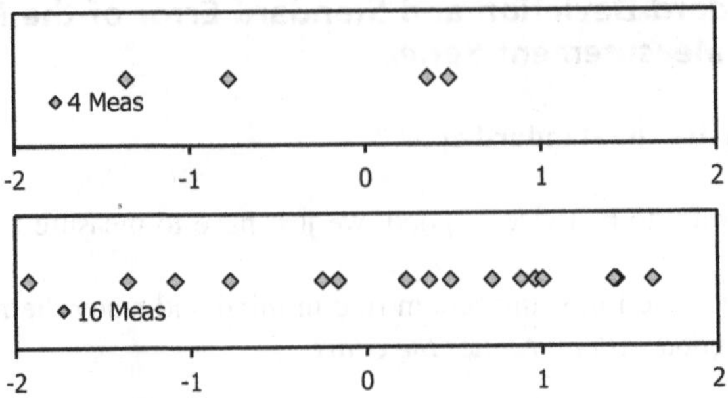

Fig. 9.5 Estimate! Estimate mean value, standard deviation, and standard error of **a** (top) a series of 4 measurements and **b** (bottom) a series of 16 measurements! The standard deviation for both series is about 1

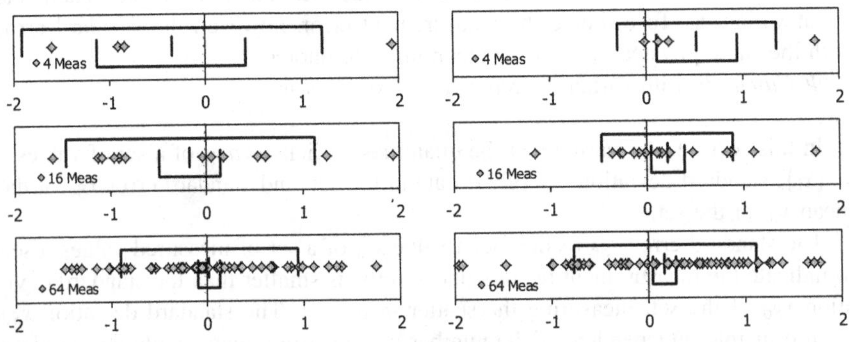

Fig. 9.6 Results for $n = 4$, 16, and 64 measurements of a physical quantity whose true value is 0; mean value x_m indicated by vertical lines, scatter $x_m \pm x_{sd}$ of the measurement series by square brackets above, and confidence interval (error range) $x_m \pm x_{se}$ of the mean value as square brackets below; the true value 0.00 is **a** (left) within the standard error range of the mean value or **b** (right) outside of the standard error range of the mean value

In the analysis and discussion of measurement series, we must clearly distinguish between

- the standard deviation x_{sd} of the data set (Eq. 9.3) and
- the standard error x_{se} of the mean value x_m (Eq. 9.4).

The standard deviation of the series characterizes the scatter of the measured values within the measurement series. A new measurement is expected to lie approximately within this range, "approximately" meaning:

Ψ Two within and one out of (the standard *deviation).*[6]

[6] Namely, with a probability of 2/3 within plus or minus the standard deviation from the mean.

The *mean value* x_m *of all measured values* can be determined by visual judgment more precisely than $\pm x_{Sd}$, as is visible in the diagrams in Figs. 9.5 and 9.6. According to theory, the *standard error* of the mean value x_{Se} is equal to the standard deviation of the series x_{Sd} divided by the square root of the number n of measurements, Eq. 9.4.

The corresponding range

$$[x_m - x_{Se}, x_m + x_{Se}] \tag{9.8}$$

called the standard error range, is indicated in the diagrams by bottom square brackets.

In the next section, we examine the meaning of the error range in more detail by determining hit rates.

▶ **Mag** What is the informative value of the standard error?

▶ **Tim** The standard error has the same physical dimension as the measured value. Therefore, it is possible to specify the measurement uncertainty with plus-minus the standard error.

▶ **Mag** Is the true value always within the standard error?

▶ **Alac** No, only if we've done everything right.

▶ **Tim** I've learned that this is only true in about 68% of cases, to be remembered with:

Ψ Two within and one out of.

▶ **Alac** Is that to say that the true value is outside of the standard error limits for every third experiment? There was never any trouble like this in our introductory lab course.

▶ **Mag** If the true value for nine lab groups lies within the standard error of the measurement $\pm x_{se}$, then three groups have cheated, I suspect.

▶ **Tim** We have also learned that the broom rule applies only if the same quantity is measured at least eight times. We never measured that often in the beginner's lab course. What is the error specification good for, then?

▶ **Mag** We will learn about that in this exercise by determining hit rates. Anyway, if you specify the number n of measurements, together with x_m and x_{Se}, you can make well-defined statements with t statistics, addressed in Exercise 9.5.

4, 16, or 64 measurements of the same quantity
Let the true value of a quantity be 0 and the standard deviation of the measurement process be 1. We simulate this process with a Gaussian generator with a mean value of 0.00 and a standard deviation of 1.00 and determine the mean value x_{m}, standard deviation x_{Sd}, and standard error x_{Se} for measurement series (samples) with different numbers of measurements. The results of the simulation are presented in Fig. 9.6.

In the first row of Fig. 9.6, the physical quantity was measured four times, in the second row, 16 times, and in the third row, 64 times. In the left subpictures, the mean value's error ranges include the true mean; in the right subpictures, they do not.

Standard deviation of the measurement series
The measured standard deviation x_{Sd} is approximately the same for all measurement series, independent of the number of measurements. So, if an additional measurement is made, the expected deviation of this new measurement from the mean value is independent of how often the measurement has been made before. The dispersion of the data set does not depend on the number of data.

How precisely can the mean value of a measurement series be determined?
In Fig. 9.6, two measurement series are plotted for each number of measurements. The mean value fluctuates more strongly for the series with a smaller number (visible here when comparing the left and right subfigures). The theory states that the expected measurement error for the mean value of a measurement series is the standard deviation of this series divided by the root of the number of measurements (Eq. 9.4). The error range (also called the confidence interval) of the mean value is also included in the diagrams mentioned. It corresponds approximately to the range in which we would locate the mean value with the eye.

▶ **Tim** Are we really twice as good with four times the effort?

▶ **Mag** The formula only states that the *standard error* of the mean calculated according to the rules decreases with $1/\sqrt{n}$, narrowing the confidence interval ("error range") correspondingly. We still have to know whether the confidence level ("hit rate") changes as well.

▶
- The standard deviation of a measurement series is in principle independent of the number of measurements; it fluctuates only by chance.
- The standard error of the mean of the measurement series decreases with the inverse of the square root of *n*, the number of measurements.
- The probablity that the standard error range captures the true value depends on the number of measurements and increases with increasing *n* up to 68%.

Hit rate by repetition of the statistical experiment
A simulation repeats the standard error range calculation and determines how often $x_m \pm x_{Se}$ captures the true value. The "hit rates" are:

0.60 (0.61) for 4 measurements,
0.66 (0.67) for 16 measurements,
0.68 (0.68) for 64 measurements,
obtained with `Python` (EXCEL) in the following sections.

9.4.2 Data Structure and Nomenclature

x_{meas}	results of 64 measurements, true value = 0
x_{m4m}	mean value of 4 measurements
x_{Sd4m}	standard deviation of 4 measurements
x_{Se4m}	standard error of the mean of 4 measurements
in_{4m}	true, if the error range captures the true value
within4m	counts the in_{4M}
	and correspondingly for 16 and 64 measurements.
in_M	3 values, $[in_{4m}, in_{16m}, in_{64m}]$.

9.4.3 Python Program

Hit rate by repetition of the statistical experiment
In the diagrams of Fig. 9.6b, the true value of the measured quantity (namely, 0.0) is outside of the standard error range. With a rep-log procedure, we examine how often this happens on average. The function *WithinSe* in Table 9.4 generates an array of 64 normally distributed random numbers, calculates, in lines 4 to 6, the characteristic parameters for the first 4 numbers of this array, and determines, in line 7, whether the standard error range captures the true value (*in4m* is `True` or `False`). The

Table 9.4 Does the standard error range capture the true value? Only the statements for 4 measurements are reported; those for 16 and 64 measurements are the same, but but with 16 or 64 instead of 4

```
1   def WithinSe():
2       x_meas = npr.randn(64)
3
4       xm4m = np.mean(x_meas[0:4])  #Entries 0, 1, 2, 3
5       xSd4m = np.std(x_meas[0:4],ddof=1)
6       xSe4m = xSd4m/np.sqrt(4)
7       in4m = xm4m- xSe4m < 0 < xm4m+xSe4m
                                 #0 is the true value
......
18      return in4m, in16m, in64m
```

Table 9.5 Rep-log procedure for calculating hit rates, *WithinSe* from Table 9.4

```
within4m =0
within16m=0
within64m=0
Ntrials=10000
for i in range(Ntrials):
    inx=WithinSe()
    within4m+=inx[0]
    within16m+=inx[1]
    within64m+=inx[2]
```

```
within4m    6018   hit rate 0.60
within16m   6632   hit rate 0.66
within64m   6759   hit rate 0.68
```

instructions for the first 16 and first 64 measurements are analogous, resulting in *in16m* and *in64m*, which are also returned in line 18.

In the numpy function np.std, a parameter ddof must be specified, meaning *Delta Degrees Of Freedom*. The divisor used in calculating the variance in Eq. 9.2 is *n* - *ddof*, where *n* is the number of data points. By default, *ddof* in np.std is zero. If we determine only the mean of a data series, as we do in this chapter, $ddof = n_P = 1$.

In the main program in Table 9.5, we call *WithinSe* several times (here, $N_{\text{trials}} = 10{,}000$) and increment the variables *within4m* and the like by 1 if the output of *WithinSe* is true. This is possible because the logical values are coded as binary numbers, 1 for *True* and 0 for *False*. We see from the bottom cell output that the hit rate increases with the number of measurements, from 0.60 for 4 measurements to 0.68 for 64.

Theoretically, the confidence interval for $\pm x_{Se}$ is 68.3% when the number of measurements approaches infinity. In other words: in more than 30% of the cases, the "true" value is also theoretically outside $\pm x_{Se}$. We simplify to the broom rule: Ψ *Two within and one out of (the standard error range).* It applies if the measurement is repeated at least 8 times. If measurements are taken less often, the hit rate is lower. For 4 measurements, it is only 60%, according to the bottom cell in Table 9.5.

9.4.4 Spreadsheet Layout for This Task

The spreadsheet layout corresponding to the Python program in Table 9.4 is shown in Fig. 9.7 (S). Columns A to H reproduce the calculation of the Python function *WithinSe* with the result in column H.

The role of the main program in Python is played in EXCEL by the VBA procedure in Fig. 9.8 (P), taking the logical values in H3:H5 as input and, if *True*, incrementing the counts in I3:I5 individually.

	A	B	C	D	E	F	G	H	I	J	K	L
1	=NORM.INV(RAND();0;1)			=AVERAGE(A3:A6)	=STDEV.S(A3:A18)	=xSd_m/SQRT(n)		=AND(xm_m-xSe_m < 0; 0<xm_m+xSe_m)	from rep-log procedure	=within_m/Ntrials		
2	**x_meas**		**n**	**xm_m**	**xSd_m**	**xSe_m**		**in_m**	**within_m**	**hitrate**	**Ntrials**	
3	2.04		4	**0.87**	1.22	0.61	4 Meas	FALSE	6116	0.612	10000	
4	0.30		16	0.12	**0.96**	0.24	16 Meas	TRUE	6738	0.674		
5	1.71		64	0.02	0.91	**0.11**	64 Meas	TRUE	6840	0.684		
66	0.02											

Fig. 9.7 (S) 64 random measurements x_{meas} are generated in column A, while subsets of these are evaluated in columns D to H. The logical values in column H are read by SUB *Protoc3* in Fig. 9.8 (P) and summed in column I after a 10,000-fold generation of new measurement series

```
Sub Protoc3()
Ntrials = Range("K3")
Range("I3:I5").ClearContents
For i = 1 To Ntrials
    Application.Calculation = xlCalculationManual
    If Range("H3") = True Then Range("I3") = Range("I3") + 1
    If Range("H4") = True Then Range("I4") = Range("I4") + 1
    If Range("H5") = True Then Range("I5") = Range("I5") + 1
    Application.Calculation = xlCalculationAutomatic
    Next i
End Sub
```

Fig. 9.8 (P) Rep-log procedure for incrementing the number of hits in I3:I5 of Fig. 9.7 (S); if a value in H3:H5 is True, the corresponding cell in column I is incremented by 1

Questions

concerning Fig. 9.7 (S)

How does the "true" value of the measurand enter the simulation of the measurement process in A3:A66?[7]

What is the formula in H5?[8]

9.4.5 How to Report a Measurement Result

► **Tim** People often say that our rule Ψ *Two within and one out of* only applies if measurements are taken more than seven times. We never measure that often in our lab course, at most, four times.

[7] As mean of a normal distribution in column H generated with NORM.INV (RANDOM(), *Mean*, *StDev*). Here, *Mean* = 0 and *StDev* = 1.

[8] See formula printed in H1; the expression is *True* if the error range does capture the true value. The formula in H5 processes entries in row 5 of *xm_m* and *xSe_m*.

▶ **Alac** That doesn't matter. Even if you measure a quantity only twice, you can apply the formulas for mean value and standard error. That's enough.

▶ **Mag** Not quite. You have to specify the number of measurements in addition to the mean value and the standard error. An example of a complete measurement protocol: The acceleration due to gravity in our laboratory was measured twice with a pendulum, resulting in an average value of 9.8 m/s^2 with a standard error of 0.2 m/s^2.

▶ **Tim** Should we report the details of the experimental setup as well? For example: How exactly we determined the length of the pendulum and the period of oscillation.

▶ **Mag** Of course, to allow for conclusions about systematic errors. The three aforementioned statistical specifications are enough to make statistically relevant statements. E.g., for the two measurements referred to by Alac, the calculated standard error must be doubled to achieve a measurement uncertainty according to Ψ *Two within and one out of*.[9]

9.5 The *t* Statistics Connects Confidence Interval with Confidence Level

The *t*-value of Student's *t* distribution relates an extended confidence interval *t* times the standard error to a confidence level (or an error probability) . For at least eight measurements, our rule for the standard error range Ψ *Two within and one out of* (the standard error range) applies.

9.5.1 Student's *t* Distribution

Practical handling
The following list explains the terms that play a role in *t* statistics:

- *Standard error range*$x_m \pm x_{Se}$ with the standard error calculated from the standard deviation of the measurement series x_{Sd} as $x_{Se} = \frac{x_{Sd}}{\sqrt{n}}$.
- *Degree of freedom* $dof = n - n_P$ with n being the number of measurements, and n_P the number of parameters estimated from the measurement series. In this chapter, $n_P = 1$, because only the mean x_m is estimated.

[9] The degree of freedom for 2 measurements is $dof = 1$ and the *t value* for this is $t = 1.84$ for a probability of error of 38% (see Exercise 9.5). For an error probability of 5%, the *t value* is 12.7.

- *C-spec error range* $x_m \pm t_C \cdot x_{Se}$, standard error multiplied by a factor t_C determined by a pre-specified confidence level C or error probability E and the degree of freedom dof.
- *Confidence level* $C(t, dof)$, corresponds to the hit rate in our statistical experiments. The error probability has the complementary value: $E(t, dof) = 1 - C(t, dof)$.

Error probability, confidence level, and t-value are obtained with the following functions, with *dof* being the degree of freedom:

EXCEL E(T,DOF) = T.DIST.2T(T;DOF)
 C(T,DOF)= 1- T.DIST.2T(T;DOF)
 T(E,DOF) = T.INV.2T(E,DOF)

```
Python  E(t,dof) = (1-sct.t.cdf(t,df=dof))*2
        C(t,dof) = 2*sct.t.cdf(t,df=dof)-1
        T(E,dof) = sct.t.ppf(1-E/2,df=dof)
```

In Fig. 9.9a, the error probability E is obtained for different degrees of freedom f and t values of 1 and 1.96. The t value for pre-specified E and *dof* is given in **b** (F:H).

For 4 measurements, the degree of freedom is 3, and a hit rate 1–0.391 ≈ 0.61 for $t = 1$ is expected. Actually, in Sect. 9.4.1, hit rates of 0.60 for the Python simulation and 0.61 for the EXCEL one come close to this expectation.

Mathematics of the *t* distribution

The entity t is defined as

$$t = (x_m - x_{True})/x_{Se} \tag{9.9}$$

	A	B	C	D	E	F	G	H	I
1				=(E-1/3)*3				=T.INV.2T(Et;dofT)	
2	dof	E				Et	dofT	t	
3	100000	0.317	=T.DIST.2T(1;dof)	-0.05		0.317	100000	1.00	
4	7	0.351	=T.DIST.2T(1;dof)	0.05		0.317	7	1.08	
5	3	0.391	=T.DIST.2T(1;dof)	0.17		0.317	3	1.20	
6	100000	0.050	=T.DIST.2T(1.96;dof)			0.05	100000	1.96	
7	7	0.091	=T.DIST.2T(1.96;dof)			0.05	7	2.36	
8	3	0.145	=T.DIST.2T(1.96;dof)			0.05	3	3.18	

Fig. 9.9 **a** (left, A:D) Error probability E for t values 1 and 1.96, each for various degrees of freedom *dof*. **b** (right, F:H) t values for E_t values of 0.317 and 0.05, as well as various degrees of freedom dof_T

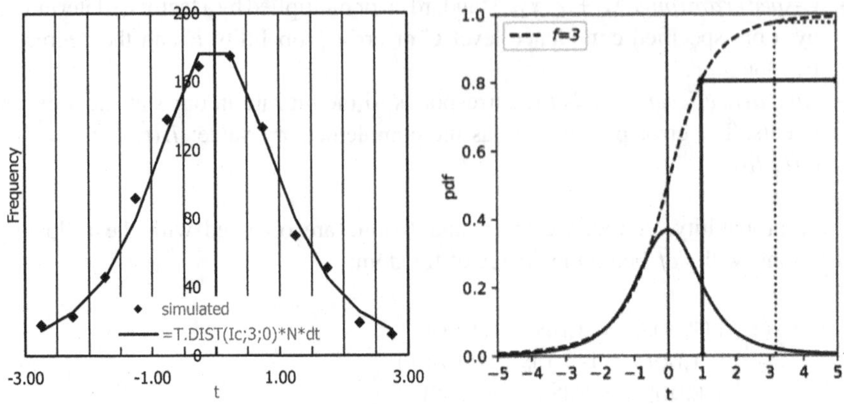

Fig. 9.10 **a** (left) Frequency of occurrence of t values obtained in a series of simulations with 4 measurements together with the theoretically predicted values for *dof* = 3; **b** (right) *ppf* and *cdf* (dashed line) of Student's t distribution for *dof* = 3 (*dof* is named *f* in the legend); the right-angled straight lines indicate `t.cdf(1,dof)` and `t.ppf (0.975,dof)` (dotted line)

describing the distance of the sample mean x_{m} to the true value x_{True}, divided by the standard error x_{Se} of the mean value.

In order to get an idea of t statistics, we:

- perform a simulation of a series of 4 measurements with a noise level between 0 and 1 from which a single value of t can be obtained,
- repeat the single experiment to get a representative set of t values, with each time a new noise level,
- and determine their frequencies of occurrence.

The result (diamonds) is shown in Fig. 9.10a together with the values predicted by the theoretical t distribution for *dof* = 3 (line).

Figure 9.10b displays the *ppf* and *cdf* of Student's t distribution for *dof* = 3; the right-angled straight lines indicate `t.cdf(1,dof)` starting at $t = 1$ and `t.ppf (0.975,dof)` (dotted line, starting at *cdf* = 0.975). We get.

$$
\begin{aligned}
&\texttt{sct.t.cdf(1,3)} = \texttt{0.8045}\\
\texttt{C(1,3)} \quad &= 2 * \texttt{sct.t.cdf(1,3)} - 1 = \texttt{0.609}\\
\texttt{T(0.05,3)} &= \texttt{sct.t.ppf(\{0.975,3\})} = \texttt{3.18\}}
\end{aligned}
$$

We see that for $t = 1$, the *cdf* is about 0.8, i.e., in about 80% of the experiments a measurement result is below the upper limit of the error range; 20% are above. As the *pdf* is symmetric to $t = 0$, likewise 20% are below the lower limit of the error range so that the error probability is $E = 0.4$ and correspondingly *C(1, 3)* $\approx$ 0.6.

For *dof* = 1,000,000, we get

`sct.t.cdf(1,1,000,000) = 0.841,`
$p = 0.16$ that $x_{True} > x_m + 1 \cdot x_{Se}$.
`C(1, 1,000,000) = 2* sct.t.cdf(1,1,000,000) - 1 = 0.683.`
`E(1, 1,000,000) = (1-0.683) = 0.317,`
$p = 0.317$ that x_{True} is outside of the error range $x_m \pm 1 \cdot x_{Se}$.
`T(0.05, 1,000,000) = Sct.t.ppf (0.975,1,000,000) = 1.96.`
a confidence level of $C = 0.95$ (error probability $0.05 = 2{\cdot}0.025$) is obtained for an error range $x_m \pm 1.96 \cdot x_{Se}$.

Set-up of the simulations to determine C-spec errors

We are simulating measurement series according to Ψ *We know everything and play stupid* by generating normally distributed random numbers with mean = 0 and standard deviation = 1, and determining the experimental hit rate, for series comprising 2 to 16 measurements, within error ranges with *t* values that correspond to error probabilities 0.317 and 0.05.

Proven or not disproven? 8 = ∞?

► **Tim** Summarizing: with the mean value and its standard deviation, and the number of measurements, statistically correct statements with confidence levels can be made.

► **Alac** Our simulation proves that.

► **Tim** No, our simulation simply gives us no reason to doubt that statement.

► **Alac** Your nitpicking's a pain in the neck.

► **Mag** But it is indispensable, because we don't do logical derivations.

► **Tim** Let's trust in statistical textbooks.

► **Alac** In the case of measurement series, we are content with a rough estimate: infinity already starts with 8 measurements.

► **Tim** 8 = ∞? Isn't that twisting the facts?

► **Mag** Discuss!

Questions

concerning Fig. 9.9 (S) By what factor t should the standard deviation of 4 measurements be increased so that the true value is within the C-spec error range with a probability of 68.3%?[10]

What is the probability of error if, for eight measurements, the standard error is specified as the measurement uncertainty?[11]

What is the error probability for four measurements if the standard error times 1.96 is used to specify the error range?[12]

9.5.2 Data Structure and Nomenclature

x_{True}	True value of a measurand
n	Number of measurements in a series
x	Series of n measurements
x_m	Mean value of the measurement series
x_{Se}	Standard error of x_m
dof	Degrees of freedom, $dof = n - 1$
C	Confidence level
E	Error probability, $E = 1 - C$
t	t-value (Student's) for the C-spec error range, determined by C and *dof*
x_{SeT}	x_{Se} multiplied with a t-value appropriate for C and E
n_T	Number of trials (repetitions of a statistical experiment)
hitRate	Counts how often the error range captures the true value
p_{Out}	(n_T-*hitRate*/n_T, probability that the error range misses the true value.

9.5.3 Spreadsheet Calculation

In column A of Fig. 9.11 (S), 32 normally distributed random numbers (mean = 0, standard deviation = 1) are generated. Subsets of 2 to 32 of them are specified by cell range addresses in column C. Indirect addressing of these ranges is used to calculate their mean (in column D) and their C-spec error for $E = 0.317$ (in column F).

Questions

concerning Fig. 9.11 (S)

[10] Figure 9.9 (S), the degree of freedom for four measurements is $dof = 3$, $t = 1.20$ in cell C5.
[11] Figure 9.9 (S), $dof = 7$, B4 = 35%.
[12] Figure 9.9 (S), $dof = 3$, B8 = 14.5%.

	A	B	C	D	E	F	G	H	I	J	K
1	=NORM.INV(RAND();0;1)		="A3:A"&B4+2	=AVERAGE(INDIRECT(Rng))	=T.INV.2T(0.317;n_-1)	=STDEV.S(INDIRECT(Rng))/SQRT(n_)*t_	=OR(xm<-xSeT;xm>xSeT)	from Rep-Log procedur			Rep-Log
2	**x**	**n_**	**Rng**	**xm**	**t_**	**xSeT**	**outSeT**	**pOut**	100,000	**t0.05**	**pOut0.05**
3	-0.44	2	A3:A4	0.34	1.84	1.45	FALSE	0.317		12.7	0.049
4	1.13	4	**A3:A6**	0.73	1.20	0.60	TRUE	0.317		3.2	0.050
5	0.33	8	A3:A10	-0.19	1.08	0.58	FALSE	0.317		2.4	0.049
6	1.88	16	A3:A18	-0.02	1.04	0.30	FALSE	0.314		2.1	0.050
7	-2.97	32	A3:A34	-0.11	1.02	0.20	FALSE	0.318		2.0	0.049
34	0.35										

Fig. 9.11 (S) Evaluation of series from column A with 2 to 32 measurements; in column F, C-spec error ranges are listed for an error probability of $E = 0.317$. The formula in column G checks whether the true value (here, 0) lies outside of the error limits. In column H, the experimental error rate is recorded after 100,000 repetitions of the statistical spreadsheet experiment (*Multiple tests for hit rates*) . The simulation is repeated for an error probability $E = p_{\mathrm{Out}}$ of 0.05 (t values in column J), with the results reported in column K

Which numbers addressed in D6 are averaged?[13]

How is the argument of the spreadsheet function AVERAGE in column D constructed?[14]

Which cells does the statement in F4 refer to?[15]

In column E, the spreadsheet function t = T.INV.2T(E;dof) reports the t value for an error probability of $E = 0.317$. The extension ".2T" indicates that deviations of x_{m} to both sides of the error range are taken into account.

In column F, the mean value's standard error is multiplied by t, so that the error rate is expected to be $p_{\mathrm{Out}} = E = 0.317$. Column G contains the logical expression [=OR(x_{m} < -x_{SeT}; xm > x_{SeT})], true if x_{mTrue}, namely, 0, is outside of the error range. In the example of Fig. 9.11 (S), this is the case (by chance) only for one measurement series. With a rep-log procedure, we perform the random experiment 100,000 times and get the error rates in column H. They are actually close to the expected 0.317.

In column J, t values that correspond to an expected error probability of $E =$ 0.05 are calculated. Repeating the simulation with the corresponding C-spec error range leads to experimental error probabilities, column K, close to the theoretical value.

[13] The mean value of the numbers in A3:A18 is calculated.

[14] The argument of AVERAGE is defined using INDIRECT from the column range with name *Rng*. From *Rng*, the value in the same row is taken.

[15] The statement in F4 (1) calculates the standard deviation of the subset A3:A6 (addressed with INDIRECT(RNG) , *Rng* in C4), (2) divides the result by $\sqrt{n}$ (B4) to get the standard error, (3) multiplies the standard error with a t (E4) for $E = 0.317$ and $dof = n - 1 = 3$, appropriate for $n =$ 4).

Table 9.6 a (top) Specification of arrays for taking up the results of simulations with *t* statistics; b (bottom) A single run of the simulation

```
1   nList=[2,4,8,16,32]
2   lnL=len(nList)
3   E=0.317
4   t=np.empty(lnL)              #t values for E
5   for n in range(lnL):
6       t[n]=sct.t.ppf(1-E/2,df=nList[n]-1)
7   out=np.zeros(lnL)
```

```
8   x=npr.randn(32)
9   xm=np.array([np.average(x[0:n]) for n in nList])
10  xSeT= np.array([np.std(x[0:n],ddof=1)/np.sqrt(n)
                                  for n in nList])*t
11  outSeT=np.logical_or((xm<-xSeT),(xm>xSeT))
12  out+=outset                   #Boolean added as 0 or 1
```

Table 9.7 Results of a run of the program in Table 9.6b

```
E         0.317
t         [1.839 1.198 1.077 1.035 1.017]
xm        [-1.026 -0.871 -0.279 -0.079 -0.04 ]
xSeT      [1.784 0.547 0.347 0.277 0.196]
outSeT    [False  True False False False]
out       [0.     1.    0.    0.    0.]
```

9.5.4 Python Program

Single experiment

The program for checking the consistency of confidence intervals and confidence levels and its result is distributed over three cells. In the first cell of Table 9.6, three lists of the same length (*nList, t, out*) are generated, with their elements specific to the 5 series with different numbers of measurements. In the second cell, the simulation is run once; all logged lists are reported in Table 9.7 A numerical array *out* is created to sum the logical values of *outSeT* as numbers 0 or 1. It counts the number of missed hits, i.e., when the true value is outside of the extended error range when the simulation is repeated.

Questions

Why does *t* have to be an array? Why is it not sufficient to specify it as a list?[16]

Replace lines 4 to 6 of Table 9.6a with a list comprehension![17]

Which columns of Fig. 9.11 (S) are equivalent to Table 9.6b?[18]

---

[16] Line 10 of Table 9.6 multiplies the array of standard errors with *t*, element-wise. This must be a numerical multiplication. Therefore, *t* has to be an array.

[17] `t = np.array ([sct.t.ppf(1-E/2,df = n-1) for n in nList])`

[18] Columns A, D, F, G; set *x* in A, x_m in D, x_{SeT} in F, *outSeT* in G.

Table 9.8 Multiple repetition of the simulation experiment; **a** (top cell) program code; **b** (middle cell) Table 9.6b changed into a function; and **c** (bottom cell) results for t values theoretically valid for an error probability of $E = 0.317$ (**b**) and $E = 0.05$ (**c**)

```
nOut=np.zeros(lnL)
nT=10000                                  #Number of trials
for m in range(nT): nOut+=Check_if_out()
np.set_printoptions(precision=3)
```

```
def Check_if_out():
    x=npr.randn(nList[-1])
    xm=np.array([np.average(x[0:n]) for n in nList])
    xSeT= np.array([np.std(x[0:n],ddof=1)/np.sqrt(n) \
        for n in nList])*t     #nList and t from table above
    outSeT=np.logical_or((xm<-xSeT),(xm>xSeT))
    return outSeT
```

```
      E   0.317    nT    100000
nOut/nT  [0.318 0.316 0.319 0.319 0.319]
```

```
      E   0.05    nT   100000
nOut/nT  [0.0501 0.05   0.05   0.0493 0.0493]
```

How do you have to change the logical query in line 11 of Table 9.6b if you want to count the hits instead of the misses?[19]

Error rates

The statistical experiment is multiply repeated in Table 9.8, in a for-loop calling the function *Check_if_out()* and summing up its logical output in a numerical array n_{Out}. The statistical experiment is performed twice, first with $E = 0.317$ in line 3 of Table 9.6a (results in Table 9.8b) and second with $E = 0.05$ (results reported in Table 9.8c). The experimentally found miss rates $p_{\text{Out}} = n_{\text{Out}}/n_{\text{T}}$ are close to the expectations, $E = 0.317$ and $E = 0.05$.

Questions

Which lines of Table 9.6 have to be integrated into the function *Check_if_out*? What variable has to be returned?[20]

9.6 Combining Results from Several Measurement Series

A combined result of two measurement series is obtained as a weighted average of the individual results with the squares of the reciprocal C-spec errors

---

[19] `inSeT = np.logical_and(-xSeT < xm, xm < xSeT)`; the true value of x_m is 0.

[20] Lines 8 to 11 with `return outSeT`. The EXCEL equivalent is a rep-log procedure that calculates p_{Out}.

as weights. The C-spec errors are derived from internally and externally consistent variances.

Ψ *Worse makes good even better. Mostly, but not always.*

9.6.1 Combining Two Measurement Results

Calculating with variances

▶ **Mag** Two research groups use different methods, A and B, to determine the value of the same measurand. They report their non-identical results with different standard errors. For example, the acceleration due to gravity at a particular location has been determined by a drop test and a pendulum. As the head of the two groups, which result do you write in the project's final report?

▶ **Alac** Clearly the one with the smaller standard error.

▶ **Tim** I simply report both results, each with its own number of measurements and standard errors. The readers can then decide for themselves.

▶ **Mag** Neither answer is correct. As a supervisor, you should know how to combine both measurements and specify a value for the measurand whose standard error is even smaller than that for the better of the two measurements.

▶ **Tim** Alright, I have found a formula in the internet. The final result x_{mAB} is a weighted mean of the two results x_{mA} and x_{mB}, with the *reciprocals of the variances* var_{A} *and* var_{B} *of the measurement series* as weights w, Eq. 9.10.

$$\mathrm{x_{mAB}} = \frac{\mathrm{w_A} \cdot \mathrm{x_m(A)} + \mathrm{w_B} \cdot \mathrm{x_m(B)}}{\mathrm{w_A} + \mathrm{w_B}}$$
$$w_A = \frac{1}{var_A} \quad w_B = \frac{1}{var_B} \tag{9.10}$$

Furthermore, the reciprocal value of the combined results' variance is calculated as the sum of the *reciprocals of the individual variances*:

$$var_{\mathrm{AB}}^{-1} = var_A^{-1} + var_B^{-1} = w_A + w_B$$
$$x_{Sd}(AB) = \sqrt{w_A + w_B} \tag{9.11}$$

▶ **Alac** Equation 9.3 is another neat formula, plus, it's trustworthy, because it fits our rule Ψ *Calculate with variances, report the standard error.*

▶ **Tim** Do we have to take *Student*'s *t*-value into account? That's the *t*-question. My internet resource does not deal with it.

► **Alac** Let's check it with *Ψ Two within and one out of*.

► **Mag** Good idea. By the way, Eq. 9.11 is called the *internally consistent* variance. The greater the variance, the lower the measurement's weight in the combined result.

Questions

Due to Eq. 9.11, will the combined result's variance be greater or smaller than the variance of the best measurement?[21]

Provided the noise is the same for two measurement series with $n = 4$ and $n = 16$, by what factor should the *standard errors* of their mean values differ?[22]

Calculation with C-spec errors

Equation 9.11 makes a statement about the variances of measurement series. We are, however, interested in confidence intervals and know already that the standard error is:

$$x_{Se} = \frac{x_{sd}}{\sqrt{n}} \tag{9.12}$$

and the C-spec error, with the t value obtained from the confidence level C and the degrees of freedom dof is:

$$x_{Ce} = x_{Se} \cdot t(C, dof) \tag{9.13}$$

Furthermore, we recall our broom rule:
Ψ Calculate with variances, report the C-spec error!
Consequently, we use in the formulas for mean and C-spec error the weights:

$$weight\ w = \frac{1}{x_{Ce}^2} \tag{9.14}$$

and check the results with hit rates.

Simulation

We set up two independent measurement series for the same measurand, one, A, with 16 individual measurements and the other one, B, with 4 individual measurements, so that they exhibit different variances. They are combined into a third series C of size 20. Furthermore, the respective results of A and B are combined into AB using the method described above.

[21] The reciprocals of the variances are added; the variance of the overall result becomes smaller.
[22] The standard errors of the mean values should differ by a factor of two (the ratio of the square roots of the numbers of measurements).

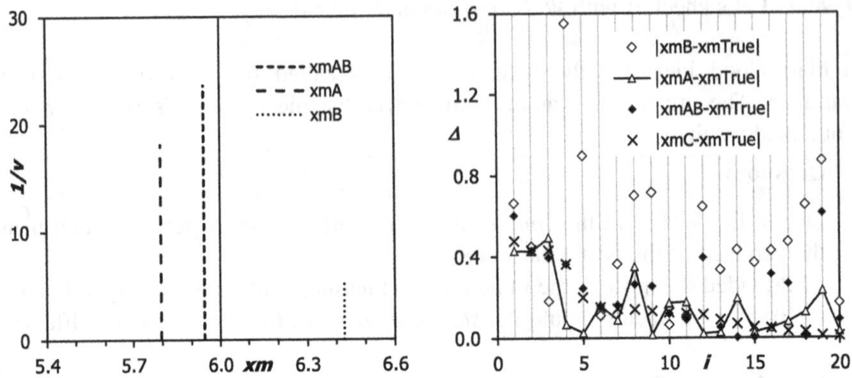

Fig. 9.12 **a** (left) A result for one measurement series each for A, B, and AB; the length of the vertical lines indicates the respective weight in the weighted sum; the true value is $x_m = 6$. **b** (right) Deviation of the mean values of the measurement series A, B, C, and AB from the true value, ordered according to the deviation $|x_{mC}-x_{mTrue}|$

We calculate the mean values x_{mA}, x_{mB}, and x_{mC} together with the squares of the C-spec errors according to Sect. 9.5.1. The combined result x_{mAB} is calculated as the weighted mean according to Eq. 9.10, together with its C-spec error according to Eq. 9.13, both with the weights in Eq. 9.14. The results for one random experiment are graphically shown in Fig. 9.12a. The "true" value of the measurand is 6. If the random experiment is repeated, different positions and heights of the vertical lines result.

The positions of the vertical lines in Fig. 9.12a correspond to the mean value x_m of the measurement series, the height to $1/x_{mCe*}{}^2$. In this example, the value of the combined measurements, x_{mAB}, is closest to the true value of 6.0.

We repeat the random experiment 20 times with a rep-log procedure and note the deviations of the four mean values for A, B, C, and AB from the true value; the result is shown in Fig. 9.12b. The experiments are ordered according to the outcome of experiment C, the direct evaluation of 20 measurements. The value for the series with four measurements (B) is usually farther away from, but sometimes closer to, the true value than the results of the series with 16 measurements (A).

Ψ *Mostly, but not always!* Fundamental rule of statistical reasoning:-).

The weighted mean values of (A) are joined with a solid line. We see that they are sometimes better and sometimes worse than those for the combined series (AB).

► **Tim** So, it would sometimes be better to report the result of experiment (A) instead of the combined result?

► **Mag** Keep in mind that we perform simulations according to Ψ *We know everything and play stupid*. In real life, the true value is unknown, and we have to be content with Ψ *Decide! Sometimes it will be wrong*. Let's have a look at the hit rates from the rep-log procedures in Table 9.9. and Fig. 9.14 (S).

Table 9.9 Internally (left two cells) and externally (right two cells) consistent evaluation of the measurement series; the first line corresponds to a variant of line 8 in Table 9.11

vABmax=vABint	vABmax=vABint	vABmax=vABext	vABmax=vABext
Theor. 0.683	Theor. 0.950	Theor. 0.683	Theor. 0.950
Hits A 0.682	Hits A 0.950	Hits A 0.682	Hits A 0.952
Hits B 0.684	Hits B 0.950	Hits B 0.681	Hits B 0.951
Hits C 0.683	Hits C 0.950	Hits C 0.681	Hits C 0.951
HitsAB 0.633	HitsAB 0.932	HitsAB 0.448	HitsAB 0.406

▶ **Alac** The hit rates of A, B, and C are close to the theoretical value of 0.683 and 0.950, but AB's hit rates are only about 0.63 and 0.93; they do not meet the expectation.

▶ **Tim** Therefore, the estimation of the standard error according to Eq. 9.11, called "internally consistent", must be wrong.

▶ **Mag** It is not wrong but does not fully satisfy our requirements. Before drawing conclusions, let's evaluate another formula to estimate a combined mean's variance called "externally consistent". Take a look at Eq. 9.15.

▶ **Tim** But that's even worse. Look at the right two cells in Table 9.9 with *HitsAB* = 0.406.

▶ **Mag** That's right. In the end, we have to calculate both variances and choose the bigger one to avoid too small an error range. In doing so, we yield hit rates *HitsAB* = 0.667 and 0.935, slightly better than those reported when taking the internally consistent variances alone.

Externally consistent variance

The *externally consistent variance* is calculated with the mean values x_{mA} and x_{mB} of series A and B and the combined mean value x_{mAB}:

$$v_{ABext} = \frac{w_A \cdot (x_m A - x_m AB)^2 + w_B \cdot (x_m B - x_m AB)^2}{(M-1) \cdot (w_A + w_B)} \tag{9.15}$$

with M being the number of measurement series whose results are to be combined, here, $M = 2$, and w_{A}, w_{B} being the weights defined in Eq. 9.14.

9.6.2 Data Structure and Nomenclature

A series A, N_{A} values
B series B, N_{B} values

C	A and B together as one series
AB	results of *A* and *B* combined into one result
x_{True}	true value of the measurand
$x_{\mathrm{mA}}, x_{\mathrm{mB}}, x_{\mathrm{mC}}$	mean values of series A and B
$x_{\mathrm{CeA}}, x_{\mathrm{CeB}}, x_{\mathrm{CeC}}$	C-spec errors errors of x_{mA}, x_{mB}, and x_{mC}
$w_{\mathrm{A}}, w_{\mathrm{B}}$	weights of series x_{mA} and x_{mB}, Eq. 9.14
x_{mAB}	mean value of x_{mA} and x_{mB} combined
v_{ABint}	internal variance of the combined result
v_{ABext}	external variance of the combined result.

9.6.3 Spreadsheet Calculation

In Fig. 9.13 (S), two data sets A and B are generated, comprising, respectively, 16 and 4 normally distributed random numbers and the union C = A|B (see cell E3) with mean x_{mTRUE} and standard deviation x_{Ns}. They are evaluated for mean values x_{m}, and C-spec errors x_{Ce}, here for an error probability $E = 0.05$.

	A	B	C	D	E	F	G	H
1	**xmTrue**	6	**xmA**	6.06	*=AVERAGE(A)*	**E**	0.05	
2	**xNs**	1	**xmB**	5.69	*=AVERAGE(B)*	**tA**	2.13	*=T.INV.2T(E;15)*
3			**xmC**	5.99	*=AVERAGE(A;B)*	**tB**	3.18	*=T.INV.2T(E;3)*
4						**tC**	2.09	*=T.INV.2T(E;19)*
5	*=NORM.INV(RAND();xAtrue;xNs)*					**xCeA**	0.55	*=STDEV.S(A)/SQRT(16)*tA*
6	**A**		**B**			**xCeB**	0.75	*=STDEV.S(B)/SQRT(4)*tB*
7	6.42		5.15			**xCeC**	0.44	*=STDEV.S(A;B)/SQRT(20)*tC*
22	4.98							

Fig. 9.13 (S) Sets A and B, normally distributed random numbers; evaluation of the data sets and C = A|B for mean value (C1:D3), as well as the C-spec error of the mean (F5:G7) for *E* in G1

	I	J	K	L	M	N	O	P
1	**wA**	3.55	*=1/xCeA^2*	**xmAB**	6.01	*=(xmA*wA+xmB*wB)/(wA+wB)*		
2	**wB**	0.51	*=1/xCeB^2*	**vAB**	0.25	*=(wA+wB)^-1*		
3	**wC**	4.62	*=1/xCeC^2*	**xsAB**	0.50	*=SQRT((wA+wB)^-1)*		
4							100000	
5				**hitA**	TRUE	*=AND(xmA-xCeA<xmTrue;xmTrue<xmA+xCeA)*	0.941	
6				**hitB**	TRUE	*=AND(xmB-xCeB<xmTrue;xmTrue<xmB+xCeB)*	0.950	
7				**hitC**	TRUE	*=AND(xmC-xCeC<xmTrue;xmTrue<xmC+xCeC)*	0.944	
8				**hitAB**	TRUE	*=AND(xmAB-xsAB<xmTrue;xmTrue<xmAB+xsAB)*	0.925	
9							**0.950**	*=1-E*

Fig. 9.14 (S) In O5:O8 are the hit rates of the experiments with the four sets A, B, C, and AB for 100,000 repetitions. The value expected from statistical theory is given in O9

Table 9.10 Defining four measurement series A, B, C, and AB; calculating their means, variances, and C-spec errors; t_A, t_B, t_C are global variables from the main program in Table 9.12; continued in Table 9.11

```
import numpy.random as npr

def measurements():
    A=xmTrue+xNs*npr.randn(16)
    B=xmTrue+xNs*npr.randn(4)
    C=np.concatenate([A,B])  #Not independent of A and B
    xmA=np.mean(A)
    xmB=np.mean(B)
    xmC=np.mean(C)
     #C-spec errors by considering t:
    xCeA=np.std(A,ddof=1)/np.sqrt(len(A))*tA
    xCeB=np.std(B,ddof=1)/np.sqrt(len(B))*tB
    xCeC=np.std(C,ddof=1)/np.sqrt(len(C))*tC
    hitA=(xmA-xCeA)<xmTrue<(xmA+xCeA)
    hitB=(xmB-xCeB)<xmTrue<(xmB+xCeB)
    hitC=(xmC-xCeC)<xmTrue<(xmC+xCeC)
```

Question

Which two of the three sets A, B, and C are pairwise independent?[23]

In Fig. 9.14 (S), first, the results of A and B are combined into one final result x_{mAB} with a calculated error x_{sAB}. Then, the statistical experiment is repeated 100,000 times with a rep-log procedure to count how often the error ranges $x_m \pm x_{Ce}$ capture the true value for the sets A, B, C, and AB. The theoretical rate is 0.950, which is closely reached for A, B, C, but with ≈ 0.925 not for AB. We conclude that the internally consistent variance is not sufficient to estimate the error range. Therefore, in the following section with Python programs, the alternative with externally-consistent error is also considered.

9.6.4 Python, Internally and Externally Consistent Error of the Combined Result

The Python function in Tables 9.10 and 9.11 simulates the measurement process. In the function in Table 9.10, the data sets A, B, and C are generated and evaluated for mean value x_m, variance v, and C-spec error x_{Ce} of the mean.

Table 9.11 continues the function *measurements* begun in Table 9.10. The combined result x_{mAB}, x_{sAB} is built, where x_{sAB} is obtained as the maximum of the

[23] *A* and *B* are independent of each other, because their members are generated in two different ranges. *C*, *A*, and *C*, *B* are not independent, because *C* is the union of *A* and *B*.

Table 9.11 Continuation of Table 9.11; line 22 is varied according to which variance is to be calculated

```
16        wA=(xCeA)**-2                   #Weight
17        wB=(xCeB)**-2
18        xmAB=(xmA*wA+xmB*wB)/(wA+wB)
19        vABint=(wA+wB)**-1              #Variances combined
20        vABext=(wA*(xmA-xmAB)**2
21                   +wB*(xmB-xmAB)**2)/(wA+wB)
22        vABmax=max(vABint,vABext)   #=vABint
23        xsAB=np.sqrt(vABmax)
           #Confid. level not yet clear. Hit rate to be determ.:
24        hitAB=(xmAB-xsAB)<xmTrue<(xmAB+xsAB)
25        return hitA,  hitB,  hitC, hitAB
```

```
hits              (True, False, True, True)
```

Table 9.12 Main program calling the function measurements in a for-loop. The values in the right cell are for the current specification E = 0.05 and for another run with E = 0.317

```
import scipy.stats as sct
E=0.05
tA =sct.t.ppf(1-E/2,df=15)
tB =sct.t.ppf(1-E/2,df=3)
tC =sct.t.ppf(1-E/2,df=19)

xmTrue=6     #True mass
xNs=1
Nrep=10000
hitA,hitB,hitC,hitAB=0,0,0,0
for rep in range(Nrep):
    hits=measurements()
    hitA+=hits[0]
    hitB+=hits[1]
    hitC+=hits[2]
    hitAB+=hits[3]
```

```
theor.   0.950
Hits A   0.951
Hits B   0.951
Hits C   0.948
Hits AB  0.936

theor.   0.683
Hits A   0.677
Hits B   0.682
Hits C   0.683
Hits AB  0.688
```

internally and externally consistent variances. Line 24 checks whether the true value x_{mTrue} is captured by $x_{mAB} \pm x_{sAB}$, and the four Boolean values are returned.

In the main program in Table 9.12, the statistical experiment is repeated $N_{rep} =$ 10,000 times for a specified error probability E, and the hit rates are determined. The results are shown in the right cell. The hit rate for AB with 0.936 is not much closer to the theoretical value of 0.950 than the 0.925 reported in Fig. 9.14 (S) for the internally consistent variance.

9.7 Propagation of Standard Deviations

You cannot prevent errors from propagating. It is in their nature. Learn to live with it! We simulate the propagation of variances and standard deviations in sums, products, and powers with statistical experiments on sets with 100 elements. Note: For products, the relative variances add up; for sums, the absolute variances. The propagation of confidence intervals is treated in Exercise 9.8.

9.7.1 Rules for Propagation of Standard Deviations

General Rule
The final result z of a measurement series is often a function of one or more measured physical quantities x: $z = f(x_1, x_2, \ldots)$. The standard theory of error propagation considers the x_i as random variables, normally distributed around the "true" value, with the result:

► The variance of the final result is a weighted sum of the variances $var(x_i)$ of the measured values of the individual variables x_i. Remember: It is the variances that propagate through the formulas to the final result. But, ultimately, the C-spec error has to be reported together with the estimate of the mean.

For two variables, x and y, the equation is

$$var_z = w_x var_x + w_y var_y \text{ with } w_x = \left(\frac{\partial z}{\partial x}\right)^2 \text{ and } w_y = \left(\frac{\partial z}{\partial y}\right)^2 \tag{9.16}$$

which can straightforwardly be extended for more variables. The variables v denote the empirical variance of the data series, Eq. 9.2. The weights w in the sum are the squares of the derivatives of z with respect to the associated variables; the greater the slope, the greater the weight.

As an example, we consider the calculation of the volume V_K of a sphere from its diameter d_K: $V_K = \frac{\pi}{6}d_K^3$

$$\text{var}(V_K) = \left(\frac{\pi}{2}d_K^2\right)^2 \cdot \text{var}(d_K) \tag{9.17}$$

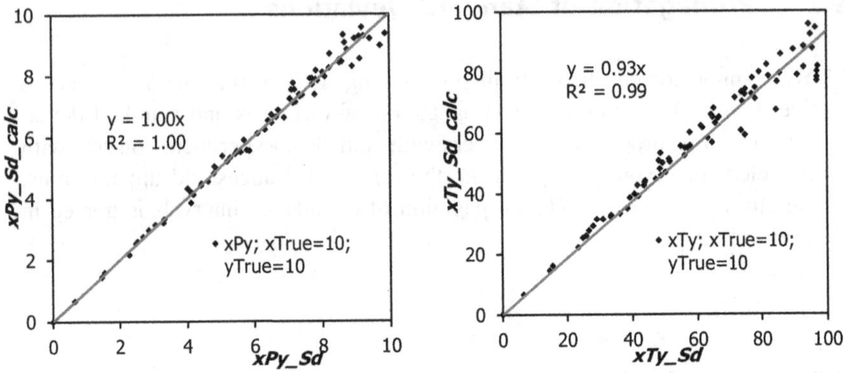

Fig. 9.15 **a** (left) Standard deviation of the sum $xPy = x + y$, calculated versus empirical values. **b** (right) Standard deviation of the product $xTy = x \cdot y$, calculated versus empirical values. The standard deviation is up to ten times greater than the mean value. X_{True}, y_{True} are the true values

Question

What are the physical units on both sides of Eq. 9.17?[24]

In the following sections, we examine the propagation of the standard deviations of sets of 100 measurement points into the final result.

Error propagation in sums
For a sum $z = x + y$, the weights in Eq. 9.16 are $w_x = 1$ and $w_y = 1$ so that $var_z = var_x + var_y$. For a difference $z = x - y$, the coefficients are the same because the partial derivatives are squared so that we get the rule:

▶ The variance of a sum or difference is the sum of the variances of the summands.

Ψ Calculate with variances, report the *C-spec error!*

For the data in Fig. 9.15a, two data sets *x* and *y* with various standard deviations for a normally distributed noise are added element-wise to get *xPy* (*x* Plus *y*). The standard deviations of the set of the sums, as derived from the standard deviations of *x* and *y*, are plotted against the empirically determined standard deviations *xPy_Sd* of the set *xPy*. The trend line has a slope 1.00, indicating accordance with the theoretical expectation. In general, that cannot be taken for granted, as

[24] The dimensional analysis $m^6 = m^4 \cdot m^2$ shows that Eq. 9.17 does not contain gross errors.

can be seen in Fig. 9.15b for xTy (x Times y), where the trend line through a corresponding plot has a slope of only 0.93.

Questions

Let the two measurands x and y have standard deviations x_{Sd} and y_{Sd}, respectively. How large is, according to the propagation rules, the standard deviation of the sum $x + y$, and how large is the standard deviation of the difference x - y?[25]

The proportionality factor in the trend line in Fig. 9.15b is, with 0.93, clearly smaller than 1. Does this contradict theory?[26]

▶ **Task** Vary the mean values and standard deviations of x and y and log the mean value and standard deviation of the sum xPy! Which regularity do you assume? Derive the expected result from Eq. 9.16!

Error propagation in products

In Fig. 9.15b, the experimentally determined standard deviation *xTy_Sd* of the set xTy is compared with the theoretical standard deviation *xTy_Sd_calc* calculated from the standard deviations of x and y. For the product $z = x{\cdot}y$, Eq. 9.16 yields the weights

$$w_x = \left(\frac{\partial z}{\partial x}\right)^2 = y^2 \text{ and } w_y = \left(\frac{\partial z}{\partial y}\right)^2 = x^2 \tag{9.18}$$

hence,

$$\begin{aligned} var_z &= y^2 \cdot \mathrm{var_x} + x^2 \cdot var_y \\ \frac{\mathrm{var}_z}{z^2} &= \frac{\mathrm{var}_x}{x^2} + \frac{\mathrm{var}_y}{y^2} \\ \left(\frac{z_{Sd}}{z_m}\right)^2 &= \left(\frac{x_{Sd}}{x_m}\right)^2 + \left(\frac{y_{Sd}}{y_m}\right)^2 \end{aligned} \tag{9.19}$$

▶ The square of the relative standard variations of a product is the sum of the relative standard variations of the factors.

Ψ Calculate with variances, report the standard error!
Ψ Even better: report confidence level and confidence interval!

[25] The standard deviations of the sum and the difference are equal: $(x + y)_{Sd} = (x - y)_{Sd} = \sqrt{x_{Sd}^2 + y_{Sd}^2}$.

[26] The formulas for error propagation, here, the square root of Eq. 9.19 for products, are based on a Taylor series development and are, therefore, valid only for small variances of the independent variables, a condition no longer satisfied here.

Standard deviations that are too large distort the result
In Fig. 9.15b, the standard deviation of the product, calculated according to the propagation formula, is plotted against the empirical one. The slope of the regression line is 0.93, significantly smaller than 1. This is because the standard deviations of the measurands are too large. The formulas for error propagation are based on a Taylor series development of the function, in Eq. 9.16 only to the first order. So, they are valid only for a small interval around the independent variables. The standard deviations in Fig. 9.15b become too large. For small standard deviations (< 60), the slope is greater than that of the trend line drawn in the figure, actually 0.99, as an appropriate check shows.

Error propagation in powers
We now investigate the power function

$$z = x^n \tag{9.20}$$

The variance of the power function is derived from Eq. 9.16 to

$$var_z = a_x\ var_x \text{ with } a_x = \left(\frac{\partial z}{\partial x}\right)^2 = \left(n\ x_m^{n-1}\right)^2 \tag{9.21}$$

Thus, the expected standard deviation of z is

$$z_{Sd} = n \cdot x_m^{n-1} \cdot x_{Sd} \tag{9.22}$$

Figure 9.16 displays the empirical standard deviation for a power $n = 2$ (in **a**) and $n = 4$ (in **b**), both for a noise level $x_{Ns} = 0.1$, as a function of the mean value x_m.

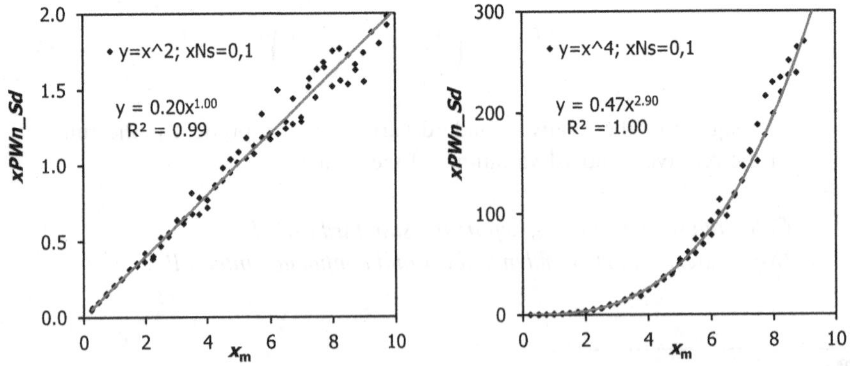

Fig. 9.16 **a** (left) Standard deviation of the second power of the random variable x with the noise level $x_{Ns} = 0.1$ as a function of the empirical mean x_t of the data set. **b** (right) Same as **a**, but with a power of $n = 4$

Question

Do the formulas for the trend lines in Fig. 9.16a and b agree with the theoretical expectation?[27]

Experimental procedure

We set up two data series (x, y) (size 100 each), defined as a true value (x_{True} or y_{True}) plus a normally distributed noise (standard deviations x_{Ns}, y_{Ns}). We then determine their mean values (x_m, y_m) and standard deviations (x_{Sd}, y_{Sd}).

We add and multiply the two sets pairwise to get the new series *xPy* ("*x* Plus *y*") and *xTp* ("*x* Times *y*") and determine their mean values xPy_m and xTy_m and standard deviations xPy_{Sd} and xTy_{Sd}, respectively. Furthermore, we create a new series *x*PW*n* by raising the elements of *x* to the power *n* and determine its mean value $x\mathrm{PW}n_m$ and standard deviation $x\mathrm{PW}n_{Sd}$.

We vary the noise levels and log the mean values and standard deviations of the sets with a rep-log procedure. This is achieved:

- in EXCEL with a VBA routine that repetitively changes the parameters of the spreadsheet calculation,
- in `Python` by outsourcing the parts of the program equivalent to the spreadsheet into a function, repetitively called by the main program.

We plot the theoretically calculated standard deviations of *xPy* and *xTy* versus the empirical ones in a scatter plot and lay a straight trend line through the data points (see Fig. 9.15). The slope of such trend lines is 1 if the theoretically calculated values correspond to the empirical ones.

We set up another set $x_{\mathrm{PW}n}$ by raising the elements of the series *x* to the *n*th power and present its standard deviation as a function of the mean value x_m of the series *x*, together with a power trend line through the data (see Fig. 9.16).

Questions

We have called the collections *x*, *y*, *x*Py, *x*Ty data series, not data sets. Why is the term "set" inappropriate here?[28]

As a result of our simulations, we plot the *standard deviations* of sums and products against the *standard deviations* of the summands or factors, without considering either the number *n* of measurements or the *t* values. Why does that not hinder the illustration of the laws of error propagation?[29]

---

[27] Yes, we expect $z_{Sd} \propto x^3$ for $f(x) \propto x^4$ and $z_{Sd} \propto x$ for $f(x) \propto x^2$ and find $y = 0.47x^{2,90}$ und $y = 0.20x^{\cdot 1.00}$.

[28] A set is an unordered collection. However, our data are ordered. The *n*th element of *xPy* and *xTy* is obtained from the *n*th elements of *x* and *y*.

[29] The sets for all terms have the same number *n* of measurements, so that the factors *n* and *t* are common to all of them and to the results.

9.7.2 Data Structure and Nomenclature

x_{True}, y_{True}	true values
x_{Ns}, y_{Ns}	noise levels (standard deviation of a normal distribution)
x, y	two series of true values plus noise
x_{m}, y_{m}	mean of x and y
x_{Sd}, y_{Sd}	standard deviations of x and y
*x*Py	series x Plus y
*x*Ty	series x Times y
*x*PWn	series x to the PoWer of n
xPy_m, xPy_Sd	mean and standard deviation of *x*Py, correspondingly for *x*Ty and *y*PW*n*
xPy_Sd_calc	standard deviation of *x*Py calculated from x_{s} and y_{S}, correspondingly for *x*Ty and *y*PW*n*.

9.7.3 Spreadsheet Calculation

In A13:B112 of Fig. 9.17 (S), we create two series x and y of normally distributed numbers, regarded as measurement values specified by their means ("true values") and standard deviations ("noise") in A1:B4. The series x and y are evaluated for mean value and standard deviation in D1:E4.

We then create three new data series in D13:E112 by adding x and y (*xPy*), multiplying x and y (*xTy*), and raising x to the power n (*x*PW*n*) specified in B5. These series are evaluated in Fig. 9.18 (S). There, the standard deviations *xPy*_Sd_calc, *xTy*_Sd_calc, and *xPWn*_Sd_calc are also reported, theoretically derived from the standard deviations of the original series.

Error propagation in sums

In Fig. 9.18 (S), the empirical standard deviation *x*Py_Sd is calculated from the series *x*Py. The theoretical standard deviation *xPy_Sd_calc* is calculated as the root of the sum of the squares of the summands' standard deviations. The two values are identical to the second decimal.

	A	B	C	D	E	F	G
1	**xTrue**	10.00		**xm**	10.00	*=AVERAGE(x)*	
2	**xNs**	0.10		**xSd**	0.11	*=STDEV.S(x)*	
3	**yTrue**	10.00		**ym**	9.54	*=AVERAGE(y)*	
4	**yNs**	7.50		**ySd**	7.56	*=STDEV.S(y)*	
5	**n**	2.00					

	A	B	C	D	E	F
11	*=NORM.INV(RAND();xTrue;xNs)*			*=x+y*	*=x*y*	*=x^n*
12	**x**	**y**		**xPy**	**xTy**	**xPWn**
13	10.06	17.89		27.95	180.02	101
112	10.00	17.31		27.31	173.07	100

Fig. 9.17 (S) Two normally distributed data series x and y of size 100, defined by their means x_{True} and y_{True} ("true" values) and standard deviations x_{Ns} and y_{Ns} ("noise").in a Gaussian

We *vary* the standard deviations of x and y independently of each other systematically with a rep-log procedure, *record* the empirical and calculated standard deviations of the sum, and *plot* both as data points in Fig. 9.15a. The trend through the data points is a straight line with a slope very close to 1. Thus, we have retrieved the theoretical statement with our simulation; at least, we cannot object to it.

Error propagation in products

The product of x and y is stored as a new series *xTy* in column E of Fig. 9.17 (S) and evaluated in column M of Fig. 9.18 (S).

We vary the standard deviations of x and y independently of each other systematically with a log procedure and record the product's empirical and calculated standard deviations. The standard deviation *xTy_Sd_calc* of the product *xTy* is calculated from those of the factors as (see Eq. 9.19)

$$xTy_{SdCalc} = z_{SdCalc} = \sqrt{(y_m\, x_{Sd})^2 + (x_m\, y_{Sd})^2} \tag{9.23}$$

The results are shown in Fig. 9.15b. The trend line has a slope of 0.93, less than the expected 1.00. The reason for this is that the standard deviations x_{Sd} and y_{Sd} are too large (see discussion in Sect. 9.7.1).

Error propagation in powers

▶ **Task** Change n in Fig. 9.17 (S) or Table 9.13 and review your interpretation!

In Fig. 9.17 (S), column F, x is raised to the nth power; the result is stored as *xPWn*. The standard deviation *xPWn_Sd* of this series is determined in Fig. 9.18 (S) in M8 together with the theoretical one *xPWn_Sd_calc* in M9.

In a rep-log procedure, the value of x_{True} is varied from 0.5 to 10 for a noise level of 0.1. The results for $n = 4$ and $n = 2$ are shown in Fig. 9.16a and b, respectively; they are approximated by a power function as a trend line.

	I	J	K	L	M	N
1		**x+y**			**x*y**	
2	**xPy_m**	20.34	*=AVERAGE(xPy)*	**xTy_m**	103.51	*=AVERAGE(xTy)*
3		20.34	*=xm+ym*		103.39	*=xm*ym*
4	**xPy_Sd**	7.79	*=STDEV.S(xPy)*	**xTy_Sd**	77.78	*=STDEV.S(xTy)*
5	**xPy_Sd_calc**	7.77	*=SQRT(xSd^2*	**xTy_Sd_calc**	77.65	*=SQRT((ym*xSd)^2*
6			*+ySd^2)*			*+(xm*ySd)^2)*
7		**x^n**				
8	**xPWn_m**	100	*=AVERAGE(xn)*	**xPWn_Sd**	1.75	*=STDEV.S(xPWn)*
9	**xm^n**	100	*=xm^n*	**xPWn_Sd_calc**	1.75	*=n*xm^(n-1)*xSd*

Fig. 9.18 (S) Continuation of Fig. 9.17 (S); evaluation of the data sets *xPy* ("*x* Plus *y*"), *xTy* ("*x* Times *y*") and *xPWn* ("*x* to the power of *n*")

Table 9.13 `Python` function, part corresponding to Fig. 9.17 (S) continued in Fig. 9.18 (S)

```
import numpy.random as npr
I=np.zeros(14)
xTrue, yTrue, n = 10, 10, 2          #True values
def simu(xNs,yNs):
      x=xTrue+xNs*npr.randn(100)     #Array with 100 items
      y=yTrue+yNs*npr.randn(100)
      xm=np.mean(x)
      xSd=np.std(x)
      ym=np.mean(y)
      ySd=np.std(y)

      xPy=x+y
      xTy=x*y
      xPWn=x**n
```

Question/task

Write the formula in N9 of Fig. 9.18 (S) in mathematical form![30]

9.7.4 Python Program

The spreadsheet calculations can be translated directly into a `Python` program, Tables 9.13 and 9.14. The program's core is the function *simu* calculating with the parameters x_{True}, y_{True}, n, x_{Ns}, y_{Ns}. The parameters x_{Ns} and y_{Ns} are chosen to be handed over in the function head, whereas x_{True}, y_{True}, and n are treated as global parameters.

The arrays x and y are created anew within the function with every call, and consequently, the scalars and arrays derived from them are recalculated. The means, empirical, and calculated standard deviations for every set are returned as an array I with 14 elements (Table 9.14).

Error propagation for sums and products

The main program for calculating the standard deviations of the sum, displayed in Fig. 9.15a, is given in Table 9.15. The standard deviations x_{Ns} and y_{Ns} are varied, whereas the true values x_{True}, y_{True} are specified before the for-loop and remain constant during the whole experiment.

Similar main programs work for the product with:
`xTy.append(I[7])`.

[30] $Z = x^n$; $Z_{Sd} = n\, x^{n-1}\, x_{Sd}$.

Table 9.14 Continuation of Table 9.13, part of the function *simu* corresponding to Fig. 9.18 (S)

```
        I[1]=np.mean(xPy)
        I[2]=xm+ym
        I[3]=np.std(xPy)
        I[4]=np.sqrt(xSd**2+ySd**2)
        I[5]=np.mean(xTy)
        I[6]=xm*ym
        I[7]=np.std(xTy)
        I[8]=np.sqrt((ym*xSd)**2+(xm*ySd)**2)
        I[9]=np.mean(xPWn)
        I[10]=xm**n
        I[11]=xSd**n
        I[12]=np.std(xPWn)
        I[13]=n*xm**(n-1)*xSd
        return I #Mean and standard deviation
                  (empirical and theoretical) of x*y, x*y, and x^n
```

Table 9.15 Main program for calculating the standard deviation of a sum $x + y$ for varying standard deviations of x and y

```
xPy_Sd=[]
xPy_Sd_calc=[]
 #Variation of the noise levels xNs and yNs
 #Constant xTrue, yTrue
xTrue=10.0
yTrue=10.0
for noiseX in np.arange(0.5,8,1):
    for noiseY in np.arange(0.5,8,1):
        I=simu(noiseX,noiseY)
        xPy_Sd.append(I[3])
        xPy_Sd_calc.append(I[4])
```

`xTy_Sd.append(I[8])`.

Table 9.16 presents a program for achieving a plot such as that in Fig. 9.15a.

Power-law trend line

A power-law trend line $y = a \cdot x^n$ is obtained with the function *fitPowLaw* in Table 9.17, by fitting a linear trend line $logy = a + m \cdot logx$ to the logarithmized data $logx = log(x)$ and $logy = log(y)$. The transformation to $y = a_R x^{p_R}$ is achieved with $p_R = m$ and $a_R = exp(a)$.

Error propagation for powers

Table 9.18 shows the main program for calculating the standard deviation of a power x^n as a function of the true value x_{True}. A program similar to Table 9.16 achieves a fit to the data points and results in a plot corresponding to Fig. 9.16.

Table 9.16 a (top) Program for producing a figure as in Fig. 9.15a, b (bottom) function to draw a straight line through (*x*, *y*)

```
FigStd('xPy_Sd',0,10,2,'xPy_Sd_calc',0,10,2)
plt.plot(xPy_Sd,xPy_Sd_calc,'kD',ms=3,
                    label='StDev Sum;\nxTrue='+str(xTrue)+\
                           ';yTrue='+str(yTrue))
PlotLin(xPy_Sd,xPy_Sd_calc,a=False)
plt.legend()
```

```
import statsmodels.api as sm

def PlotLin(x,y,a=False): #Linear regression line
    if a==False:              #No y-axis intercept
        model=sm.OLS(y,x)
        results=model.fit()
        mR=results.params[0]
        yLin=mR*np.array(x)
        plt.plot(x,yLin,'k-',label="y=%.2f*x"%mR)
    if a==True:               #y-axis intercept allowed
        xx=sm.add_constant(x)
        model=sm.OLS(y,xx)
        results=model.fit()
        aR=results.params[0]
        mR=results.params[1]
        yLin=aR+mR*np.array(x)
        lbl="y=%.2f+%.2fx"%(aR,mR)
        plt.plot(x,yLin,'k-',label=lbl)
```

Table 9.17 Plotting a power-law trend line by a straight line through logarithmized data

```
def fitPowLaw(x,y):
    logx=np.log(x)
    logx=sm.add_constant(logx)     #y-axis intercept allowed
    logy=np.log(y)
    model=sm.OLS(logy,logx)        #Linear regression
    results=model.fit()
    aR=np.exp(results.params[0])   #Amplitude
    pR=results.params[1]           #Power
    R2=results.rsquared
    #print(pR)    #Only in test phase
    yPow=aR*x**pR
    lbl="y=%.2f*x**%.2f,R²=%.2f"%(aR,pR,R2)
    plt.plot(x,yPow,'k-',label=lbl)
```

Table 9.18 **a** (top) Main program for the power series *xPWn, simu* from Table 9.13; **b** (bottom) Plotting the original data and a power trendline through the data

```
xPWn=[]
xPWn_Sd=[]
xtp=[]
n=2
 #Variation of xTrue for constant noise level
xNs=0.1
for x_1 in np.arange(0.25,10,0.25):
    for rep in np.arange(2):
        xTrue=x_1
        I=simu(0.1,0.1)
        xtp.append(I[10]**(1/n))
        xPWn.append(I[12])
        xPWn_Sd.append(I[13])
```

```

FigStd('xTrue',0,10,2,'xPWn_Sd',0,2,0.5)
plt.plot(xtp,xPWn_Sd,'kD',ms=3)
fitPowLaw(xtp,xPWn_Sd)  #Function in table above
plt.legend()
```

9.8 Propagation of Confidence Intervals

We learn how to get from variances of two data sets to a confidence interval for a combined result, sum or product of the two measurands. We use a statistical simulation for the last step to get confidence-specified errors ("C-spec errors").

9.8.1 From Variance to Confidence

In Exercise 9.7, we investigate the propagation of standard variations in sums, product, and powers by means of series of 100 elements. In real-world experiments, usually much less measurements are performed so that t factors have to be taken into account. Furthermore, the number of measurements can be different for the different operands. As a consequence, the C-spec error of the result cannot be deduced straightforwardly from the standard deviation of the result; we have to calculate with C-spec errors also for the operands.

Measurement series for single measurands

In Fig. 9.19 (S), two measurement series x_{Ai} and x_{Bi} are simulated, with the true values x_{A} and x_{B} and the noise levels ns_{A} and ns_{B}, specified in rows 1 and 2,

	A	B	C	D	E	F	G	H	I	J	K
1	xAi	xBi		xA	4			xB	2		
2	3.81	1.85		nsA	0.2			nsB	0.1		
3	3.98	1.89									
4	3.83	1.98		nA	4	=COUNT(xAi)		nB	10	=COUNT(xBi)	
5	3.84	1.92		xAm	3.87	=AVERAGE(xAi)		xBm	1.94	=AVERAGE(xBi)	
6		1.98		xAse	0.04	=STDEV.S(xAi)/SQRT(nA)		xBse	0.03	=STDEV.S(xBi)/SQRT(nB)	
7		1.98									
8		2.12		Con	0.683						
9		1.83		tA	1.20	=T.INV.2T(1-Con;nA-1)		tB	1.06	=T.INV.2T(1-Con;nB-1)	
10		1.97		xAce	0.05	=xAse*tA		xBce	0.03	=xBse*tB	
11		1.88									
12					FALSE	6852			FALSE	6865	
13					=AND(xAm-xAce<xA;xA<xAm+xAce)				=AND(xBm-xBce<xB;xB<xBm+xBce)		
14						10000				10000	

Fig. 9.19 (S) Confidence intervals x_{Ace} and x_{Bce} for a confidence level *Con* for two measurement series x_{Ai} and x_{Bi}

entering our standard function x_{Ai} = [=NORM.INV(RAND();XA;NSA)] and x_{Bi} = [=NORM.INV(RAND();XB;NSB)] for the simulation of noisy measurements.

In rows 4 to 6, the number of elements (n_{A}, n_{B}), the mean values (x_{Am}, x_{Bm}), and the standard errors (x_{ASe}, x_{BSe}) of x_{Ai} and x_{Bi} are determined. Remember: the *standard error of the mean* is obtained as square root of the *variance of the series* (their standard deviation) divided by the square root of the number of elements in the series, Eq. 9.12.

To come from the standard error x_{Se} of the mean to a *confidence interval* x_{Ce} around the mean, we have to specify a *confidence level*. This is done in E8 with *Con* = 0.683, valid for the standard error when the number of measurements is large. Student's *t*-values for this confidence level are determined in line 9 and multiplied in line 10 with the standard error to yield the C-spec errors x_{Ace} and x_{Bce} of the confidence intervals.

In E12 and I12, we check in the usual way whether the confidence intervals capture the true values. Furthermore, with a rep-procedure, we check how often that is the case for 10,000 repetitions of the statistical experiment (reported in F12 and J12). The numbers are close to the expected values.

9.8.2 Sum and Product of Two Measurands

In Fig. 9.20 (S), we determine product *AtP* ("A times B") and sum *ApB* ("A plus B") of the means x_{Am} and x_{Bm} of the two measurement series x_{Ai} and x_{Bi} and calculate C-spec errors AtB_{ce} and ApB_{ce} for the errors obtained according to the rules of error propagation explained in Exercise 9.7. They include tentative *t* factors $t\mathrm{AxB}_{\mathrm{emp}}$ and $t\mathrm{ApB}_{\mathrm{emp}}$, first set to 1 in N5 and S5. With these intervals, we get hit rates of more than 7000 for 10,000 repetitions of the statistical experiment,

	M	N	O	P	Q	R	S	T	U	V
1	**AtBtrue**	8.00	=xA*xB			**ApBtrue**	6.00	=xA+xB		
2	**AtBemp**	8.10	=xAm*xBm			**ApBemp**	6.01	=xAm+xBm		
3	**varA**	0.001	=(xAce/xAm)^2			**varAp**	0.01	=xAce^2		
4	**varB**	0.000	=(xBce/xBm)^2			**varBp**	0.00	=xBce^2		
5	**tAtBemp**	0.959				**tApBemp**	0.986			
6	**AtBce**	0.26				**Apbce**	**0.12**			
7							=SQRT(varAp+varBp)*tApBemp			
8		=AND(AtBemp-AtBce<AtBtrue;AtBtrue<AtBemp+AtBce)								
9		**TRUE**	6805	10000			TRUE	6833	10000	
10			1.03	=T.INV.2T(1-O9/P9;16)				1.00	=T.INV.2T(1-T9/U9;1000)	
11			1.03	=T.INV.2T(1-Con;16)				1.00	=T.INV.2T(1-Con;1000)	
12			1.01	=O11/O10				1.00	=T11/T10	
13			0.96	=O12*tAxBemp				0.99	=T12*tApBemp	

Fig. 9.20 (S) Product *AtB* ("times") and Sum *ApB* ("plus") of the results of the two measurement series of Fig. 9.19 (S)

significantly larger than expected from a confidence level $Con = 0.683$. The reason for this discrepancy is that the rules for error propagation set up in Exercise 9.7 are mathematically correct only if the number of measurements is big enough.

C-spec error by adapting the *t* factor

We do not delve deeper into the mathematical derivation but reduce the hit rates by adapting the tentative *t* factors. To do so, we calculate Student's *t* value:

- for the empirically obtained hit rates (O12 and T12); they should approach the confidence level after some iterations,
- for the envisaged confidence level *Con* (O11 and T11); they do not change in the course of the simulation.

We build the ratio of the two *t* values and multiply with the current tentative *t* values (results in P13, T13) and copy/paste the results into $t\text{AtB}_{emp}$ and $t\text{ApB}_{emp}$ (N5, S5). We repeat the statistical simulation and get hit rates 6738 (with $t = 0.940$ for A_B) and 6708 (with $t = 0.962$ for ApB). After a second iteration, the results are 6805 (with $t = 0.959$) and 6833 (with $t = 0.986$), indeed close to the values expected for the specified confidence level. The quantities AxB_{ce} and ApB_{ce} are called the confidence-specified errors, or short *C-spec errors*.

9.9 Mass of a Thin Film on a Glass Substrate

Our task is to determine the mass of a thin film on a glass plate. To this end, the glass plate is weighed several times on a microbalance, before and after being coated. The mass of the film is obtained with recursion to error propagation and t statistics.

9.9.1 Instructions for Use for Accurate Measurements and Their Results

A measuring process ...

In an ideal coating process, a glass substrate of mass $m_{Sub} = 1$ g is coated with a thin film of mass $m_F = 1$ mg. To determine the film's mass, the substrate is weighed before and after coating. The balance's nominal accuracy is 1 μg, which means that the mass is displayed in grams, with six digits after the decimal point. However, the measurement's actual accuracy is significantly lower, e.g., due to disturbances by air currents or building vibrations.

Questions

What is the ratio of the masses of the substrate and the thin film?[31]

How many digits after the decimal point can the layer's mass be determined nominally, if given in mg?[32]

By what percentage does the coating increase the mass of the sample?[33]

... and its simulation

The substrate is weighed n times before and n times after being coated. The process is simulated by series $m_{Bef} = m_{Sub} + noise$ and $m_{Aft} = m_{Sub} + m_F + noise$, where the noise is normally distributed with m_{Ns} as the standard deviation.

EXCEL quotes each number with 15 digits, something like 0.569410526368089, `Python` even with 16 digits. Therefore, the results of the simulated weighing have to be rounded to six digits after the decimal point to mimic the display resolution of the balance: ROUND(...;6) = 0.569411 and `np.round(..., 6).`

The following explanations are valid for the preceding simulation but also for real measurements.

Immediate evaluation of the measurement results

The *standard errors* of the masses of the uncoated and coated substrate are calculated as the standard deviation of the measurement series divided by the square root of the number n of measurements in the series. In order to get the *C-spec errors* m_{BefCe} and m_{AftCe} (half the width of the *confidence intervals*), the standard error has to be

[31] 1 g/1 mg = 1000; the substrate is 1000 times heavier than the film.

[32] Displaying an accuracy of 1 μg, the films' mass could naively be specified as, e.g., 1.001 mg.

[33] 1 mg of 1 g, corresponding to 0.1%.

multiplied by Student's t-factor that is determined by the degree of freedom ($dof = n - 1$) and the specified confidence level Con. In this exercise, we choose $Con = 0.683$ which corresponds to the confidence level of the standard error if the number n of measurements is big enough.

The mass of the film is estimated as the difference in the means of the two measurement series,

$$m_{\text{fest}} = m_{\text{aft}} - m_{\text{Befm}}$$

Confidence interval of the estimated mass of the film

In a first approach, we estimate an error range $\pm\, m_{\text{Fc0}}$ of the film's mass with the rule for sums and differences as square root of the sum of the squares of the *C-spec errors* of the summands:

$$m_{Fc0} = \sqrt{m_{BefCe}^2 + m_{AftCe}^2} \tag{9.24}$$

We know from Exercise 9.8, that the confidence level of this estimate does not necessarily correspond to the confidence level Con of the estimated masses of the substrate before and after coating. In order to correct that, we have to multiply m_{Fc0} with a t factor that is to be determined through a statistical simulation.

For this simulation, we take as parameters the estimated masses m_{BefM} of the uncoated substrate and m_{Fest} of the film, and a noise level m_{Ns} estimated as the standard deviation of the measurement series of the uncoated and coated substrate. In this exercise, these values are the result of a simulation but in laboratory praxis such are the results of real measurements.

The number of trials in the simulation must be big enough so that the fluctuations of the hit rate for the same t value are sufficiently small to avoid too large jumping of the hit rate and assure improvement in the iterative adaptation of the t value.

Rounding the numerical results to relevant digits

The layer's mass and its C-spec error are to be determined from the simulated measurement data and written as a final result with rounded numbers. To give an example: If the layer's mass is determined as the difference of the weighings to be 0.0009748 g with an error of 6.57×10^{-5}, the final result is $9.7(7) \times 10^{-4}$ g or $9.7 \pm 0.7 \times 10^{-4}$ g:

```
0.0009748   0.00097
±0.0000657 ±0.00007
```

This means that the result can only be expressed sensibly with two digits and that the uncertainty in the last digit is ± 7. An alternative notation is 0.97(7) mg. For completeness, the *confidence level of the reported error* must be noted.

Questions

You get the following results for a single measurand: Mean value 0.0010818 g, standard error of the mean value 4×10^{-5} g. How do you report the final result?[34]

Is 1.033 ± 0.037 more precise than $1.03 \pm 0{,}04$?[35]

"Official" provisional measurement result

In the report of the measurement results, you should state the result, here the film's estimated mass, and its C-*spec error*, and the level of confidence:

$$\texttt{Mass of the film (C} = \texttt{0.683)} = \texttt{1.01} \pm \texttt{0.06mg}$$

If you do not calculate the C-spec error, you must state all values necessary for the simulation.

Question

Which results of a measurement series for determining the film's mass do you have to state if you do not calculate the C-spec error?[36]

Syntactical differences between Excel and Python

In EXCEL, [= ROUND(NORM.INV(RAND();MSUB;MNS);DSP)] is written into $n = 5$ cells with range name m_{Bef}, whereas in PYTHON, there is only one statement:

```
mBef = np.round(mSub + mNs * npr.randn(nBef), dsp)
```

using the function `npr.randn` to generate normally distributed random numbers that have to be multiplied by m_{Ns} to get the right scale of the measurement noise.

EXCEL uses the function STD.S to get the variance of a sample, whereas `Python` uses `np.std` with `ddof`, the deduction of the degrees of freedom, to be specified as a key argument; $ddof = 1$, in our case, e.g., `np.std (mBef,ddof = 1)`.

---

[34] 1.08×10^{-3} g; 0.04×10^{-3} g $\rightarrow$ (1.08 ± 0.04) mg; uncertainty of measurement given as one standard error. Ψ *Two inside and one out of* applies if enough repetitions of the measurement have been made or the error is extended by a t factor.

[35] No, the standard error is also only an estimate and is affected by statistical error.

[36] (1) Estimated masses of the uncoated substrate and the film, (2) the noise level of the measurement series. These are the parameters necessary to perform a simulation for getting the C-spec error.

9.9.2 Data Structure and Nomenclature

m_{Sub}	true mass of the substrate
m_{F}	true mass of the film
m_{Ns}	measurement noise (standard deviation of a normal distribution)
dsp	display precision of the scales
n_{Bef}, n_{Aft}	number of measurements before and after coating
m_{Bef}	array of the results of the weighing before coating
m_{Aft}	array of the results of the weighing after coating
m_{BefM}, m_{AftM}	means of m_{Bef} and m_{Aft}
m_{BefSe}, m_{AftSe}	standard errors of m_{Bef} and m_{Aft}
Con	confidence level
m_{BefCe}, m_{AftCe}	C-spec errors of m_{BefM} and m_{AftM} for Con
m_{Fest}	Estimated mass of the film, $= m_{\text{AftM}} - m_{\text{BefM}}$
m_{FSe}	standard error of m_{Fest}
m_{Fce}	C-spec error of m_{Fest}.

9.9.3 Spreadsheet Solution

Simulation of the weighing process

The parameters of the exercise are specified and the measurement series simulated in Fig. 9.21 (S). All values are stored in named cells, with the names listed in column A and in D1:E1. The uncoated substrate's mass is measured as $m_{\text{Bef}} = m_{\text{Sub}} +$ noise, that of the coated substrate as $m_{\text{sub}} + m_{\text{F}} +$ noise. The noisy measurement is simulated with the spreadsheet function NORM.INV(RAND();MSUB;MNS), with mean value m_{Sub} (or $m_{\text{Sub}} + m_{\text{F}}$ for the coated substrate) and standard deviation m_{Ns}.

Evaluation of the simulated or real-world weighing process

The calculation with these parameters is shown in Fig. 9.22 (S). The values in D:E could also be the results of real measurement. The following evaluation and simulation would be the same.

	A	B	C	D	E	F
1	**mSub**	1	g	**mBef**	**mAft**	
2	**mF**	1.0E-03	g	0.999934	1.000833	
3	**mNs**	1.0E-04	g	**1.000047**	1.001021	*=ROUND(NORM.INV(RAND();mSub;mNs);dsp)*
4				0.999899	1.001115	
5	**dsp**	6		0.999952	**1.000912**	*=ROUND(NORM.INV(RAND();mSub+mF;mNs);dsp)*
6				1.000154	1.000854	

Fig. 9.21 (S) A:C, "True" values of the masses of the substrate and the film and an assumed value for the noise of the measurement process; dsp = number of digits in the scale display; D:E, simulated results of measurement series for the uncoated and coated substrate.

	H	I	J	K	L	M
1	mBefM	0.999997	=AVERAGE(mBef)			
2	mAftM	1.000947	=AVERAGE(mAft)			
3	nBef	5	=COUNT(mBef)	Con	0.683	
4	nAft	5	=COUNT(mAft)	tBef	1.14	=T.INV.2T(1-Con;nBef-1)
5				tAft	1.14	=T.INV.2T(1-Con;nAft-1)
6	mBefSe	4.62E-05	=STDEV.S(mBef)/SQRT(nBef)	mBefCe	5.28E-05	=mBefSe*tBef
7	mAftSe	5.32E-05	=STDEV.S(mAft)/SQRT(nAft)	mAftCe	6.08E-05	=mAftSe*tAft
8	noise	1.11E-04	=(STDEV.S(mBef)+STDEV.S(mAft))/2			

Fig. 9.22 (S) Evaluation of five weighing processes each of the uncoated and the coated substrate, here simulated in Fig. 9.21 (S). The results m_{BefSe} and m_{AftSe} are standard errors. The report could also be the result of a real-world experiment

	O	P	Q	R	S	T
1	varBef	2.79E-09	=mBefCe^2	TRUE	=AND(mFest-mFce<mF;mF<mFest+mFce)	
2	varAft	3.69E-09	=mAftCe^2	HitRate	673	From procedure
3	mFc.0	8.05E-05	=SQRT(varBef+varAft)	Trials	1000	
4	tAd	0.891		tHR	1.12	=T.INV.2T(1-HitRate/Trials;4)
5	mFce	7.17E-05	=mFc.0*tAd	tF	1.14	=T.INV.2T(1-Con;4)
6	mFest	9.50E-04	=mAftM-mBefM	tAd.New	0.91	=tAd*tF/tHR

Fig. 9.23 (S) Setting an error m_{Fce} and determining its hit rate and therefrom its t factor t_{HR}. The t factor t_F for the confidence level *Con* is calculated in S5

Questions

Why are the measurement results rounded to six decimal places in Fig. 9.21 (S)?[37]

Which parameter in Fig. 9.22 (S) characterizes the measuring accuracy of the weighing process?[38]

Confidence interval by statistical simulation

In Fig. 9.23 (S), we first calculate an error m_{Fc0} for the film mass (P3) with the rule of error propagation for sums and differences and then multiply it with a tentative t-value t_{Ad} to calculate a C-spec error m_{Fce} (P5).

In the beginning, we had set $t_{Ad} = 1$ and obtained a hit rate 731 for 1000 trials, more than the 683 expected from the confidence level *Con*. For the first iteration, we replaced t_{Ad} with t_{AdNew} to get a hit rate of 673, close enough to the expected 683 to state the resulting m_{Fce} = 7E-05 as the C-spec error for the confidence level *Con* = 0.683. The old tentative C-spec error is multiplied by the ratio of the t value t_F for the desired confidence level and the t value for the current hit rate.

[37] This corresponds to the *display* accuracy of the balance.

[38] The accuracy of the current measuring process is characterized by the noise during weighing. In the case of Fig. 9.22 (S) I8, it is 1.11E-4 g, estimated as the average standard deviation of the measurement series.

Table 9.19 Parameters chosen for the simulation

```
mSub=1.00      #[g] True mass of the substrate
mF=1.0e-3      #[g] True mass of the film
mNs=1.0e-4     #[g] Measurement noise
dsp=6          #    Number of displayed digits
nBef=5         #    Number of measurements
nAft=5

Con=0.683      #    Confidence level
```

9.9.4 Python Program

The parameters for the simulation of the weighings are specified in Table 9.19.

The main program to determine the hit rates of the confidence intervals is given in Table 9.20. It is essentially a for-loop calling a function *InRange* which takes the t factor t_{Ad} as input and returns a Boolean *inRa* that states whether the confidence interval captures the true value, and the C-spec error m_{Fce}.

The function *InRange* is given in Table 9.21. It recurs to two functions *MassBefore* and *MassAfter* reported in Table 9.22. *MassBefore* simulates the measuring process and returns the estimated mass m_{BefM} of the substrate before coating together with its C-spec error m_{BefC}. *MassAfter* does the equivalent for the mass of the substrate after coating.

The result of the for-loop is a hit rate that is used to calculate a new t value t_{AdNew} expected to be related to the specified confidence level. For the next iteration, the value of t_{AdNew} is inserted into line 3 by hand.

The final result is rounded with the function *FinRes* from Sect. 9.3.3 to the relevant number of digits (see Table 9.23).

Table 9.20 Main program to determine the hit rate and the C-adjusted t value t_{AdNew}. For a new run to iteratively improve t, insert t_{Adnew} manually in line 3

```
1   Trials=10000
2   HitRate=0
3   tAd=1                              #To be adjusted for chosen C
4   for i in range(Trials):
5       inRa,mFce=InRange(tAd)         #Defined in table below
6       if inRa==True: HitRate+=1
7
8   tF=sct.t.ppf(1-(1-Con)/2,nAft-1)   #Target value of t
9   tHR=sct.t.ppf(1-(1-HitRate/Trials)/2,nAft-1)
10  tAdNew=tAd*tF/tHR
```

```
tAd      1.00  HitRate  7171
tF       1.14
tHR      1.24
tAdNew   0.92  HitRate  6725
```

Table 9.21 Function to determine the C-spec error m_{Fce} of the film mass and the logical value whether the confidence interval captures the true value, recurs to functions in Table 9.22 that perform the simulations

```
def InRange(tAd):
#Estimate mass of the film!
    mBefM,mBefCe=MassBefore()          #In table below
    mAftM,mAftCe=MassAfter()           #In table below
    mFest=mAftM-mBefM                  #Film mass
    varBef=mBefCe**2
    varAft=mAftCe**2
    mFc0=np.sqrt(varBef+varAft)
    mFce=mFc0*tAd                      #C-spec error
    inRa=mFest-mFce<mF<mFest+mFce
    return inRa,mFce
```

Table 9.22 Simulation of the weighing of the substrate before and after being coated

```
import scipy.stats as sct

def MassBefore():
#Mass of the substrate before coating:
    mBef=np.round(mSub+mNs*npr.randn(nBef),dsp)
    mBefM=np.average(mBef)             #Mean
    mBefSe=np.std(mBef,ddof=1)/np.sqrt(nBef)
    tBef=sct.t.ppf(1-(1-Con)/2,nBef-1)
    mBefCe=mBefSe*tBef                 #C-spec error
    return mBefM,mBefCe

def MassAfter():
#Mass of the substrate after coating:
    mAft=np.round(mSub+mF+mNs*npr.randn(nAft),dsp)
    mAftM=np.average(mAft)             #Mean
    mAftSe=np.std(mAft,ddof=1)/np.sqrt(nAft)
    tAft=sct.t.ppf(1-(1-Con)/2,nAft-1)
    mAftCe=mAftSe*tAft                 #C-spec error
    return mAftM,mAftCe
```

```
mAftM  1.000913  mAftCe  8.06e-05
mBefM  0.999904  mBefCe  4.08e-05
mFest  0.001009
```

Table 9.23 Rounding the result to the relevant number of digits with the function defined in Exercise 9.3

```
Result=FinRes('Mass of the film (C=0.683) ',
                    mFest*1000,mFc*1000," mg")
```

```
Mass of the film (C=0.683) =1.01±0.06 mg
```

9.10 Questions and Tasks

Explain the broom rules:

1. Ψ *Twice as good with four times the effort.*
2. Ψ *Two within and one out of.*
3. Ψ *Worse makes good even better.*
4. Ψ *Mostly, not always.*
5. Ψ *Report the C-spec errors but calculate with their squares!*

Evaluation of a measurement series

A quantity x was measured 9 times, with the resulting mean value $x_m = 10$ and standard deviation of the nine individual measurements $x_{Sd} = 1.8$.

6. Which spreadsheet and which `Python` function do you use to simulate this measurement series?
7. What is the standard error of the mean value?
8. A measurement series yields, as the mean value, $x_m = 7.12546 \times 10^4$ and, as the standard error of the mean value, $x_{Se} = 6.28743 \times 10^2$. How do you specify the measurement result correctly rounded?
9. Does the formula x_1 =*ROUND(0.847; 1) - ROUND(0.155; 1)* yield the same value as x_2 = *ROUND(0.847--0.155; 1)*?
10. In a departing airplane, the acceleration a is measured with two different methods, ten times each. The measurement results are $a_1 = 20 \pm 1$ m/s^2 and $a_2 = 21 \pm 0.5$ m/s^2. Which value and which measurement uncertainty should the crew report to the ground station?

Error propagation

11. A quantity z is the difference of two quantities $z = s_1 - s_2$. The results of the summands *are* $s_1 = 10 \pm 2.24$; $s_2 = 20 \pm 2$. How large are the difference z and its error z_{Se} calculated from the standard errors of the summands?
12. A quantity is the product of two quantities, $p = p_1 \cdot p_2$. Measurements with many repetitions result in $p_1 = 10 \pm 1$ and $p_2 = 20 \pm 3.5$. How big are the product p and its error p_{Se} calculated from the standard errors of the factors?
13. What is the difference between the C-spec error of a quantity that is calculated as a function of several measurands and its error calculated from the standard error of the measurands?

Fitting Trend Curves to Data Points 10

We create points on user-specified functions, transform them into measurement points by adding noise to their y-values, and then fit appropriate trend lines to the noisy data. Confidence intervals of the trend lines' coefficients are obtained using t statistics. In doing so, we learn how far we can trust the parameters of the trend lines. We use functions for linear regression, LINEST of Excel and OLS of the `statsmodel` library of `Python`, and nonlinear regression, OLVER and `curve_fit`.

10.1 Introduction: Linear and Nonlinear Regression

Solutions of Exercises 10.2 (Excel), 10.3 (Python), 10.4 (Excel), and 10.6 (Python) can be found at the internet adress: go.sn.pub/26leyH.

10.1.1 Straight Line Through Data Points by Sight

From the beginner's physics lab course, we are familiar with a procedure for getting physical parameters from measured data by suitably plotting them and drawing a straight line through the data points with a ruler. This way, the two characteristic parameters of a straight line, slope and y-axis intercept, can be obtained. Below are two examples.

(a) The Curie–Weiss law for the temperature dependence of the magnetic susceptibility χ_m of a ferromagnet above the Curie–Weiss temperature Θ,

D. Mergel, *Physics with Excel and Python*,
https://doi.org/10.1007/978-3-030-82325-2_10

$$\chi_m = \frac{C}{T - \theta} \tag{10.1}$$

leads to a linear plot of the reciprocal of the measured data, $1/\chi_{\mathrm{m}}$, versus T

$$\frac{1}{\chi_m} = \frac{1}{C} \cdot T - \frac{\Theta}{C} \tag{10.2}$$

from which the parameters C (Curie–Weiss constant) and Θ can be obtained.

(b) The reaction rate R of a chemical reaction as a function of the absolute temperature T, with k being the Boltzmann constant

$$R = R_0 \cdot \exp\left(-\frac{E_0}{kT}\right) \tag{10.3}$$

is logarithmized and plotted over $1/T$ (Arrhenius plot):

$$\ln R = \ln R_0 - \left(\frac{E_0}{k}\right) \cdot \frac{1}{T} \tag{10.4}$$

from which the collision rate R_0 of the reactants and the activation energy E_0 of the reaction can be obtained.

These are examples of linear regression.

10.1.2 Multilinear Regression

One method for finding an optimum linear function through data points is linear regression. The linear regression models in EXCEL and `Python` apply to functions with several independent variables x_i. They are based on the general *multilinear form*:

$$y = a + m_1 x_1 + m_2 x_2 + m_3 x_3 + \ldots \tag{10.5}$$

We will apply multilinear regression to points scattered around a straight line, a parabola $y = a + b{\cdot}x + c\,x^2$, and an exponential function $\ln(y) = \ln(A) + m{\cdot}x$.

Trend lines in charts, spreadsheets, Python
In EXCEL, you can insert trend lines of the linear, exponential, potential, logarithmic, polynomial, or power type into *xy*-diagrams ("scatter diagrams"). Coefficients of a multilinear trend line are obtained, together with their standard errors, with the LINEST function of EXCEL or the corresponding `OLS` ("ordinary least squares") in `Python`.

Both methods, LINEST and `ols`, output a *standard error* for each coefficient whose exact meaning depends on the number of data points and the number of parameters estimated from the data set, generally speaking, on the degrees of freedom. In our exercises, we check, with multiple repetitions of a random experiment, whether we have correctly recognized the degrees of freedom and correctly set the confidence limits. If this is the case, our broom rule Ψ *Two within and one out of (the standard error range)* must hold.

This check is possible because our data points are generated with "true" values on a curve and newly blurred with random noise for every repetition; then, the trend parameters are again estimated from the noisy data set. Ψ *We know everything and play stupid.*

Procedure of a random experiment
In order to check whether the hit rate corresponds to an expected confidence level, we

- generate "true" data points (x, y) from x-values and a function $y = f(x)$,
- blur them with normally distributed noise,
- transform the noisy values into a linear form,
- calculate the coefficients of a linear trend line through the noisy data, together with their standard errors extended by a t-value,
- check whether the error ranges capture the true values,
- determine the hit rate, i.e., how often in a series of simulations the error range of the trend line's coefficients does capture the true values.

10.1.3 Nonlinear Regression

If the measurements are subject to background noise, a constant must be added to the trend function. Then, no straight-line equations, such as those in Eq. 10.2 or Eq. 10.4, can be obtained by coordinate transformations. In these cases, nonlinear regression must be applied. Likewise, if the function itself is not linearizable by a coordinate transformation as, e.g., $y = \cos \omega_1 t + \cos \omega_2 t$.

In such cases, we generate the trend function to be adjusted with preselected parameters and vary them so that the function passes through the measured data. This can be done by hand and visual judgment, but also automatically with the EXCEL functions GOAL SEEK and SOLVER, or with the functions `curve_fit` from `scipy.stats` or `minimize` from the `scipy.optimize` library of `Python`.

For fitting trend lines to measurement data, there are special program packages, e.g., *Origin* for the general case, SCOUT for fitting model dielectric functions to optical spectra, and other programs for Rutherford backscatter spectra or impedance spectroscopy. However, it has proven to be very useful for improving their judgment when the students can organize simple adjustments in spreadsheets or computer programs themselves. This way, they are experiencing nonlinear regression behavior with examples in which they have all of the parameters in their own hands.

10.1.4 Coefficient of Determination R^2

Explained and residual variance

The coefficient of determination for any trend line is defined as

$$R^2 = \frac{explained\ variance}{total\ variance} = \frac{(v_{Tot} - v_{Res})}{v_{Tot}} \tag{10.6}$$

The terms in this formula are explained by means of Fig. 10.1 for a linear trend line through the data points y_{Ns}. In part **a**, the distances to the horizontal $y = y_{\mathrm{m}}$ (mean y) are marked with vertical lines. The sum of the square of these distances is $v_{\mathrm{Tot}} = 61.4$, entering Eq. 10.6 as *total variance* v_{Tot} *of the data set*. In part **b**, the trend line within the chart is plotted, together with its formula and the value of the coefficient of determination $R^2 = 0.71$ in the legend. Here, it is the distance of the data points to the trend line that is marked with vertical lines. The sum of the squares of these distances is $v_{\mathrm{res}} = 17.9$, entering Eq. 10.6 as *residual variance* v_{Res} *of the trendline*.

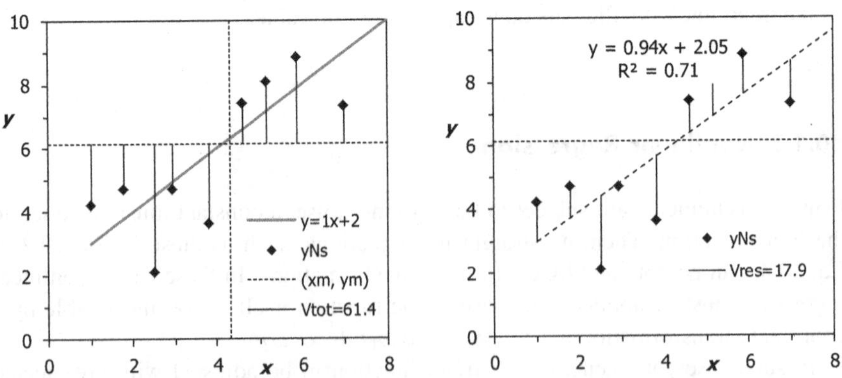

Fig. 10.1 **a** (left) Original curve and noisy data points; distances of the data points to the line $y = y_{\mathrm{m}}$ are marked by vertical lines. **b** (right) The same data as in (**a**), together with a linear trend line. Distances of the data points to the trend line are marked with vertical lines

The *variance explained by the trend line* is the difference between the residual variance and the total variance. In our example, Eq. 10.6 results in $R^2 = (61.4 - 17.9)/61.4 = 0.71$, the same value as reported in Fig. 10.1b.

Adjusted R^2
For the above calculation, the same degree of freedom has been supposed for v_{Tot} and v_{Res}, namely, $N - 1$, the number of data points minus 1. This is justified for v_{Tot} because only one parameter, the mean value, is estimated from the data set. Still, it is questionable for v_{Res} because more parameters describing the trend line are estimated from the data set, two for a straight line (*ddof* = 2 in `Python`). If the variances are calculated with the actual degree of freedom $N - ddof$, adjusted R^2 values result:

$$R^2_{adj} = \frac{v_{Tot} - v_{Res} \cdot \left(\frac{N-1}{N-ddof}\right)}{v_{Tot}} \tag{10.7}$$

leading to an expression with the easily obtainable entities R^2, N, and *ddof*.

$$R^2_{adj} = 1 - \left(1 - R^2\right) \cdot \left(\frac{N-1}{N-ddof}\right) \tag{10.8}$$

The term *ddof* stands for "delta degrees of freedom". For a set of N data points, we have the following degrees of freedom:

$f = N - 1$ for the variance of the data set,
$f = N - 2$ for the coefficients of a linear trend line (and an exponential trend line; see Exercise 10.4),
$f = N - 3$ for the coefficients of a parabolic trend line.

10.1.5 C-spec Error with Iterative *t* Adaptation

Real-world data
We can apply the procedure of a random experiment presented in Sect. 10.1.2 (Ψ *We know everything and play stupid*) to real-world data when making two steps beforehand:

- taking the experimental trend line through the data points as the "true" line and
- estimate the noise level as standard deviation of the residuals. The residuals are the difference between the y values of the data points and the trend line.

For multilinear trend lines, we can check whether our assumptions about the confidence level are correct.

Experimental *t* factors
The solver function in EXCEL does not return standard errors of the optimized coefficients, contrary to the function `curve_fit` of `scipy.stats`. Therefore, in Sect. 10.7.6, we determine C-spec errors ("confidence-specified") by a random experiment with tentative error ranges that are iteratively adapted to yield a hit rate that corresponds to the desired confidence level.

10.2 Linear Trend Line

We generate a set of points on a straight line ("generating line"), add normally distributed noise to their *y*-values, and then lay a straight trend line through these "data points". The coefficients of such a line and their standard errors are determined with the spreadsheet function LINEST. In `Python`, this is achieved with the function OLS ("ordinary least squares") of the library `Statsmodels`. The didactically most important message: Ψ *Two within and one out of*, meaning that in about 1/3 of the cases, the "true" coefficients of the generating line lie outside the *standard error range* of the estimated coefficients.

10.2.1 Creating Data Points and Evaluating Them

A linear trend line is to be laid through data points, minimizing the square deviation of the *y*-values. We generate the data points ourselves by randomly choosing *x*-values within a specific range, calculating the corresponding *y*-values on a straight-line ("true values"), and blurring them with normally distributed noise. Then, we estimate the coefficients of an optimum linear trend line through the data points and check whether the coefficients' error ranges capture the true values.

Data points are generated
We generate a set of data points scattered around a straight line defined by

$$y = a + m \cdot x \tag{10.9}$$

with m being the slope and a the y-intercept. The independent variable x is randomly equi-distributed between 0 and x_{Max}. The values of y are blurred with a normally-distributed noise specified by a standard deviation Ns:

$$y_{Ns} = y + \boldsymbol{noise} \tag{10.10}$$

The formula for *noise* is:

- Ns * randn(n) in Python, n being the size of the array generated, and
- NORM.INV(RAND; 0; Ns) in EXCEL, to be written into a range comprising n cells.

In EXCEL, y_{Ns} could also be specified as a whole as

$y_{Ns}(x) =$ NORM.(RAND(); y; Ns) .

The set (x, y_{Ns}) is our data cloud.

Linear trend line through the data points

Our task is to determine a linear trend line

$$y_R = m_R \cdot x + a_R \tag{10.11}$$

through the data cloud, which minimizes the sum of the quadratic deviations of y_{Ns} to this line. This procedure is called *linear regression*.

Two examples are shown in Fig. 10.2, together with the linear trend line (by definition, going through the center (x_m, y_m) of the data cloud) and two straight lines with coefficients with + or – their standard errors. The equations and the coefficients of determination R^2 of the trend line are reported in the legend. The trend line is called the "linear regression line". Its slope m_R is called the *regression coefficient*.

We start with the "true" function $y = 1x + 2$ to finally get the trend line for the noisy data:

$y_r = 1.16x + 0.98$ with $R^2 = 0.78$ for a noise level $Ns = 1.4$,

$y_r = 0.89x + 2.42$ with $R^2 = 0.28$ for a noise level $Ns = 2.0$.

Hit rates and degree of freedom *dof*

Statistical theory states that the "true" value is captured by the standard error range, plus-minus standard error from the estimated value, in 68.2% of the cases. This

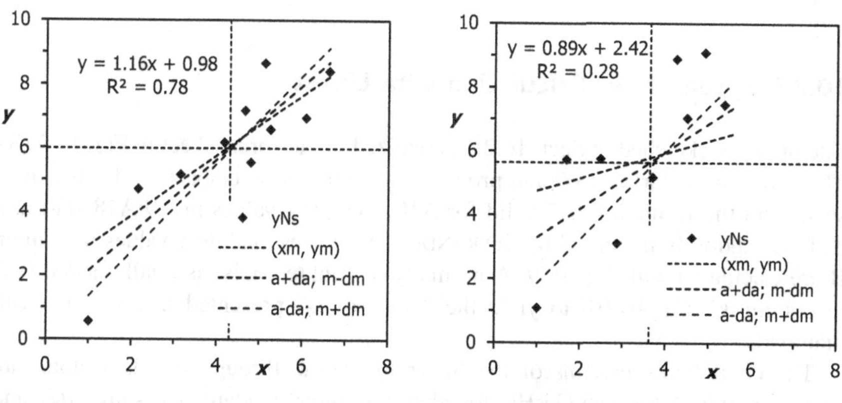

Fig. 10.2 Noisy data (diamonds) around the "true" line ($y = 2 + x$) with linear trend lines (solid) and lines for coefficients + or – their standard errors. **a** (left) Noise level 1.4. **b** (right) Noise level 2.0; noise level enters as standard deviation into a normal distribution

statement applies exactly only if the number of measurements goes to infinity, but our broom rule Ψ *Two within and one out of* applies as of 9 measuring points on a straight line.

▶ **Tim** Why now suddenly as of 9 measurement points? For measurement series, we have learned that the rule holds as of 8 measurements.

▶ **Mag** That's because of a different degree of freedom. For measurement series, the degree of freedom is the number of measurements -1, because only one parameter, the mean value, is estimated from the measurement series. For straight lines, however, two parameters are estimated, namely, slope m_R and *y*-axis intercept a_R. The degree of freedom $dof = 7$ is thus reached with 9 data points.

10.2.2 Data Structure and Nomenclature

x	set of independent variables starting at 1 and randomly increased by a value < dx
a, m	coefficients of a straight-line equation, "true values"
y	$y = a + m \cdot x$
Ns	noise level entering as the standard deviation in a normal distribution
y_{Ns}	y blurred with noise
x_m, y_m	average of the sets of x and y_{Ns}
a_R, m_R	coefficients estimated from a trend line through y_{Ns}
a_{Se}, m_{Se}	standard error of a_R and m_R
R^2	coefficient of determination
in_A, in_M	TRUE, if the error range captures the true value
hitA, *hitM*	number of hits, counting how often in_A, in_M are true.
dof	degrees of freedom, number of points – 2 for a linear trend line.

10.2.3 Spreadsheet Calculation with Linest

The program sketched in Sect. 10.2.1 is realized in the spreadsheet in Fig. 10.3 (S). The values of x start at 1.00 and progress with random interspaces ≤ dx, as can be seen from the formula in A7 valid for A10. The 10 x-values in A9:A18 start at 1 and are randomly increased by dx*RAND(). The corresponding y-values in column B are calculated with Eq. 10.9. A normally-distributed noise is finally added (y_{Ns} in column C, Eq. 10.10) to yield the "data points" presented in Fig. 10.2 with diamonds.

The coefficients m_R, a_R of the linear trend line through the data cloud are calculated with LINEST in G4:H6, together with their standard errors m_{Se}, a_{Se}, and the coefficient of determination R^2. With in_M and in_A (F9:G9), we check whether the error range captures the true values of m and a. For in_M, we take the standard

	A	B	C	D	E	F	G	H	I	J
1	*y intercept*	**a**	2.00	**y=1x+2**			*=LINEST(yNs;x;;1)*			
2	*Slope*	**m**	1.00				**mR**	**aR**		
3	*0<= x <=xMax*	**dx**	1.40				**mSe**	**aSe**		
4	*Noise*	**Ns**	2.00			Estimated	0.864	3.362		
5						Standard error	0.194	0.883		
6						**R²**	0.713	1.304		
7	*=A9+dx*RAND()*	*=m*x+a*	*=y+NORM.INV(RAND();0;Ns)*	*=AVERAGE(x)*	*=AVERAGE(yNs)*	*=AND(mR-mSe<m;m<mR+mSe)*	*=AND(aR-t.95*aSe<a;a<aR+t.95*aSe)*	10000	10000	
8	**x**	**y**	**yNs**	**xm**	**ym**	**in_m**	**in_a**	**hitM**	**hitA**	
9	1.00	3.00	4.55	4.03	6.85	TRUE	TRUE	6502	9474	
10	**1.67**	3.67	5.98			**t.95**	2.31	*=T.INV.2T(0.05;8)*		
18	8.17	10.17	9.36							

Fig. 10.3 (S) A linear function $y = m{\cdot}x + a$ is blurred with noise. LINEST in G4:H6 determines the coefficients of a linear trend line through the data cloud. In F9:G9, we check whether the standard error range captures the true values of m or the t-extended error range the true values of a. *Hit*M and *hit*A are returned from a VBA routine repeating the spreadsheet calculation 10,000 times

error m_{Se} whereas for in_A, we take the confidence interval for an error probability $E = 0.05$. Hit_M and hit_A are returned from a VBA rep-log procedure running the spreadsheet calculation 10,000 times.

Question

Questions concerning Fig. 10.3 (S):

What is the formula for y_{Ns}, the y-values of the data points, in C9:C18?[1]

Which parameter of the normal distribution is set by the parameter *Ns* in this formula?[2]

Trend line in the diagram

In Fig. 10.2, linear trend lines have been drawn through the "data points", together with their equations and coefficients of determination R^2. This is achieved by activating the data points and going through DESIGN/ADD CHART ELEMENT/TRENDLINE/LINEAR (EXCEL 2019). Proceed further with activating the trend line, going through FORMAT/SERIES 1 TRENDLINE 1/ and setting check-marks in DISPLAY EQUATION ON CHART and DISPLAY R- SQUARED VALUE ON CHART.

Each time the spreadsheet is modified (for example, by "clearing" the contents of an already empty cell), the noise is recalculated, and, accordingly, all data points in the graph change, yielding a new trend line.

[1] C9:D18 = [= y + NORM.INV (RAND(); 0; *Ns*)].

[2] *Ns* indicates the standard deviation of the normal distribution.

Linest as matrix function

What are the standard errors in the coefficients of the trend line? They are not reported in the legend for the trend line. To get them, we apply the LINEST spreadsheet function implemented in G4:H6 of Fig. 10.3 (S), performing linear regression. Its syntax is

$$\text{LINEST (known\_ys; [known\_xs]; [const]; [stats])} \tag{10.12}$$

LINEST is a matrix function taking, as input, the Y-values (here, y_{Ns} from Fig. 10.3 (S)) and the X-values (here, x from Fig. 10.3 (S)) and outputting a matrix with 4 rows and 2 columns, of which only the first three rows are of interest to us here.

As a reminder: To insert a matrix function into a spreadsheet, first, mark the range into which the results are to be written; in Fig. 10.3 (S), it is the (3R × 2C) matrix G4:H6 (highlighted in grey). Then, write [= LINEST (. After the parenthesis "(", the function window opens, indicating the expected input, Eq. 10.12. After entering the ranges for the Y and X-values, you may enter control parameters that determine whether the trend line should go through the origin (CONST = FALSE or 0; default is TRUE) or whether statistical characteristics should be output (STATS = TRUE or 1; default is FALSE).

If CONST = 0 or FALSE, then the straight line is drawn through the origin. If STATS = TRUE, additional regression characteristics are output, e.g., the standard errors of the coefficients and the coefficient of determination R^2. In our case, the trend line should not be forced to go through zero; therefore, the corresponding site remains empty. Regression characteristics are to be output; therefore, we set a 1 as the last argument, thus the function is called as LINEST (y_{Ns}; x;; 1).

▶ **Mag** How do you confirm the input of a matrix function?

▶ **Alac** With the magic chord![3]

In the first row of the output matrix of LINEST are the estimates m_R and a_R for m and a (note the order!). In the second row are the standard errors m_{Se} and a_{Se} of these coefficients. In the third row on the left is the coefficient of determination R^2, and on the right, the prediction error. The coefficients m_R and a_R, and R^2 correspond exactly to those of the trend line in a diagram with the same values for x and y_{Ns}.

The R^2 values reported in the figures and by LINEST are not adjusted for the reduced degree of freedom. To get R^2_{adj}, use Eq. 10.8!

Hit rates and degree of freedom

How often does the standard error range capture the true value?

With a rep-log procedure, we count how often *in_m* and *in_a* in F9:G9 are true. They are true if the respective error intervals ($m_R - m_E$, $m_R + m_E$) and ($a_R - a_E$,

[3] CTL + SHIFT + RETURN.

$a_R + a_E$) capture the true coefficients. We get hit rates of $hitM = 6502$ for 10,000 trials, closely corresponding to the theoretical value C = 0.653 for 8 degrees of freedom obtained with 1-T.DIST.2T(1; 8) = 0.653 (`2 * sct.t.cdf(1, 8)-1` in `Python`). The 8 degrees of freedom are given by 10, the number of data points, diminished by 2, because the 2 parameters m_R and a_R are estimated from the data set.

The hit rate for a is calculated with the standard error multiplied with 2.31, the t factor for $E = 0.05$. It corresponds closely to the expected 9500.

Rounding to a meaningful number of digits
In the diagrams, the coefficients of the trend line are given with 2 digits. By comparison with the standard errors from LINEST, we see that this assumes an accuracy that is greater than the uncertainty allows. The equation of the straight trend line is now, according to the results of LINEST in G4:H4 of Fig. 10.3 (S) and rounded to the number of significant digits,

$$y_R = (0.9 \pm 0.2)x + 3.4 \mp 0.9$$

We have rounded by inspection and written the result down by hand. However, this can also be achieved with mathematical formulas specified in the spreadsheet calculation in Exercise 9.3 or with the help of the user-defined function *FinRes* in Sects. 9.3.3 and 9.3.4. For more information about LINEST, see EXCEL HELP for this feature.

10.2.4 Python Program

The `Python` program in Table 10.1 generates 10 noisy data points (x, y_{Ns}) around a straight line (cell **a**) and evaluates them with `sm.OLS` (cell **b**). If OLS is called as = `sm.OLS(yNs, x)`, a trend line $\mathrm{y_R} = \mathrm{m_R}\ \mathrm{x}$ is fitted to the data. To get a fit to $\mathrm{y_R} = \mathrm{a_R} + \mathrm{m_R}\ \mathrm{x}$, we have to create a new set of independent variables `xx = sm.add_constant(x)` . The output in cell **c** is arranged so that it resembles the output of LINEST in EXCEL.

A figure corresponding to Fig. 10.2, as obtained with the program presented in Table 10.2.

Question

Where are the results in lines 13 and 14 in Table 10.1 needed? [4]
In Fig. 10.4a, $R^2 = 0.74$ is reported. How big is $R^2{}_{\mathrm{adj}}$? [5]

[4] The coordinates x_m and y_m of the center of the data cloud are not needed in the presented program. They are necessary for a figure like Fig. 10.2.
[5] With Eq. 10.8 and $N = 10$ and `ddof` = 2, we get $R^2{}_{\mathrm{adj}} = 0.7075 \approx 0.71$.

Table 10.1 **a** (top) Generation of noisy data points (x, y_{Ns}) around a straight line; **b** (bottom left) Linear regression line through the data points; **c** (bottom right) Results of the linear regression OLS

```
import numpy.random as npr
import statsmodels.api as sm
a,m = 2.0,1.0
x=np.zeros(10)
dx=1.4
Ns=1.4          #Noise level
x[0]=1
for i in range(1,10):
    x[i]=x[i-1]+dx*npr.rand() #Uneven spacing of x values

y=m*x+a
yNs=y+Ns*npr.randn(len(x))
```

```
xm=np.average(x)
ym=np.average(yNs)
xx=sm.add_constant(x)
model=sm.OLS(yNs,xx)
results=model.fit()
aR=results.params[0] #y=a+mx
mR=results.params[1]
aE=results.bse[0]      #Errors
mE=results.bse[1]
r2=results.rsquared
r2_ad=results.rsquared_adj
in_a=(aR-aE < a) \
          and (a < aR+aE)
in_m=(mR-mE < m < mR+mE)
```

```
xx [[ 1.00   1.00]
    [ 1.00   1.18]
       . . .
    [ 1.00   3.86]
    [ 1.00   4.54]]

               mR     aR
est            0.98   2.05
mE,aE          0.21   0.57
r²             0.74
r²_adj         0.70
in_m,in_a      True True
```

Table 10.2 Python program for plotting Fig. 10.4a, variables specified and results reported in Table 10.1

```
lblT="y="+str(np.round(mR,2))\
    +"x+"+str(np.round(aR,2))\
    +"\n"+"R²="+str(np.round(r2,2))

FigStd('x',0,6,2,'y',0,8,2)
plt.plot(x,yNs,'kx')
ylin=x*mR+aR
plt.plot(x,ylin,'k-', label = lblT)
plt.legend()
plt.savefig("PhEx 9-2 trend line",dpi=1200)
```

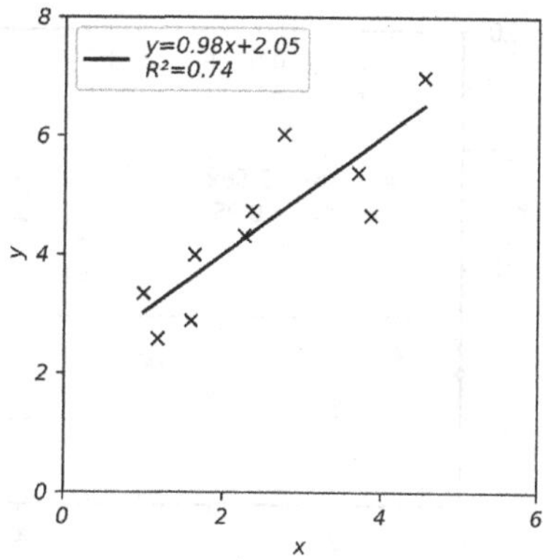

Fig. 10.4 Chart with trend line and formula produced with the Python program in Table 10.2

10.3 Fitting a Polynomial Trend Line to Data Points with Multilinear Regression

We generate points on a parabola and add normally distributed noise to their *y*-values. A polynomial of the second degree is fitted to these simulated data, first, as a trend line within a diagram, and then with the spreadsheet function LINEST in EXCEL and with `OSL` of the `statsmodel` library of `Python`. These functions return, in addition to the coefficients of the polynomial, their standard errors. - Confidence intervals and confidence levels are interrelated through *Student*'s *t* statistics. - Experience teaches us that measurement points must extend beyond the parabola vertex for its coefficients to be estimated reliably.

10.3.1 Introduction

Noisy data points around a parabola
The formula of a parabola is

$$y = a + bx + cx^2$$
$$y = ax^0 + bx^1 + cx^3 \quad (10.13)$$

The second variant emphasizes the powers of x.

Figure 10.5 shows 9 data points around a parabola generated with Eq. 10.13 and blurred with normally-distributed noise to yield values y_{Ns}, in **a** for a noise level (standard deviation of a normal distribution) of $Ns = 1$ and in **b** for $Ns = 2$.

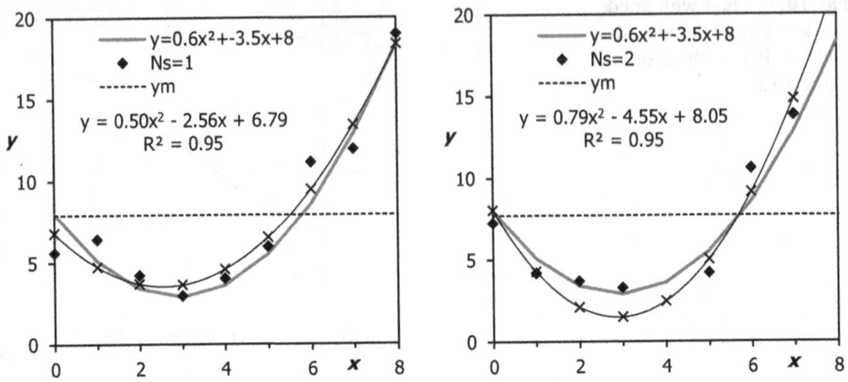

Fig. 10.5 Nine data points around a parabola (grey lines), together with polynomial trend lines (thin black lines) of degree 2. **a** (left) Noise level $Ns = 2$. **b** (right) Noise level $Ns = 4$

Trend parabola

Trend parabolas $y_R = a_R \cdot x^0 + b_R \cdot x^1 + c_R \cdot x^2$ are obtained by multilinear regression, in which the function is regarded as linearly depending on three variables x^0, x^1 and x^2 that formally have to be stacked together in a matrix with three columns, e.g., in `Python`, with `xxx = np.array([1, x, x²])`. The multilinear regression functions return the values a_R, b_R, c_R, together with their standard errors a_{Se}, b_{Se}, c_{Se}, and the coefficient of determination R^2.

In EXCEL diagrams, polynomial trend lines can be fitted to the data, with their equations and R^2 being reported in its legend, as in Fig. 10.5. Attention: R^2 is unadjusted for a reduced degree of freedom.

Questions

Consider two data sets A and B, with the same number of points around the same parabola, but with the noise level of B being higher than that of A. What do you expect for the R^2 values for a fit with a parabola to A and B? Which one is higher?[6]

What is the degree of freedom for regression analysis of our parabola with 9 data points?[7]

The coefficient of determination R^2 is reported in the figures, together with the equation of the trend parabolas. Describe how R^2 is calculated with the data points (x, y_{Ns}) and the points (x, y_R) on the trend line! Consider the different variances, total, residual, explained![8]

---

[6] The R^2 for a fit to the data set with a lower noise level is expected to be higher, contrary to Fig. 10.5. Average values of 0.97 for **a** and 0.90 for **b** are to be expected.

[7] Degree of freedom $dof = 9$ (number of data points) – 3 (parameters estimated from the data set) $= 6$.

[8] Consult Eq. 10.6! Total variance $v_{Tot} = var(y_{Ns}) = (y_{Ns}\text{-}y_m)/dof$, residual variance $v_{Res} = sum(y_{Ns}\text{-}y_{Trend})^2/dof$, $R^2 = (v_{Tot}\text{-}v_{Res})/v_{Tot}$ = explained variance/ total variance.

What are the degrees of freedom to be set for adjusted R^2?[9]

Hit rates, with and without the *t* factor

The hit rates for, e.g., ($a_R - a_E < a < a_R + a_E$) are around 0.64, as will be seen later in Fig. 10.7, and thus significantly smaller than suggested by our rule Ψ *Two within and one out of*. This is because the degree of freedom for the parabola with 9 (points) – 3 (coefficients of the parabola) = 6 is too small for this rule to apply. For a degree of freedom $f = 6$, a hit rate of 0.644 (= 1- T.DIST.2 T(1; 6) in EXCEL or `1 - (1 - 2 * sct.t.cdf(1, 6)` in `Python`) is actually expected according to Student's *t*-distribution.

▶ **Tim** I'm confused. What should I state as the result?

▶ **Alac** Don't bother. Nobody knows about degrees of freedom and the *t*-value anyway.

▶ **Mag** If you specify the coefficients' standard error and the number of data points, an expert can derive statistically relevant statements. You can become such an expert yourself with our exercises.

The *t*-value for six degrees of freedom is $t = 1.091$ obtained with T.INV.2S(0.318; 6)] (EXCEL) or `sct.t.ppf (1-0.318/2,6) (Python)` for an error probability of 0.318 (confidence level 0.682) and $t = 2.45$ for an error probability of 0.05 (confidence level 0.95). Tests with an uncertainty given by the standard error multiplied by the *t*-values ("*t*-extended standard error") indeed give hit rates that correspond to the confidence levels.

▶ **Mag** What is the degree of freedom when calculating the variance of y_{Ns} or ($y_{\text{Ns}} - y_{\text{Trend}}$)?

▶ **Alac** I would again say 6, but I'm not sure.

▶ **Tim** To calculate the variance of a set, we have to estimate the mean of this set. So, I conclude that the degree of freedom is 9 – 1 = 8.

▶ **Mag** Tim is correct. Calculating $v_{Tot} = \sum(y_{Ns} - y_m)^2/8$ and $v_{Res} = \sum(y_{Ns} - y_R)^2/8$ yields the same value for $R^2 = (v_{Tot} - v_{Res})/v_{Tot}$ as that reported in the figures or obtained with the multilinear regression functions.

▶ **Alac** OK, that's clear.

[9] *Dof* = $N - 1$ for the total variance, *dof* = $N - 3$ for the residual variance, because 3 parameters are estimated from the data set.

▶ **Mag** But take care! For the adjusted R^2 value, you have to adjust the degrees of freedom, see Sect. 10.1.4!

Reliable fit of a parabola to the data points

In Fig. 10.6, trend parabolas are drawn through 9 data points. In Fig. 10.6a, the trend parabola differs more from the original parabola than in Fig. 10.6b, although $R^2 = 0.99$ is the same for both. The reason is that the characteristic feature of a parabola, the region around an extremum, is less well represented in the data points in **a**.

Questions

Why do the coefficients of the trend parabola fitted to the data points in Fig. 10.6a differ so much (≈20%) from the true values (see the legend), although $R^2 = 0, 99$ is achieved?[10]

Why does the trend curve in Fig. 10.6b capture the "true"parabola better than the one in Fig. 10.6a?[11]

Higher-order polynomials

One can insert polynomials of higher degree as a trend line in diagrams. This is also possible with LINEST in the spreadsheet or with `OLS` in `Python`. You have to create column vectors x, x^2, …, and x^n, and enter the whole set as an argument for KNOWN_X'S in LINEST. For OLS, you have to extend the matrix with x^0, i.e., a sequence of ones. This procedure can be generalized to other functions of x.

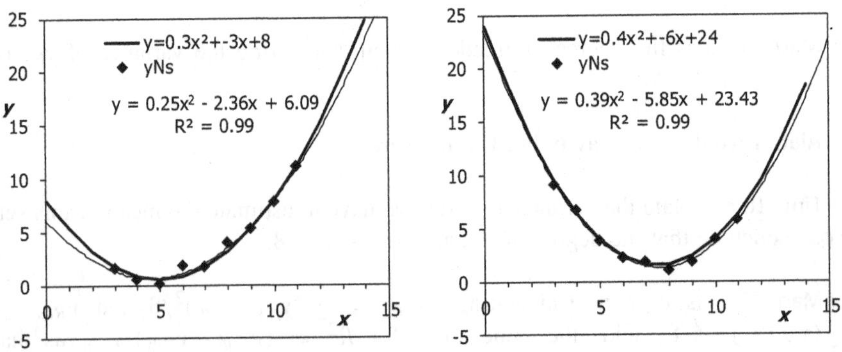

Fig. 10.6 a (left) Original parabola, simulated data points and trend line (second-degree polynomial) for unfavorably located data points. **b** (right) As in **a**, but for favorably located data points

[10] The differences in the coefficients are so big because the data points do not capture the minimum well.

[11] In Fig. 10.6b, the characteristic property of the parabola, an extremum, is captured.

	A	B	C	D	E	F	G	H	I	J
1	**a**	8.00			=LINEST(yNs;xx;;1)					
2	**b**	-3.50			**cR**	**bR**	**aR**	Estimated		
3	**c_**	0.60			**cSe**	**bSe**	**aSe**	Stand. Error		
4		$y=0.6x^2+-3.5x+8$			0.66	-4.33	10.14	Estimated		
5	**Ns**	1.00			0.04	0.34	0.58	Stand. Error		
6		Ns=1		R^2	0.98	0.71	#N/A			
7	**dx**	1.00		**N**	**NinC**	**NinB**	**NinA**	**t.32**	1.09	=T.INV.2T(0.317;6)
8				1000	654	949	956	**t.05**	2.45	=T.INV.2T(0.05;6)
9	=A11+dx	=x^2	=a+b*x+c_*x2p	=y+Ns*NORMINV(RAND();0;1)	=AND(cR-t.32*cSe<c_;c_<cR+t.32*cSe)	=AND(bR-t.05*bSe<b;b<bR+t.05*bSe)	=AND(aR-t.05*aSe<a;a<aR+t.05*aSe)	=AVERAGE(yNs)		
10	**x**	**x2p**	**y**	**yNs**	**inC**	**inB**	**inA**	**ym**		
11	0.00	0.0	8.00	10.06	FALSE	FALSE	FALSE	7.86		
12	**1.00**	1.0	5.10	6.46						
19	8.00	64.0	18.40	18.30						

Fig. 10.7 (S) In columns A:D, data points around a parabola $y = a + b{\cdot}x + c{\cdot} \times 2$ are created. In E4:G6, the noisy data $y_{Ns} = f(x)$ are fitted with a parabola. The results in E8:G8 are reported by a rep-log procedure that counts how often in_C, in_B, in_A are true for N repetitions of the experiment. The formulas for in_C, in_B, and in_A are different! N_{inC} is for $t = 1$, N_{inB} for $t = 1.09$, N_{inC} for $t = 2.45$. Data points and the trend line are shown in Fig. 10.5

10.3.2 Data Structure and Nomenclature

a, b, c	"true" coefficients of a parabola $y = a + b \cdot x + c{\cdot}x^2$
x	array of independent variables
x_{2p}	x to the second power
XX	$[x, x_{2p}]$ for EXCEL
XXX	$[1, x, x_{2p}]$ for Python
y	$y(x, x_{2p})$, function of two variables, values of the parabola for x and x_{2p}
Ns	noise level, entering as standard deviation in a normal distribution
y_{Ns}	y blurred with normally distributed noise
a_R, b_R, c_R	coefficients of a regression (trend) line through the data points (x, y_{Ns})
m_{Se}, a_{Se}. c_{Se}	standard error of m_R, a_R, and c_R
in_A, in_B, in_C	TRUE, if the error range captures the true value
N_{inA}, N_{inB}, N_{inC}	number of hits, counting how often in_A, in_B, in_C are true
dof	degrees of freedom = number of points – 3 for a parabolic trend line
t_{32}, t_{05}	t values for confidence levels 0.68 and 0.95.

The argument structure is different for EXCEL and Python:

EXCEL LINEST (yNs;xx;;1)
Python `sm.OLS(yNs, xxxT)` with *xxx*T being the inverse of *xxx*.

10.3.3 Spreadsheet Solution

Generating the data set
In Fig. 10.7 (S), a spreadsheet layout for fitting a parabola trend line to data points is presented. The formula for y accesses the two variables $x_1 = x$ and $x_2 = x^2$ ($=x_{2p}$), as well as the constant a. The y-values are masked with Gaussian noise with standard deviation Ns and stored in y_{Ns}. Now, our data set is (x, y_{Ns}).

Evaluating the data set
The noisy data set is evaluated in E4:G6 with

LINEST(*yNs;xx;;1*).

to obtain the estimated coefficients c_R, b_R, a_R, together with their standard errors c_{Se}, b_{Se}, a_{Se}. Note the argument for the independent variables in the formula reported in E1! The independent variables x and x_{2p} have to be put together in one range, here, in a range A11:B19 named *xx*.

We calculate the error range of the coefficients for different confidence levels, i.e., we have to multiply the standard error with an appropriate t-value, and count, with a rep-log procedure, how often the error ranges capture the true values. The counts are stored in variables named $N_{in}C$, $N_{in}B$, $N_{in}A$.

The t-value for a confidence level of 0.95 is obtained with T.INV(0.05; *dof*), where *dof* is the degree of freedom; see the formula in J8, valid for I8.

Question

concerning Fig. 10.7 (S):

Which are the independent variables in LINEST?[12]

Where are the t-values calculated, and what is their value for inC, inB, inA?

What is the confidence level of in_C, in_B, in_A? Are the numbers of TRUE in N_{in}C, N_{in}B, N_{in}A in agreement with expectation?[13]

---

[12] The independent variables are x in A11:A19 and x^2 in B11:B19, put together as argument xx = A11:B19 for LINEST.

[13] In_C is calculated with t.32, valid for a confidence level of 1–0.317 = 0.683 close to the empirical $N_{in}C/N$ = 0.654; in_A with t.05 for a confidence level of 1–0.05 = 0.95 close to $N_{in}A/N$ = 0.956.

How do you use LINEST to determine the coefficients of a trend line for the function $y = a \cdot x_1^5 + b \cdot \log(x_2)$?[14]

Trend line in the diagram
To lay a parabola through the nine data points, DESIGN/ADD CHART ELEMENT/TRENDLINE/ FORMAT TRENDLINE/TRENDLINE OPTIONS/ offers, among others, POLYNOMIAL/ ORDER 2. With appropriate clicks, the obtained regression equation and the coefficient of determination R^2 of the fit are reported in the diagram.

Hit rates
We apply a rep-log procedure to the spreadsheet calculation reading the values TRUE or FALSE in E11:G11 N times and returning the counts into $N_{in}C$, $N_{in}B$, $N_{in}A$ (E8:G8). The hit rates 6422/10000 = 0.642, 6837/10000 = 0.684, 9501/10000 = 0.95 correspond to the confidence levels expected for the different t-values applied for in_C, in_B, in_A (see Footnote 13).

10.3.4 Python Solution

Generation of noisy data
In Table 10.3, noisy data y_{Ns} are generated around a parabola, the coefficients of which are specified all together in a list *abc*, not separately as *a, b, c*. The variables are specified as three row vectors x_0, x, and x_{2P}, with x_0 being a list of ones and x_{2P} the squares of x; they are gathered in a matrix *xxx* that is transposed to *xxxT* to become a list of column vectors (see lines 7–11) required as entry into the OLS function, together with y_{Ns}.

Polynomial through the noisy data
In Table 10.4, the data set (*xxxT*, y_{Ns}) is evaluated with the OLS (ordinary least squares) function of `scipy.stats`, returning the results into *est* = sm.OLS(...) consisting of

- a list of the estimated parameters (est.params),
- a list of their standard errors (est.bse),
- some characteristics of the fit (e.g., est.rsquared).

The results for a run are shown in the bottom cell of Table 10.4.

[14] A two-column range is calculated for $x_1{}^5$ and $\log(x_2)$ and entered as KNOWN_X'S in LINEST.

Table 10.3 Python program, generating the data set (*xxxT*, y_{Ns}) from (*x*, *y*)

```
import numpy as np
import numpy.random as npr

abc=[8.0,-3.5,0.6]                          #Coeffs. of parabola
Ns=1
dx=1
x=np.linspace(0,8,9)
x2p=x**2
x0=np.ones(len(x))
xxx=np.array([x0,x,x2p])
xxxT=xxx.transpose(1,0)

y=abc[0]*x0+abc[1]*x\+abc[2]*x2p  #True parabola
yNs=y+Ns*npr.randn(len(y))                  #Blurred with noise
```

```
yNs:
 [ 7.44  6.15  1.76  5.14  3.42  5.13  8.74  12.32  16.80]
xxxT:
 [[ 1.00  0.00  0.00]
 [ 1.00  1.00  1.00]
 [ 1.00  2.00  4.00]
 ...
 [ 1.00  6.00  36.00]
 [ 1.00  7.00  49.00]
 [ 1.00  8.00  64.00]]
```

Table 10.4 Evaluation of the data set (*xxxT*, y_{Ns}) with OLS (ordinary least squares) of scipy.stats

```
import statsmodels.api as sm
import scipy.stats as sct

est=sm.OLS(yNs,xxxT).fit()
t05 = sct.t.ppf(0.975,6)     #t value for E = 0.05
inABC=np.zeros(3)            #For the three coeffs.
for i in range(3):
    inABC[i]=(est.params[i]-t05*est.bse[i] <
              abc[i] < est.params[i]+t05*est.bse[i])
                             #In error range?
par=list(est.params)
err=list(est.bse)
rSq=est.rsquared
```

```
abc     8.00  -3.50   0.60
par     7.38  -3.39   0.60
err     0.87   0.51   0.06
rSq     0.97
t05     2.447
in         1      1      1
```

Table 10.5 Program lines of Tables 10.3 and 10.4 are assembled into a function returning the logical list `in ABC`

```
t05 = sct.t.ppf(0.975,6)              #t value for E=0.05
t32 = sct.t.ppf(0.8415,6)             #t value for E=0.318
t=[1.0,t32,t05]
def FitToPol():
    yNs=y+Ns*npr.randn(len(y))
    est=sm.OLS(yNs,xx).fit()          #Linear regression
    inABC=np.zeros(3)
    for i in range(3):                #Coeffs. in error range?
        inABC[i]=(est.params[i]-t[i]*est.bse[i]
                 < abc[i] <
                 est.params[i]+t[i]*est.bse[i])
    return inABC
```

Questions

How do you calculate Student's t value for a parabola's coefficients for a confidence level of 2/3 and 12 data points?[15]

How is the regression line in Table 10.4 (coefficients in bottom cell) sensibly reported with uncertainties in the coefficients?[16]

Which lines in Tables 10.3 and 10.4 have to be gathered in a function, to be called in a loop to get hit rates for $E = 0.05$? Which variable has to be returned and then summed up in the main program?[17]

Hit rates

In Table 10.5, generation of y_{Ns} and evaluation of the data set are transferred into a function that returns the logical values in_{ABC}. This function is called in the main program of Table 10.6 in a loop in order to determine the hit rates. We have chosen three different t values for the three coefficients, 1, t_{32}, and t_{05}, corresponding to confidence levels of 0.644, 0.683 ($= 1 - (1 - 0.8415) * 2$), and 0.95, respectively.

In a loop in Table 10.6, we check whether the error ranges capture the true values. To do this efficiently, parameters concerning the coefficients a, b, c are put together in lists of shape comparable to the list in `est`:

- a, b, c in abc, see Table 10.3,
- the logical values in_{A}, in_{B}, in_{C} in in_{ABC}, see Table 10.4,

[15] `sct.t.ppf (1-1/6, 12-3) = 1.02.`

[16] $y_{\mathrm{R}} = 7.4(9) - 3.4(5)\,\mathrm{x} + 0.60(6)\cdot x^2$.

[17] Lines 14, 18–23 have to be gathered in a function; in_{ABC} has to be returned and summed up in the main program, see Table 10.5.

Table 10.6 Main program for getting hit rates for *a*, *b*, *c*; `NinA = NinABC[0]`, etc.

```
nRep=10000
NinABC=np.zeros(3)
for n in range(nRep):
    inABC=FitToPol()
    NinABC[0]+=inABC[0]
    NinABC[1]+=inABC[1]
    NinABC[2]+=inABC[2]
```

t	1.00	1.09	2.45
N	NinA	NinB	NinC
10000	6438	6888	9524

and are addressed in Tables 10.4, 10.5 and 10.6 with their indices in these lists, e.g., `NinABC[0]+ = inABC[0]`.

The *t*-value for a confidence level of 0.95 is calculated according to

`t05= sct.t.ppf (0.975, 6).`

`t.ppf` returns the value for a one-sided test. As we are checking whether the true values are larger than the lower error limit and smaller than the upper error limit, we have to enter an error probability 0.05/2 = 0.025 instead of 0.05. The above formula returns $t_{05} = 2.45$.

The hit rates in the bottom cell of Table 10.6 are, with 0.64, 0.69, and 0.95, close to the theoretical confidence levels 0.642, 0.69, and 0.950 (see also Footnote 13).

10.4 Exponential Trend Line

We generate noisy data points (x, y_{Ns}) around an exponential and fit a trend line to the data, (1) as an exponential trend line in an EXCEL diagram, (2) with LINEST, and `OLS` of the `statsmodel library` of `Python` as a fit of a straight line through the logarithmized y_{Ns} data, and (3) with the spreadsheet matrix function LOGEST with a corrected standard error for the amplitude. For the exponential function to be clearly distinguishable from a parabola, the data points must cover a sufficiently large *x*-range.

10.4.1 Exponential and Logarithm

An exponential curve is defined by

$$y = g \cdot \exp(h \cdot x) \tag{10.14}$$

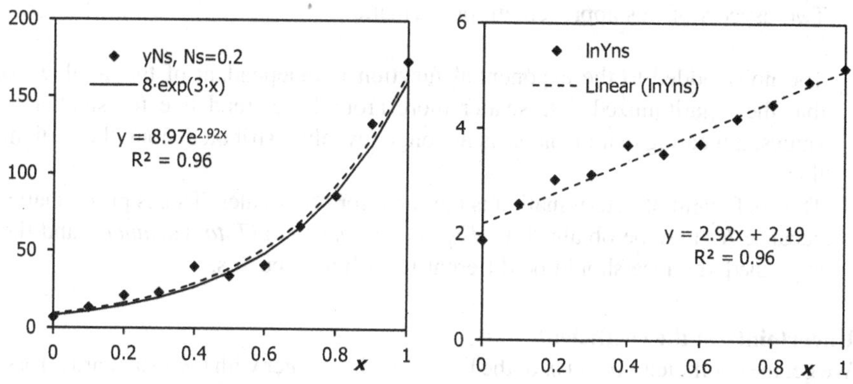

Fig. 10.8 **a** (left) Data points obtained from an exponential function blurred with noise on the *y* values, together with an exponential trend line. **b** (right) Natural logarithm of the values in **a**, together with a linear trend line

with *g* being the amplitude and *h* a characteristic growth parameter.

To simulate measured data, we add normally distributed noise to the function values *y*. A fit to the noisy data is most easily obtained by adding an exponential trend line in an EXCEL diagram, with the formula and coefficient of determination R^2 displayed as a legend. An example is shown in Fig. 10.8a.

Linear trend of the logarithmized data

By logarithmizing *y*, Eq. 10.14 can be converted into a straight-line equation of the form $z = a + m{\cdot}x$:

$$\ln(y) = \ln(g) + h \cdot x \tag{10.15}$$

with $z = ln(y)$, $a = ln(g)$, and $m = h$.

The logarithmized noisy y_{Ns} values of Fig. 10.8a are shown in Fig. 10.8b, together with a linear trend line. The output in the legend is essentially the same as in **a**, considering that $a = \exp(2.19) = 8.97$. This indicates that the exponential trend line is indeed obtained by a least-squares fit to the logarithmized data.

Question

When fitting an exponential trend line in a diagram, sometimes, the following error message appears: TRENDLINE CAN NOT BE CALCULATED FOR NEGATIVE VALUES OR FOR ZERO. When and why is that the case?[18]

[18] The y_{Ns} values are logarithmized within the function that computes the exponential trend line. An error occurs when a *y*-value is negative or zero what occurs if the noise value is negative and its absolute value bigger than the signal.

Two aspects of this approach are problematic:

- The noise added to the exponential function is independent of the y-value, so that the logarithmized data scatter more around the trend line for smaller y-values, indicating that the noise is no longer evenly distributed along the straight line.
- The coefficient of determination is the same for both scales. This is problematic, because R^2 may be obtained as (*Explained variance*)/(*Total variance*), and the explained variance should be different for different curves.

Uncertainty in the coefficients

We get the coefficients a and m of the linear trend, together with their standard errors, as usual, with LINEST and OLS and have to transform them into the coefficients g and h of the exponential function with $h = m$ and $g = \exp(a)$.

As the coefficient m is identical to h, so is the standard error $h_{\mathrm{Se}} = m_{\mathrm{Se}}$. The standard error g_{Se}, however, must be calculated according to the error-propagation law. We have $g = \exp(b)$ and $dg_E/db = \exp(b)$ so that

$$g_{Se} = \exp(b_R) \cdot b_{Se} \tag{10.16}$$

Due to the problems arising from logarithmizing the noisy data and deriving the uncertainty in the coefficient of the exponential from those of the linear trend of the logarithmized data, we have to be careful with the confidence level.

Influence of noise on the confidence level

For reasonable estimates of the standard error, the hit rate should be about 0.657 (Ψ *Two within and one out of*). With a rep-log procedure repeating the random experiment, "Eleven points around an exponential" 1000 times, we check how often this is actually the case for our experiment. The result can be seen in Fig. 10.9, below the label "const. noise".

For the fitting of an exponential trend to 11 data points, the degree of freedom is $dof = 11 - 2 = 9$. This results in a confidence level for the standard error range of 0.657, calculated as 1-T.VERT.2S(1; 9) (EXCEL) or `2 * sct.t.cdf(1,9)-1`

	const. noise				prop. noise	
	h=3	h=1.7	h=0.5	h=0.1	h=3	h=1.7
Total	1000	1000	1000	1000	1000	1000
Hits h	460	547	651	652	650	651
Hits g	383	463	601	641	643	659

Fig. 10.9 How often are the true values of h and g captured by the error range, for signals with constant noise level (*const. noise*) and noise proportional to the y-values of the exponential (*prop. noise*)?

(Python). With 1000 repetitions of the random experiment, one, therefore, expects that, in 657 cases, the error range of the estimated coefficients will capture the true value. However, in Fig. 10.9, the greater the coefficient h in the exponent, the less the theoretical value of 657 is reached, independent of the noise level.

Figure 10.8b indicates the reason for this. The deviations from the "true" curve are unevenly distributed on the log-scale. For large values of x, they are smaller, so that the trend line is closer to the "true" curve than for small values of x. This is a consequence of logarithmization. However, the unequal distribution of the deviations from the "true" values does not correspond to the mathematical model on which linear regression is based.

If we apply a noise proportional to the y-value, e.g., with

YNS = Y + NORM.INV(RAND();0;NS) * Y/4(EXCEL)
or `yNs = y + Ns * randn(len(y)) * y/4 (Python)`,

then the deviations in the logarithmic values are more evenly distributed over the range of t (see Fig. 10.10b). Now, the deviations in the linear representation, Fig. 10.10a, are distributed unevenly. The number of hits in Fig. 10.9 (under "prop. noise") deviates only slightly (less than 4%) from 657, also independent of the noise level.

► **Tim** When fitting an exponential trend, probably nothing fits at all; not even the noise is reliable.

► **Alac** I don't see it that way. You can always give reasonable values for the coefficients.

► **Mag** In principle, that's true. For low noise, the uncertainties and the differences among the various types of noise are not so big.

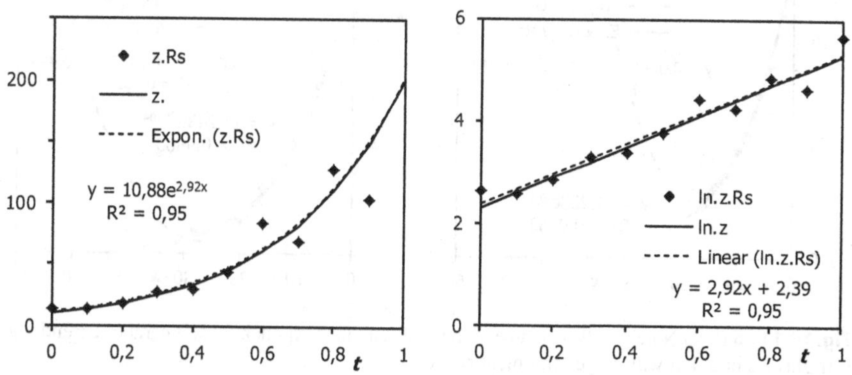

Fig. 10.10 **a** (left) Exponential curve with noise proportional to the y-value. **b** (right) Logarithmized data of **a**, together with a linear trend line

▶ **Tim** What exactly does "not so big" mean?

▶ **Alac** In a given case, we just figure it out with a simulation based on our method Ψ *We know everything and play stupid* to get the hit rate.

C-spec error by simulation-based *t* adaptation
We take the regression curve as the "true" curve, estimate the noise level from the data as well, and perform a statistical simulation with these parameters, similar to Sect. 10.7.6.

10.4.2 Exponential or Polynomial?

We fit both an exponential and a polynomial to the same set of data points generated around an exponential function. Figures 10.11a, b show two examples.

In Fig. 10.11a, both trend lines describe the experimental data equally well, with coefficients of determination R^2 of 0.9942 and 0.9953. When repeating the random experiment (new measurement points are randomly generated), the parabola sometimes gets an even higher R^2 value, even though the "true" original curve is exponential. There are not enough measurement points to clearly identify the exponential character. In Fig. 10.11b, a larger argument range was chosen. Even now, the parabolic trend line does not differ significantly from the measurement points; but it would predict completely wrong *y*-values for *x* greater than 35.

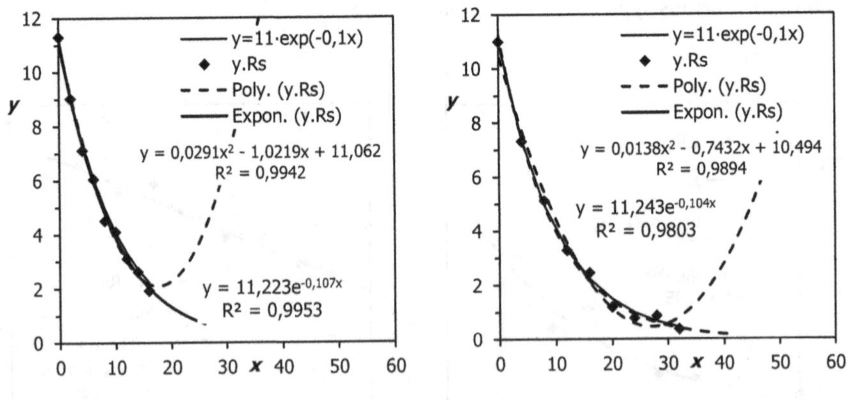

Fig. 10.11 **a** (left) Nine points near an exponential function; exponential and parabolic trend line. **b** (right) As in **a**, but with larger argument range

Question

Why is a parabola not an appropriate choice for the data in Fig. 10.11?[19]

► **Alac** In any case, the exponential trend line in Fig. 10.11a yields a good value of the decay constant.

► **Mag** If you are convinced that the underlying physical process is actually exponential, you may determine the decay constant from Fig. 10.11a.

► **Tim** What does it mean to be "convinced"?

► **Mag** That there is a theory of the physical process that predicts an exponential.

► **Alac** …and then we remember "not disproved but also not proven" …

► **Tim** …but preliminarily accepted. This time, that makes sense to me.

► **Mag** That's the best one can get out of the experiment. Discuss!

10.4.3 Data Structure and Nomenclature

x	series of independent variables
y	exponential values of x; $y = g{\cdot}exp(h{\cdot}x)$
Ns	noise level entering as a standard deviation into a normal distribution
y_{Ns}	y blurred with normally distributed noise
$\ln Y_{Ns}$	y_{Ns} logarithmized
m_R, aR	coefficient of a straight trend line through (x, lnY_{Ns})
m_{Se} a_{Se}	standard error of m_R, a_R
h_R, g_R	coefficients of the exponential trend line (regression line)
h_{Se}, g_{Se}	standard error of h_R, g_R, to be calculated from the results of the linear trend.

10.4.4 Python Program

The noisy data points around an exponential are generated in the top cell of Table 10.7 and logarithmized in the middle cell. The logarithmized data are fitted with a linear trendline, the coefficients thereof being presented in the bottom cell.

In Table 10.8, the coefficients of the linear trend fitted to the logarithmized

[19] The characteristic of a parabola, its vertex, is not captured (see Sect. 10.2.2 on parabolas).

Table 10.7 **a** (top) Generation of noisy data points around an exponential; **b** (middle) Linear fit to the logarithmized data with OLS; **c** (bottom) Results of OLS

```
g,h= 10.0, 3.0                  #y = g·exp(h·x)
Ns=4                            #Noise level
x=np.linspace(0,1,11,endpoint=True)
y=g*np.exp(h*x)
yNs=y+Ns*np.random.randn(len(x))
```

```
import statsmodels.api as sm

lnYns=np.log(yNs)               #Natural logarithm
xx=sm.add_constant(x)           #y intercept allowed
model=sm.OLS(lnYns,xx)          #Linear regression
results=model.fit()
mR=results.params[1]            #Optimized coeffs of y=a+mx
aR=results.params[0]
mE=results.bse[1]               #Standard error
aE=results.bse[0]
r2=results.rsquared
```

```
         m     a
est.    2.78, 2.45
error   0.22, 0.13
r²=        0.95
```

Table 10.8 Transformation of the coefficients of the linear trend into those of an exponential trend

```
hR=mR
hE=mE
gR=np.exp(aR)
gE=aE*gR
inG=(gR-gE<g<gR+gR)
inH=(hR-hE<h<hR+hE)
```

	h	g
true	3.00,	10.00
est.	2.78,	11.61
error	0.22,	1.49
inG, inH	False	True

data are transformed into coefficients of the exponential trend. The Boolean variables *inH* and *inG* check whether the estimated standard error ranges of the estimated coefficients capture the true values of *h* and *g*.

To get the hit rates, we have to integrate lines 5, 8–16 and 17–22 into a function that returns in_G and in_H and have a loop in the main program running over that function and counting the `Trues` in in_G and in_H.

10.4.5 Spreadsheet Solution

LINEST with the Logarithmized Data

In Fig. 10.12 (S) , noisy data (x, y_{Ns}) are created, logarihmized, and fitted with a linear trend line with LINEST in E3:F5. The coefficients m_R and a_R of the linear trend and their standard errors are transformed in H3:I5 into the values of the exponential trend. H5:I5 contains the usual check as to whether the error ranges capture the true values. The numerical values of the coefficients h_R, g_R correspond to those of an exponential trend line in a figure.

We get the hit rates by applying a rep-log procedure that repeats the statistical experiment by writing a number into a cell to refresh the spreadsheet calculation and counts the TRUEs in H5 and I5.

Exponential trend with the spreadsheet function logest

EXCEL provides a spreadsheet function LOGEST that determines the parameters of an exponential trend and their uncertainties, much like LINEST does for a linear trend. The equation of the curve to be fitted is

$$y = a \cdot o^x = a \cdot \exp(h)^x = a \cdot \exp(hx) \tag{10.17}$$

Comparing with Eq. 10.14, the following equations are valid:

$$h = \ln(o) \text{ and } g = a.$$

	A	B	C	D	E	F	G	H	I	J
1	**g**	8			**mR**	**aR**		**hR**	**gR**	
2	**h**	3			**mSe**	**aSe**		**hSe**	**gSe**	
3	**Ns**	0.2		Estimated	3.08	2.10		3.08	**8.19**	*=EXP(aR)*
4				Standard error	0.20	0.12		0.20	**0.96**	*=aSe*EXP(aR)*
5				R²	0.96	0.21		**TRUE**	TRUE	
6		*=g*EXP(h*x)*	*=y+NORMINV(RAND();0;Ns)*	*=LN(yNs)*	*↑{=LINEST(lnYns;x;;1)}*			*↑=AND(hR-hSe<h;h<=hR+hSe)*		
7	**x**	**y**	**yNs**	**lnYns**						
8	0	8.00	7.25	1.98				**0.657** *=1-T.DIST.2T(1;9)*		
9	0.1	10.80	12.78	2.55						
18	1	160.68	207.88	5.34						

Fig. 10.12 (S) The noisy data y_{Ns} are created in columns A:D and logarithmized in D. A linear trend line through (x, y_{Ns}) is calculated with LINEST (formula in E6). The coefficients of the linear trend are transformed into those of the exponential trend; formula for H5 in H6

Fig. 10.13 (S) Use of the spreadsheet function LOGEST to determine the coefficients of an exponential trend and their standard errors; the matrix formula in E21 applies to E24:F26

	E	F	G	H	I
21	{=LOGEST(yNs;x;;1)}			=LN(oR)	
22	**oR**	**gR**		**hR**	**gR**
23	**oSe**	**gSe**		**hSe**	**gSe**
24	21.86	8.19		**3.08**	8.19
25	0.20	0.12		0.20	~~0.12~~
26	0.96	0.21			

For multiple independent variables, the function to be fitted is extended to

$$y = a \cdot o_1^{x1} \cdot o_2^{x2} \cdot \ldots \tag{10.18}$$

In Fig. 10.13 (S), LOGEST is applied to the data of Fig. 10.12 (S). The same value $R^2 = 0.96$ is obtained for the coefficient of determination and the same values for $h_R = \ln(o_R) = 3.08$ and $g_R = 8.19$ (calculated in I24) are returned. However, LOGEST outputs a wrong standard error for g_R. Comparison with E4 in Fig. 10.12 (S) shows that the error of the ordinate intercept of the linear trend is reported as untransformed. The same transformation as in I4:J4 of Fig. 10.12, based on Eq. 10.16, should be done.

10.5 Solving Nonlinear Equations

This exercise introduces the nonlinear optimization algorithms SOLVER of EXCEL and `minimize` of the `Python` library `scipy.optimize`, determining, as an example, the intersections of a polynomial with a straight line. This technique is applied in nonlinear regression.

10.5.1 Intersection of Straight Lines with a Parabola

SOLVER of EXCEL is an analysis tool that varies up to 200 independent variables ("adjustable parameters") so that the value of a target variable (must be a scalar) as a function of these parameters becomes optimum, maximum, minimum, or close to a given value, depending on the setting. The function `minimize` of the `scipy.optimize` library solves the same tasks. We apply it to determine the intersections of a parabola with straight lines. For the task in this exercise we apply the function `fsolve`, also of the `scipy.optimize` library, that finds the roots of an equation. `Minimize` of this library is applied in Exercise 10.6.

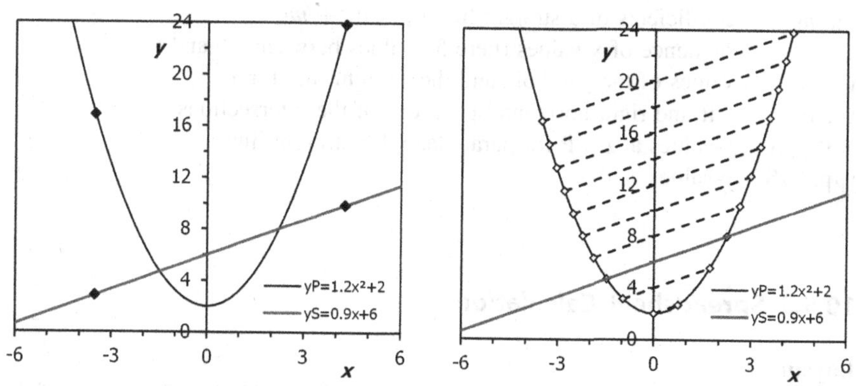

Fig. 10.14 **a** (left) The intersections of the straight line with the parabola are to be found. The currently selected x-values are not yet the solution. **b** (right) The intersections of ten straight lines with the parabola were determined with a solver algorithm

In Fig. 10.14a, a parabola and a straight line are shown whose intersections are to be found. This problem leads to a quadratic equation that can be solved analytically (*pq* solutions of the reduced quadratic equation). Therefore, the numerical method with a solver algorithm is not necessary here, but is very convenient, and also works for more complicated problems that cannot be solved analytically, e.g., for the intersections of a cosine function with a third-degree polynomial.

For the currently selected x-values, x_L (left of x = 0) and x_R (right of x = 0), the y-values on the straight line y_LS and y_RS are different from the corresponding y_PL and y_PR on the parabola. The x-values are to be modified so that the y-values become equal.

The equations for a straight line, and a parabola symmetric to $x = 0$ are

$$y_S = c_S + m_S \cdot x \tag{10.19}$$

$$y_P = b_P + a_P \cdot x^2 \tag{10.20}$$

We can manually adjust the x-values of the two points, e.g., with sliders, so that the y-values on the straight line and the parabola match. The same can be done with SOLVER by minimizing the quadratic deviations of the two y-values. Figure 10.14b shows the solutions for 10 straight lines with different ordinate intercept c_S. The two intersections for a particular straight line are found in one run by minimizing the sum of the two quadratic deviations of the y values. The 10 solutions are obtained in a rep-log procedure for the spreadsheet and in a for-loop in the main `Python` program.

10.5.2 Data Structure and Nomenclature

a_P, b_P coefficients of a parabola $y_P = a_P \cdot x^2 + b_P$

c_S, m_S	coefficients of a straight line $y_S = c_S + m_S \cdot x$
x	sequence of x-values (here 51 values between—6 and 6)
y_P, y_S	values of the parabola and the straight line for x
x_L, x_R	left and right horizontal positions of the intersections
y_LP, y_LS	y-values at x_L, P for parabola, S for straight line
y_RP, y_RS	y-values at x_R.

10.5.3 Spreadsheet Calculation

Layout

A spreadsheet layout for the task is presented in Fig. 10.15 (S). The graphs of the parabola $y_P(x)$ and the straight line $y_S(x)$ are calculated in D11:F61 for 51 points. The intersections are calculated in D7:F9, first, by adjusting x_L and x_R with sliders, and then with SOLVER. Later, we will use a rep-log procedure, triggered with the button "Intersections", to get the intersections for the ten straight lines in Fig. 10.14b.

Activating Solver in Excel

The solver function must be activated with FILE/ OPTIONS/ ADD- INS/ SOLVER ADD-IN, then we find GO and click (see Fig. 10.16a). When we call DATA/SOLVER, the table in Fig. 10.16b pops up. The solver function optimizes the target cell's value (SET OBJECTIVE) by varying the parameter cells' values (BY CHANGING VARIABLE CELLS). The target cell is linked to parameter cells through a set of formulas. If the value of a parameter cell is changed, the value of the target cell also changes.

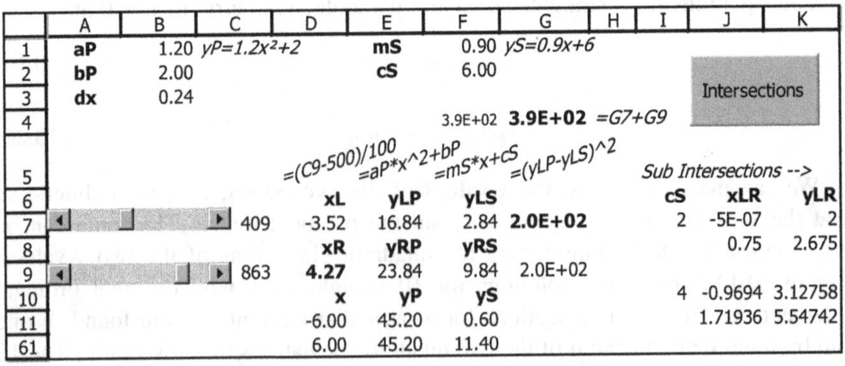

A B C D E F G H I J K
aP 1.20 yP=1.2x²+2
mS 0.90 yS=0.9x+6
bP 2.00
cS 6.00
dx 0.24
Intersections
3.9E+02 3.9E+02 =G7+G9
=(C9-500)/100
=aP*x^2+bP
=mS*x+cS
=(yLP-yLS)^2
Sub Intersections -->
xL yLP yLS
cS xLR yLR
409 -3.52 16.84 2.84 2.0E+02
2 -5E-07 2
xR yRP yRS
0.75 2.675
863 4.27 23.84 9.84 2.0E+02
x yP yS
4 -0.9694 3.12758
-6.00 45.20 0.60
1.71936 5.54742
6.00 45.20 11.40

Fig. 10.15 (S) The parameters for the parabola are in A1:B2, those for the straight line in E1:F2; arguments and function values in D11:F61; in the range D7:G9, the two intersection points are to be determined; in columns I to K, the intersection points for ten different straight lines with various y-axis intercepts c_S are stored. The button "Intersections" triggers the procedure *Intersectio* in Fig. 10.20 (P)

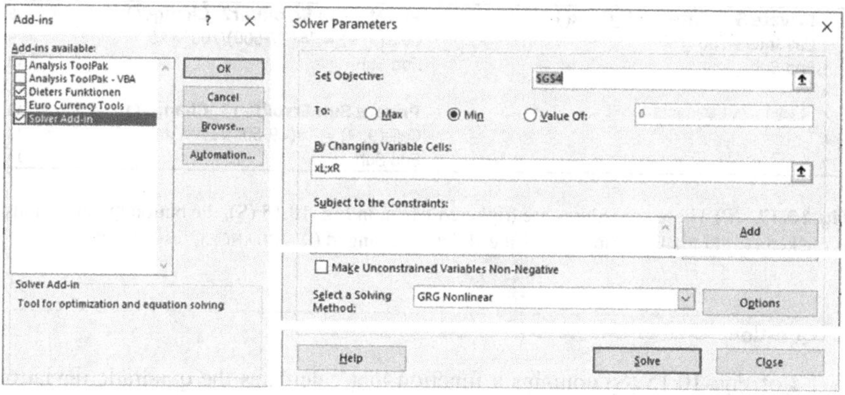

Fig. 10.16 **a** (left) Possible add-ins that can be activated in the EXCEL options. "Dieters Funktionen" are user-defined functions that have been saved as add-ins (see Sect. 4.9.1). **b** (right) Window after calling the solver function with DATA/SOLVER (in the Analysis Group); SET OBJECTIVE: the target value in the target cell is maximized (MAX), minimized (MIN), or adjusted to a given value (VALUE OF) by varying the values in the changeable cells (BY CHANGING VARIABLE CELLS)

Often, the option ☑ make unconstrained variables non-negative is activated as default. This check-mark must be removed, because the variable x to be optimized can also be negative for our task.

Solver determines intersections
The intersections are now determined with the SOLVER function. To do this, open the DATA tab (see Fig. 1.1 in Sect. 1.7) and click on the SOLVER button at the far right of the Analysis group, DATA/ SOLVER. A window opens, as in Fig. 10.16b. First, we take on only one intersection point and enter G7 (of Fig. 10.15 (S)), containing the left point's square deviation of the y values of the parabola and the straight line, as the target cell in the SOLVER tab, and the x-value in cell D7 as the changeable cell. After pressing the SOLVE button, the two points on the parabola and the straight-line slide together.

We could now determine the right intersection in the same way; but we choose a different solution, with both intersections determined simultaneously, by summing up the square deviations of the two points in G4. We enter G4 as the target cell to be minimized, and the x-values in D7 and D9 as variable cells showing up with their names x_L and x_R. When we press SOLVE, the two intersections are determined together in one run.

```
Private Sub Intersect_Click()
Call Intersectio
End Sub

Private Sub ScrollBar2_Change()
Cells(7, 4) = "=(C7-500)/100" 'xL
End Sub

Private Sub ScrollBar3_Change()
Cells(9, 4) = "=(C9-500)/100" 'xR
End Sub
```

Fig. 10.17 (P) These procedures are triggered when, in Fig. 10.15 (S), the button "Intersections" is clicked (…_CLICK) or when one of the sliders is changed (…_CHANGE), respectively

Question

F4 of Fig. 10.15 (S) contains a function that calculates the quadratic deviation of the two intersections without accessing column G. Which function can that be?[20] See Section 5.10 (Mathematical and Trigonometric Functions) for advice!

▶If the option ☑ MAKE UNCONSTRAINED VARIABLES NON- NEGATIVE is activated, SOLVER cannot find any negative x-values. This option must be unchecked if this is inadequate for the problem under consideration, as it is in this exercise.
There are two ways to select the changeable cells:

- The solver algorithm varies the values in cells D7 and D9 and overwrites the original formula. However, throughout this exercise, we want to change these values with the slider again. To be able to do so, we reinsert the formula with a macro, that is always triggered when the slider is operated; see SUB SCROLLBAR2_CHANGE und SUB SCROLLBAR3_CHANGE in Fig. 10.17 (P).
- We can have SOLVER vary the values in C7 and C9. Then, the formulas in D7:D9 remain unchanged.

▶ **Mag** Let's choose the first variant!

▶ **Alac** Why would we start there? The second variant is easier, because no macro is needed. Are we following the motto: Why be straightforward when we can complicate things?

▶ **Tim** Well, the first one teaches us how to link a macro to a control element.

▶ **Mag** Exactly, sometimes we learn via detours.

Sometimes, SOLVER declares the same x-value as the solution for the two intersections. In such cases, the initial x-values have been unfavorably chosen, e.g., both

20 =SUMXMY2(E7:E9; F7:F9) "Sum x minus y squared"

```
Sub Macro1()
Sheets("calc").Select
SolverOk SetCell:="$G$4", MaxMinVal:=2, ValueOf:=0, ByChange:="$D$7,$D$9", _
    Engine:=1, EngineDesc:="GRG Nonlinear"
SolverSolve
End Sub
```

Fig. 10.18 (P) Recorded macro when initializing the SOLVER function as in Fig. 10.16b

to be greater than zero. Such fallacies are possible in all optimization programs, because, primarily, only local optima are found.

► Use the sliders to change the changeable cells' initial values so that an approximate solution is reached before you start SOLVER!

Questions

Why does the formula in D7 or D9 in Fig. 10.15 (S) need to be re-entered after SOLVER has run?[21]

How do you re-enter the formula in D7 or D9?[22]

According to Fig. 10.20, what are the initial values of x for a y-axis intercept of 20?[23]

Calling Solver from a VBA procedure

The SOLVER function can be called from a macro. To get the corresponding commands, we turn DEVELOPER/RECORD MACRO on before calling SOLVER. The result for our example is presented in Fig. 10.18 (P).

Before starting the program, we must activate the reference to the solver function in the VBA editor with TOOLS/REFERENCES/SOLVER (see Fig. 10.19).

We insert the commands recorded in Fig. 10.18 (P) into the intended procedure as in Fig. 10.20 (P).

In a loop in SUB *Intersectio*, the ordinate intercept of the straight line is incremented from 2 to 20 in steps of 2 (line 11), and each time, SOLVERSOLVE calls the SOLVER function (line 16). The addition USERFINISH:=TRUE causes the solution proposed by SOLVER to be accepted immediately. If this addition is missing, a window pops up after each proposal of the SOLVER function in which the user can click OK.

[21] Because, in the current variant of the optimization process, the formulas in these cells are overwritten by the SOLVER function.

[22] By introducing macros like SUB SCROLLBAR2_CHANGE in Fig. 10.17 (P).

[23] The initial values are $x = -3$ and $+3$. They are the same for all y-axis intercepts, because they are always reset within the loop "cg=" in lines 13 and 14 in Fig. 10.20 (P). A better solution could be to take the previously optimized x-values as the start, because they are closer to the expected optimized value.

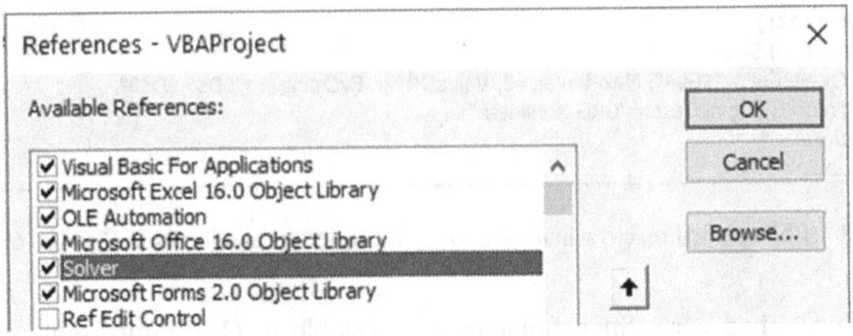

Fig. 10.19 Activating the SOLVER function in the VBA editor via TOOLS/REFERENCES

```
5 Sub Intersectio()
6 r2 = 7                  'Row index
7 sp2 = 9                 'Column index
8 Range("I7:K400").Clear
9 SolverOk SetCell:="$G$4", MaxMinVal:=2, _
10    ValueOf:="0", ByChange:="$D$7;$D$9"
11 For cS = 2 To 20 Step 2
12    Cells(2, 6) = cS
13    Cells(7, 4) = -3 'left x to start
14    Cells(9, 4) = 3 'right x to start
15
16    SolverSolve Userfinish:=True
17    Cells(r2, sp2) = cS
18    'ordinate section of the straight line
19    Cells(r2, sp2 + 1) = Cells(7, 4) 'xL
20    Cells(r2, sp2 + 2) = Cells(7, 6) 'yLS
21    r2 = r2 + 1
22    Cells(r2, sp2 + 1) = Cells(9, 4) 'xR
23    Cells(r2, sp2 + 2) = Cells(9, 6) 'yRS
24    r2 = r2 + 2
25 Next cS
26 End Sub
```

Fig. 10.20 (P) SUB *Intersectio*, working on Fig. 10.15 (S) specifies the parameters for the SOLVER function in lines 9 and 10 and selects the initial value c_S of the ordinate intercept of the straight line in the loop (FOR $cS =$), calls the SOLVER function in line 16, and saves the coordinates of the intersections in the spreadsheet in the range below I6:K6, starting with $r_2 = 7$ (line 6). The specifications in lines 9 and 10 have to be defined only once, after which they apply to all subsequent calls of SOLVER

The initial values are $x = -3$ and $+3$. They are the same for all y-axis intercepts, as they are always reset within the loop (FOR cS=) in Fig. 10.20 (P). If they were set before the loop, the previously optimized x-values would be the next y-axis intercept's initial values. This could be better, because the values are closer to the x-values expected for the next y-axis intercept.

Question

What effect does the instruction r2 = r2 + 2 in line 24 of Fig. 10.20 (P) have on range I7:K11 of Fig. 10.15 (S)?[24]

[24] Empty cells are introduced between the coordinates of the points, e.g., J9:K9, so that the data are displayed in the diagram as separate points.

10.5.4 Python Program

Lambda functions

In Table 10.9, function tables of the parabola (x, y_{Px}) and the straight line (x, y_{Sx}) are set up using the lambda functions *Prb* and *Str*. A lambda function is a small function associated with a variable. It may take any number of arguments, but must only contain a single executable expression. Another example is *make_0* in Table 10.10.

Furthermore, in Table 10.9, two x values, x_L, and x_R, are specified, together with their y-values y_{RP} and y_{RS}, serving later as initial values for the intersections of the parabola with the straight line.

Table 10.9 Function tables of the parabola and the straight line, and two points on each of them

```
x=np.linspace(-6,6,51,endpoint = True)
aP,bP=1.2, 2.0                      #Parabola
Prb=lambda x: aP*x**2+bP
yPx=Prb(x)
mS,cS=0.9, 20.0                     #Straight line
Str=lambda x: mS*x+cS
ySx=Str(x)

xL=-5                               #Initial x left
yLP,yLS=Prb(xL),Str(xL)
xR=5                                #Initial x right
yRP,yRS=Prb(xR),Str(xR)
```

Table 10.10 Plotting initial points and optimized intersections

```
FigStd('x',-6,6,2,'y',0,50,10)
plt.plot(x,yPx,'k--',label="parabola")
plt.plot(x,ySx,'k-',label="straight line, cS="+str(cS))
plt.plot([xL,xL,xR,xR],[yLS,yLP,yRS,yRP],'ks',
    ms=5, fillstyle='none')                   #Initial points

from scipy.optimize import fsolve
make_0 = lambda x : Prb(x)-Str(x)

xL=fsolve(make_0, xL)                         #Optimized left x
yLS=Str(xL)
xR=fsolve(make_0, xR)                         #Optimized right x
yRS=Str(xR)
plt.plot([xL,xR],[yLS,yRS],'ks',ms=5) #ms, Marker size
plt.legend()
```

Function fsolve

In Table 10.10, first, the curves and the initial points are plotted, after which the intersections are determined by `fsolve` and then plotted in the same figure. The function `scipy.optimize.fsolve (func, x0, ...)`. finds the roots of a function of the (in general non-linear) equations defined by `func(x) = 0` given a starting estimate x_0. In our case, the lambda function *make_0* is the root-defining function.

10.6 Temperature Dependence of the Saturation Magnetization of a Ferromagnet

The nonlinear Langevin equation relates the saturation magnetization of a ferromagnet with the temperature. We solve it with SOLVER, called from a macro, and with the function `minimize` of the `scipy.optimize` library of `Python`.

10.6.1 Langevin Function

Langevin equation

The Langevin equation describes the temperature dependence of the saturation magnetization M of a ferromagnet:

$$M = N\mu \cdot \tanh\left(\frac{\mu\lambda M}{k_B T}\right) \tag{10.21}$$

with:

μ magnetic moment of a magnetic element, e.g., an electron,
N density of magnetic elements,
λ $\mu\lambda N =$ molecular field,
k_B Boltzmann's constant,
T absolute temperature.

The saturation magnetization M thus appears on both sides of the equation. This is physically justified because a magnetic dipole aligns itself in the entire field and also contributes to the entire field.

Equation 10.21 can be simplified by introducing reduced variables

$$m := \frac{M}{N}\mu \text{ and } t := \frac{k_B T}{N\mu^2\lambda} \tag{10.22}$$

to

$$m = \tanh\left(\frac{m}{t}\right) \tag{10.23}$$

The reduced magnetization m is proportional to the magnetization M. The reduced temperature t is proportional to the temperature T. The solution $m = m(t)$ is called the *Langevin function.*

Graphical and numerical solution

The Langevin equation, Eq. 10.23, cannot be solved analytically. Still, it can be solved graphically by plotting $y = \tanh(m/t)$ for a given value of t as a function of m and by determining the point of intersection with the straight line $y = m$ (see Fig. 10.21a). It is solved numerically by keeping t fixed and varying m with a solver function so that the square deviation $(m - \tanh(m/t))^2$ is minimized.

We determine 18 points on the curve $m(t)$ by implementing two peculiarities: (a) all points are optimized simultaneously, and (b) all points are initially on a quarter circle.

As to (a): The Langevin equation can be solved for each reduced temperature t independently of the other temperatures. However, we solve it simultaneously for all points on the curve by adding up the quadratic deviations for the 18 cases and minimizing this sum by varying the 18 values of m.

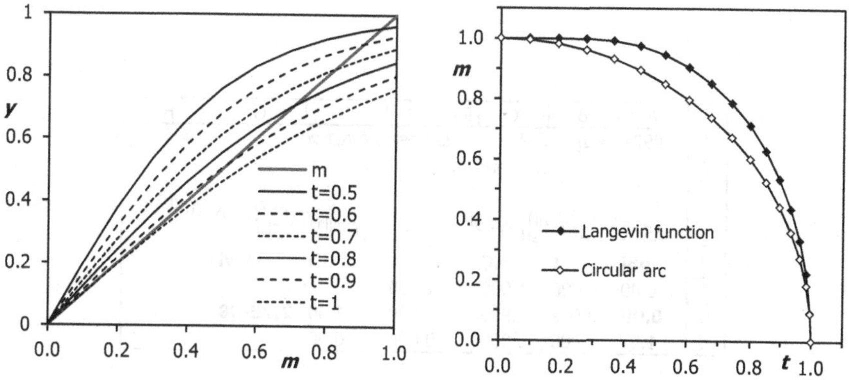

Fig. 10.21 **a** (left) Graphical solution of the Langevin equation; the y-axis is valid for m and $\tanh(m/t)$. **b** (right) Langevin function as a varied circular arc

As to (b): Since we expect a strongly bent curve, steeply going towards zero at $t = 1$, we select starting points (before minimizing) on a circular arc (see Fig. 10.21b).

Physical interpretation
The Langevin function in Fig. 10.21b is, with respect to the circular arc, flatter for small t and steeper for t approaching 1. The magnetic moments first stabilize each other in parallel alignment, but above a specific temperature, the order collapses.

10.6.2 Data Structure and Nomenclature

ϕ array of 18 angles of a quarter circle
t $\sin(\phi)$
m reduced magnetization to be optimized
m_C $\cos(\phi)$, initial values of m
th tan*hyp*(m/t), tangens hyperbolicus.

The entries $m(0) = 1$ and $m(1) = 0$ are held fixed; $th(0)$ is not defined.

10.6.3 Spreadsheet Layout

A spreadsheet solution of the Langevin equation is presented in Fig. 10.22 (S). The initial values m_C are on a circular arc and copied (with PASTE/PASTE VALUES) into m (E5:E22) before optimization. The solver algorithm is then used to vary these m-values so that they equal *tanh* according to the Langevin equation. If something goes wrong with the optimization, we can start again by copying m_C into m.

	A	B	C	D	E	F	G	H
1	0.0924	=PI()/17/2			Langevin function			
2			Circular arc					
3	=A5+\$A\$1	=SIN(phi)	=COS(phi)			=TANHYP(m/t)	=SUMXMY2(m;th)	
4	**phi**	**t**	**mC**		**m**	**th**	**devi**	
5	0.00	0.000	1.000		1.000			
6	**0.09**	0.092	0.996		1.000	1.00	**2.2E-08**	
22	1.57	1.000	0.000		0.000	0.00		

Fig. 10.22 (S) Solving the Langevin equation; reduced magnetization: initial positions m_C on a circular arc, m after optimization; m is defined as [E6:E22]

Table 10.11 Optimizing the Langevin function

```
from scipy.optimize import minimize

phi=np.linspace(0.01,np.pi/2,18)
t=np.sin(phi)                    #Values on quarter arc (not varied)
mC=np.cos(phi)                   #Initial values on quarter arc

def objective(m):
    th=np.tanh(m/t)
    return sum((m-th)**2)

m=minimize(objective,mC,method='SLSQP')

FigStd('m',0,1.0,0.2,'y',0,1,0.2)
plt.plot(mC,t,'k+-')
plt.plot(m.x,t,'kx-')   #m.x, optimized values
plt.axis('scaled')
```

Question

What are the target (objective) and the adjustable variables for SOLVER in Fig. 10.22 (S)?[25]

10.6.4 Python

In Table 10.11, the Langevin equation is solved with the scipy function minimize. The syntax is

`m = minimize (objective,mC, method = 'SLSQP')`

where

m	list of variables to be varied; shape determined by m_C
m_C	list of initial values for m
objective	name of a function with m as the argument and a scalar as the output to be minimized
SLSQP	*Sequential Least-Squares Programming*, type of solver to be applied.

The optimization works although $m[0]$ and $m[-1]$ are not excluded from the optimization because $m_C = \tanh(m_C/t)$ is already fulfilled.

[25] The target is *devi* (G6) and the variable cells are m, except its first value (E6:E22), because the magnetization m must be 1 at $t = 0$ and be kept fixed during optimization.

10.7 Fitting Gaussians to Spectral Lines with Nonlinear Regression

We fit a sum of two Gaussians to two overlapping EDX spectral lines, taking advantage of the additional knowledge that, for physical reasons, both lines must have the same width. The EXCEL tool is SOLVER; the Python function is `curve_fit` from the `scipy.stats` library. The C-spec errors of the SOLVER solution are obtained through simulation-based t adaptation (Student's t) using the parameters obtained from the fit.

10.7.1 Fitting the Sum of Two Gaussians to Data Points

In this exercise, we use the SOLVER tool of EXCEL and the function `curve_fit` from the `scipy.stats` library to fit functions with several fit parameters to measurement data, generated artificially in the by now well-proven manner.

Generation of two spectrally overlapping EDX signals
In Fig. 10.23a, a spectrum of an EDX analysis is shown. EDX is the abbreviation for "energy dispersive X-ray analysis". It means the energy-resolved analysis of characteristic X-ray radiation with silicon detectors after excitation of the sample with an electron beam. The count rate y of a detector is recorded as a function of the energy x of the photons arriving at the detector.

Here, the mapped spectra are artificially generated, but are very similar to the actual data for a $SrTiO_3$ coated Si wafer. The generation is done by specifying two Gaussian bell curves with the parameters maximum value, centre, and standard deviation as $A = 600$, $x_A = 1.75$, $x_{Ad} = 0.05$ and $B = 150$, $x_B = 1.85$, $x_{Bd} = x_{Ad} = 0.05$, respectively, adding them together with a constant background noise level y_C.

General bell curve
The normal distribution for medium x_m and standard deviation x_{Sd} is defined as

$$N(x, x_m, x_{Sd}) = \frac{1}{x_{Sd}\sqrt{2\pi}} \cdot \exp\left(-\frac{1}{2}\left(\frac{x - x_m}{x_{Sd}}\right)^2\right) \qquad (10.24)$$

The pre-factor is chosen so that the integral over the function is 1, as it must be for a probability distribution. The functions for normal distributions are used to model the spectra, and later to provide fits to the noisy data. To this end, it is advantageous that maximum A and width x_d be chosen independently of each other. The fit function is now $f(x, x_m, x_{Sd}) = A_N \cdot N(x, x_m, x_{Sd})$. Consequently, A_N for $N(x, x_m, x_{Sd})$ becomes

$$A_N = A \cdot x_{Sd} \cdot \sqrt{2\pi} \tag{10.25}$$

When A and x_{Sd} are independent of each other, the initial parameters can be set with the eye more intuitively.

Question

What is the area under the curve $f(x, x_m, x_{Sd}) = A_N \cdot N(x, x_m, x_{Sd})$?[26]

Two different fits to the spectrum
The sum of two bell curves, G_a and G_b, and a constant background y_{Ar} are fitted to a simulated spectrum by means of a solver algorithm.

Two different fits to the same measurement data can be seen in Fig. 10.24. For Fig. 10.24a, the standard deviations of the two bell curves have both been varied to obtain an optimum fit. For Fig. 10.24b, the two standard deviations have been forced to be equal. Both fits appear to be equally good, with a very high $R^2 = 0.99$. Therefore, the fit is not unique.

Due to physical considerations, a fit with equal spectral width is more likely, because the width of the lines is determined by the resolving power of the detector and should be approximately constant within the considered energy range. So, we decide to interpret the experimental spectrum with the fit in Fig. 10.24b.

Caution when ftting with many parameters!
The described method of fitting curves with a set of parameters to measurement data is useful, but also dangerous. If you have enough fit parameters, you can fit almost any data series without the parameters having a meaningful physical interpretation. Gain experience! Our exercises offer an excellent practice field for this, as we invent "true" data ourselves.

▶ **Alac** Sure, according to the motto: Ψ *We know everything and play stupid.*

▶ **Tim** And thereby learn statistics.

Uncertainty in the coefficients
The function `curve_fit` of the `scipy.optimize` library returns, in addition to the parameters of the regression curve, the covariance matrix of the optimized coefficients from which the standard errors can be obtained as the square root of the diagonal elements.

[26] The area under this curve is A_N.

The solver function of EXCEL returns only the optimized parameters. The standard errors in the fitted curves' parameters can also be determined in a mathematically exact way in spreadsheets; see, e.g., E. Joseph Billo, *EXCEL for Chemists*, Wiley–VCH (1977) *ISBN 0-471-18,896-4*, Chap. 17 or John Wiley (2011) ISBN 978–0470-38,123-6, Chap. 15.

In practice, a coefficient can be changed individually by hand until there is no longer a good fit upon visual inspection. The deviation from the optimum value of the coefficient may then be reported as the uncertainty. An example can be seen in Fig. 10.23b, where the center of the left bell curve has been changed from 1.75 to 1.755. Evidently, the total curve G_{abc} no longer fits the data points. So, 0.005 is a rough estimate of the error in x_{Ar}.

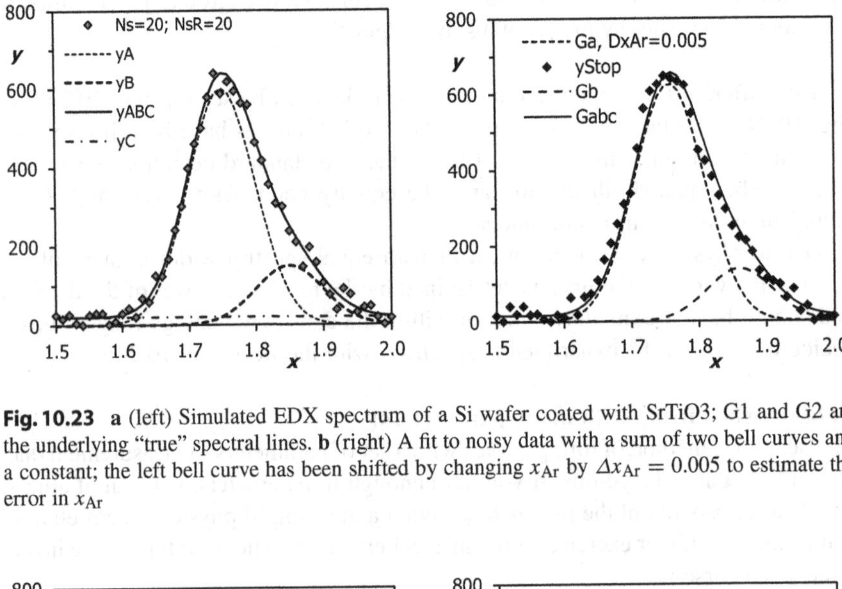

Fig. 10.23 **a** (left) Simulated EDX spectrum of a Si wafer coated with SrTiO3; G1 and G2 are the underlying "true" spectral lines. **b** (right) A fit to noisy data with a sum of two bell curves and a constant; the left bell curve has been shifted by changing x_{Ar} by $\Delta x_{Ar} = 0.005$ to estimate the error in x_{Ar}

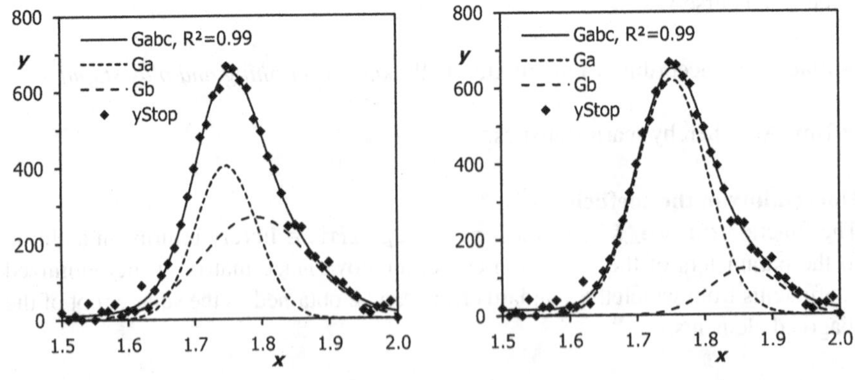

Fig. 10.24 Fit to an EDX spectrum with two Gaussian curves. **a** (left) The two standard deviations are varied independent of each other. **b** (right) The two standard deviations are forced to be equal

▶ **Tim** Is that allowed, and is it accurate enough?

▶ **Alac** Better a rough estimate of the uncertainties of the coefficients than none at all. In my experience, many labs engage in that practice. The researchers feel that a report of the optimum coefficients without uncertainty is, in any case, useless. Nevertheless, even this is done in many laboratories, as I have also observed.

▶ **Mag** Alac is right. However, we will do better with a simulation described in the next section by determining the hit rates for the assumed errors in the coefficients and obtaining the hit rates to the C-spec errors for a specified confidence level by simulation-based *t* adaptation.

10.7.2 C-spec Errors of the Coefficients by a Statistical Simulation

Figure 10.25 (S) lists standard errors of the parameters of the two bell curves fitted to noisy data, obtained with various methods. The values for `Python` from row 10 on are obtained as the square root of the diagonal elements of the covariance matrix that is returned by the function `curve_fit` of the `scipy.optimize` library.

The C-spec errors ("confidence-specified") for EXCEL are obtained with a statistical simulation in six steps:

(1) Starting from a fit to the spectrum and taking the resulting parameters of the optimized bell curves (row 2) as values of a "true" noise-less spectrum, as well as the residual noise of the fit as the "true" noise level. The residual noise is

	A	B	C	D	E	F	G
1	**Ar**	**xAr**	**xAsdr**	**Br**	**xBr**		
2	625	1.752	0.050	151	1.86		
3	**DyAr**	**DxAr**	**DAsdr**	**DyBr**	**DxBr**		
4	20	0.0050	0.0020	20	0.002	by eye	
5	10	0.0019	0.0013	16	0.008	D from row 4	
6	11	0.0019	0.0015	16	0.006		
7	10	0.0021	0.0015	16	0.007		
8	11	0.0016	0.0014	16	0.007	new D from row 7	
9							
10	Python						
11	10	0.0014	0.0014	13	0.006		
12	11	0.0013	0.0016	15	0.007		
13	12	0.0015	0.0015	19	0.005		

Fig. 10.25 (S) [A1:E2]: Results of a fit of two bell curves to experimental data; standard errors of the coefficients—obtained in EXCEL by eye (row 4) or by adjusting the values from row 4 with a *t*-value (rows 5–8)—in rows 11 to 13 with `Python` as the square root of the diagonal elements of the covariance matrix

calculated as the standard deviation of the difference between the noisy data and the fitted curve.

(2) Assuming *initial errors* in the coefficients by changing them by hand until the deviation from the data becomes visible (Figs. 10.24 and 10.25 (S) row 4).
(3) Simulating noisy data with the parameters from (1) and (2) to obtain a *fit to these data* with our model "two bell curves plus constant background noise".
(4) Repeating the statistical experiment (3) 100 times or more and so determining the *hit rates* for the initial error ranges from (2).
(5) Estimating the confidence level from the hit rates in (4), and from that, the *t values* for the desired confidence level.
(6) Adjusting the *C-spec errors* by dividing the errors assumed in (2) by the *t* values from (5).

Row 4 of Fig. 10.25 contains the initial values of the errors obtained by visual inspection. The next three rows report the results of statistical simulations with these errors. After that, the initial values of row 4 were replaced with those of the fit in row 7 to obtain the values in row 8. They are within the error range obtained with `Python`.

So, our method Ψ *We know everything and play stupid* is also successful for a task occurring in real laboratory life.

10.7.3 Data Structure and Nomenclature

Generation of a noisy spectrum

A, x_A, x_{Asd}	amplitude, center, standard deviation
y_A	bell curve with the above parameters
B, x_B, x_{Bsd}, y_B	second bell curve
y_C	background signal level
Ns	noise level
$y_{ABC}ns$	$y_A + y_B + y_C$ +*noise*, noisy spectrum.

Best fit to noisy spectrum

A_r, x_{Ar}, x_{Asdr}	amplitude, center, standard deviation of the fit curve
y_{Ar}	bell curve with the above parameters
$B_r, x_{Br}, x_{Bsdr}, y_{Br}$	bell curve with the above parameters
y_{Cr}	background signal level
$y_{ABC}new$	$= y_{Ar} + y_{Br} + y_{Cr}$
N_{SR}	noise level estimated as standard deviation of ($y_{ABC}ns$ – $y_{ABC}new$).

Visually estimated standard errors of the fit parameters

Dy_{Ar}, Dx_{Ar}, D_{Adr} estimated standard errors of A_r, x_{Ar}, x_{Adr}
Dy_{Br}, DxB_r, D_{Bdr} estimated standard errors of B_r, x_{Br}, x_{Bdr}.

C-spec errors of the fit parameters obtained from a statistical experiment

p_{out} error rate for the estimated standard errors
t Student's*t* value calculated from p_{out}
st_E adapted standard error calculated from p_{out} and t, to get a confidence level 0.68.

10.7.4 Python

For the `Python` program in Table 10.12, three user-defined functions are needed:

- a bell curve (`def gauss`) with parameters amplitude A, center x_m, and standard deviation x_{Sd},
- a function (`def gauss2`) adding two bell curves with independent parameters,
- a function (`def gauss3`) adding two bell curves with the same standard deviation.

In Table 10.13, the function `gauss` is used to generate two bell curves $y_A(x)$ and $y_B(x)$ intended to mimic ideal spectral lines. A noisy spectrum is simulated by adding these two curves and a constant background level y_C to get y_{ABC}, which is then blurred with normally-distributed noise to become $y_{ABC}ns$. To avoid negative values that do not occur in reality, a *maximum* function is applied in line 13.

Table 10.12 Defining a Gaussian and the sums of two Gaussians

```
import scipy.stats as sct

def gauss(x,A,xm,xd):
    f=sct.norm(0,1).pdf(0)
    return A/f*xd*sct.norm(xm,xd).pdf(x)

def gauss2(x,A,xA,xAd,B,xB,xBd,yC):
    G2=gauss(x,A,xA,xAd)+gauss(x,B,xB,xBd)+yC
    return G2

def gauss3(x,A,xA,xAd,B,xB,yC): #Same width
    G3=gauss(x,A,xA,xAd)+gauss(x,B,xB,xAd)+yC
    return G3
```

Table 10.13 Generation of a noisy EDX signal with two spectral lines

```
import numpy.random as npr

A,xA,xAd =600,1.75,0.05 #Amplitude, position, width
B,xB,xBd=150,1.86,0.05
yC=20                        #Background level
Ns=20                        #Noise level
dx=0.01

x=np.arange(1.5,2+dx,step=dx)
yA=gauss(x,A,xA,xAd)
yB=gauss(x,B,xB,xBd)
yABC=yA+yB+yC
yABCns=np.maximum(yABC+Ns*npr.randn(len(x)),0)
```

In Table 10.14, a function of type gauss2 is fitted to the noisy data by a nonlinear least-squares fit with curve_fit, returning the optimized coefficients (here, stored in p_{opt}) and the covariance matrix (here, stored in cov). The standard errors D of the coefficients are obtained as square roots of the diagonal elements of the covariance matrix. The current fit looks like Fig. 10.24a with two bell curves with different widths, $x_{Asd} = 0.047$ and $x_{Bsd} = 0.67$, as reported in the bottom cell of Table 10.14.

Table 10.14 Fitting gauss2 (independent widths of the two bell curves) to noisy data; "error" means standard error

```
from scipy.optimize import curve_fit

p0=[400,1.7,0.05,200,1.9,0.05,130]          #Initial guess
popt, cov = curve_fit(gauss2, x, yABCns,p0)
D=np.zeros(7)
for i in range(7):
    D[i]=np.sqrt(cov[i][i])      #√Trace of covariance matrix
```

	A	xA	xAd		B	xB	xBd	yC
true	600	1.750	0.050	true	150	1.860	0.050	20
estim.	534	1.745	0.047	estim.	180	1.828	0.067	21
error	189	0.006	0.004	error	100	0.062	0.027	6

Table 10.15 Fitting gauss3 (equal widths of the two bell curves) to noisy data; "error" means standard error

```
coef=[A,xA,xAd,B,xB,yC]
inC=np.ndarray(6,dtype=bool)
yABCns=np.maximum(yABC+Ns*npr.randn(len(x)),0)
ini=[400,1.7,0.05,200,1.85,200]    #Initial guess
popt3, cov = curve_fit(gauss3, x, yABCns,p0=ini)
D=np.zeros(6)
for i in range(6):
    D[i]=np.sqrt(cov[i][i])
    inC[i]=popt3[i]-D[i]<coef[i]<popt3[i]+D[i]
```

	A	xA	xAd
true	600	1.750	0.050
estim.	597	1.752	0.054
error	10	0.0015	0.0015
in	True	False	False

	B	xB	xBd	yC
true	150	1.860	0.050	20
estim.	144	1.869	0.054	12
error	14	0.007	0.001	5
in	True	False	False	False

Question

How can lines 5-7 in Table 10.14 be formulated with list comprehension in one line?[27]

A fit with *gauss3*, Table 10.15, supposing equal widths of the bell curves, results in coefficients close to the "true" ones. In order to determine the confidence levels of the calculated standard errors in the coefficients, we check whether the error ranges capture the true values. The variables "in are reported in the bottom cell. To do that efficiently in a loop, the coefficients are put together in a list *coef,* to be addressed with the loop index. The list is of the same size as the estimated coefficients *popt3* and their errors *D*, and the result of the logical check is stored in a list in_C of the same size.

Hit rates

Based on the logical check in Table 10.15, we can estimate the hit rates of the error ranges (see Table 10.16). The statements for a fit with *gauss3* are assembled into a function *hitRates* that returns the Boolean array in_C, stating, for every coefficient, whether the true value is captured by the standard error range. A loop over `rep = 1000` repetitions counts the number of Trues in N_{in}CC, with the result being reported in the bottom cell of Table 10.16. The hit rates are close to 668, so that we have no reason to doubt that `curve_fit` returns the standard errors of the estimated coefficients.

[27] `D = [np.sqrt(cov[i][i]) for i in range(len(cov))].`

Table 10.16 Determining the hit rates for the standard errors obtained in Table 10.15; "function name" should be "variable namefit with gauss3, according"

```
ini=[400,1.7,0.05,200,1.85,200]

def hitRates():
    inC=np.ndarray(6,dtype=bool)
    yABCns=np.maximum(yABC+Ns*npr.randn(len(x)),0)
    popt3, cov = curve_fit(gauss3, x, yABCns,p0=ini)
    D=np.zeros(6)
    for i in range(6):
        D[i]=np.sqrt(cov[i][i])
        inC[i]=popt3[i]-D[i]<coef[i]<popt3[i]+D[i]
    return inC

coef=[A,xA,xAd,B,xB,yC]
NinCC=np.zeros(6)
rep = 1000
for r in range(rep):
    inCC=hitRates()
        #Without () only another variable name
    NinCC+=inCC
```

```
inC:
[ 664.00  679.00  683.00  669.00  680.00  691.00]
```

Questions

What is the difference between the statements `inCC = hitRates` and `inCC = hitRates()`?[28]

What is the degree of freedom for a fit with gauss3, according to Sect. 10.7.2?[29]

10.7.5 Spreadsheet

Generation of a noisy spectrum

The ideal spectrum y_{ABC} without noise is generated in Fig. 10.26 (S) with the parameters specified in B1:E5. A standard normal distribution is used for the noise, addressed as NORM.DIST(*x*; xA,. xAsd;0) where "0" stands for pdf. Its amplitude is calculated

[28] The statement `inCC = hitRates()` calls the function `hitRates()`, whereas `inCC = hitRates` simply assigns another name to this variable.

[29] From lines 11 of Table 10.12 and 9 of Table 10.13: *dof* = 51 (points) – 6 (parameters) = 45.

	A	B	C	D	E	F	G	H	I	J	K
1	Amplitude	**A**	600	**B**	150				varTot	46158	
2	Center	**xA**	1.75	**xB**	1.85				varRes		
3	StD	**xAsd**	0.05	**xBsd**	0.05				R²		
4	Offset	**yC**	20								
5	Noise level	**Ns**	20	**nD**	0.399	*=NORM.DIST(0;0;1;0)*					
6		**dx**	0.01				Ns=20; NsR=22				
7		*=B9+dx*	*=A/nD*xAsd*NORM.DIST(x;xA;xAsd;0)*	*=B/nD*xBsd*NORM.DIST(x;xB;xBsd;0)*	*=yA+yB+yC*	*=NORM.INV(RAND();0;Ns)*	*=MAX(yA+yB+yC+Noise;0)*	*=SQRT(SUMXMY2(yABCns;yABC)/50)*			
8		**x**	**yA**	**yB**	**yABC**	**Noise**	**yABCns**	**NsR**			
9		1.5	0.00	0.00	20.00	-7	13	22.01			
10		**1.51**	0.01	0.00	20.01	-22	0				
59		2	0.00	1.67	21.67	-4	17				

Fig. 10.26 (S) Generation of a noisy spectrum y_{ABC}ns; step (1) of Sect. 10.7.2

in E5 as n_D, guaranteeing that the integral over the function is 1. As we want to handle the maxima A and B of the curves as parameters independent of the widths, the pre-factors for y_A and y_B are calculated corresponding to Eq. 10.25. With A and the width x_{Asd} (or B and B_{sd}), we have two parameters at hand that have an immediate visual meaning and can independently be adjusted by hand for the bell curves to fit the experimental curves.

The ideal "true" spectrum y_{ABC} is blurred with noise to become $y_{ABC}ns$ in column G. The signal, a count rate, is never negative in reality; this is guaranteed with MAX(*; 0). The residual noise Ns_R is calculated as the standard deviation of (y_{ABC}ns - y_{ABC}). Its value is close to the pre-specified "true" noise level Ns.

Question

To generate the noisy spectrum in Fig. 10.26 (S), two functions are used: NORM.DIST (reported in C7) and NORM.INV (reported in F7). Which roles do they play?[30]

Fitting bell curves to the spectrum

In Fig. 10.27 (S), the sum of two bell curves, G_a plus G_b plus an offset y_{Cr}, is fitted to the noisy spectrum created in Fig. 10.26 (S). This is done in a separate spreadsheet into which the vectors x and $y_{ABC}ns$ are copied from Fig. 10.26 (S). The variables for SOLVER are the 7 coefficients in C1:F4 with index r. They are changed in each iteration of SOLVER so that y_{ABC}ns is calculated anew. To have the "experimental" spectrum fixed, the contents of $y_{ABC}ns$ have been value-copied (COPY, PASTEVALUES) into y_{Stop}. The variable $y_{ABC}ns$ is no longer needed in the current fit. The target cell for SOLVER is Ns_R in G9, the residual noise level for (y_{Stop} − G_{abc}). The coefficient of determination r_{sq} of the fit is calculated in H1:H3 according to Eq. 10.6.

[30] NORM.DIST produces a smooth Gaussian curve as a function of energy. NORM.(RAND();0;NS) produces the normally distributed noise of the y-values.

	A	B	C	D	E	F	G	H	I	J
1		Amplitude	**Ar**	405.20	**Br**	268.20	**varTot**	45495	*=VAR.S(yABCns)*	
2		Center	**xAr**	1.75	**xBr**	1.79	**varRes**	316	*=SUMXMY2(yABCns;Gabc)/50*	
3		Std	**xAsdr**	0.046	**xBsdr**	0.080	**rSq**	0.99	*=(varTot-varRes)/varTot*	
4		Offset	**yCr**	10.14						
5		Noise level	**NsR**	16.3	*=G9*	Gabc, R²=0.99				
6										
7	*=x*	*=yABCns*	*427*	*=Ar/nD*xAsdr*NORM.DIST(x;xAr;xAsdr;0)*	*=Br/nD*xBsdr*NORM.DIST(x;xBr;xBsdr;0)*	*=Ga+Gb+yCr*	*=SQRT(SUMXMY2(yStop;Gabc)/50)*			
8	**x**	**yABCns**	**yStop**	**Ga**	**Gb**	**Gabc**	**NsR**			
9	1.5	21	19	0.0	0.4	10.5	16.3			
59	2	26	0	0.0	8.7	18.8				

Fig. 10.27 (S) New spreadsheet calculation: fit of the sum of two bell curves to the noisy data of Fig. 10.26 (S); y_{Stop} is the "frozen" result of the fit procedure, reported in Fig. 10.24a

Questions

What is the adjusted coefficient of determination $r_{Sq}Adj$ for Fig. 10.27 (S)?[31]
Is the procedure reported in Fig. 10.27 (S) a least-squares fit?[32]

The results for two different fits with the sum of two bell curves, (a) with individual widths and (b) with identical widths, are shown in Fig. 10.28 (S). Both fits yield a very high coefficient of determination, $R^2 = 0.99$, so there is no good reason to prefer one particular fit. However, we know that, for physical reasons, the width of the spectral lines is determined by the resolution of the detector, and therefore should be the same for both lines; thus, (b) is preferred.

10.7.6 C-spec Error of the Optimized Coefficients by Simulation-Based *t* Adaptation

Fig. 10.29 (S), based on Fig. 10.28 (S), presents a spreadsheet calculation for obtaining the errors of the optimized coefficients.

Row 4 lists the optimized coefficients of a fit of the sum of two bell curves to noisy experimental data. They are now regarded as the true values for a *new statistical experiment* and used as parameters to generate a spectrum in the calculation model of Fig. 10.26 (S) that is then entered into Fig. 10.27 (S). The noise level Ns_R, calculated as the standard deviation of the difference between the noisy spectrum and the fitted curve in H9 of Fig. 10.26 (S), is value-copied into G9 of Fig. 10.27 (S).

[31] $R^2_{adj} = 1 - (1 - R^2) \cdot \frac{N-1}{N-ddof}$ (Eq. 10.8), here, $N = 51$ (Fig. 10.26 (S)), $ddof = 7$ and $R^2 = 0.99$ (Fig. 10.27 (S)), so that $R^2_{adj} = 0.989$. In this example, the difference between R^2 and R^2_{adj} is unimportant.

[32] Strictly speaking, not because the target is the square root of the squares of the deviations. However, *sqrt* is a strictly monotonously increasing function so that the minimum in *sqrt* is also the minimum in the squares.

	A	B	C	D	E	F	G	H	I	J
1		Amplitude	**Ar**	405.20	**Br**	268.20	**varTot**	45495	*=VAR.S(yABCns)*	
2		Center	**xAr**	1.75	**xBr**	1.79	**varRes**	316	*=SUMXMY2(yABCns;Gabc)/50*	
3		Std	**xAsdr**	0.046	**xBsdr**	0.080	**rSq**	0.99	*=(varTot-varRes)/varTot*	
4		Offset	**yCr**	10.14						
5		Noise level	**NsR**	16.3	*=G9*	Gabc, R²=0.99				

	A	B	C	D	E	F	G	H	I	J
1		Amplitude	**Ar**	621.70	**Br**	122.50	**varTot**	44929	*=VAR.S(yABCns)*	
2		Center	**xAr**	1.75	**xBr**	1.86	**varRes**	284	*=SUMXMY2(yABCns;Gabc)/50*	
3		Std	**xAsdr**	0.052	**xBsdr**	0.052	**rSq**	0.99	*=(varTot-varRes)/varTot*	
4		Offset	**yCr**	15.51						
5		Noise level	**NsR**	16.1	*=G9*	Gabc, R²=0.99				

Fig. 10.28 (S) Copy of instances from Fig. 10.27 (S); results of fitting with bell curves with **a** (top) individual widths, **b** (bottom) identical widths; F3 = [=xAsdr]

	K	L	M	N	O	P	Q	R	S
3		**Ar**	**xAr**	**xAsdr**	**Br**	**xBr**			
4		405	1.75	0.046	268	1.79			
5		**DyAr**	**DxAr**	**DAsdr**	**DyBr**	**DxBr**			
6		10	0.0021	0.0015	16	0.0071			
7		*=AND(Ar-DyAr<A;A<Ar+DyAr)*	*=AND(xAr-DxAr<xA;xA<xAr+DxAr)*	*=AND(xAsdr-DAdr<xAsd;xAsd<xAsdr+DAdr)*	*=AND(Br-DyBr<B;B<Br+DyBr)*	*=AND(xBr-DxBr<xB;xB<xBr+DxBr)*			
8		**inA**	**inxA**	**inxAd**	**inB**	**inxB**			
9		FALSE	FALSE	FALSE	FALSE	FALSE			
10	100	67	79	71	66	67			
11		0.67	0.79	0.71	0.66	**0.67**	*=P10/$K$10*		
12	**pOut**	0.33	0.21	0.29	0.34	0.33	*=1-P11*		
13	**t**	0.99	1.27	1.07	0.96	0.99	*=T.INV.2T(pOut;44)*		
14	**stE**	11	0.0016	0.0014	16	0.007	*=DxBr/t*		

Fig. 10.29 (S) Continuation of Fig. 10.28 (S); spreadsheet calculation for determining the standard errors of the optimized coefficients; L4:P4 copied from D1:F3

The deviation values reported in row 6 are obtained by changing the optimized parameters until the trend curve differs visibly from the noisy spectrum to be pre-supposed as errors for the coefficients obtained in *new statistical experiments* with spreadsheet calculations as in Fig. 10.26 (s) and Fig. 10.27 (S) in order to determine whether the supposed confidence intervals capture the true values.

With a rep-log procedure, the *new statistical experiment* is repeated 100 times to obtain the hit rates reported in row 10 of Fig. 10.29 (S). From that, we get the confidence levels in row 11 and the error probability p_{Out} into row 12; therefrom, the corresponding t-value in row 13; and finally, the standard errors *stE* for a confidence level of 0.68 estimated as initial errors (row 6) divided by t.

This simulation with 100 trials is repeated with the *stE* copied manually as standard errors in row 6 in the following run. The results of this iterative procedure are reported in Fig. 10.25 (S).

10.8 Questions and Tasks

1. How do you obtain a linear mapping of the function

$$y = \mathrm{A} \cdot \exp\left(\left(\frac{x}{\Delta x}\right)^2\right)$$

and what meaning do the y-axis intercept and slope at $x = 0$ have?

2. In Fig. 10.30a (S), a straight line is adapted to measurement points with LINEST. What is the meaning of the numbers in D1:E2 and D3?
3. A parabola is fitted to 7 data points with LINEST. How many degrees of freedom does the fitted parabola have? Does our rule: Ψ *Two within and one out of*? apply to the coefficients of the parabola?
4. In Fig. 10.30b, a polynomial $y_C(x)$ is fitted to data points (x, y_S) with SOLVER. Which is the target cell, and which are the adjustable cells? Which mathematical formula is behind the spreadsheet formula in S5?
5. In Fig. 10.31 (S), a parabola is fitted with LINEST to 7 data points (x, y). A rep-log procedure counts 10,000 times how often the true values of a, b, and c are outside of the error range. Which data are in B7:C13? What is the meaning

	D	E	F	G
1	2.97	1.35	=LINEST(y;x;;1)	
2	0.19	1.20		
3	0.97	1.76		

	N	O	P	Q	R	S	T	U
5					=aS*x^2+bS*x+cS	=SUMXMY2(yC;yS)		
6	x			yC	yS			
7	0	aS	1.11	1.74	2.49	4.06		
8	1	bS	1.66	6.34	5.25			
9	2	cS	2.49	10.84	10.24			
13	6			52.38	52.41			

Fig. 10.30 a (left, S) Output of LINEST. b (right, S) A polynomial y_S is fitted to the data points (x, y_C) with SOLVER

	A	B	C	D	E	F	G	H	I	J	K	L	M
1	a	1			1.11	1.66	2.49						
2	b	2			0.11	0.69	0.88	{=LINEST(A7:A13;B7:C13;;1)}					
3	c	3			1.00	1.01	#N/A						
4	Ns	1						10000					
5					=AND(E1-E2<a;a<E1+E2)			=T.DIST.2T(1;4)		=AND(F1-2*F2<b;b<F1+2*F2)		=T.DIST.2T(2;4)	
6	y	x	x²		a.1	b.1	c.1		a.2	b.2	c.2		
7	1.74	0	0		FALSE	TRUE	TRUE		TRUE	TRUE	TRUE		
8	6.34	1	1		0.378	0.379	0.383	0.374	0.119	0.120	0.119	0.116	
9	10.84	2	4										
13	52.38	6	36										

Fig. 10.31 (S) A parabola is fitted to the 7 data points (x, y) with LINEST. The values in E8:G8 and I8:K8 are the result of a rep-log procedure

Table 10.17 Python code snippet for performing a linear least-square fit to data points (x, y_{Ns}) in arrays x and y_{Ns}

```
1  #Data points (x,y) in lists x and yNs
2  model=sm.OLS(yNs,xx)
3  results=model.fit()
4  aR=results.params[0]
5  aE=results.bse[0]
6  r2=results.rsquared
7  r2_ad=results.rsquared_adj
```

Fig. 10.32 (S) For determining the intersection of two straight lines

	A	B	C	D	E
1	a.1	1		a.2	3
2	m.1	2		m.2	-2
3					
4	x	y.1	y.2		
5	1	3	1	4	

of the values in E2:G2 and the arguments of the formula for T.DIST.2T in H8 and L8? How have the values in E8:G8 and in I8:K8 been obtained?

6. Table 10.17 shows a Python code snippet for performing a linear least-square fit to data points (x, y_{Ns}) in arrays x and y_{Ns}. A linear trend line is fitted to the data points with the OLS function of the statsmodel.api library. What is the array *xx* in the argument of OLS? How do you get the coefficients of the linear trend line and their standard errors? What is the difference between the two quantities queried in lines 6 and 7?
7. Figure 10.32 (S) shows the spreadsheet layout for determining the intersection of two straight lines with SOLVER. What are the formulas in B5, C5, and the target cell D5? What are the objective (target) cell and the variable cells in SOLVER?
8. Figure 10.33a shows a spreadsheet layout for calculating the intersections of three straight lines displayed in **b.** What are the formulas for y_1, y_2, and y_3? The target cell to be minimized is E4. What are the formulas in E6, E7, and E8?
9. Table 10.18 shows a Python program for solving the problem of Fig. 10.33, namely finding the intersections of three straight lines with a solver algorithm. It is incomplete. Introduce the 6 missing statements!

	A	B	C	D	E	F	G
1	**a.1**	2	**a.2**	-1	**a.3**	8	
2	**m.1**	1.5	**m.2**	-1	**m.3**	-4	
3							
4					**8.9E-13**	*=SUM(E6:E8)*	
5	**x**	**y.1**	**y.2**	**y.3**			
6	-1.20	0.20	0.20	12.80	3.3E-13		
7	1.09	3.64	-2.09	3.64	5.7E-13		
8	3.00	6.50	-4.00	-4.00	0.0E+00		

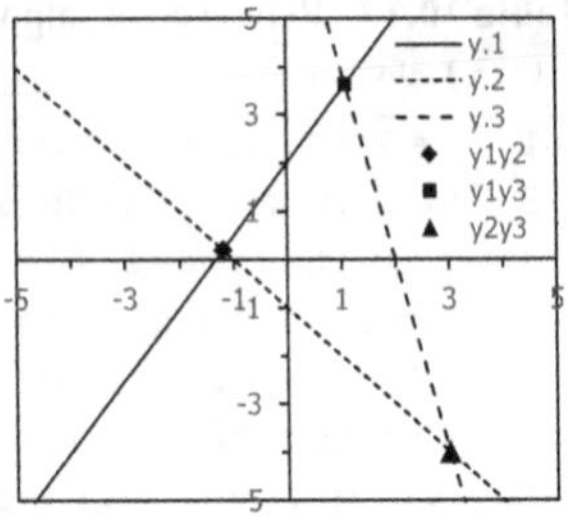

Fig. 10.33 **a** (S, left) spreadsheet layout for calculating the intersections of y_1, y_2, and y_3 using SOLVER, the column vector x contains the optimized x-values of the intersections. **b** (right) Straight lines of **a**, together with their intersections

Table 10.18 Python program for solving the problem in Fig. 10.33; 6 lines are omitted

```
a1, m1 = 2, 1.5
y1 = lambda x: a1+m1*x
def target(x):
    y12=(y1(x[0])-y2(x[0]))**2
    return y12+y13+y23
xI=[-1,1,2.5]
x=minimize(target,xI,method='SLSQP')
print(x.x)
```

Index

D. Mergel, *Physics with Excel and Python*,
https://doi.org/10.1007/978-3-030-82325-2

Printed in the United States
by Baker & Taylor Publisher Services